Bioinorganic Photochemistry

Bioinorganic Photochemistry

GRAŻYNA STOCHEL, MAŁGORZATA BRINDELL,
WOJCIECH MACYK, ZOFIA STASICKA,
KONRAD SZACIŁOWSKI

Faculty of Chemistry, Jagiellonian University, Poland

WILEY
A John Wiley & Sons, Ltd., Publication

This edition first published 2009

Registered office
John Wiley & Sons Ltd, The Atrium, Southern Gate, Chichester, West Sussex, PO19 8SQ, United Kingdom

For details of our global editorial offices, for customer services and for information about how to apply for permission to reuse the copyright material in this book please see our website at www.wiley.com.

Library of Congress Cataloging-in-Publication Data

Bioinorganic photochemistry / Grażyna Stochel ... [et al.].
p. cm.
Includes bibliographical references and index.
ISBN 978-1-4051-6172-5 (cloth : alk. paper)
1. Bioinorganic chemistry. 2. Photobiochemistry. I. Stochel, Grażyna.
QP531.B545 2009
5721′.435–dc22

2008044387

A catalogue record for this book is available from the British Library.

978-1405-161725

Set in 10 on 12 pt Times by SNP Best-set Typesetter Ltd., Hong Kong
Printed in and bound in Singapore by Fabulous Printers Pte Ltd

Acknowledgements

The authors wish to thank colleagues from the Coordination and Bioinorganic Physicochemistry group at the Jagiellonian University for their friendly support. Special thanks to Zygmunt Wołek, Ewa Kuliś, Przemysław Łabuz, Joanna Kuncewicz, Łukasz Orzeł, Agnieszka Jańczyk, Sylwia Gawęda and Agnieszka Podborska for their assistance in preparation of the manuscript.

The completion of this book was made possible with the encouragement and love of our families.

Contents

Preface

Bioinorganic photochemistry is a new branch of modern science dealing with the interaction of light with inorganic matter, which has a huge impact in all forms of life on the Earth from its origin up to the present: it is responsible for the origin and maturation of biosphere, its environment, and sustainable development.

The photobioinorganic interactions may be found not only now and here but also in the Universe far from Earth and in times dated more than 4 billion years ago, when the first hints of life probably emerged on Earth.

Small inorganic molecules under the influence of light are able to convert and assemble to form a variety of organic compounds, which not only are a life supplement but also may be treated as responsible for primordial life forms. Photochemistry of the inorganic species made its contribution to the creation of the world and played a fundamental role in the evolution of life. Both primordial and present life are protected from the destructive action of the high-energy part of solar radiation by the stratospheric photochemical processes involving oxygen and ozone.

The maintenance of life on Earth is possible only as a result of photosynthesis, which takes place in green plants and did in the past, when it was followed by decomposition and formation of deposits of coal, oil, and natural gas, currently used as fossil fuels. One of the future developments of bioinorganic photochemistry appears to be a pathway to enable the creation of new sources of energy that are both cheap and environmentally friendly.

A lot of photoreactions occurring in the atmosphere, hydrosphere, and soils ensure the health, comfort, and welfare of human beings, creatures, and the environment. These processes are mostly driven by coordination compounds of transition metals, which play the role of (photo)catalysts or (photo)sensitizers. There is also increased understanding of the role of supramolecular inorganic systems in their interaction with light and the great variety of processes that may ensue.

Development of artificial light sources, and especially the introduction of lasers, brought about an enormous increase in research on light–biomatter interactions. Thus the application of inorganic photochemistry and photophysics generates challenging new areas in bioscience and biotechnology.

Recently, nanotechnology and nanomaterials have been revolutionizing important areas in biomedical photonics, especially diagnostics and therapy at the molecular and cellular levels. Once again, inorganic species offer unique possibilities for practical applications.

Despite the rapidly growing knowledge in bioinorganic photochemistry, there is no single book devoted to this new interdisciplinary branch of science. The information of some specific problems from bioinorganic photochemistry is spread throughout various books devoted to bioinorganic chemistry, inorganic photochemistry, photobiology, environmental photochemistry, or bioanalytical and biomedical applications. Therefore the goal of this book is to provide a comprehensive overview of bioinorganic photochemistry taken as a new interdisciplinary branch of science. We hope that the book that arose from the review paper published in Chemical Reviews (2005;105:2657–94) will serve as a guide for newcomers in the field, as well as the first source of information for more involved readers. After introductory remarks on bioinorganic photochemistry as a new area of interdisciplinary science, the second part contains essential information from the field of photochemistry and especially inorganic photochemistry. The next part of the book is devoted to bioinorganic solar photochemistry, from the origin and maturation of the biosphere to the sustainable development of its environment. Parts IV and V focus on artificial light interactions with biomatter both in the context of application (medical, biomedical, environmental) and as models of important biochemical and biophysical phenomena.

Grażyna Stochel, Małgorzata Brindell, Wojciech Macyk, Zofia Stasicka, Konrad Szaciłowski

Abbreviations

[12]aneN_4	1,4,7,10-tetraazacyclotetradecane
8-oxo-G	7,8-dihydro-8-oxoguanine
A	adenine
ACT	antimicrobial chemotherapy
ADP	adenosine-5′-diphosphate
AETE	absorption/energy-transfer/emission
AM	air mass
AOP	advanced oxidation process
AOT	advanced oxidation technique
APDT	antimicrobial photodynamic therapy
ATP	adenosine-5′-triphosphate
BChl	bacteriochlorophyll
bet	back electron transfer
bphb	4-[4-(2,2′-bipyridin-4-yl)phenyl]-2,2′-bipyridine
bpip	2-(4′-benzyloxy-phenyl)imidazo[4,5-*f*]-1,10-phenanthroline
bpy	2,2′-bipyridine
bpy′	4-(4′-methyl-2,2′-bipyridin-4-yl)butanamide
bpz	2,2′-bipyrazine
Car	carotenoid
CB	conduction band
CFT	crystal field theory
Chl	chlorophyll
chrysi	chrysene-5,6-diylidenediamine
cnoip	2-(2-chloro-5-nitrophenyl)imidazo[4,5-*f*]-1,10-phenanthroline
COX	cytochrome oxidase
Cp	cyclopentadienyl
CT	charge transfer
CTTS	charge transfer-to-solvent
cyclam	1,4,8,11-tetraazacyclotetradecane
cyt	cytochrome
ddz	dibenzo[*h,j*]dipyrido[3,2-*a*:2′,3′-*c*]phenazine

dicnq	dicyanodipyrido quinoxaline
dip	4,7-diphenyl-1,10-phenanthroline
dmb	4,4′-dimethyl-2,2′-bipyridine
dmso	dimethyl sulfoxide
dpb	2,3-bis(2-pyridyl)benzo[*g*]quinoxaline
dpp	2,3-dipyridin-2-ylpyrazine
dppz	dipyrido[3,2-*a*:2′,3′-*c*]phenazine
dpq	dipyrido[3,2-*d*:2′,3′-*f*]quinoxaline
ed3a	ethylenediaminetriacetate
edta	ethylenediaminetetraacetate
en	1,2-diaminoethane
ESR	electron spin resonance
ET	electron transfer
FAD	flavine adenine dinucleotide
FRET	Förster resonant energy transfer
fttp	tetrakis(4-trifluoromethylphenyl)porphyrin
G	guanine
G^{ox}	oxidized guanine
GMP	guanosine monophosphate
GOD	glucose oxidase
h^+	hole
hat	1,2-diaminoethane
hat	1,4,5,8,9,12-hexaazatriphenylene
Hb	haemoglobin
hnaip	2-(2-hydroxy-l-naphthyl)imidazo[4,5-*f*]-1,10-phenanthroline
hnoip	2-(2-hydroxy-5-nitrophenyl)imidazo[4, 5-*f*]-1,10-phenanthroline
HOMO	highest occupied molecular orbital
hpip	2-(2-hydroxyphenyl)imidazo[4,5-*f*]-1,10-phenanthroline
HS	humic substance
IC	internal conversion
IFET	interfacial electron transfer
IL (or ILCT)	intra-ligand charge transfer
ip	imidazo[4,5-*f*]-1,10-phenanthroline
IPCT	ion-pair charge transfer
IR	infrared
ISC	intersystem crossing
IT (or IVCT)	intervalence transfer
L	ligand
LC	ligand centred
LDH	lactate dehydrogenase
LED	light emitting diode
LF	ligand-field
LHC	light-harvesting centre
LLCT	ligand-to-ligand charge transfer
LMCT	ligand-to-metal charge transfer
LSPR	localized surface plasmon resonance

LUMO	lowest unoccupied molecular orbital
MBCT	metal-to-band charge transfer
MC	metal centred
Me_2dppz	11,12-dimethyl-4,5,9,14-tetraazabenzo[*b*]triphenylene
mgp	*N*-(1,10-phenanthrolin-4-ylmethyl)guanidine
MLCT	metal-to-ligand charge transfer
MMCT	metal-to-metal charge transfer
MPCT	metal-to-particle charge transfer
MRH	nitromerocyanine
MRI	magnetic resonance imaging
mRNA	messenger ribonucleic acid
NADPH	the reduced form of nicotinamide adenine dinucleotide phosphate
nc	naphthalocyanine
NHE	normal hydrogen electrode
NIR	near infrared
MTHF	5,10-methenyltetrahydrofolylpolyglutamate
NADPH	the reduced form of nicotinamide adenine dinucleotide phosphate
NCPs	nucleosome core particles
NOS	nitric oxide synthase
NP	nanoparticle
OAc	acetate
OEC	oxygen-evolving complex
oep	octaethylporphyrin
OSCT	outer-sphere charge transfer
PACT	photodynamic antimicrobial chemotherapy
PAN	peroxyacetyl nitrate
pc	phthalocyanine
PCT	photoinduced charge transfer
PD	photodiagnosis
PDD	photodynamic diagnosis
PDI	photodynamic inactivation
PDT	photodynamic therapy
pdta	3-(pyridine-2-yl)-as-triazino[5,6-*f*]acenaphthylene
pdtb	3-(pyridine-2-yl)-5,6-diphenyl-as-triazine
pdtp	3-(pyridine-2-yl)-as-triazino[5,6-*f*]phenanthroline
PET	photoinduced electron transfer
Ph	phenyl or phosphorescence
phehat	1,4,5,8,9,10,17,18-octaazaphenanthro[9,10-*b*]triphenylene
phen	1,10-phenanthroline
Pheo	pheophytin
phi	phenanthrene-9,10-diylidenediamine
phzi	benzo[*a*]phenazine-5,6-diylidenediamine
pip	2-phenylimidazo[4,5- 2 2*f*]-1,10-phenanthroline
PMCT	particle-to-metal charge transfer

POM	polyoxometallate
poq-Nmet	2-{2-[(7-chloroquinolin-4-yl)methylamino]ethylsulfanyl}-*N*-[1,10]-phenantrolin-5-yl-acetamide
ppip	2-(4′-phenoxy-phenyl)-imidazo-1,10-phenantroline
PQ	plastoquinone
PSI	photosystem I
PSII	photosystem II
PTT	photothermal therapy
pydppz	3-(pyrid-2′-yl)dipyrido[3,2-*a*:2′,3′-*c*]phenazine
qdppz	naptho[2,3-*a*]dipyrido[3,2-*h*:29,39-*f*]phenazine-5,18-dione
QD	quantum dot
qpy	2,2′:4′,4″:2″,2‴-quaterpyridine
RC	reaction centre
RNOS	reactive NO species
ROS	reactive oxygen species
SCF	supercritical fluid
SEM	semiconductor
Sens	sensitizer
SOD	superoxide dismutase
SP	nitrosporopyran
SPE	single-photon excitation
SSCT	second-sphere charge transfer
T	thymine
tap	pyrazino[2,3-*f*]quinoxaline
TAP	1,4,5,8-tetraazaphenanthrene
TEOA	2,2′,2″-nitrilotriethanol; triethanolamine
tex	texaphyrin
TON	turnover number
tmtp	tetra(4-methylphenyl)porphyrin
TPE	two-photon excitation
tpp	tetraphenylporphyrin
tpy (terpy)	2,2′:6′,2″-terpyridine
tren	triethylenetetramine
TS	transition state
U	uracil
UV	ultraviolet light
VB	valence band
VR	vibrational relaxation
XOD	xanthine oxidase

Part I
Introduction

1

Philosophy of Bioinorganic Photochemistry

The most important thing in science is not so much to obtain new facts as to discover new ways of thinking about them.
Sir William Bragg

Bioinorganic photochemistry is a rapidly growing and evolving new interdisciplinary research area integrating inorganic photochemistry with biological, medical, and environmental sciences (Figure 1.1) [1]. The role of light and inorganic species in natural systems and the possibility of their application in artificial systems of medical or environmental importance are in the limelight of bioinorganic photochemistry. From the earliest times humans have been aware of the influence that solar radiation exerts on matter and life; however, it is mainly during the last century that a systematic understanding of this phenomena has been developed [2–9]. Photochemistry of the inorganic species had its contribution in the creation of the world and has played a fundamental role in the evolution of life. Photosynthesis and many photoreactions proceeding in the atmosphere, hydrosphere and soil, involving inorganic species, ensure life on Earth. Bioinorganic solar photochemistry deals with the interaction of sunlight with inorganic matter, which has a huge impact on all forms of life on the Earth from its origin until now.

Sunlight supplies energy to the whole terrestrial environment: atmosphere, hydrosphere, lithosphere and biosphere. The spectral range of sunlight reaching our planet has varied with time. Atmospheric oxygen appeared owing to photosynthesis around 2.7 billion years ago. Atomic oxygen produced by short-wavelength ultraviolet (UV) irradiation (<240 nm) reacted then with molecular dioxygen to form an ozone layer shielding the Earth's surface from the most harmful UV. Four hundred million years ago the concentration of ozone reached 10% of the present level and allowed living systems to evolve from aquatic to terrestrial life. Today this ozone layer, with a maximum concentration in the stratosphere at 25 km above sea level, absorbs solar UV at wavelengths shorter than 290 nm. The radiation energy effective

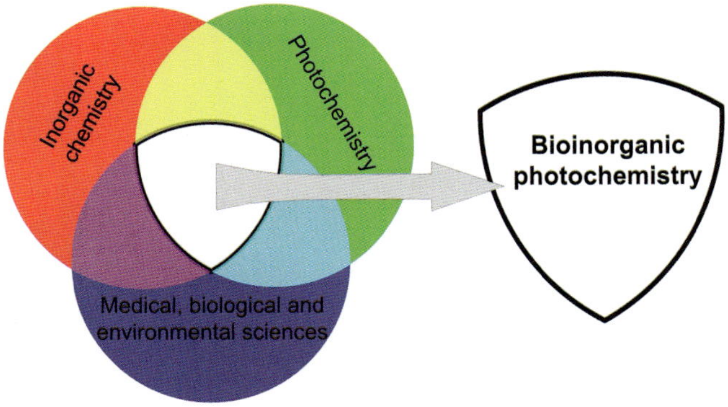

Figure 1.1 *Bioinorganic photochemistry connects inorganic photochemistry with biological, medical, and environmental sciences.*

for photobiology lies between 300 and 900 nm. Practically all photobiological behaviour of plants and animals, photosynthesis, phototropism, phototaxis, photoperiodism, and visions utilize this range of radiation [2–4, 10].

Natural photobiochemical processes, as a result of evolution, follow essentially the biologically desirable pathways [11]. In contrast, adventitious photobiochemical processes are likely to follow a multiplicity of pathways and usually find a variety of targets [11]. The mechanisms that underline both types of photoprocesses are highly complex, and their elucidation requires knowledge of the physics of light, the chemistry and structure of a photoacceptor molecule and its microenvironment, as well as physical and chemical processes leading to the final effect. Light is composed of energy packets called photons, which at the same time are energy quanta and bits of information. All the phenomena related to the interaction between light and matter, and the great number of photochemical and photophysical applications in science and technology, can ultimately be traced back to these two aspects of light [12]. Sun is the main light source, but nowadays light can also be provided by various artificial light sources. Introduction of lasers has caused an enormous increase of research on the interaction of light with biomatter [13–15].

The results obtained from the interaction of light with matter depend on the degree of organization of the receiving matter. To be useful for solar energy harvesting, organic synthesis, degradation of pollutants, therapeutic or diagnostic processes, etc the systems activated by light must fulfil various requirements [4, 12, 16–19]. In this context, inorganic photochemistry has recently attracted much attention because it offers feasible solutions [2, 12, 18–36]. Metal ions and other inorganic species can be involved in both natural and adventitious processes. There are numerous systems containing various metal ions and other inorganic species that are flexible enough to drive their photochemistry or photophysics towards desired actions [1, 13, 37, 38].

There is better understanding of the natural systems and phenomena that speeds the design of various molecular devices of medical, biochemical, or environmental importance. Metal complexes exhibit a high level of organization, so they are quite useful as components of molecular level photochemical devices. Moreover, a variety of transition metal-based supramolecular systems, or heterogeneous nanoassemblies supplemented with transition metal compounds, can be carefully designed to perform desired functions such as energy conversion, molecular sensing, labelling, switching, catalysis, etc [16, 17, 20, 29, 39–44]. Photochemical or photophysical processes induced by sunlight or artificial light sources are often damaging to biological systems, especially when suitable photoprotective mechanisms are absent or insufficient – typical examples are photocarcinogenesis and the photoinduced generation of pollutants. In many cases it is possible to take advantage of the damaging action of light to obtain beneficial effects. Achieving this goal requires a detailed knowledge of the mechanisms involved in a given photoprocess, so that their progress can be strictly controlled and intermediates and ultimate actions directed towards defined targets.

Photochemistry is the chemistry of excited electronic states. The change in electron distribution caused by photon absorption can cause substantial modifications in the chemical and physical properties of a molecule. Among these properties the energy, molecular geometries, polarizability, dipole and magnetic moments, and related redox and acid-based properties can change on passing to excited states [44]. Over the past decades inorganic photochemistry, which extends from simple Werner transition-metal complexes through supramolecular and multimetallic systems to homogeneous and heterogeneous nanoassemblies, has attracted increasing interest in various fields of science and technology, including bioscience and biotechnology [2–4, 11, 12, 16–36, 45–62]. One of the tremendous advantages of photochemical activation of transition-metal complexes is the generation of electronic excited states with moderate energy consumption. Transition-metal complexes distinguish themselves from organic compounds by both the number of accessible electronic excited states and their spin multiplicity. Consequently, depending on the wavelength (energy) of irradiation, optical excitation leads to various electronic excited states of different reactivity. In some favourable circumstances this behaviour allows tuning of photochemical reactivity and switching between various pathways such as electron transfer (preferably due to the population of diverse charge transfer states – CT), dissociation/substitution/rearrangement reactions (caused by excited metal-centred states – MC), and ligand-centred reactivity caused by the population of intraligand states (ligand centred – LC).

The great variety of available electronic excited states may be used for photogeneration of coordinatively unsaturated species, transition-metal compounds with changed formal oxidation numbers, as well as free ligands and ligand radicals. Such species generated photochemically not only can take part in stoichiometric processes but also open new pathways into both light-induced catalytic reactions and chain processes [21, 52]. Two limiting cases of photocatalysis, photoinduced catalytic and photoassisted reactions, can be distinguished. Photoinduced catalysis is the photogeneration of a catalyst that subsequently promotes a catalyzed reaction. Photoassisted reactions enable conversion of solar energy into useful energy. The activation of transition-metal complexes by visible and UV light provides definite advantages compared with the usual thermal activation. Catalysts usually have unique properties and are generated with high selectivity. The strategic synthetic design of transition-metal complexes or organometallic compounds, as well as the choice of the irradiation wavelength (ligand field, charge transfer, and intraligand excitation), allows modelling of photoinduced catalytic or photoassisted reactions. Photocatalytic reactions can be carried out at ambient temperature and pressure. To utilize the broadband solar energy spectrum it is necessary to involve some spectral sensitization [21, 52].

Application of inorganic photophysics generates challenging new areas in bioscience and biotechnology. Optical and, especially, fluorescence spectroscopy are widely used research tools in biochemistry, molecular biology, and environmental studies [31, 63]. Fluorescence has also become the dominant method enabling the revolution in medical diagnostics, DNA sequencing, and genomics. To date, all fluorescence observables, including spectral shifts, anisotropies, quantum yields, and lifetimes, have found both scientific and analytical applications [31]. New opportunities in fluorescence and radiation decay engineering, eg modification of fluorophore emissions by changing non-radiation decay rates has been described [32]. Transition-metal complexes have many potential advantages including numerous excited states of long lifetimes and high luminescence quantum yields [32, 33, 58–60, 64]. Luminescent polynuclear transition-metal complexes containing multichromophoric ligands with extended conjugation have been extensively studied in recent years, partly because of their potential use as sensors, labels, and probes in (bio)chemical systems [13, 35, 58, 65–70]. Many of them are easily excited by visible light with low-cost light-emitting diodes (LEDs) or inexpensive diode lasers. They show large spectral shifts between the excitation and emission bands that minimize the difficulty of isolating the excitation and emission wavelengths. Long lifetimes allow efficient time discrimination from the ubiquitous background fluorescence. Importantly, the longer lifetimes also allow the excited state ample time to sample its environment, making these materials particularly sensitive reporters [27, 33, 59, 60].

Supramolecular systems are constituted of a number of discrete molecular components with definite individual properties held together by various interactions. In natural systems the molecular components are frequently assembled by intermolecular forces (hydrogen bonds, donor–acceptor interactions, van der Waals' forces, etc), whereas in artificial systems covalent or coordination bonds are used to achieve a better control of the supramolecular structure [12, 16, 30, 40, 42, 56, 71–73]. Supramolecular systems began the concept of molecular devices, assemblies of molecular components designed to achieve specific functions, such as photoinduced electron and energy transfer in solar energy conversion, electron collection, photosensitization, antenna effect, photoswitching of electric signals, light-energy conversion, and photoinduced structural changes in switch on/off applications (photoisomerizable systems, molecular wires, and sensors) [12, 16, 17, 21, 29, 71].

The development of supramolecular chemistry has allowed construction of structurally organized and functionally integrated chemical systems capable of elaborating energy and information input photons so that they can perform complex functions [40, 41, 72, 74–83]. Over the past decade research on transition metal supramolecular systems has experienced extraordinary progress. In terms of bonding strength, the moderate coordination bonds between transition metals and ligands are between strong covalent bonding in carbon-based systems and weak interactions in biological systems. Some advantages of employing transition metals to build supramolecular systems include the following [30, 44, 58, 84]:

- Involvement of *d* orbitals, which offer more bonding modes and geometric symmetries than simple organic molecules
- A range of electronic and steric properties that can be fine-tuned by employing various ancillary ligands
- Easily modified size of the desired supermolecules by utilizing various lengths of bridging ligands
- Incorporation of their distinct spectral, magnetic, redox, photophysical, and photochemical properties.

Moreover, the diverse bonding angles imported by the transition-metal centres and the high directionality of the bonding between the ligands and metals also provide superior features over weak electrostatic, van der Waals', and π–π interactions. Another interesting aspect is that thermodynamically driven spontaneous self-assembly of individual molecular components into well-defined molecular structures in solution is expected to be rather similar for both coordination chemistry and biology, and this enables transition metal complexes to be valuable mimics of the more complicated biological systems [58, 71].

Recently, nanoscience and nanotechnology, which involve research on materials and species at length scales between 1 to 100 nm, have been revolutionizing important areas in environmental protection and biomedical photonics, especially diagnostics and therapy at the molecular levels [13, 16, 85–87]. The combination of photonics and nanotechnology has already led to new generation of methods and devices for probing the cell machinery and elucidating intimate life processes occurring at the molecular level. This will open the possibility of detecting and manipulating atoms and molecules using nanodevices, which have the potential for a wide variety of medical uses at the cellular level.

Colloidal-metal nanoparticles have found use in biology and medicine in the last 20 years, but semiconductor nanocrystals have only recently been used for biological labelling [13, 88–90].

A great trend in biotechnology is the development of multiplex sensing and ultrasensitive imaging technologies for the rapid molecular profiling of cells, tissues, and organs. These probes are traditionally based on organic dyes conjugated to biomolecules. As a result of their complex molecular structures, however, organic fluorophores often exhibit unfavourable absorption and emission properties such as photobleaching, environmental quenching, broad and asymmetric emission spectra, and the inability to emit various colours at a single wavelength excitation. These problems can be overcome by exploiting the unique optical properties of metal and semiconductor nanoparticles. The ongoing research attempts on bioconjugation of quantum dots to peptides, proteins, oligonucleotides and other biomolecules have demonstrated their applications in assembling new materials, in homogeneous bioassays, and multicolour fluorescence imaging and detection [13, 88, 91–93].

Nanostructured materials offer many new opportunities to study fundamental processes in a controlled manner, and this in turn leads to fabrication of numerous photonic and optoelectronic devices [13, 16, 17, 87, 94–98]. The design of photochemical molecular devices requires the ability to organize molecules on a nanometric scale with the fine control of their arrangement/distribution, mobility, and spectral and redox properties. Several types of heterogeneous multiphase systems have been proposed and tested. Mesoporous membrane-type films with a large surface area can be prepared from nanosized colloidal semiconductor dispersions. Nanocrystalline oxide or chalcogenide semiconductor thin films and particles such as TiO_2, ZnO, ZnS, CdS, and CdSe have been used for that purpose [13, 17, 85, 99, 100]. By suitable molecular engineering the metal complexes can be readily attached to the surface. These films with anchored photoactive complexes find increasing use in energy-conversion devices such as optical sensors [17, 54, 55]. For biophotonic applications the photoactive component has to meet several stringent requirements: the first is the intensity and spectral range of light absorption in the UV, visible, and near-infrared (IR) regions; the second is tunability of the absorption band; the third are photophysical properties: types and number of accessible excited states and their lifetimes and quantum yields of radiative and non-radiative decays; and the fourth are redox properties of the ground and excited states. For redox sensitizers or redox mediators there are further requirements of stability of both the redox forms and the reversibility of the redox processes. The interaction with biomatter is of primary importance. In all cases transition-metal complexes with polypyridines or polypyrrolic macrocycles as ligands come out clearly as the sensitizers of preferred choice [1, 13, 17].

Some specific aspects of light and inorganic compound interactions in bioscience and biotechnology have been reviewed [1, 12, 19, 22–28, 32–34, 36, 58–60]. Future perspectives of bioinorganic photochemistry will depend on both development of bioinorganic solar photochemistry and progress in understanding, as well as application of artificial light interaction with biomatter.

References

1. Szaciłowski K, Macyk W, Drzewiecka-Matuszek A, Brindell M, Stochel G. Bioinorganic photochemistry: Frontiers and mechanisms. *Chem Rev* 2005;**105**:2647–94.
2. Milgrom LR. *The Colours of Life*. Oxford: Oxford University Press, 1997.
3. Wolken JJ. *Light and Life Processes*. New York: Van Nostrand Reinhold Co., 1998.
4. Suppan P. *Chemistry and Light*. Cambridge: Royal Chemical Society (RSC), 1994.
5. Renger G. *Primary Processes of Photosynthesis* – Part 1. Cambridge: RSC, 2008.
6. Renger G. *Primary Processes of Photosynthesis* – Part 2. Cambridge: RSC, 2008.
7. Batschauer A. *Photoreceptors and Light Signalling*. Cambridge: RSC, 2003.
8. Frausto da Silva JJR, Williams RJP. *The Biological Chemistry of the Elements. The inorganic chemistry of life*. Oxford: Oxford University Press, 2001.
9. Blankenship R. *Molecular Mechanisms of Photosynthesis*. St Louis, MO: Washington University, 2002.
10. Sarna T, ed. *Advanced Course in Photobiology and Photomedicine*. Kraków: UJ Press, 1999.
11. Bonnett R. Photosensitizers of the porphyrin and phthalocyanine series for photodynamic therapy. *Chem Soc Rev* 1995;**24**:19.

12. Balzani V, Credi A, Venturi M. Photoprocesses. *Curr Opin Chem Biol* 1997;**1**:506–13.
13. Vo-Dinh T. *Biomedical Photonics Handbook*. New York: CRC Press, 2003.
14. Kraatz H-B, Metzler-Nolte N, eds. *Concepts and Models in Bioinorganic Chemistry*. Weinheim: Wiley, 2006.
15. Saleh BEA, Teich MC. *Fundamentals of Photonics*. New York: John Wiley & Sons, Inc., 1991.
16. Balzani V, Venturi M, Credi A. *Molecular Devices and Machines – A journey into nanoworld*. Weinheim: Wiley-VCH, 2003.
17. Kalyanasundaram K, Grätzel M. Applications of functionalized transition metal complexes in photonic and optoelectronic devices. *Coord Chem Rev* 1998;**77**:347–414.
18. Nalwa HS, ed. *Handbook of Photochemistry and Photobiology*, Vol 4. *Photobiology*. New York: American Scientific Publishers, 2003.
19. Litter MI. Heterogeneous photocatalysis. *Appl Catal B: Environ* 1999;**23**:89–114.
20. Balzani V, Scandola F. *Photochemical and Photophysical Devices*. Oxford: Pergamon Press, 1996.
21. Kalyanasundaram K, Grätzel M, eds. *Photosensitization and Photocatalysis Using Inorganic and Organometallic Compounds*. Dortrecht: Kluwer Academic Press, 1993.
22. Schance KS, Schmehl RH. Applications of inorganic photochemistry in the chemical and biological sciences. *J Chem Educ* 1997;**74**:633–5.
23. Bonnett R. Metal complexes for photodynamic therapy. In: McCleverty JA, Meyer TJ, eds. *Comprehensive Coordination Chemistry II*. Amsterdam: Elsevier, 2003: 945.
24. Faulkner S, Matthews JL. Fluorescent complexes for biomedical applications. In: McCleverty JA, Meyer TJ, eds. *Comprehensive Coordination Chemistry II*. Amsterdam: Elsevier 2003: 913.
25. Stochel G, Wanat A, Kuliś E, Stasicka Z. Light and metal complexes in medicine. *Coord Chem Rev* 1998;**171**:203–20.
26. Stochel G, Drzewiecka A, Kuliś E, et al. Photochemistry and photophysics of coordination compounds in medicine. In: Melnik M, Sirota A, eds. *Challenges for Coordination Chemistry in the New Century*. Bratislava: Slovak Technical University Press, 2001: 301–6.
27. Yam VW-W, Lo KK-W. Recent advances in utilization of transition metal complexes and lanthanides as diagnostic tools. *Coord Chem Rev* 1998;**184**:157–240.
28. Ali H, van Lier JE. Metal complexes as photo- and radiosensitizers. *Chem Rev* 1999;**99**:2379–450.
29. Balzani V, Scandola F. *Supramolecular Photochemistry*. New York: Ellis Horwood, 1991.
30. Balzani V, Credi A, Venturi M. Photochemistry and photophysics of coordination compounds. An extended view. *Coord Chem Rev* 1998;**171**:3–16.
31. Lakowicz JR. *Principles of Fluorescence Spectroscopy*. New York: Kluwer Academic/Plenum Publishers, 1999.
32. Lakowicz JR. Radiative decay engineering: biophysical and biomedical applications. *Anal Biochem* 2001;**298**:1–24.
33. Demas JN, DeGraff BA. Application of luminescent transition platinum group metal complexes to sensor technology and molecular probes. *Coord Chem Rev* 2001;**211**:317–51.
34. Cieśla P, Kocot P, Mytych P, Stasicka Z. Homogeneous photocatalysis by transition metal complexes in the environment. *J Mol Catal A: Chem* 2004;**224**:17–33.
35. Balzani V, Juris A, Venturi M, Compagna S, Serroni S. Luminescent and redox-active polynuclear transition metal complexes. *Chem Rev* 1996;**96**:759.
36. Burdette SC, Lippard SJ. ICCC34 – golden edition of coordination chemistry reviews. Coordination chemistry for the neurosciences. *Coord Chem Rev* 2001;**216–217**:333–61.
37. McCleverty JA, Meyer TJ, eds. *Comprehensive Coordination Chemistry II*. Amsterdam: Elsevier, 2003.
38. Balzani V, Campagna S, eds. *Photochemistry and Photophysics of Coordination Compounds*. Berlin: Springer-Verlag GmbH, 2007.

39. Balzani V, Credi A, Venturi M. *Molecular Devices and Machines. Concepts and perspectives for the nanoworld*. Weinheim: Wiley-VCH 2008.
40. Balzani V, Credi A, Venturi M. Molecular machines working on surfaces and at interfaces. *ChemPhysChem* 2008;**9**:202–20.
41. Balzani V, Credi A, Venturi M. Processing energy and signals by molecular and supramolecular systems. *Chem Eur J* 2008;**14**:26–39.
42. Balzani V, Credi A, Venturi M. Photochemical conversion of solar energy. *ChemSusChem* 2008;**1**:26–58.
43. Vos JG, Forster RJ, Keyes TA, eds. *Interfacial Supramolecular Assemblies*. Chichester: John Wiley & Sons Ltd, 2003.
44. Turro NJ. From molecular chemistry to supramolecular chemistry to superdupermolecular chemistry. Controlling covalent bond formation through non-covalent and magnetic interactions. *Chem Commun* 2002:2279–92.
45. Balzani V, Carassiti V. *Photochemistry of Coordination Compounds*. London: Academic Press, 1970.
46. Adamson AW. *Concepts of Inorganic Photochemistry*. New York: Wiley-VCH, 1975.
47. Fox MA. *Photoinduced Electron Transfer*. Amsterdam: Elsevier, 1988.
48. Sykora J, Sima J. Photochemistry of coordination compounds. *Coord Chem Rev* 1990;**107**:1.
49. Ramamurthy V, Schance KS. *Multimetallic and Macromolecular Inorganic Photochemistry*. New York: Marcel Dekker, 1999.
50. Nalwa HS, ed. *Handbook of Photochemistry and Photobiology*, Vol 1. *Inorganic Photochemistry*. Stevenson Ranch: American Scientific Publishers, 2003.
51. Roundhill DM. *Photochemistry and Photophysics of Metal Complexes*. New York: Plenum Press 1994.
52. Hennig H. Homogeneous photocatalysis by transition metal complexes. *Coord Chem Rev* 1999;**182**:101–23.
53. Grätzel M. *Energy Resources through Photochemistry and Catalysis*. New York: Academic Press, 1983.
54. Coleman JP, Lynch AT, Madhukar P, Wagenknecht JH. Printed, flexible electrochromic displays using interdigitated electrodes. *Sol Energy Mater Sol Cells* 1999;**56**:395–418.
55. Bonhôte P. Nanocrystalline electrochromic displays. *Displays* 1999;**20**:137–44.
56. Nalwa HS, ed. *Handbook of Photochemistry and Photobiology*, Vol 3. *Supramolecular Photochemistry*. New York: American Scientific Publishers, 2003.
57. Bonnett R. *Chemical Aspects of Photodynamic Therapy*. Amsterdam: Gordon Science Publishers, 2000.
58. Sun S-S, Lees AJ. Transition metal based supramolecular systems: synthesis, photophysics, photochemistry and their potential applications as luminescent anion chemosensors. *Coord Chem Rev* 2002;**230**:171–192.
59. Bergonzi R, Fabbrizzi L, Licchelli M, Mangano C. Molecular switches of fluorescence operating through metal centred redox couples. *Coord Chem Rev* 1998;**170**: 31–47.
60. Valeur B, Leray I. Design principles of fluorescent molecular sensors for cation recognition. *Coord Chem Rev* 2000;**205**:3–40.
61. Ariga K, Kunitake T. *Supramolecular Chemistry – Fundamentals and Applications*. Heidelberg: Springer Verlag, 2006.
62. Stroscio MA, Dutta M, eds. *Biological Nanostructures and Applications of Nanostructures in Biology*. New York: Kluwer Academic Publishers.
63. Valeur B. *Molecular Fluorescence. Principles and applications*. Weinheim: Wiley-VCH, 2002.
64. Vogler A, Kunkely H. Luminescent metal complexes: Diversity of excited states. *Top Curr Chem* 2001;**213**:143–82.
65. de Silva AP, Eilers J, Zlokarnik G. Emerging fluorescence sensing technologies: from photophysical principles to cellular applications. *Proc Natl Acad Sci USA* 1999;**96**: 8336–7.

66. de Silva AP, Fox DA, Huxley AJM, McClenaghan ND, Roiron J. Metal complexes as components of luminescent signalling systems. *Coord Chem Rev* 1999;**185–186**: 297–306.
67. de Silva AP, Fox DA, Huxley AJM, Moody TS. Combining fluorescence, coordination chemistry and electron transfer for signalling purposes. *Coord Chem Rev* 2000;**205**:41–57.
68. de Silva AP, Fox DB, Moody TS, Weir SM. The development of molecular fluorescent switches. *Trends Biotechnol* 2001;**19**:29–34.
69. de Silva AP, Gunaratne HQN, Gunnlaugsson T, et al. Signaling recognition events with fluorescent sensors and switches. *Chem Rev* 1997;**97**:1515–66.
70. de Silva AP, McClean GD, Moody TS. Fluorescent sensors. In: *Encyclopedia of Supramolecular Chemistry*. New York: Marcel Dekker, Inc., 2004: 572–8.
71. Lehn J-M. *Supramolecular Chemistry: Concepts and perspectives*. Weinheim: VCH, 1998.
72. Balzani V, Credi A, Venturi M. Light-powered molecular-scale machines. *Pure Appl Chem* 2003;**75**:541–7.
73. Balzani V, Sabbatini N, Scandola F. 'Second-sphere' photochemistry and photophysics of coordination compounds. *Chem Rev* 1986;**86**:319–37.
74. Balzani V, Ceroni P, Maestri M, Vicinelli V. Light-harvesting dendrimers. *Curr Opin Chem Biol* 2003;**7**:657–65.
75. Balzani V, Credi A, Raymo FM, Stoddart JF. Machines at the molecular level. *Angew Chem Int Ed* 2000;**39**:3348.
76. Balzani V, Credi A, Silvi S, Venturi M. Artificial nanomachines based on interlocked molecular species: recent advances. *Chem Soc Rev* 2006;**35**:1135–49.
77. Balzani V, Credi A, Venturi M. Molecular logic circuits. *ChemPhysChem* 2003;**3**:49–59.
78. Credi A. Artificial molecular motors powered by light. *Aust J Chem* 2006;**59**:157–69.
79. Credi A. Molecules that make decisions. *Angew Chem Int Ed* 2007;**46**:5472–5.
80. de Silva AP, McClenaghan ND. Molecular-scale logic gates. *Chem Eur J* 2004; **10**:574–86.
81. de Silva AP, Uchiyama S, Vance TP, Wannalerse B. A supramolecular basis for molecular logic and computation. *Coord Chem Rev* 2007;**251**:1623–32.
82. Szaciłowski K. Digital information processing in molecular systems. *Chem Rev* 2008;**108**: 3481–548.
83. Atwood JL, Davies JED, MacNicol DD, Vögtle F, Lehn J-M, eds. *Comprehensive Supramolecular Chemistry*. Oxford: Pergamon, 1996.
84. Balzani V, Bergamini G, Campagna S, Puntoriero F. Photochemistry and photophysics of coordination compounds: overview and general concepts. In: Balzani V, Campagna S, eds. *Photochemistry and Photophysics of Coordination Compounds I*. Berlin: Verlag 2007: 1–36.
85. Ozin GA, Arsenault AC. *Nanochemistry. A chemical approach to nanomaterials*. Cambridge: Royal Chemical Society, 2005.
86. Dupas C, Houdy P, Lahmani M, eds. *Nanoscience. Nanotechnologies and nanophysics*. Berlin: Springer Verlag, 2007.
87. Fahrner WR, ed. *Nanotechnology and Nanoelectronics*. Berlin: Springer Verlag, 2005.
88. Mirkin CA, Taton TA. Semiconductors meet biology. *Nature* 2000;**405**:626–7.
89. Daniel M-C, Astruc D. Gold nanoparticles: assembly, supramolecular chemistry, quantum size-related properties, and applications toward biology, catalysis, and nanotechnology. *Chem Rev* 2004;**104**:293–346.
90. Medintz IL, Clapp AR, Mattoissi H, Goldman ER, Fisher B, Mauro JM. Self-assembled nanoscale biosensors based on quantum dot FRET donors. *Nature Mat* 2003;**2**:630–8.
91. Niemeyer CM. Nanoparticles, proteins and nucleic acids: biotechnology meets material science. *Angew Chem Int Ed Eng* 2001;**40**:4128–58.
92. Mirkin CA. Programming the assembly of two and three-dimensional architectures with DNA and nanoscale inorganic building blocks. *Inorg Chem* 2000;**39**:2258–72.
93. Rosi NL, Mirkin CA. Nanostructures in biodiagnostics. *Chem Rev* 2005;**105**:1547–62.

94. Dragoman M, Dragoman D. *Nanoelectronics. Principles and devices*. Norwood, MA: Artech House, Inc., 2006.
95. Minolli D. *Nanotechnology Applications to Telecommunications and Networking*. Hoboken, NJ: Wiley-Interscience, 2006.
96. Regis E. *Nano: The emerging science of nanotechnology*. Newport Beach, CA: Back Bay Books, 1996.
97. Bhushan B, ed. *Handbook of Nanotechnology*. Berlin: Springer Verlag, 2004.
98. Lyshevski SE, ed. *Nano and Molecular Electronics*. London: CRC Press, 2007.
99. Kamat PV. *Composite Semiconductor Nanoclusters. Studies in surface science and catalysis*. Amsterdam: Elsevier, 1996.
100. Di Ventra M, Evoy S, Heflin JR Jr, eds. *Introduction to Nanoscale Science and Technology*. New York: Kluwer Academic Publishers, 2004.

Part II

Fundamentals

2

Light and Matter

Shadow is a colour as light is, but less brilliant; light and shadow are only the relation of two tones.
Paul Cezanne

2.1 Nature of Light

Photochemical processes are activated by interaction between light and matter. Light, similar to other electromagnetic radiations, is a form of energy that exhibits both particle-like and wave-like properties. An elementary particle of electromagnetic energy, in the sense of the wave–particle duality, is called a quantum. A photon is the quantum of electromagnetic energy (E) at a given frequency of the radiation (ν):

$$E = h\nu = hc/\lambda \tag{2.1}$$

where h is Planck's constant, λ wavelength of the radiation and c the velocity of light [1].

Visible light is characterized by its wavelengths between 400 and 750 nm corresponding to energy of 300–160 kJ/einstein (mole of photons), respectively. Vision is stimulated by the photochemical transformation of an eye pigment containing a moiety related to retinal, which is the aldehyde derived from vitamin A (retinol). The eye receptors are sensitive to the light energy:

- 300–267 kJ/einstein (400–450 nm) photons are received as violet
- 267–240 kJ/einstein (450–500 nm) as blue
- 240–209 kJ/einstein (500–575 nm) as green

- 209–203 kJ/einstein (575–590 nm) as yellow
- 203–185 kJ/einstein (590–650 nm) as orange
- 185–160 kJ/einstein (650–750 nm) as red.

The photochemical reaction becomes possible when a molecular entity is in an electronically excited state, ie the state where energy is higher than that of the ground state by about 150–1200 kJ/einstein (called the actinic radiation range). The relevant radiant energy involves not only the visible range, but also near-infrared (IR) (750–800 nm, 160–150 kJ) and ultraviolet (UV) radiation of wavelengths between 100 and 400 nm (1200–300 kJ/einstein). The photochemically active radiation in visible, near-UV and near-IR ranges is colloquially called 'light' [2, 3].

2.2 Accessible Light Sources

There is a large variety of light sources used in photochemistry; these can be: natural (sun, stars) or artificial (lamps, lasers); incoherent (sun, lamps) or coherent (lasers), monochromatic or polychromatic; operating continuously or pulsed. The most important light source, not just for natural photochemical processes, is the sun, the radiation spectrum of which spreads widely from radiowave to cosmic radiation (Figure 2.1). Passing through the atmosphere sunlight loses the most energetic radiation and only sunlight of wavelengths longer than 290 nm reaches sea level, ie apart from visible light it contains near UV, as UV-A (315–400 nm) and UV-B (290–315 nm). Every second the sun emits 2×10^{11} MJ of radiant energy towards the Earth, about 60% of which reaches sea level; this makes $>3 \times 10^{18}$ MJ/year. Unfortunately, only a small portion of this energy (about 2×10^{15} MJ/year) is converted to chemical energy by photosynthesis in plants and microorganisms [4].

Artificial sources of incoherent radiation functioning continuously involve:

- incandescent lamps (a tungsten filament incandescent in an inert or halogen atmosphere)
- electrical discharge lamps such as mercury lamps (medium pressure and high pressure), xenon, antimony–xenon or mercury–xenon lamps

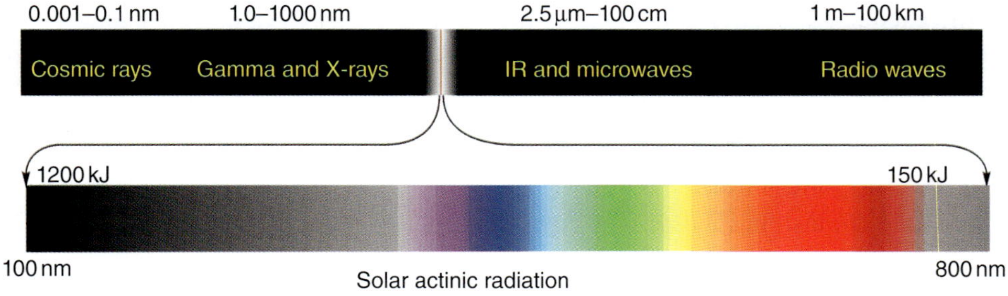

Figure 2.1 *Solar actinic radiation against its whole electromagnetic radiation spectrum*

- resonance lamps (low-pressure mercury lamps, lamps filled with pure vapour of other elements, eg Cd, Zn, Na, Kr, Xe)
- light emitting diodes (LEDs).

The spectral characteristic of this light depends on the kind of the source: the incandescent lamps are used primarily as sources of polychromatic visible light characterized by a continuous spectrum; electrical discharge lamps produce band or line spectra in UV, visible and near-IR regions, whereas resonance lamps emit resonance radiation of atoms and their ions in the form of line spectra and can be used as sources of monochromatic light, eg Hg (184.9 and 253.7 nm), Cd (228.8 and 643.8 nm), Na (589.0 nm), Zn (213.8, 330.0, 334.5, and 636.2 nm), Kr (116.5 and 123.6 nm), Xe (129.6 and 147.0 nm) [1].

Sources of pulsed incoherent radiation are so-called flash lamps, which produce short light pulses of broadband continuous-wave characteristic. Commonly, an intense pulse of short duration is used to generate sufficient concentration of a transient species suitable for spectroscopic observation.

Sources of coherent radiation in UV, visible, or IR range are lasers, which produce **l**ight **a**mplification by **s**timulated **e**mission of **r**adiation (from which the word laser is derived). Laser can be used as pulsed or non-pulsed source of electromagnetic radiation (marked as CW or continuous wave). Depending on the active medium (lasing species) there are different kinds of lasers, eg argon, helium–neon, helium–cadmium, krypton, nitrogen, CO_2, copper vapour, dye, semiconductor, neodymium, ruby laser, etc. They emit single lines or can be tunable lasers. Depending on the operation mode free-running, mode-locked, or Q-switched lasers can be distinguished [1, 5].

2.3 Interaction between Light and Matter

When a light beam passes a matter the different phenomena are observed, which are schematically illustrated in Figure 2.2. These encompass: reflection, diffusion, transmission, refraction, scattering, absorption, and luminescence.

For photochemical purposes only absorption and luminescence are of importance: absorption is the main method of excited state generation, whereas luminescence belongs to photophysical processes, which compete with the photoreactions in the excited state deactivation.

Quantitatively, absorption is often expressed as absorbance (A), which is the logarithm to the base 10 of the ratio of the spectral radiant power of incident, essentially monochromatic (of λ wavelength), radiation $\left(P_\lambda^0\right)$ to the radiant power of transmitted radiation (P_λ):

$$A = \log P_\lambda^0 / P_\lambda = -\log T \tag{2.2}$$

where T is the internal transmittance. The absorbance of a beam of collimated monochromatic radiation in a homogeneous isotropic medium is proportional to

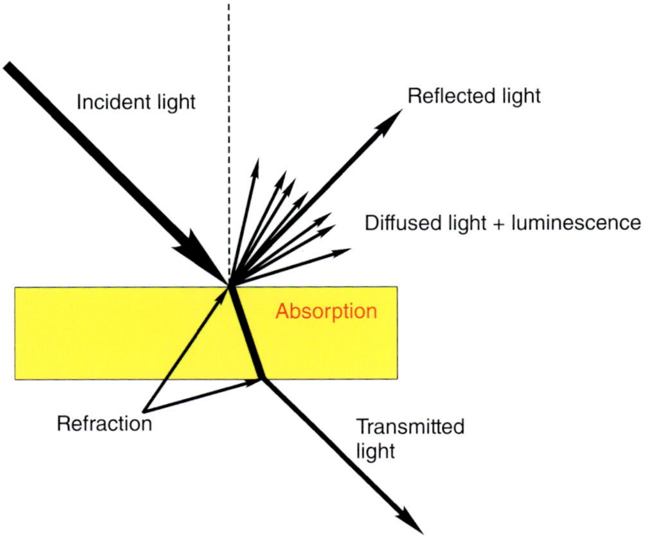

Figure 2.2 *Phenomena observed during interaction of light beams with matter*

the absorption pathlength, l, and to the concentration, c, or – in the gas phase – to the pressure of the absorbing species. This law, called the Beer–Lambert law can be expressed as:

$$A = \log(P_0/P) = \varepsilon c l \tag{2.3}$$

where the proportionality constant, ε, is called the molar absorption coefficient [1].

References

1. Braslavsky SE, Houk KN, Verhoeven JW. *Glossary of Terms used in Photochemistry*. International Union of Pure and Applied Chemistry, Organic Chemistry Division, Commission on Photochemistry, 1996.
2. Scandola F, Balzani V. Interaction between light and matter. In: Serpone N, Pelizzetti E, eds. *Photocatalysis Fundamentals and Application*. New York: Wiley-Interscience, 1989: 9–44.
3. Balzani V, Scandola F. *Supramolecular Photochemistry*. New York: Ellis Horwood Ltd, 1991.
4. Parmon VN, Zamaraev KI. Photocatalysis in energy production. In: Serpone N, Pelizzetti E, eds. *Photocatalysis Fundamentals and Applications*. New York: Wiley-Interscience, 1989: 565–602.
5. Krausz E, Riesen H. Laser spectroscopy. In: Solomon EI, Lever ABP, eds. *Inorganic Electronic Structure and Spectroscopy*. Oxford: Wiley-Interscience, 1999: 307–52.

3

Formation and Properties of Electronic Excited States

Everything in the future is a wave, everything in the past is a particle.
Lawrence Bragg

The interaction of light with a molecular system is generally an interaction between one molecule and one photon, which can be written:

$$R + h\nu \rightarrow {}^*R \tag{3.1}$$

where R and $^*R$ denote reagent molecule in its ground and electronically excited states and $h\nu$ the absorbed photon.

The energy transfer from an electromagnetic field to a molecular entity occurs through absorption of a photon by the entity in its ground state. The light absorbed by the sample is converted into the excitation energy of the species, which in turn is transformed into chemical energy and/or dissipated in the form of radiant or heat energy. As distinct from the ground-state entity, a molecule in its excited state lives only a very short time. The lifetime of the excited molecule (ie time needed for a concentration of the entity to decrease to $1/e$ of its original value) ranges from femtoseconds to milliseconds, depending on the number and rates of the deactivation processes. The lifetime depends on the molecule nature, energy, and type of the excited state, as well as on the composition and properties of its surrounding. Possible relaxation processes rely on (1) ground state regeneration and loss of the electronic excitation energy with or without a photon emission or (2) conversion of the excitation energy into a chemical energy. The former loss is classified as photophysical radiative or radiationless transition, respectively, whereas the latter is named a photochemical reaction.

The short lifetime is a consequence of the high energy of the molecule in excited state, which exceeds that of the ground state, mostly by 150–600 kJ/mol. Besides energy, the excited-state molecule differs from that in the ground state in many properties: electronic configuration, electronic density distribution, charge distribution, geometric structure, bond lengths and strengths, redox behaviour, acid–base behaviour, magnetic properties, etc. These differences not only make the excited-state entities short-lived, but often their reactivity would be completely unlike that of the ground-state species.

The molecule in its excited state cannot only be generated by light absorption, but also the excitation energy can be obtained in some chemical, biochemical, electrochemical processes, or by conversion of yet another kind of energy, eg ultrasonic or mechanical. The excited-state behaviour is, however, independent of its origin.

3.1 Wave Mechanics and Quantum Numbers

The quantitative treatment of absorption, electronic excitation, and deactivation of the excited state falls within the domain of wave mechanics.

The theory for a particle having a wavelength is represented by the Schrödinger equation, which, for the particle confined to a small region of space (such as an electron in an atom or molecule) can be solved only for certain energies, ie the energy of such particles is quantized or confined to discrete values. Moreover, some other properties, eg spin or orbital angular momentum, are also quantized.

The theory gives rise to a concept of atomic or molecular orbital, ie the wavefunction, which depends explicitly on the spatial coordinates of only one electron and the quantum numbers that define energy, spin, orbital momentum, and symmetry properties of the two last wavefunctions. Quantum numbers of the wavefunctions in lower and upper states determine the possible interaction of the entity with electromagnetic radiation [1].

Individual electrons possess both spin angular momentum and orbital angular momentum. The vector addition of these momentums generates the vector properties associated with a molecular entity. The spin moment of a single electron, s, is equal to ½. The overall molecular spin, S, which is the vector summation of spin moments of individual electrons, is important in defining the electronic states of the molecular entity. The number of possible spin orientations, calculated as $2S + 1$, corresponding to a given total spin quantum number S, for the same spatial electronic wavefunction, is called spin multiplicity [2]. It is common practice to specify electronic states by the spin multiplicity, rather than by the spin itself. For the molecular entity, in which all electrons are paired, $S = 0$ and $2S + 1 = 1$, whereas for two unpaired electrons $S = 1$, $2S + 1 = 3$. Depending on their spin properties, the electronic states receive special names and in some cases also symbols, eg singlet (S) state means that all electrons are paired, doublet (D) is used for one unpaired electron, triplet (T) – for two unpaired electrons (Table 3.1).

In the case of organic compounds and most simple inorganic molecules the ground state is characterized by the possible lowest number of unpaired electrons; thus, in most cases, it is a singlet state. The triplet states are commonly the excited states of the entities with singlet ground states (the spectacular exception is molecular oxygen). A simple scheme for describing electronic states of the molecular entity consists of a list of states numbered in order of their energy, eg S_0, S_1, S_2, etc, and

Table 3.1 Spin multiplicity of the electronic states

Number of unpaired electrons	Total spin quantum number, S	Multiplicity, 2S + 1	Name, symbol
0	0	1	Singlet (S)
1	1/2	2	Doublet (D)
2	1	3	Triplet (T)
3	3/2	4	Quartet
4	2	5	Quintet
5	5/2	6	Sextet

T_1, T_2, T_3, etc (for the case with S_0 ground state). For transition metal complexes, with central atoms that may have up to five unpaired electrons, the ground states that occur are of multiplicity from singlet to sextet (see Table 3.1); for six and seven unpaired electrons in *f* electronic elements, the states are called septet and octet (for f^6 and f^7, respectively).

Spin is of particular significance because, according to selection rules, the electronic transitions that do not involve a change in the spin part of the wavefunction are transitions of a high probability, called spin allowed, whereas the others (spin forbidden), although sometimes observed, are significantly less probable than transitions between states of the same multiplicity. Upon transfer of electronic energy between an excited molecular entity and other molecular entity in its ground or excited state, the overall spin angular momentum of the system should not change (Wigner's spin conservation rule) [2].

In a similar way as individual spin moments (s_1, s_2, s_3, etc) couple to give a total spin number S, also the individual angular momenta (denoted l_1, l_2, l_3, etc) couple to give a resultant total angular momentum, L. It is common practice to represent an electronic state by the symbol of total angular momentum and the spin multiplicity, denoted ^{2S+1}L.

The total angular momentum makes its contribution to the selection rules, according to which not only spin but also the symmetries of the wavefunctions of the initial and final states are important. Therefore, additional information is needed about the effect that symmetry operations have on the sign of the wavefunction – in particular, the effect of inversion through a centre of symmetry or reflection by a plane of symmetry passing through the molecule defines a state as 'even' or 'odd', denoted *g* (German *gerade*) or *u* (German *ungerade*), or as '+' or '–', respectively. For monophotonic radiative transitions in centrosymmetrical systems, the only non-vanishing electric-dipole transition moments are those that connect an even term (*g*) with an odd term (*u*) (Laporte's rule) [2].

3.2 Electronic Excitation

In diatomic or linear molecules, the electronic state may still be defined in part by the orbital electronic angular momentum. In the case of more complex molecules,

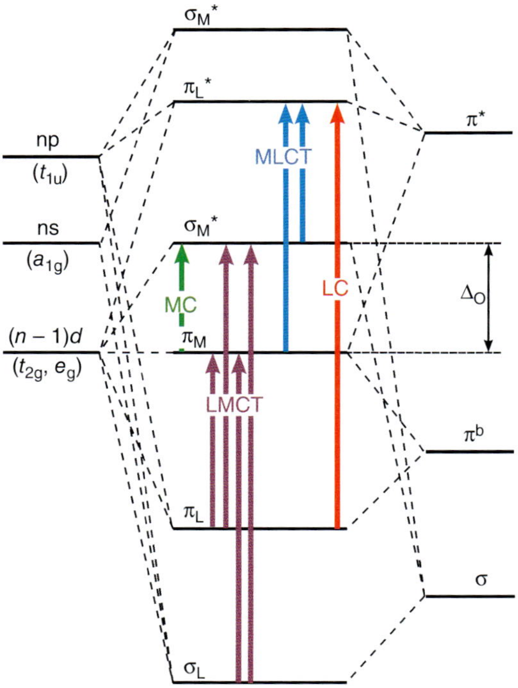

Figure 3.1 *Electronic transitions for an octahedral transition metal complex. LC, ligand centred; LMCT, ligand-to-metal charge transfer; MC, metal centred; MLCT, metal-to-ligand charge transfer*

symmetry elements may be non-existent and the state may be then described in terms of the individual molecular orbitals. The latter are characterized according to their role in bonding character: σ and π denote bonding orbitals, σ^* and π^* denote antibonding orbitals, and n denotes a non-bonding orbital.

Excitation in general consists in promotion of an electron from one of the highest-energy occupied molecular orbitals (HOMOs) to achieve one of the lowest-energy unoccupied molecular orbitals (LUMOs) of the ground-state molecule. This may be expressed, for example, by the terms (n, π^*), (n, σ^*), (π, π^*), or (π, σ^*), whereas the multiplicity is shown as a prefix, eg $^1(n, \pi^*)$. Population of the antibonding excited states thus usually results in bond weakening.

In coordination entities, the various electronic states are classified according to distribution of their electron density (Figure 3.1) [3]:

(1) ligand-centred, strongly σ-bonding orbitals (σ_L)
(2) ligand-centred, π-bonding orbitals (π_L)
(3) metal-centred π-bonding orbitals (π_M), predominantly metal d-orbitals (t_{2g})
(4) metal-centred σ-antibonding orbitals (σ_M^*), predominantly metal d-orbitals (e_g)

(5) ligand-centred π-antibonding orbitals (π_L^*)
(6) metal-centred, strongly σ-antibonding (σ_M^*) orbitals.

Accordingly, the electronic transitions are classified into those in which charge redistribution takes place within a central atom (of angular character) or within ligands, or those with charge transfer between the central atom and ligands (radial charge redistribution). At relatively low energies (visible and near-UV range), three main types of electronic transitions may be expected (Figure 3.1) [3–6]:

1. Transitions from essentially non-bonding to antibonding *d*-orbitals localized in central atom, called metal-centred (MC) or ligand field (LF) or *d–d* transitions (characterized by the ligand field splitting energy, Δ)
2. Transitions from π-bonding to π-antibonding ligand orbitals, called ligand-centred (LC) or internal ligand (IL)
3. Transitions with a fairly large radial charge transfer: ligand-to-metal charge transfer (LMCT) and metal-to-ligand charge-transfer (MLCT) transitions.

Under special conditions still further transitions can be observed as (1) metal-to-metal (MMCT) or intervalence (IT) charge transfers characteristic for polynuclear complexes and (2) charge transfer to solvent (CTTS) or ion-pair-charge-transfer (IPCT) detectable in cases of fairly strong external interactions [3].

In investigation of the nature and fate of the excited-state the special very fast techniques had to used [6–9].

References

1. Wayne CE, Wayne RP. *Photochemistry*. Oxford: Oxford University Press, 2001.
2. Braslavsky SE, Houk KN, Verhoeven JW. *Glossary of Terms used in Photochemistry*. International Union of Pure and Applied Chemistry, Organic Chemistry Division, Commission on Photochemistry, 1996.
3. Balzani V, Scandola F. *Supramolecular Photochemistry*. New York: Ellis Horwood Ltd, 1991.
4. Grills DC, Huang KW, Muckerman JT, Fujita E. Kinetic studies of the photoinduced formation of transition metal–dinitrogen complexes using time-resolved infrared and UV–visible spectroscopy. *Coord Chem Rev* 2006;**250**:1681–95.
5. Vogler A, Kunkely H. Excited state properties of organometallics based on interligand interactions between aromatic ligands. *Coord Chem Rev* 2005;**249**:1511–16.
6. Vlcek A, Jr. Mechanistic roles of metal-to-ligand charge-transfer excited states in organometallic photochemistry. *Coord Chem Rev* 1998;**177**:219–56.
7. Suppan P. *Chemistry and Light*. London: The Royal Society of Chemistry, 1994.
8. Turner JJ. Infrared vibrational band shapes in excited states. *Coord Chem Rev* 2002;**230**:213–24.
9. Mataga N, Chosrowjan H, Taniguchi S. Ultrafast charge transfer in excited electronic states and investigations into fundamental problems of exciplex chemistry: Our early studies and recent developments. *J Photochem Photobiol C: Photochem Rev* 2005;**6**:37–79.

4

Photophysical Deactivation of Electronic Excited States

In order for the light to shine so brightly, the darkness must be present.
Francis Bacon

4.1 Spontaneous Deactivation

Excited-state relaxation can proceed spontaneously in monomolecular processes or can be stimulated by a molecular entity (quencher) that deactivates (quenches) an excited state of another molecular entity, by energy transfer, electron transfer, or a chemical mechanism [1]. The quenching is mostly a bimolecular radiationless process (the exception is a quencher built into the reactant molecule), which either regenerates the reactant molecule dissipating an energy excess or generates a photochemical reaction product (Figure 4.1).

Photoexcitation by photon absorption and subsequent events that lead from one to another state of a molecular entity through radiation and radiationless transitions without any chemical change are called photophysical processes. The processes are classified as radiative and radiationless ones, depending on the photon emission (or absorption) and energy loss without any photon emission; according to the kinetic aspects the monomolecular (spontaneous) and bimolecular (quenched) processes are distinguished (see Figure 4.1).

Spontaneous emission of radiation from an electronically and/or vibrationally excited species not in thermal equilibrium with its environment is called luminescence: when the emission is accompanied by the formation of a molecular entity of the same spin multiplicity it is called fluorescence (F), whereas luminescence involving change in spin multiplicity (typically from triplet to singlet) is called phosphorescence (Ph) [1].

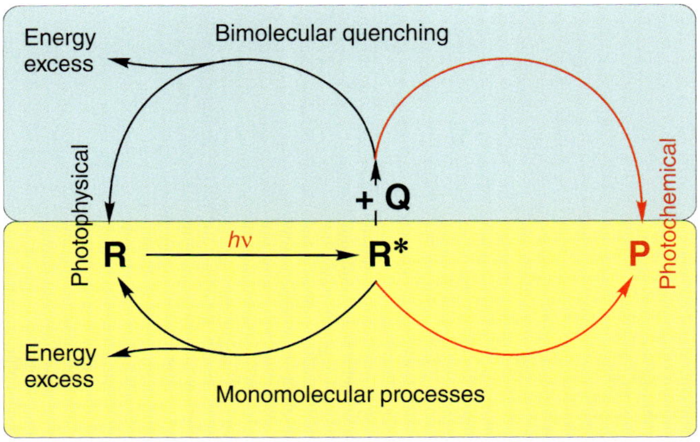

Figure 4.1 *Excitation by light absorption (hν) and main pathways of the excited state deactivation: R and R*, the reactant molecule in its ground and excited state, respectively; P, photochemical reaction product; Q, quencher molecule*

Radiationless deactivation processes involve vibrational relaxation (VR), internal conversion (IC), intersystem crossing (ISC), and quenching (Q). Vibrational relaxation is the loss of vibrational excitation energy by a molecular entity through energy transfer to the environment caused by collisions. Internal conversion is an isoenergetic radiationless transition between two electronic states of the same multiplicity. Typically, the transition results in a vibrationally excited molecular entity of the target electronic state; the internal conversion is followed by instantaneous vibrational relaxation, leading to the lowest vibrational level. Intersystem crossing is an isoenergetic radiationless transition between two electronic states with different multiplicities. It often results in a vibrationally excited molecular entity in the lower electronic state, which then usually deactivates to its lowest vibrational level.

The best tool for interpreting the nature of the photophysical processes is the so-called Jabłoński diagram (Figure 4.2), constructed in the first half of the twentieth century by Polish physicist Aleksander Jabłoński (1898–1980) to show that the fluorescent state of a molecular entity is the lowest excited state from which the transition to the ground state is allowed, whereas the phosphorescent state is a metastable state below the fluorescent state, reached by radiationless transition. Currently, modified Jabłoński diagrams are frequently used and are actually state diagrams in which molecular electronic states, represented by horizontal lines displaced vertically to indicate relative energies, are grouped according to multiplicity into horizontally displaced columns. Excitation and relaxation processes that interconvert states are indicated in the diagram by arrows. Radiative transitions are generally indicated with straight arrows (→), whereas radiationless transitions are generally indicated with wavy arrows (⤳) [2].

A photon absorption leads to population of the electronic excited state of the same multiplicity, eg for the singlet S_0 ground state it is the S_1, S_2, or S_n state, depending on the absorbed photon energy. When excitation leads to higher vibrational

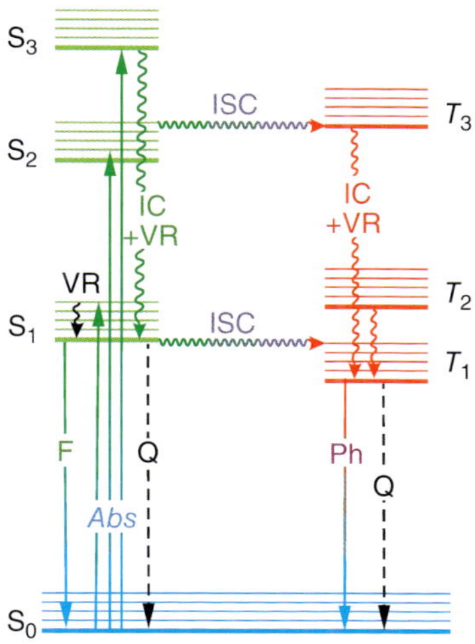

Figure 4.2 *Jabłoński diagram for a molecule in the singlet ground state (S_0). Abs, absorption; F, fluorescence; IC, internal conversion; ISC, intersystem crossing; Ph, phosphorescence; Q, quenching; VR, vibrational relaxation*

states of S_n, then the molecular entity relaxes fast into vibrational equilibrium with its environment. The excited state deactivation may also proceed via internal conversion or intersystem crossing. Population by absorption of the states that differ from the ground state because of multiplicity (eg triplet states) is forbidden by the selection rule, although in metal complexes this rule is not obeyed precisely. These states are normally populated via intersystem crossing. The lowest excited states are the relatively long lived, especially those that have different multiplicity from the ground state; thus they are the source of luminescence and, moreover, they are most often responsible for a photochemical behaviour.

4.2 Quenching

Not only do the excited states deactivate spontaneously, but their decay may be induced by a resonant perturbing electromagnetic field (stimulated emission) or the quenching effect (see Figure 4.1). An external environmental influence or even substituent in the excited molecule (external or internal quencher) may also induce its deactivation through a non-radiative inter- or intramolecular process, respectively. As stated earlier a quencher is a molecular entity that deactivates (quenches) an excited state of another molecular entity, by energy transfer, electron transfer, or a chemical mechanism [1]. Quenching is especially important in the liquid phase, where collisions are very frequent. The quenching processes are relevant in such important fields as photosensitization and photocatalysis (see Chapter 6).

The term 'energy transfer' is used to describe a photophysical process in which an excited state of one molecular entity (donor, e.g. AB*) is deactivated to a lower-lying state by transferring energy to a second molecular entity (acceptor, Q), which is thereby raised to a higher energy state:

$$AB^* + Q \rightarrow AB + Q^* \tag{4.1}$$

A quencher should have an excitation energy lower than that of the donor species and the appropriate electronic configuration. Transfer of excitation energy proceeds by radiative or radiationless deactivation of the donor molecular entity. Radiative energy transfer (also called trivial energy transfer) consists of light emission by the donor molecule and reabsorption of the emitted light by the acceptor molecular entity.

Radiationless energy transfer can proceed via an electron exchange mechanism (called Dexter excitation transfer – Figure 4.3). It requires an overlap of the wavefunctions of the energy donor and the energy acceptor, and the overall spin angular momentum of the system should not change (spin conservation rule). The simultaneous electron exchange needs the electronic configuration enabling the exoergic electron transfer. Although both singlet and triplet excited states can be quenched, the dominant mechanism is the triplet–triplet energy transfer [3].

The energy transfer can also proceed through an alternative mechanism, which can occur between molecular entities separated by distances considerably greater than the sum of their van der Waals' radii (eg in molecular monolayers). The electron shift to less energy in AB* and the increase in energy of electron in Q are coupled by their electric fields. It is described in terms of an interaction between the transition dipole moments (a dipolar mechanism, called Förster excitation transfer, Figure 4.3).

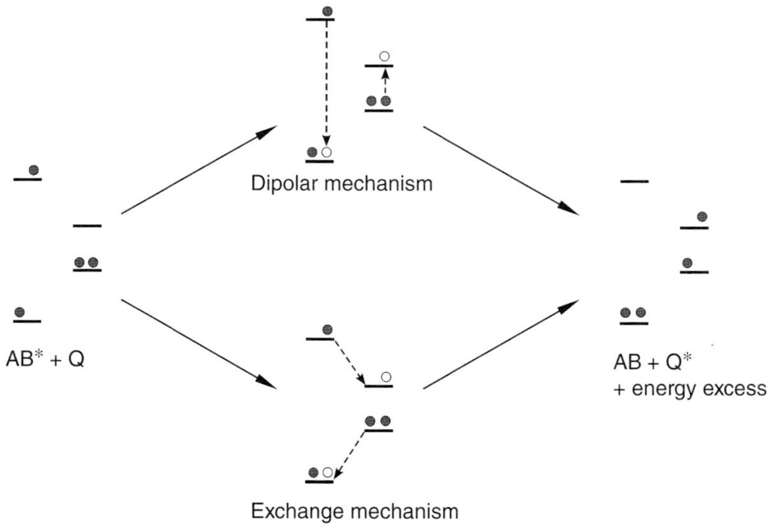

Figure 4.3 Scheme of the alternative mechanisms of energy transfer from the excited AB* donor molecule to the quencher Q. (Adapted from Scandola et al. [4])

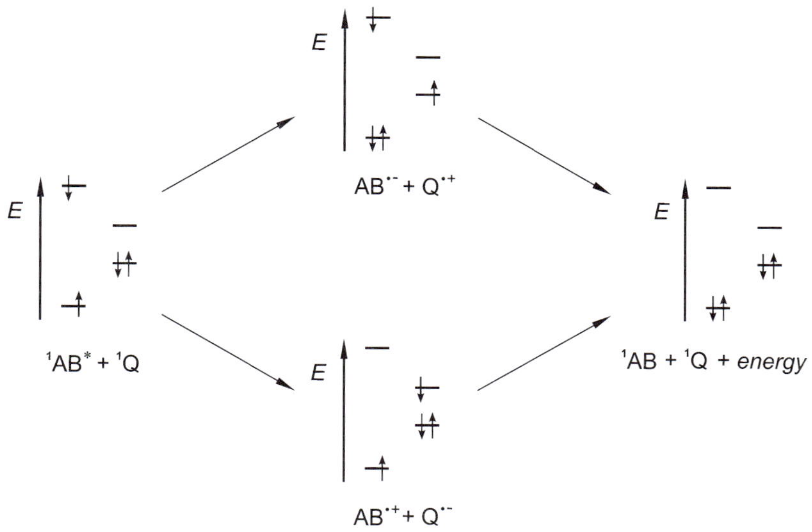

Figure 4.4 *Excitation quenching by electron transfer between excited* $^1AB^*$ *molecule and a quencher in its singlet ground state,* ^1Q

Excited states may be quenched as well via an electron transfer between the excited and quencher molecular entities. The electron can be transferred by two alternative ways, generating a radical anion and cation as a transient species (Figure 4.4). These then react thermally: when the reaction leads to reproduction substrate AB and quencher Q in their ground states, the photophysical deactivation occurs; when radical ions react with other medium components generating new species, the process belongs to photochemical redox reactions (see Chapter 6).

Photophysical processes are commonly encountered and can participate in such important phenomena as energy harvesting (antenna effect), which is the basis for natural photosynthesis and artificial systems mimicking it [4–21] (see below).

4.3 Coordination and Organometallic Compounds

The Jabłoński diagram presented in Figure 4.2 exemplifies electronic states for a molecule in the singlet ground state, which is typical for organic compounds and the most common for simple inorganic molecules. In the case of coordination and organometallic entities, however, the multiplicities of their ground states range from singlet to sextet (or even octet in the case of *f* elements), depending on the nature and configuration of the central atom, as well as the nature and field symmetry of the ligands. The same concerns the excited states of these compounds, and as opposed to organic compounds the multiplicity of excited states can be lower than that of the ground state. Chromium(III) complexes may exemplify this case, for which the ground state is quartet, whereas its excited states are quartets and doublets (Figure 4.5).

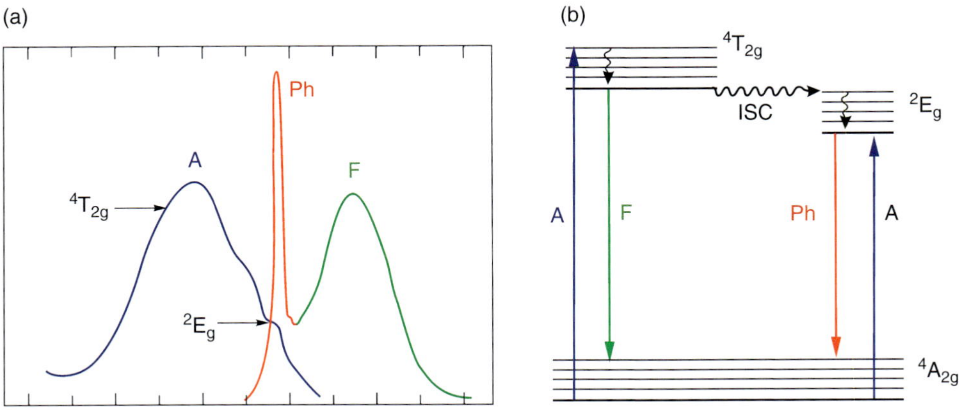

Figure 4.5 *(a) Spectra of the Cr^{III} complex with urea [Cr(urea)$_6$]$^{3+}$: absorption (A), phosphorescence (Ph) and fluorescence (F); (b) the relevant electronic transitions (ISC, intersystem crossing)*

Nevertheless, in coordination and organometallic compounds the same photophysical processes can be observed as those presented in Figure 4.2 [4, 18, 22–31], which can be exemplified by Cr^{III} complexes with various ligands, for which the intersystem crossing and other photophysical processes have been studied most efficiently [24, 26, 32–34] (Figure 4.5).

The broad emission band (fluorescence) is a mirror image of the absorption band generated by the spin-allowed transition (between the quartet ground state of the Cr^{III} complex, $^4A_{2g}$, and the lowest excited quartet state, $^4T_{2g}$), which is shifted towards long waves. The narrow emission (phosphorescence) derives from the forbidden $^4A_{2g} \rightarrow {}^2E_g$ transition. The difference (usually in frequency units) between the spectral positions of the band maxima of the absorption and luminescence arising from the same electronic transition is called Stokes shift. Generally, the luminescence occurring at a longer wavelength than the absorption is stronger than the opposite, which is called an anti-Stokes shift. Similar to other systems, fluorescence and phosphorescence differ in their lifetimes: phosphorescence as derived from a forbidden transition is a relatively long-lived process; phosphorescence lifetimes of metal complexes at 77 K are of the millisecond order, whereas at room temperature they are of the micro- or nanosecond order.

Luminescence of metal complexes is widely applied (eg to sensors or luminescent probes [17, 24, 25, 29, 35, 36]); the effects of components (chromophores) and ambient medium (solvents) on its lifetime have been intensively studied. To prolong luminescence lifetime, metal complexes have recently been built into supramolecular systems [24, 37, 38].

Luminescence of the metal complexes can also be a result of the excited state population in an alternative way to absorption, eg as seen in the outcome of the chemical reaction (chemiluminescence) or electrochemical processes (electrochemiluminescence), which are often applied in analytical chemistry [24, 25, 37, 39–44].

References

1. Braslavsky SE, Houk KN, Verhoeven JW. *Glossary of Terms used in Photochemistry*. International Union of Pure and Applied Chemistry, Organic Chemistry Division, Commission on Photochemistry, 1996.
2. McNaught AD, Wilkinson A. *Compendium of Chemical Terminology. IUPAC recommendations*. Oxford: Blackwell, Science, 1997.
3. Grills DC, Huang KW, Muckerman JT, Fujita E. Kinetic studies of the photoinduced formation of transition metal–dinitrogen complexes using time-resolved infrared and UV–visible spectroscopy. *Coord Chem Rev* 2006;**250**:1681–95.
4. Scandola F, Chiorboli C, Prodi A, Iengo E, Alessio E. Photophysical properties of metal-mediated assemblies of porphyrins. *Coord Chem Rev* 2006;**250**:1471–96.
5. Zhang X, Xiao Y, Qian X. Highly efficient energy transfer in the light harvesting system composed of three kinds of boron-dipyrromethene derivatives. *Org Letters* 2008;**10**: 29–32.
6. Chen Y, Zhuang X, Zhang W, et al. Synthesis and characterization of phthalocyanine-based soluble light-harvesting CIGS complex. *Chem Mater* 2007;**19**:5256–61.
7. Balzani V, Bergamini G, Marchioni F, Ceroni P. Ru(II)-bipyridine complexes in supramolecular systems, devices and machines. *Coord Chem Rev* 2006;**250**:1254–66.
8. Elias B, Kirsch-De Mesmaeker A. Photo-reduction of polyazaaromatic Ru(II) complexes by biomolecules and possible applications. *Coord Chem Rev* 2006;**250**:1627–41.
9. Roger C, Muller MG, Lysetska M, Miloslavina Y, Holzwarth AR, Wurthner F. Efficient energy transfer from peripheral chromophores to the self-assembled zinc chlorin rod antenna: a bioinspired light-harvesting system to bridge the 'green gap'. *J Am Chem Soc* 2006;**128**:6542–3.
10. Ward MD. $[Ru(bipy)(CN)_4]^{2-}$ and its derivatives: Photophysical properties and its use in photoactive supramolecular assemblies. *Coord Chem Rev* 2006;**250**:3128–41.
11. Baitalik S, Wang X, Schmehl RH. A Trimetallic mixed Ru(II)/Fe(II) terpyridyl complex with a long-lived excited state in solution at room temperature. *J Am Chem Soc* 2004;**126**:16304–5.
12. Wang P, Klein C, Humphry-Baker R, Zakeeruddin SM, Grätzel M. A high molar extinction coefficient sensitizer for stable dye-sensitized solar cells. *J Am Chem Soc* 2005; **127**:808–9.
13. Amadelli R, Argazzi R, Bignozzi CA, Scandola F. Design of antenna-sensitizer polynuclear complexes. Sensitization of titanium dioxide with $[Ru(bpy)_2(CN)_2]\ Ru(bpy(COO)_2)_2^{2-}$. *J Am Chem Soc* 1990;**112**:7099–103.
14. Balzani V, Campagna S, Denti G, Juris A, Serroni S, Venturi M. Designing dendrimers based on transition-metal complexes. Light-harvesting properties and predetermined redox patterns. *Acc Chem Res* 1998;**31**:26–34.
15. Balaban TS. Tailoring porphyrins and chlorins for self-assembly in biomimetic artificial antenna systems. *Acc Chem Res* 2005;**38**:612–23.
16. Li X, Sinks LE, Rybtchinski B, Wasielewski MR. Ultrafast aggregate-to-aggregate energy transfer within self-assembled light-harvesting columns of zinc phthalocyanine tetrakis (perylenediimide). *J Am Chem Soc* 2004;**126**:10810–11.
17. Yersin H, Kratzer C. Energy transfer and harvesting in $[Ru_{1-x}Os_x(bpy)_3](PF_6)_2$ and {L-$[Ru(bpy)_3]\Delta$-$[Os(bpy)_3]\}(PF_6)_4$. *Coord Chem Rev* 2002;**229**:75–93.
18. Campagna S, Di Pietro C, Loiseau F, et al. Recent advances in luminescent polymetallic dendrimers containing the 2,3-bis(2′-pyridyl)pyrazine bridging ligand. *Coord Chem Rev* 2002;**229**:67–74.
19. Sommovigo M, Denti G, Serroni S, et al. Polynuclear polypyridine complexes incorporating Ru(II), Os(II), and Pt(II): decanuclear dendrimeric antennas. *Inorg Chem* 2001;**40**: 3318–23.
20. Kalyanasundaram K, Grätzel M, Nazeeruddin MK. Luminescence and intramolecular energy-transfer processes in isomeric cyano-bridged rhenium(I)-rhenium(I) and

rhenium(I)-ruthenium(II)-rhenium(I) polypyridyl complexes. *Inorg Chem* 1992;**31**: 5243–53.
21. Fleming CN, Maxwell KA, DeSimone JM, Meyer TJ, Papanikolas JM. Ultrafast excited-state energy migration dynamics in an efficient light-harvesting antenna polymer based on Ru(II) and Os(II) polypyridyl complexes. *J Am Chem Soc* 2001;**123**:10336–47.
22. Adamson AW. Properties of excited states. *J Chem Educ* 1983;**60**:797–802.
23. Demas JN. Photophysical pathways in metal complexes. *J Chem Educ* 1983;**60**:803–8.
24. Lees AJ. Quantitative photochemistry of organometallic complexes: insight to their photophysical and photoreactivity mechanisms. *Coord Chem Rev* 2001;**211**:255–78.
25. Sun SS, Lees AJ. Transition metal based supramolecular systems: synthesis, photophysics, photochemistry and their potential applications as luminescent anion chemosensors. *Coord Chem Rev* 2002;**230**:171–92.
26. Wang X, Del Guerzo A, Schmehl RH. Photophysical behavior of transition metal complexes having interacting ligand localized and metal-to-ligand charge transfer states. *J Photochem Photobiol C: Photochem Rev* 2004;**5**:55–77.
27. Jukes RTF, Adamo V, Hartla F, Belser P, De Cola L. Electronic energy transfer in a dinuclear Ru/Os complex containing a photoresponsive dithienylethene derivative as bridging ligand. *Coord Chem Rev* 2005;**249**:1327–35.
28. Prodi A, Indelli MT, Kleverlaan CJ, Alessio E, Scandola F. Energy transfer pathways in pyridylporphyrin metal adducts and side-to-face arrays. *Coord Chem Rev* 2002;**229**: 51–8.
29. Zhao S-B, McCormick T, Wang S. Ambient-temperature metal-to-ligand charge-transfer phosphorescence facilitated by triarylboron: boron and its metal complexes. *Inorg Chem* 2007;**46**:10965–7.
30. Tears DKC, McMillin DR. Exciplex quenching of photoexcited platinum(II) terpyridines: influence of the orbital parentage. *Coord Chem Rev* 2001;**211**:195–205.
31. Omary MA, Mohamed AA, Rawashdeh-Omary MA, Fackler JP Jr. Photophysics of supramolecular binary stacks consisting of electron-rich trinuclear Au(I) complexes and organic electrophiles. *Coord Chem Rev* 2005;**249**:1372–81.
32. Porter GB. Introduction to inorganic photochemistry. *J Chem Educ* 1983;**60**:785–90.
33. Forster LS. Intersystem crossing in transition metal complexes. *Coord Chem Rev* 2006;**250**:2023–33.
34. Forster LS. Thermal relaxation in excited electronic states of d^3 and d^6 metal complexes. *Coord Chem Rev* 2002;**227**:59–92.
35. Demas JN, DeGraff BA. Applications of luminescent transition platinum group metal complexes to sensor technology and molecular probes. *Coord Chem Rev* 2001;**211**: 317–51.
36. Rogers CW, Wolf MO. Luminescent molecular sensors based on analyte coordination to transition-metal complexes. *Coord Chem Rev* 2002;**233–234**:341–50.
37. McClenaghan ND, Leydet YM, Indelli MT, Campagna S. Excited-state equilibration: a process leading to long-lived metal-to-ligand charge transfer luminescence in supramolecular systems. *Coord Chem Rev* 2005;**249**:1336–50.
38. Amendola V, Fabbrizzi L, Foti F, et al. Light-emitting molecular devices based on transition metals. *Coord Chem Rev* 2006;**250**:273–99.
39. Noffsinger JB, Danielson ND. Generation of Chemiluminescence upon reaction of aliphatic amines with tris(2,2′-bipyridine)ruthenium(III). *Anal Chem* 1987;**59**:865–8.
40. Sato K, Tanaka S. Determination of metal ions by flow injection analysis with peroxyoxalate chemiluminescence detection. *Microchem J* 1996;**53**:93–8.
41. Yıldız G, Demiryurek AT. Ferrous iron-induced luminol chemiluminescence: a method for hydroxyl radical study. *J Pharm Toxic Methods* 1998;**39**:179–84.
42. Voloshin AI, Shavaleev NM, Kazakov VP. Water enhances photoluminescence intensity of europium(III), terbium(III) and samarium(III) tris-β-diketonates in toluene solutions and chemiluminescence intensity of europium(III) and samarium(III) tris-β-diketonates in the reaction with dioxetane. *J Photochem Photobiol A: Chemistry* 2000;**136**:203–8.

43. Barnett NW, Hindson BJ, Jones P, Smith TA. Chemically induced phosphorescence from manganese(II) during the oxidation of various compounds by manganese(III), (IV) and (VII) in acidic aqueous solutions. *Anal Chim Acta* 2002;**451**:181–8.
44. Brown AJ, Lenehan CE, Francis PS, Dunstan DE, Barnett NW. Soluble manganese(IV) as a chemiluminescence reagent for the determination of opiate alkaloids, indoles and analytes of forensic interest. *Talanta* 2007;**71**:1951–7.

5

Kinetics of the Excited-State Decay

Science never solves a problem without creating ten more.
George Bernard Shaw

Photophysical and photochemical processes are characterized quantitatively by the quantum yield value (Φ_λ), which determines the number of defined events occurring per photon absorbed by the system (λ is the wavelength of absorbed radiation) [1]. Integral quantum yield is defined by:

$$\Phi_\lambda = \frac{\text{number of events}}{\text{number of photons absorbed}} \tag{5.1}$$

where both quantities relate to the same volume and time interval. In addition to the integral the differential quantum yield is also used:

$$\Phi_\lambda = n\mathrm{d}x/\mathrm{d}t \tag{5.2}$$

where $\mathrm{d}x/\mathrm{d}t$ is the rate of change of a measurable quantity, and n is the amount of photons (mole or its equivalent, einstein) absorbed per unit time. Φ_λ can be used for photophysical processes or photochemical reactions, eg quantum yield of fluorescence, phosphorescence, or photochemical reaction. In the last case, Φ_λ can characterize the primary photochemical process (Φ_1, primary quantum yield), generation of a selected product (Φ_P, product quantum yield), or photoreaction quantum yield (quantum yield of substrate decay).

The quantum yield for the photochemical reaction is determined by:

$$\Phi_\lambda = \frac{\text{amount of reactant consumed or product formed}}{\text{amount of photons absorbed}} \tag{5.3}$$

or by differential quantum yield $\Phi_\lambda = n$ dc/dt, where dc/dt means rate of the photochemical reaction expressed by the substrate decay ($-$dc_R/dt) or product formation (dc_P/dt) [1].

The rate of photochemical conversion of reactant R into product(s), P:

$$\mathrm{R} \xrightarrow{h\nu} \mathrm{P} \tag{5.4}$$

depends on the amount of the monochromatic light absorbed per unit time and volume ($I_{a\lambda}$), and the quantum yield (Φ_λ):

$$\mathrm{d}c/\mathrm{d}t = \Phi_\lambda I_{a\lambda} \tag{5.5}$$

Absorbed light intensity ($I_{a\lambda}$) controls the rate of the excited state generation (*R):

$$\mathrm{R} \xrightarrow{I_a} {}^*\mathrm{R} \tag{5.6}$$

whereas its decay depends on rates of all deactivation processes, both photophysical and photochemical. When deactivation proceeds spontaneously its kinetics can be analyzed in terms of general models of unimolecular processes:

$${}^*\mathrm{R} \xrightarrow{k_{nr}} \mathrm{R} \tag{5.7}$$

$${}^*\mathrm{R} \xrightarrow{k_r} \mathrm{R} \tag{5.8}$$

$${}^*\mathrm{R} \xrightarrow{k_{cr}} \mathrm{P} \tag{5.9}$$

where k_{nr}, k_r, k_{cr} are the first-order rate constants of radiationless, radiative, and chemical processes, respectively. The reagent in excited state R* is formed by absorption and decays fast in photophysical and photochemical processes; thus changes in its concentration can be expressed by:

$$\mathrm{d}c_{R^*}/\mathrm{d}t = I_{a\lambda} - k_{nr}c_{R^*} - k_r c_{R^*} - k_{cr}c_R \tag{5.10}$$

The produced R* thus disappears with the rate constant k^o, which is the sum of the first-order rate constants for all applicable processes:

$$k^o = k_{nr} + k_r + k_{cr} \tag{5.11}$$

As, under steady irradiation conditions, the changes in the concentration R* (dc_{R^*}/dt) are negligible, they are assumed to approximate zero (called photostationary state approximation) [2, 3]. Then:

$$I_{a\lambda} = (k_{nr} + k_r + k_{cr})c_{R^*} \tag{5.12}$$

and

$$c_{R^*} = \frac{I_{a\lambda}}{k_{nr} + k_r + k_{cr}} \tag{5.13}$$

As rate of product, P, formation $dc_P/dt = k_{cr}\ c_{R^*}$, and $dc_P/dt = \Phi_P\ I_{a\lambda}$, its quantum yield:

$$\Phi_P = dc_P/dt\ I_{a\lambda} = k_{cr}\ c_{R^*} = \frac{k_{cr}}{k_{nr} + k_r + k_{cr}} \tag{5.14}$$

When quencher (Q) is present, a bimolecular reaction:

$$^*R + Q \xrightarrow{k(q)} R + Q\ (\text{or}\ ^*Q) \tag{5.15}$$

should be considered with the second-order rate constant (k_q). As usual concentration of Q is much larger than that of *R, the quenching reactions are assumed to be the pseudo-first order. Then, the overall rate constant (k) is the sum of all first- and pseudo-first-order constants:

$$k = k_{nr} + k_r + k_{cr} + k_q \tag{5.16}$$

The lifetime of the excited state (τ^0) is equal to the reciprocal of the sum of the (pseudo)unimolecular rate constants of all processes that cause the decay:

$$\tau^0 = \frac{1}{\sum_n k_i} \tag{5.17}$$

where k_i is the rate constant of one of the nth first or pseudo-first order decay processes [2, 3]. When the excited state decay is only via radiative deactivation the term 'radiative lifetime' is used [1].

In the case of second-order processes the lifetime depends on the initial concentration of the reactant, or quencher (τ). Then, the ratio of lifetimes in absence and presence of quencher is given by:

$$\frac{\tau^0}{\tau} = \frac{k_{nr} + k_r + k_{cr} + k_q[Q]}{k_{nr} + k_r + k_{cr}} = 1 + k_q \tau^0 [Q] \tag{5.18}$$

The $k_q\ \tau^0$ product, designated K_{SV} and called the Stern–Volmer constant, can be determined as the slope of the linear plot of τ^0/τ versus [Q].

In the presence of a quencher, variations of quantum yields of photophysical processes or photochemical reaction with the concentration of the quencher are also observed. In the simplest case, a plot of Φ^0/Φ vs concentration of quencher, [Q], is linear, obeying the equation

$$\Phi^0/\Phi = 1 + K_{SV}[Q] \tag{5.19}$$

The Stern–Volmer equation applies when a quencher inhibits either a photochemical or a photophysical process by a single reaction [1].

The kinetic aspects shown so far suggest that the photochemical reaction rate is independent of the reactant concentration, ie that the photoreactions follow zero-order kinetics:

$$-dc_R/dt = \Phi_\lambda \cdot I_{a\lambda} = \text{const} \cdot \Phi_\lambda I_{0\lambda} \tag{5.20}$$

where $I_{0\lambda}$ is the incident light intensity of wavelength λ. The thesis is really true only when the reactant absorbs the incident light almost completely, ie its absorbance ($A_\lambda = \varepsilon_\lambda c_R l$) is >2 ($\varepsilon_\lambda$ = molar absorption coefficient, l = optical pathlength). In other cases, the rate depends on the substrate concentration in a rather complex way; only when absorbance is very low ($A < 0.1$) does the relationship approach first-order kinetics:

$$-dc_R/dt = \Phi_\lambda I_{a\lambda} c_R = k_1 c_R \tag{5.21}$$

where k_1 depends on both Φ_λ and $I_{0\lambda}$.

Quantum yield of a single excited-state decay, ie sum of all photophysical and photochemical processes of this state, amounts to unity: $\Phi_i = 1$. Thus, the larger the share of photophysical processes, the smaller the contribution of photochemical reactions to the excited-state decay. It also means that the quantum yield of the initial photochemical reaction cannot exceed unity: $0 \le \Phi_i \le 1$. Moreover, the contribution of photophysical processes may change the energy and nature of the photochemically reactive excited state (see Figure 4.2), and thereby the course of chemical reaction and its quantum yield. This is especially important for coordination compounds, because, for most organic substances, the photochemically active state is the lowest excited singlet or triplet state. In the case of the latter the quantum yield value is independent of the absorbed light energy, whereas for coordination entities it usually depends on the energy of the reactive excited state.

Quantum yields of product formation, Φ_P, can range from $<10^{-6}$ to $>10^6$, depending on the reaction system; a quantum yield >1 suggests the secondary reaction courses. A secondary step allows another reactant molecule to be consumed by the primary photoproduct and then Φ_P could reach 2. The quantum yield >2 suggests a chain reaction mechanism. A good example of this is the photochemical synthesis of HCl, for which $\Phi_P = 10^6$, which means that absorption of one photon by a Cl_2 molecule results in the production of a million molecules of HCl.

If the absorbing species is in a condensed phase, intermolecular interactions may arise that control the fate of the excited-state entity, eg when the excited molecule is in a solvent cage and the initial photochemical reaction relies on the homolytic or heterolytic bond fission, diffusion of the initial products (radicals) from the same precursor is inhibited. Instead the products remain in the cage for several vibration periods which enables the back reaction to form the substrate or its isomeric form [4–6].

The coordinating properties of solvents often control the yield and nature of the photosubstitution of transition metal complexes, as in the *trans*-$[Cr(NCS)_4(NH_3)_2]^-$ complex, which is inactive photochemically in nitromethane ($\Phi = 0$), whereas in water it generates the $[Cr(NCS)_3(H_2O)(NH_3)_2]$ complex with quantum yield, $\Phi = 0.3$ [2, 3].

To determine the chemical nature, concentration, and kinetics of reactive intermediates, time-resolved techniques are used. To detect short-lived species, an inert matrix at extremely low temperature [7], an extremely high-intensity light source, extremely sensitive detection method, or combination of these methods is used. The method using an intensive light source, called flash photolysis, is a technique of transient spectroscopy and transient kinetic studies in which a light pulse is used to produce transient species. Commonly, an intense pulse of short duration is used to produce sufficient concentration of a transient species for spectroscopic observation. The method can be applied to follow concentrations of substrates, intermediates, and products as a function of time after the flash, which enables in the elucidation of photochemical reaction mechanisms (kinetic spectroscopy) [8, 9].

The flash lamps used in archetypal time-resolved techniques produced an intense pulse of short duration and broadband continuous-wave characteristic, and provided information in a millisecond timescale. Gradual improvements, and especially use of laser sources, allow decrease of resolution times to micro-, nano- and femtoseconds [2, 10–12].

Spectroscopic observation within the UV–visible range is the main detection method, but recently have also been successfully applied the IR and ESR (electron spin resonance) methods [13–16].

References

1. Braslavsky SE, Houk KN, Verhoeven JW. *Glossary of Terms used in Photochemistry*. International Union of Pure and Applied Chemistry, Organic Chemistry Division, Commission on Photochemistry, 1996.
2. Kutal C, Adamson AW. Photochemical processes. In: Wilkinson G, Gillard RD, McCleverty JA, eds. *Comprehensive Coordination Chemistry*. Oxford: Pergamon Press 1987:385–414.
3. Yıldız G, Demiryurek AT. Ferrous iron-induced luminol chemiluminescence: a method for hydroxyl radical study. *J Pharm Toxic Methods* 1998;**39**:179–84.
4. Braden DA, Parrack EE, Tyler DR. Solvent cage effects. I. Effect of radical mass and size on radical cage pair recombination efficiency. II. Is geminate recombination of polar radicals sensitive to solvent polarity? *Coord Chem Rev* 2001;**211**:279–94.
5. Cunningham CT, Cunningham KLH, Michalec JF, McMillin DR. Cooperative substituent effects on the excited states of copper phenanthrolines. *Inorg Chem* 1999;**38** 4388–92.
6. Cummings SD, Eisenberg R. Tuning the excited-state properties of platinum(II) diimine dithiolate complexes. *J Am Chem Soc* 1996;**118**:1949–60.
7. Brozik JA, Crosby GA. Photochemical reactions of rhodium(III) diimine complexes in solid glycerol matrices. Ligand-field influences on activation parameters. *Coord Chem Rev* 2005;**249**:1310–15.
8. Wayne CE, Wayne RP. *Photochemistry*. Oxford: Oxford University Press, 2001.
9. Suppan P. *Chemistry and Light*. London: The Royal Society of Chemistry, 1994.
10. McCusker JK. Femtosecond absorption spectroscopy of transition metal charge-transfer complexes. *Acc Chem Res* 2003;**36**:876–87.
11. Krausz E, Riesen H. Laser spectroscopy. In: Solomon EI, Lever ABP, eds. *Inorganic Electronic Structure and Spectroscopy* 1999: 307–52.

12. Ford PC, Bridgewater JS, Massick SJM. Photochemical strategies for investigating organometallic intermediates relevant to catalysis mechanisms. *Catalysis Today* 1999; **49**:419–30.
13. Grills DC, Huang KW, Muckerman JT, Fujita E. Kinetic studies of the photoinduced formation of transition metal–dinitrogen complexes using time-resolved infrared and UV–vis spectroscopy. *Coord Chem Rev* 2006;**250**:1681–95.
14. Kawai A, Shibuya K. Electron spin dynamics in a pair interaction between radical and electronically-excited molecule as studied by a time-resolved ESR method. *J Photochem Photobiol C: Photochem Rev* 2006;**7**:89–103.
15. Hirota N, Yamauchi S. Short-lived excited triplet states studied by time-resolved EPR spectroscopy. *J Photochem Photobiol C: Photochem Rev* 2003;**4**:109–24.
16. Murai H. Spin-chemical approach to photochemistry: reaction control by spin quantum operation. *J Photochem Photobiol C: Photochem Rev* 2003;**3**:183–201.

6

Photochemical Reactions

Never fear shadows. They simply mean that there's a light somewhere nearby.
Ruth E. Renkei

Reactions initiated from ground electronic states are, in photochemistry, called dark reactions (because they go on in the absence of light) or thermal reactions (because their rates depend on temperature, as distinct from the photochemical reaction rates). Chemical reactions lead to product generation only when thermodynamic and kinetic requirements are satisfied.

Photochemical reactions start from excited states and, distinct from the photophysical processes, convert the substrate molecule into product(s). Generally the initial photochemical products are unstable and undergo secondary thermal and/or photochemical reactions, yielding more stable photoproduct(s). Their chemistry is usually different from that of the parent ground-state species.

Energy of the electronically excited entity amounting 150–600 kJ/mol exceeds a typical activation energy ($E_a \leq 30$ kJ/mol), which opens some new reactive channels (Figure 6.1). Conversion of an excited reactant into a product needs only negligible activation energy (E_a^* usually within 8–17 kJ/mol) and its intermediate product(s) (IP) can be precursor of the following:

(1) final products the same, as those formed in thermal reactions (P_t); many ground state reactions have their photochemical counterparts
(2) products obtainable only in photochemical reactions via new electronic arrangement – not possible for the ground electronic state, $P_{ph}(1)$
(3) products generated only in photochemical reactions by overcoming an activation barrier unattainable in thermal processes, $P_{ph}(2)$.

Photoproducts can thus be the same as, or quite different from, the products of the dark reactions. In the former case the desired products can be obtained

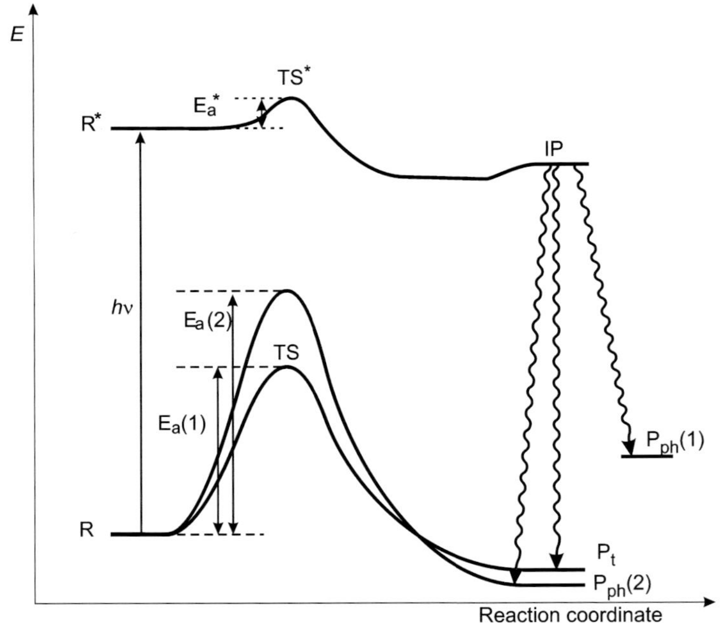

Figure 6.1 *Comparison of thermal and photochemical reaction courses; R and R* mean the reactant, TS and TS* – transition states, E_a and E_a^* – activation energies in the ground and excited states, respectively; IP means intermediate photochemical product; P_t and P_{ph} signify products of the thermal and photochemical reactions, respectively. (Adapted from Stasicka [158])*

without any heating, ie under mild, environment-friendly conditions. The photoproducts, P_{ph}, different from those of dark reactions, are often unique species, which could not be formed in any other way, or are the species richer in internal energy, than the substrate, ie the products that can store the absorbed radiant energy. The unique photoproducts (often radicals) are irreplaceable catalysts in photocatalytic processes, whereas generation of the energy rich species enables solar energy conversion into the chemical energy (see below).

The factors that contribute to the reactivity of an excited state include not only the energy excess, but also the intrinsic reactivity of the specific electronic arrangement and the relative efficiencies of the different competing pathways. The factors are all correlated and therefore the reaction pathways allowed by the correlation rules are often different from those of the ground-state partners [1].

6.1 Photochemical Reaction Channels

Angular or spatial electron shift induced by excitation results, in general, in weakening of some bonds within the molecule. The probability of the effective photochemical

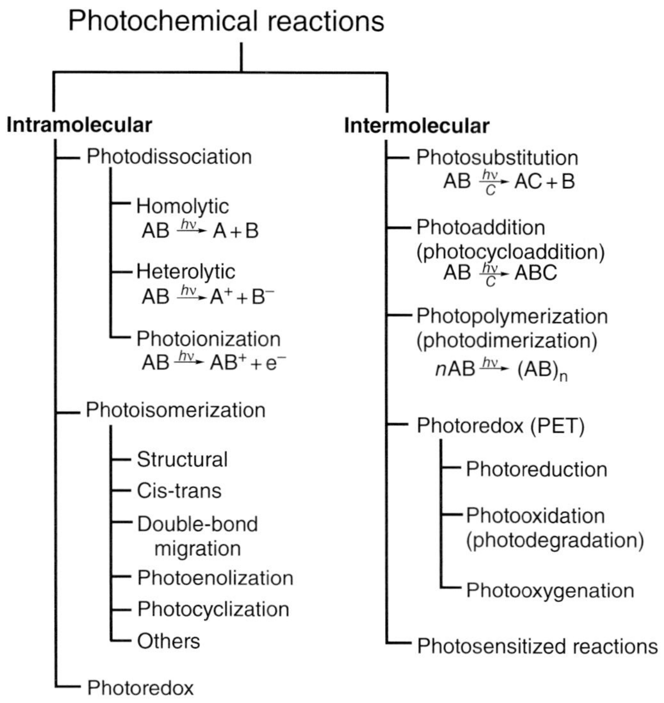

Figure 6.2 *Main types of photochemical reactions*

decay is the higher, the lower the contribution of the photophysical processes, so photochemical reactions of most organic and simple inorganic compounds proceed from the excited states of the longest lifetimes, ie from the lowest triplet (T_1) or singlet (S_1) states (see Jabłoński diagram – Figure 4.2). In the case of coordination and organometallic compounds, the photochemical transformations can also be initiated from higher excited states.

There are many ways in which the excited species can react; in general the photochemical reactions may be classified as intramolecular and intermolecular, depending on their mono- and bimolecular pathways, respectively (Figure 6.2). The former processes predominate in diluted phases, whereas the latter are observed mostly in condensed phases.

6.2 Intramolecular Photoreactions

The most important monomolecular photoreactions are photodissociation and photoisomerization. The former leads to a fragmentation in a molecular species by a photon of sufficient energy to rupture a bond, which may be of homo- or heterolytic character. The light-induced bond cleavage is often called photolysis, especially when flash techniques are used (flash photolysis).

In more complex or large systems intramolecular redox photoreactions are also observed; this pathway can be illustrated by photoinduced electron transfer in the compound containing Cu^{II} and Mo^{IV} centres connected by the CN^- bridges

$$[Cu^{II}(tren)]_6[Mo^{IV}(CN)_8](ClO_4)_2 \xrightarrow{h\nu} Cu^{I}Mo^{V} \tag{6.1}$$

which is accompanied by a distinct change in its magnetic properties (photomagnetism) [2, 3]. The phenomenon became the basis of the development of new magnetic materials and techniques, in which magnetic properties can be controlled by optical stimuli. Optically switchable magnetic materials are becoming increasingly important in the field of high-density information storage media, because the photon mode allows us to access a variety of different types of materials with high speed and superior resolution. In particular, the design of molecular compounds that exhibit photoinduced magnetization behaviour has attracted great attention [4–10].

The photoinduced electron transfer (PET) is especially important in the case of large or giant molecules (supermolecules), ie systems made up of molecular components in the same way as molecules are made up of atoms [11–19]. As the systems are made up of a number of discrete components held together by different but not always exactly specified forces (covalent bonds, electrostatic interactions, hydrogen bonds, or other intermolecular interactions), the photoinduced electron transfer or energy transfer in these systems may be formally treated as intermolecular [20].

6.2.1 Photodissociation and Photoionization

Several mechanisms are recognized for dissociation processes; the three main routes involve optical dissociation, predissociation, and induced predissociation. In the case of diatomic molecules, for which photodissociation is the principal mode, the processes are easily illustrated by the familiar energy curves, the shapes of which are characteristic for specific electronic states. Optical dissociation may come when absorption energy overcomes dissociation energy of the excited species if absorption occurs to a bound (Figure 6.3a) or unbound, repulsive state (Figure 6.3b). Predissociation occurs when absorption energy to a bound excited state is lower than dissociation energy, but the state crosses the unbound dissociative state, enabling the passage through a potential energy barrier (Figure 6.3c). When dissociation is an after-effect of perturbing influences such as collisions or external magnetic or electric fields, the phenomenon is referred to as induced predissociation [1].

In the photochemistry of larger molecules the same physical principles apply and the sequence of events is essentially the same, although descriptions are more complex and less precise. Opposite to diatomic, the polyatomic molecules can yield a multitude of different sets of products. To establish the photodissociation mechanism, the nature of the elementary chemical process undergone by an electronically excited molecular entity (primary photoreaction) yielding primary photoproducts should be known, eg in the case of alkanes both the radical fission

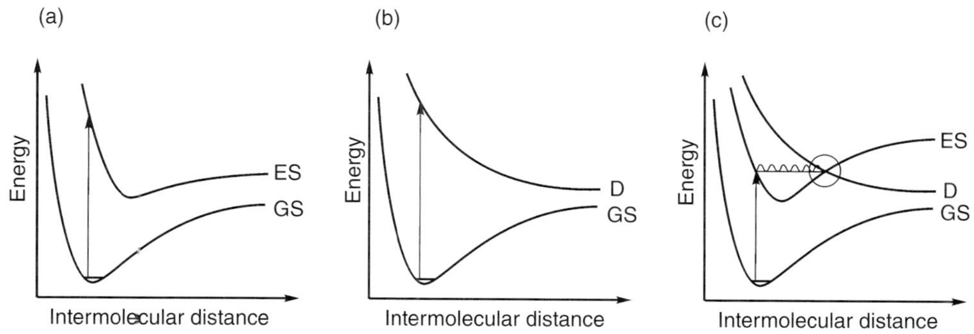

Figure 6.3 *(a, b) Absorption leading to optical dissociation and (c) absorption resulting in predissociation. GS, ground state; ES, bound excited state; D, unbound dissociative state. (Adapted from Wayne and Wayne [1])*

$$C_nH_{2n+2} \xrightarrow{h\nu} C_nH_{2n+1} + H^{\bullet} \quad (6.2)$$

and hydrogen elimination modes

$$C_nH_{2n+2} \xrightarrow{h\nu} C_nH_{2n} + H_2 \quad (6.3)$$

are possible, whereas nitrous oxide can photodissociate yielding alternatively oxygen

$$N_2O \xrightarrow{h\nu} N_2 + O \quad (6.4)$$

or nitrogen atoms:

$$N_2O \xrightarrow{h\nu} NO + N \quad (6.5)$$

Photodissociation in the gas phase is of considerable importance in various aspects of photochemistry, especially many of the radicals, ions, and electrons present in the Earth's atmosphere which are formed as a result of these pathways [21, 22]:

$$NO_2 \xrightarrow{h\nu} NO + O \quad (6.6)$$

or

$$O_2 \xrightarrow{h\nu} 2O \quad (6.7)$$

In the condensed phase the close proximity of neighbouring molecules can have a considerable effect on the subsequent fate of the excited species. Solvation can reduce molecular energies, affect the selection rules, and greatly increase the number of collisions undergone by each molecule [23]. The last effect can increase the photophysical decay and decrease the photoreaction quantum yield. Moreover, the excited molecules can be removed by chemical quenching in bimolecular reactions or the reactant molecules can be effectively trapped within cages of solvent molecules (cage effect). The fragment separation by normal diffusion is then hindered and their germinal recombination becomes more probable. The fragments (called a geminate pair) make many mutual collisions before they are eventually liberated from the cage, so they may preferentially react with each other, leading to geminate recombination. The reaction can either be a back reaction that restores the substrate in its ground-state or a reaction that leads to a new bond formation or bond reorganization. The geminate recombination usually, but not necessarily, results in reformation of the precursor species, although just occasionally it leads to generation of other products.

Photodissociation may also result from ejection of an electron into a surrounding medium induced by the absorption of electromagnetic radiation from a neutral or positively charged molecular entity; then it is called photoionization. Ionization processes in the gas phase are generally of high energy and so require short-wavelength radiation. They are, however, of great importance in upper atmospheric regions, where short-wavelength UV radiation from the sun can lead to appreciable ionization of the chemical species present, eg nitric oxide undergoes ionization:

$$NO \xrightarrow{h\nu} NO^+ + e^- \tag{6.8}$$

induced by absorption of radiation at λ <134 nm [1].

Photoionization can also proceed in solutions of the coordination compounds containing low-valence, readily oxidizable centres, eg V^{2+}, Cr^{2+}, Fe^{2+}. The process proceeds at the expense of moderate excitation energy and relies on solvated electron generation, with concomitant oxidation of the relevant metal species [24, 25] (see below).

6.2.2 Photoisomerization

Another pathway of intramolecular deactivation of the excited-state species consists of photoisomerization. Examples of this kind of step are afforded by structural isomerizations, *cis–trans* isomerizations, double-bond migrations, photoenolizations, etc. The excitation facilitates the transformations as a result of changes in electronic configurations, which remove the constraints holding the precursor molecule, eg the *cis–trans* isomerizations of olefins is induced by an electron promotion from the highest filled bonding π orbital to the lowest unfilled antibonding π^* orbital, relaxing fixed planar geometry and enabling free rotation around the carbon–carbon axis.

Similarly, changes in electronic configuration of a complex entity, eg $[Ru^{III}X_2L_4]$ implicate a ligand release and conversion of inert octahedral geometry into a labile square pyramidal or trigonal bipyramidal geometry (Figure 6.4), which enables the *cis* ↔ *trans* photoisomerization.

An example of more complex structural photoisomerization may be the photo-Fries rearrangement leading to a photorearrangement of *O*-acyl-phenols or *N*-acyl-anilines to give the [1,3]- or [1,5]-rearranged products, as shown in Figure 6.5.

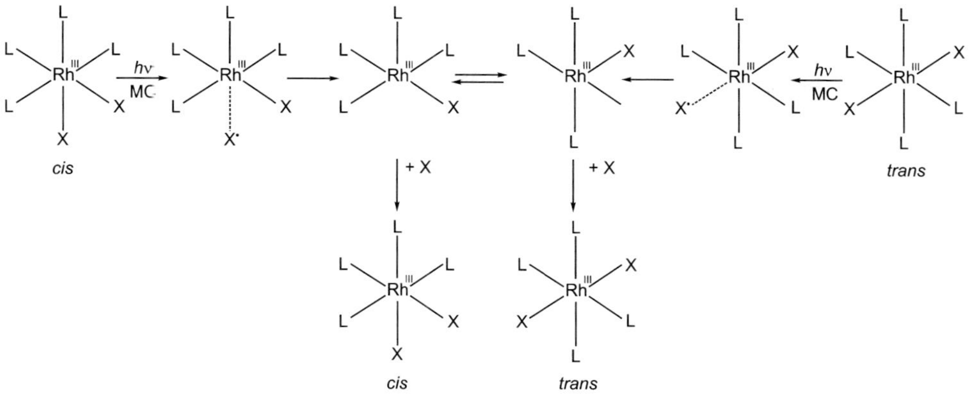

Figure 6.4 Mechanistic scheme proposed for photoisomerization of the [$Ru^{III}X_2L_4$] complexes. (Adapted from Endicott [33] and Lees [41])

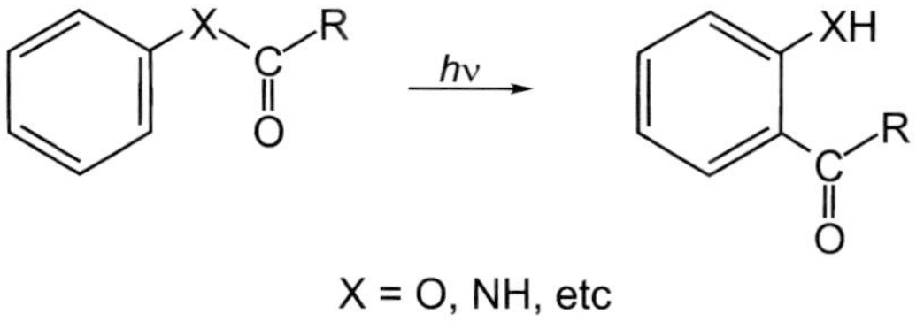

Figure 6.5 Photo-Fries rearrangement

The *trans-to–cis* photoisomerization of stilbene-like organic species can be carried out upon their coordination to the Re^I centre to form the *fac*-$[Re(CO)_3(NN)(trans\text{-}L)]^+$ complexes [26]. Another spectacular example of photoisomerization is generation of several different isomers by photochemical rearrangement of norbornadiene in the presence of different metal catalysts, such as Cu^I complexes or carbonyls of Fe, Ni, and Cr [27, 28].

6.3 Intermolecular Photoreactions

The deactivation of an excited molecular entity through a non-radiative process can also occur as a result of an external environmental influence. A molecular entity that deactivates (quenches) an excited state of another molecular entity, by energy transfer, electron transfer, or a chemical mechanism is called a quencher [29]. Photophysical processes (energy or electron transfer) were described in Chapter 4; the present discussion is confined to the chemical consequences of quenching.

The intermolecular systems represent situations in which excitations are used to promote reactions that are not possible for the ground electronic states. Their main types can be specified: photosubstitution, photoaddition (photocycloaddition),

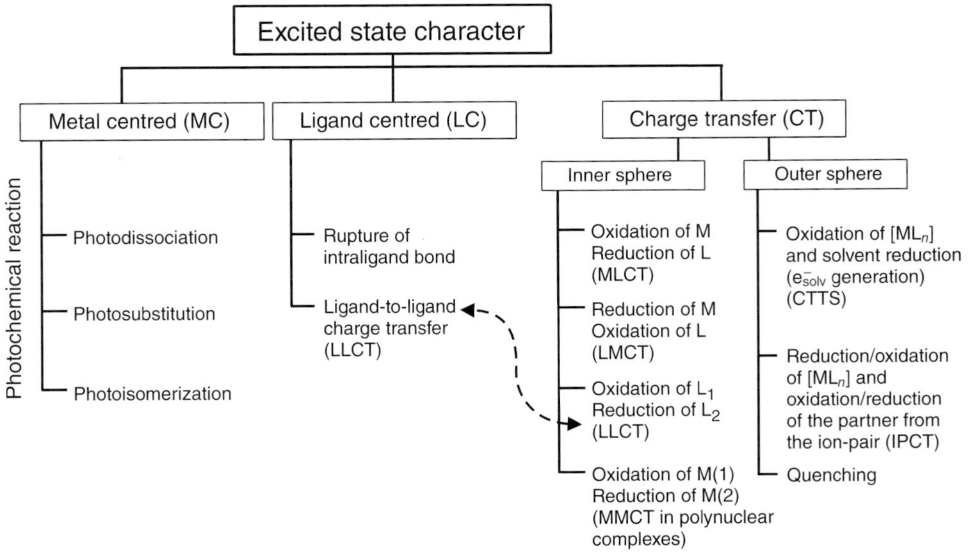

Figure 6.6 *The typical relationships between the reactive excited states and nature of direct photoreactions in coordination compounds*

photopolymerization (photodimerization), photoredox processes (photoreduction, photo-oxidation or photo-oxygenation), and photosensitization (see Figure 6.2).

A polymerization process requiring a photon for the propagation step is called photopolymerization or photoinduced polymerization, when polymerization of a monomer by a free radical or ionic chain reaction is initiated by photoexcitation.

Photoredox processes include both photoreduction and photo-oxidation of the excited species. An electron transfer that results from an electronic state produced by the resonant interaction of electromagnetic radiation with matter is called photoinduced electron transfer (PET) [30–32]. This can be done by either a direct or a photosensitized process (see below and Figure 6.6).

Reduction reactions by light (photoreductions) can proceed by either addition of one or more electrons to an excited species or the photochemical hydrogenation of a substance. Reactions in which the substrate is not electronically excited are sometimes called photoinitiated reductions.

Oxidation reactions induced by light (photo-oxidations) consist of loss of one or more electrons from a chemical species as a result of its photoexcitation. Oxidation also may be achieved with molecular oxygen participation. Reactions in which neither the substrate nor the oxygen is electronically excited are sometimes called photoinitiated oxidations.

When molecular oxygen is incorporated into a molecular entity and remains in the product, the process is also called photo-oxygenation, and there are three common mechamisms of them:

1. The reaction of triplet molecular oxygen with radicals formed photochemically
2. The reaction of photochemically produced singlet molecular oxygen with molecular entities, giving rise to oxygen-containing molecular entities
3. The third mechanism proceeds by electron transfer, producing superoxide anion as the reactive species.

Photo-oxidation often leads to transformation of a molecule into the lower-molecular-weight fragments; this is called photodegradation, and is a term widely used in the destruction of pollutants via their oxidation by means of solar radiation or UV-based processes.

Energy or electron transfer between an excited-state species and a non-absorbing quencher induces the quencher to undergo photochemical and/or photophysical transformations. The process of activation of a non-absorbing substrate is called photosensitization and the species absorbing and transferring the radiant energy is called a photosensitizer [29] (for details see Photosensitization below).

6.4 The Coordination Compound Specificity

Photochemical reactions of coordination compounds, similar to other species, may be classified into two categories, ie intra- and intermolecular processes with similar reaction types in each category (see Figure 6.2). A unique feature of coordination compounds is that their excited states differ between them not only in energy but also in charge distribution: The electronic transitions (see Figure 3.1) can be divided into metal-centred (MC) and charge transfer (CT) transitions, depending respectively on the angular and radial charge shift accompanying excitation. As a consequence, the different excited states may initiate various chemical effects.

The photochemical transformations after the radial MC charge shift are effective in weakening of the M–L bond between the central atom and ligand, which usually results in the photodissociation, photosubstitution, or photoisomerization mode (Figure 6.6).

Population of the ligand-centred (LC) excited state in a complex ligand (eg AB) can lead to weakening and finally to rupture of the intraligand bond (A–B), or the charge can be transferred to another ligand, ensuring a similar effect to ligand-to-ligand charge transition (LLCT).

Charge-transfer excited states are able to undergo different redox reactions, which can occur via either an inner-sphere or an outer-sphere mechanism (Figure 6.6). The detailed course depends mostly on the type of the reactive excited state: oxidation of the central atom and ligand reduction may be a consequence of the metal-to-ligand (MLCT) excitation, the central atom reduction and ligand oxidation can result from ligand-to-metal (LMCT) excitation, whereas oxidation of one ligand at the expense of another is often due to LLCT excitation. The metal-to-metal charge transfer (MMCT) in polynuclear complexes can bring about redox behaviour of the different centres [11, 33–37].

The outer-sphere CT processes result in the photoinduced electron transfer (PET) modes, which are mostly the consequence of charge-transfer-to-solvent

(CTTS) or ion-pair-charge-transfer (IPCT) excitations, although other CT processes can be also assigned to the PET modes.

The dependence of the reaction type on the nature of excited reactive state cannot be treated as an absolute rule, especially as not only do CT excited state species have increased redox properties, but they can also be treated as highly diagnostic.

6.4.1 Ligand Field Photochemistry

Energy of the MC excited states is generally not higher than approximately 400 kJ/mol, and charge redistribution accompanying the MC transition within the *d* electron shell is of an angular character; these features result in a weakening of the M–L bond, often followed by:

- photodissociation [38]:

$$[ML_n] \xrightarrow{h\nu} [ML_{n-1}] + L \tag{6.9}$$

- photosubstitution:

$$[ML_n] + L' \xrightarrow{h\nu} [ML_{n-1}L'] + L \tag{6.10}$$

- and photoisomerization of the complex, eg

$$cis\text{-}[ML_4L'_2] \xrightarrow{h\nu} trans\text{-}[ML_4L'_2] \tag{6.11}$$

- or photoisomerization of ligand.

The most frequent case of photosubstitution is photosolvation (photoaquation), eg photoaquation of the octahedral Cr^{III} complex:

$$[Cr(NH_3)_6]^{3+} + H_2O \xrightarrow{h\nu} [Cr(NH_3)_5(H_2O)]^{3+} + NH_3 \tag{6.12}$$

is a consequence of the ${}^4A_{2g} \rightarrow {}^4T_{1g}(t_{2g}^3 \rightarrow t_{2g}^2 e_g^1)$ excitation (see Figure 4.5), which populates the antibonding e_g orbital shifting thereby the *d* electron density, towards the M–NH_3 bonds [39].

Ligand substitution is often preceded by photodissociation which leads to generation of a coordinately unsaturated complex, able to coordinate another accessible ligand, L′. This can be exemplified by photodissociation of carbonyl complexes, eg hexacarbonylchromium [28]:

$$[Cr(CO)_6] \xrightarrow{h\nu} [Cr(CO)_5] + CO \tag{6.13}$$

which is over within 250 fs and then may be followed by addition of solvent molecule to the coordinately unsaturated $[Cr(CO)_5]$:

$$[\mathrm{Cr(CO)_5}]+\mathrm{solv}\rightarrow[\mathrm{Cr(CO)_5(solv)}] \quad (6.14)$$

lasting about 15–20 ps. Instead of a solvent molecule, another ligand may fill the free coordination site. It could also be a small inactive molecule, such as N_2:

$$[\mathrm{ML(CO)}_x]\xrightarrow{h\nu}[\mathrm{ML(CO)}_{x-1}]+\mathrm{CO} \quad (6.15)$$

$$[\mathrm{ML(CO)}_{x-1}]+\mathrm{N_2}\rightarrow[\mathrm{ML(CO)}_{x-1}(\mathrm{N_2})] \quad (6.16)$$

This coordination was observed in the case of several central atoms (M=W, Cr, Fe, Mn) and phosphine co-ligands L [40]. As the entering ligand changes its characteristics, coordination to a metal centre is often used to transform selected substrate into the desirable product, especially in organic synthesis; among other things it may be used to activate small molecules (see Chapter 6).

The M–L bond weakening in the MC excited state could also lead to photoisomerization, which can yield different categories of isomeric complexes, eg geometrical, linkage, optical, spin isomers. The most popular geometrical photoisomerization has already been discussed above in section 6.2.2, and illustrated in Figure 6.4 [33, 41].

The first reported example of photochemical linkage isomerism was based on the observation of a colour change of the red $[Co(NH_3)_5(NO_2)]Cl_2$ nitro derivative, stable at room temperature, to the yellow $[Co(NH_3)_5(ONO)]Cl_2$ nitrate(III) derivative. Subsequently, photochemical linkage isomerization was observed for a number of ligands including thiocyanate, dimethylsulphoxide and sulphur dioxide [42–44]. Low temperature photolysis studies of transition metal nitrosyl compounds revealed the existence of metastable states of nitrosyl complexes that turned out to be nitrosyl linkage isomers, in which the NO ligand is coordinated by either an oxygen atom or the π bond [42].

Relaxation of the MC excited state undergoing intersystem crossing in some complexes may lead to a change in coordination number. This is the case of Ni^{II} complexes, which are able to form tetrahedral, square pyramidal, or octahedral, high-spin complexes and low-spin four-coordinate planar complexes. In these complexes the two spin states (and different coordination geometries) are usually not very different in energy. When the low-spin four-coordinate planar complex is excited to its singlet state, which undergoes intersystem crossing (ISC) to triplet state, in fluid solution it can be transformed into a metastable high-spin six-coordinate complex. Thus, the photoinduced change of spin multiplicity results in a change of coordination number, which, in turn, stabilizes the excited state configuration [33].

6.4.2 Photochemistry from LC or LLCT States

LC transitions concern only ligands that are composed of two or more atoms, eg 2,2′-bipyridine, 1,10-phenantroline, or porphyrins. LC excitation entails a population of ligand-antibonding orbitals weakening the intra-ligand bonds, and consequently leading to bond cleavage:

$$[L_nM(AB)] \xrightarrow{h\nu\,(LC)} [L_nM(AB)^*] \rightarrow [L_nMA] + B \qquad (6.17)$$

This photochemical mode was observed in oxalato-Co^{III} complexes and in azido complexes of Rh^{III} and Ir^{III}, in which heterolytic splitting of C–C or N–N yields CO_2 or N_2 and formate (bonded by the C atom), or nitrene intermediates, respectively [27, 45], eg:

$$[M(N_3)(NH_3)_5]^{2+} \xrightarrow{h\nu\,(LC)} [M(NH)(NH_3)_5]^{2+} + N_2 \qquad (6.18)$$

where M=Rh^{III} or Ir^{III}.

Chemical deactivation of the LC excited states sometimes leads to ligand isomerization, as in the case of *fac*-$[Re(CO)_3(NN)($*trans*-L$)]^+$ (NN = polypyridyl, L = stilbene), when *trans*-stilbene is transformed into *cis*-stilbene [26].

Excitation to the LC states may also result in population of the CT-excited states, especially the LLCT states. These phenomena are frequently encountered in complexes containing both π-acceptor (eg 1,10-phenantroline or 2,2′-bipyridyl) and π-donor ligands (eg aromatic thiols). Then the LC excitation can induce charge transfer between these ligands through central atom (LLCT) that leads to a photoredox reaction. Such reactions were reported in the case of heteroleptic organometallic compound $[Rh^{III}(C_7H_7)(C_5H_5)]^{3+}$ [37], heteroleptic Re^I complex *fac*-$[Re^I(L)(CO)_3(bpy)]^{n+}$ [46] and metal–carbon-bonded platinum complexes [47].

The redox reactions were observed also in the case of Ti^{IV}, Mo^{VI}, or Re^{VII} complexes with the O_2^{2-} ligand, which is a reductant as well as an oxidant and can serve both as CT donor and CT acceptor. The LC excitation was followed by interligand charge transfer (LLCT) within the O_2^{2-} moiety, leading in consequence to photodismutation to $2O^{2-}$ and O_2 [48].

In analogy to coordination compounds, which simultaneously contain a reducing and an oxidizing ligand and thereby are characterized by low-energy LLCT transitions, the CT interactions between donors and acceptors in organic zwitterions were recently discussed [49].

6.4.3 Inner-Sphere Charge Transfer Photochemistry

Excitation to inner-sphere charge transfer (ISCT) transitions induces radial charge shift within the coordination entity, which may result in redox reactions including the central atom and ligands. The charge redistribution increases the complex susceptibility towards protonation, isomerization, redox processes, or nucleophilic or electrophilic attack.

To compete effectively with the photophysical processes, the chemical reactions from the highly energetic CT states should be very fast, otherwise the MC states become populated. On the other hand, redox processes may sometimes occur from other than CT excited states. The phenomenon is a consequence of redox potential changes after excitation, which make the entity in any excited state a much stronger oxidant and a much stronger reducer than the ground state complex, eg the standard oxidation potential of the $[Fe(bpy)_3]^{2+}$ complex is 1.05V in the ground state and

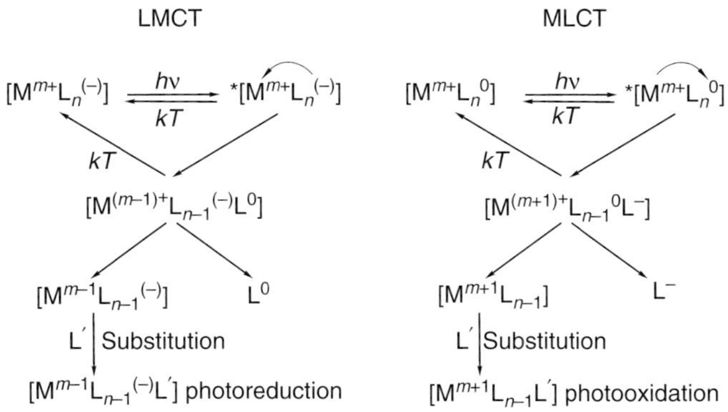

Figure 6.7 *Inner-sphere redox processes induced by different CT (charge transfer) excited states in transition metal complexes. LMCT, ligand-to-metal charge transfer; MLCT, metal-to-ligand charge transfer*

approximately 0.1V in the MC state, and its reduction potentials are –1.26 and –0.4V, respectively [50]. Similar differences in redox potentials are observed for other transition metal complexes.

Typically, however, the redox reactions are induced by the CT states. The reactions can proceed via an inner- or outer-sphere mechanism (see Figure 6.6), irrespective of the initial mode; the redox processes are usually accompanied by reactions with exterior molecules, induced by the decreased coordination ability of the centre, or ligand, or both. This manifests in successive reactions with solvents or other accessible ligands present in the medium [51, 52] (Figure 6.7).

The CT inner-sphere, photoinduced redox reaction results in oxidation (or reduction) of the central atom at the expense of the respective reduction (or oxidation) of ligand. The direction of the redox process depends on the CT state character (see Figure 3.1 and section 3.2 in Chapter 3): the central atom reduction is a consequence of the LMCT transition, whereas central atom oxidation is a result of the MLCT transition (Figure 6.7). The photochemical oxidation of one ligand at the expense of another (LLCT) can also be rated among the inner-sphere redox processes (see above).

Reactions of the LMCT States

The radial charge redistribution resulting from LMCT excitation towards the central atom can be effective in its reduction and ligand oxidation. The redox reaction is usually followed by M–L bond cleavage, ligand radical release from the coordination entity, and formation of a coordinately unsaturated complex. To complete its coordination sphere the central atom coordinates the accessible ligand (most often a solvent molecule). Such reaction patterns are observed in complexes of metals at their high oxidation numbers (eg Cu^{II}, Fe^{III}, Ru^{III}, Co^{III}, Rh^{III}, Pt^{IV}) with relatively good electron donors as ligands (eg heavier halogens, CN^-, NCS^-, N_3^-, O_2^{2-}, NO^-, S^{2-}, RS^-, $NORS^-$ (RS^- = thiolate), EDTA, its derivatives and analogues) [45, 48, 51–60]. In many of the organometallic compounds, various important organic ligands can also function as CT donors (eg cyclopentadienyl anions, allyl anions, carbenes). Accordingly, organometallic complexes with these ligands are frequently characterized by LMCT absorptions, and CT excitation can lead to emission or photoreduction [35, 36].

Among transition metal complexes the iron compounds are those for which LMCT photochemistry is of crucial environmental relevance [51, 61, 62]. Their photoreductions yield the Fe^{II} species and ligand radicals ($L^{\bullet}$), eg:

$$[Fe^{III}(L)(H_2O)_5]^{2+} \xrightarrow{h\nu\,(LMCT)} \xrightarrow[H_2O]{} [Fe^{II}(H_2O)_6]^{2+} + L^{\bullet} \tag{6.19}$$

One of the most important effects is generation of $OH^{\bullet}$ radicals from the hydroxo complexes of Fe^{III}; the reaction is the basis of the so-called photo-Fenton processes [63–78] (for details see Chapter 21).

Reactions of the MLCT States

The radial charge redistribution caused by MLCT excitation outside the central atom can be effective in its oxidation and reduction of the ligand or even an external solvent molecule. These reaction modes are observed in complexes of metals at their low oxidation numbers (eg Cu^{I}, Au^{I}, Fe^{II}, Ru^{II}, Mo^{IV}) with relatively good π-acceptor ligands, eg CN^-, NCS^-, NO^+, O_2^{2-}, pyridine, bipyridine, 1,10-phenanthroline, methylene blue) [48, 53, 79]. Often the final product is really not the oxidized, but only the substituted complex, eg:

$$[M(CO)_4(\alpha\text{-diimine})] + L \xrightarrow{h\nu\,(MLCT)} fac\text{-}[M(CO)_3(L)(\alpha\text{-diimine})] + CO \tag{6.20}$$

where M=Cr, Mo, W, and L is a solvent molecule or a nucleophile such as phosphine, phosphite, nitrile, or pyridine [80, 81].

Also in many of the organometallic compounds various important organic ligands can function as CT acceptors (eg ethines, carbenes, arenes, or cyclohexadienyl or cycloheptatrienyl cations). Accordingly, organometallic complexes with such ligands are frequently characterized by the MLCT absorptions and their CT excitation can lead to emission or photo-oxidation [35, 36].

A relatively frequent mode is mixing of MLCT and LLCT characters in low-lying electronic transitions and excited states; recently it was reported in the case of *fac*-$[Re(NCS)(CO)_3(bpy)]$ [48, 49, 82].

A recent study on the luminescence lifetime prolongation of metal polypyridine complexes in supramolecular systems revealed rapid and reversible electronic energy transfer between lowest-lying isoenergetic triplet excited states of a pyrenyl-appended Ru^{II} polypyridine complex, eg a triplet metal-to-ligand charge-transfer state (^{3}MLCT), and a triplet $^3\pi$–π^* state localized on a separate organic chromophore [83].

Reactions of the MMCT States

MMCT excitations are observed in polynuclear complexes in which two or more metal centres are bound directly (M–M) or by the ligand bridge (M–L_b–M). The latter case of remote CT resembles the LLCT transition (L–M–L), because electron donor and acceptor do not interact directly but via a metal or ligand centre [49].

The charge transfers in polynuclear complexes with metal centres ready to undergo redox reactions are called intervalence charge transfer (IT). These compounds are often characterized by low-energy M–M transitions and thereby have intense colour (eg Prussian blue and its analogues).

The photochemical consequence of the MMCT excitation is usually splitting of the M–M or M–L_b–M bond, as was proved in the case of the $[M(\eta^5\text{-}C_5H_5)(CO)_2]_2$ (M=Fe or Ru) dimers [52, 84, 85].

The primary photochemical reaction of the $[MM'(CO)_{10}]$ (M, M′=Mn, Re) complexes in solution has been shown to result in (1) homolytic fission of the metal–metal σ-bond to form $^{\bullet}M(CO)_5$ radicals, (2) dissociation of a CO ligand to form $[MM'(CO)_9]$, and (3) non-dissociative relaxation to the electronic ground state [86]. Similar behaviour is shown by other dinuclear carbonyl complexes [87].

The MMCT excitation is often accompanied by redox reaction, eg [2, 3, 35, 88]:

$$[Cp_2Ti^{IV}\{\mu\text{-}NCRu^{II}(CN)_5\}_2]^{6-} \xrightarrow{h\nu\,(MMCT)} [Cp_2Ti^{III}\{\mu\text{-}NCRu^{II}(CN)_5\}]^{3-} + [Ru^{III}(CN)_6]^{3-} \quad (6.21)$$

where Cp = cyclopentadiene,

$$[Hg^{II}Co^{I}_2(CN)_{10}]^{6-} \xrightarrow{h\nu\,(MMCT)} Hg^0 + 2[Co^{III}(CN)_5]^{3-}, \quad (6.22)$$

or

$$\{[Cu^{II}(tren)]_6[Mo^{IV}(CN)_8]\}^{6+} \xrightarrow{h\nu\,(MMCT)} \{[Cu^{II}(tren)]_5[Cu^{I}(tren)][Mo^{V}(CN)_8]\}^{6+} \quad (6.23)$$

The changes in magnetic properties accompanying the redox reactions are relevant to switchable magnetic properties [4–10] (see Introduction to section 6.2).

Photoinduced charge and energy transfers in polynuclear macromolecular systems containing bipyridine, porphyrins, phthalocyanines, fullerenes, and metal centres (eg Ru, Os, Zn) enable construction of systems for long-range energy or electron transport applicable to solar energy conversion, optoelectronic devices, and photonic switching systems [11, 14, 17, 18, 89, 90].

6.4.4 Outer-Sphere Charge Transfer Photochemistry

In addition to the inner-sphere (or intramolecular) CT transitions, outer-sphere CT (OSCT) transitions are observed. These OSCT interactions include transitions between complex and solvent (CTTS transitions), complex and its ion-pair partner (IPCT transitions), and complex and a non-bonded quencher. Frequently, OSCT excitation leads to photochemical reactions [27, 30, 50, 91, 92] (see Figure 6.6). As a consequence, the complex undergoes oxidation at the expense of the solvent (solvated electron generation as a result of CTTS), or a redox reaction between the excited coordination entity and its ion pair partner (following IPCT) or other external quencher. The driving force for the electron transfer is determined by the redox potentials of the excited state species and its reaction partner.

Reactions of the CTTS States

Distinct from other transitions, the energy of the OSCT bands depends on medium parameters; in the case of CTTS such parameters as solvent nature and its polarity, composition of the solution, and temperature should be considered.

Usually, the products of the CTTS excitations are solvated electrons and oxidized substrates:

$$[\mathrm{M}^m\mathrm{L}_n] \xrightarrow{h\nu\,(\mathrm{CTTS})} [\mathrm{M}^{m+1}\mathrm{L}_n] + \mathrm{e}^-_{\mathrm{solv}} \qquad (6.24)$$

This photochemical mode is observed in the case of electron-rich metal centres ligated by poor electron acceptors; cyanide complexes, eg $[Fe^{II}(CN)_6]^{4-}$, $[Ru^{II}(CN)_6]^{4-}$, $[Mo^{IV}(CN)_8]^{4-}$, and $[W^{IV}(CN)_8]^{4-}$ are known as the most efficient in the e^-_{solv} production. A typical reaction is photo-oxidation of hexacyanoferrate(II):

$$[\mathrm{Fe^{II}(CN)_6}]^{4-} \xrightarrow{h\nu} [\mathrm{Fe^{III}(CN)_6}]^{3-} + \mathrm{e}^-_{\mathrm{solv}} \qquad (6.25)$$

which can reach quantum yield as high as 0.89 at $\lambda \leq 229$ nm [24, 25, 27, 28, 33, 51].

Solvated electrons are generated as well in photo-oxidation of halogen and amine Cu^I complexes, $[CuX_2]^-$, $[CuX_3]^{2-}$, and aqua complexes of strongly reducing centres, such as Co^+, Ni^+, Zn^+, Cd^+, V^{2+}, Cr^{2+}, and Fe^{2+} [24, 93]. In acidic solutions hydrated electrons are scavenged by protons, yielding molecular hydrogen. This reaction pathway was considered as usable for H_2 production from acidified water (see below).

Reactions of the IPCT States

Redox processes following the IPCT were studied repeatedly [30, 91, 92, 94]. Coordination compounds are usually charged species that can give rise to ion-association phenomena. Although ion pairs are loosely bond species, their spectroscopic, photochemical, and photophysical properties differ markedly from those of the two isolated partners. Under favourable conditions additional IPCT optical transitions may be observed. If the ion pair consists of a relatively strong reducing and oxidizing agents, the IPCT transition can occur at quite low energies, eg at 14,000 cm^{-1} for the ion pair $\{[Ru(NH_3)_6]^{3+}\cdot[Fe(CN)_6]^{4-}]\}$ [30, 94]. The chemical consequences of the photoinduced charge transfer are well documented, the typical example may be photoreduction of a Co^{III} complex at the expense of an external ion, eg [91]:

$$\{[\mathrm{Co(NH_3)_6}]^{3+}\cdot\mathrm{I}^-\} \xrightarrow{h\nu\,(\mathrm{IPCT})} [\mathrm{Co(NH_3)_6}]^{2+}\cdot\mathrm{I}^\bullet\} \rightarrow \mathrm{Co^{II}aq} + 6\mathrm{NH_3} + 1/2\,\mathrm{I_2} \qquad (6.26)$$

When two associated ions are coordination compounds, the IPCT transitions may lead to charge transfer between both centres, between ligands of the two complexes, and between metal and ligand of different complexes [11].

CT State Quenching

A second sphere interaction may result in quenching of the CT excited states by energy transfer processes, leading to luminescence quenching, eg the $MLCT_{Re \rightarrow bpy}$ luminescence of Re^{I} monomeric and polymeric complexes is quenched by Cu^{II} species and the sacrificial electron donor 2,2′,2″-nitrilotriethanol (TEOA) [95].

Changes in the electronic and molecular structures after CT excitation can also result in chemical bond formation between the excited complex {AB}* and another molecule (Q) of the system, yielding an encounter complex {AB–Q}*. The complex formed by interaction of an excited molecular entity with a ground state partner of the same structure is called an excimer, whereas an electronically excited dimer of definite stoichiometry, formed by interaction of an excited molecular entity with a ground state partner of a different structure, is named an exciplex [29]. Both excimers and exciplexes have their own properties and can thus be regarded as new chemical species. Their generation is well documented by the concentration effect on the fluorescence of some solutes or by flash photolysis measurements [11, 24].

Although the phenomenon is more common in organic photochemistry, a coordination entity can also act in the process of excimer or exciplex formation as an excited molecule {AB}* or quencher (Q). The second-sphere donor–acceptor interaction with an acceptor quencher causes oxidative quenching of {AB}*, whereas interaction with a donor quencher yields reductive quenching.

Chromate(VI) has been reported to undergo reduction to Cr^{V} as a result of PET between its LMCT excited state and an external electron donor. In the study carried out for several aliphatic alcohols (methanol, ethanol, propan-2-ol, butan-1-ol, butan-2-ol, 2-methyl-propan-2-ol) two pathways of PET were identified: one-electron transfer for intermolecular and two-electron transfer for intramolecular systems [96, 97]. The intermolecular mechanism of the Cr^{VI} excited state quenching was also found for phenol or its derivatives [98], whereas in the case of an anion donor (such as oxalate) an effect of external cations was observed [99].

Some CT excited states may be quenched by both electron-donor and electron-acceptor quenchers; this behaviour can be exemplified by the Fe^{III} complex with EDTA: the LMCT excited state was found to be reductively quenched, leading to metal centre reduction:

$$[Fe^{III}(edta)(H_2O)]^{-} \xrightarrow{h\nu\,(LMCT)} [Fe^{II}(edta^{\bullet})(H_2O)]^{-} \xrightarrow{D} [Fe^{II}(edta)(H_2O)]^{2-} + D^{\bullet +} \quad (6.27)$$

by donors (D) such as methanol, ethanol, propan-2-ol, butan-2-ol, nitrate(III), sulphate(IV), and external EDTA. The same complex undergoes oxidative quenching, which results in the EDTA ligand oxidation:

$$[Fe^{II}(edta^{\bullet})(H_2O)]^{-} \xrightarrow{A} [Fe^{III}(ed3a)(H_2O)] + A^{\bullet -} + CO_2 + HCHO \quad (6.28)$$

in the presence of electron acceptor (A = molecular oxygen or chromate(VI), ed3a-ethylenediaminetriacetate) [61, 62, 100].

Sometimes the MLCT transition results in an increased electron density at a ligand atom, enhancing its basic properties and yielding a dinuclear complex, eg

$$*[Ru(bpy)_2(dpp)] + [MCl_6]^{n-} \rightarrow [Ru(bpy)_2(dpp)MCl_4]^{2-n} + 2Cl^- \quad (6.29)$$

where dpp = 2,3-bis(2-pyridyl)pyrazine, and M = Pt^{IV}, Rh^{III}, Pd^{IV} [101].

The Ru^{II} complexes that contain polyaza-aromatic ligands, such as TAP (1,4,5,8-tetra-azaphenanthrene) and HAT (1,4,5,8,9,12-hexa-azatriphenylene) manifest the very high oxidation potential of their ^{3}MLCT states. Therefore, under illumination, they are able to abstract an electron even from rather poor electron donors, eg biopolymers such as DNA, oligonucleotides, and oligopeptides. Other effects of the photoinduced electron transfer may be photoadduct formation and photo-crosslinking via metal complexes [102].

The photoinduced electron transfer reaction rates depend strongly on the molecular structures of the donor and acceptor entities [16]. The nature of the linker between them influences the charge or energy transfer, which is facilitated considerably when both partners are connected by bridging ligands [13, 17, 31].

The photoinduced electron and energy transfers occurring in coordination compounds, especially when visible light is a driving force, make them very useful photosensitizers. These processes are of importance in many fields, eg biological diagnosis, gene phototherapy, photosynthesis, vision processes, self-assembling systems, light harvesting systems, storage of solar energy, sensors, charge separation devices, and information storage systems [11, 12, 15–17, 19, 102–104].

Besides the direct photoreactions reported to date (see Figure 6.6), coordination entities also undergo the photosensitized modes in which they play the role of non-absorbing substrate. The application is usually connected with solving some mechanistic problems, as in the early study of cobalt(III) complexes with organic sensitizers, such as biacetyl, benzophenone, benzaldehyde, and naphthalene, which provided valuable mechanistic information about the excited states responsible for the photophysical, photoredox, and photosubstitution processes observed in cases of these complexes [105].

6.5 Photosensitized Reactions

The excited-state species can also reduce its energy by transfer to another molecule (quencher), which may diffuse it or change it into chemical energy. The process by which a photochemical or photophysical alteration occurs in one molecular entity, as a result of initial absorption of radiation by another molecular entity, is called photosensitization.

The species absorbing and transferring the radiant energy is called the photosensitizer. In mechanistic photochemistry the term is limited to cases in which the photosensitizer is not consumed in the reaction [29].

The phenomenon, illustrated in Figure 6.8, consists of three stages:

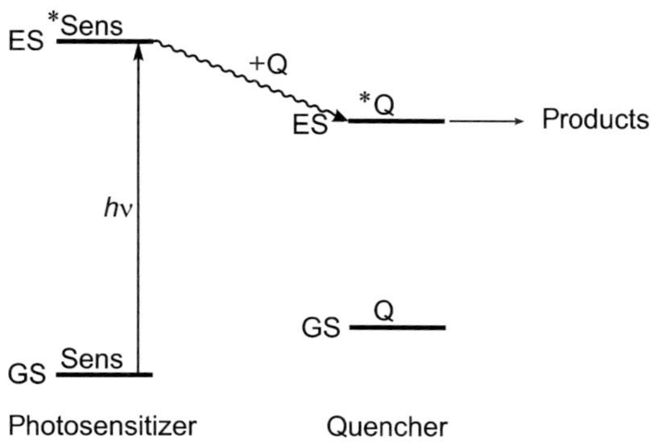

Figure 6.8 *Photosensitization reaction pathway*

1. Excitation of a photosensitizer molecule (Sens) by light absorption:

$$\text{Sens} \xrightarrow{h\nu} {}^*\text{Sens} \tag{6.30}$$

2. Quenching of the sensitizer excited state by a quencher molecule (Q):

$$^*\text{Sens} + \text{Q} \rightarrow \text{Sens} + {}^*\text{Q} \tag{6.31}$$

3. Photochemical reaction of the excited quencher molecule:

$$^*\text{Q} \rightarrow \text{products} \tag{6.32}$$

When the reaction of a non-absorbing quencher is induced by electron transfer (not energy transfer) using an excited light-absorbing sensitizer, the process is called electron transfer photosensitization. Then interaction between sensitizer and quencher consists of photoinduced electron transfer:

$$^*\text{Sens} + \text{Q} \rightarrow \text{Sens}^+ + \text{Q}^- \tag{6.33}$$

or

$$^*\text{Sens} + \text{Q} \rightarrow \text{Sens}^- + \text{Q}^+ \tag{6.34}$$

and, depending on the action of the excited sensitizer as electron donor or acceptor, the sensitization is called reductive or oxidative. The conceivable mechanisms of photosensitized electron transfer are compared with the direct photoinduced electron transfer processes in Figure 6.9.

The opposite of direct photoinduced electron transfer, but similar to photosensitization by energy transfer, the overall process must be such that the sensitizer is recycled. This can be done using a sacrificial electron donor, D:

$$\text{Sens}^+ + \text{D} \rightarrow \text{Sens} + \text{D}^+ \tag{6.35}$$

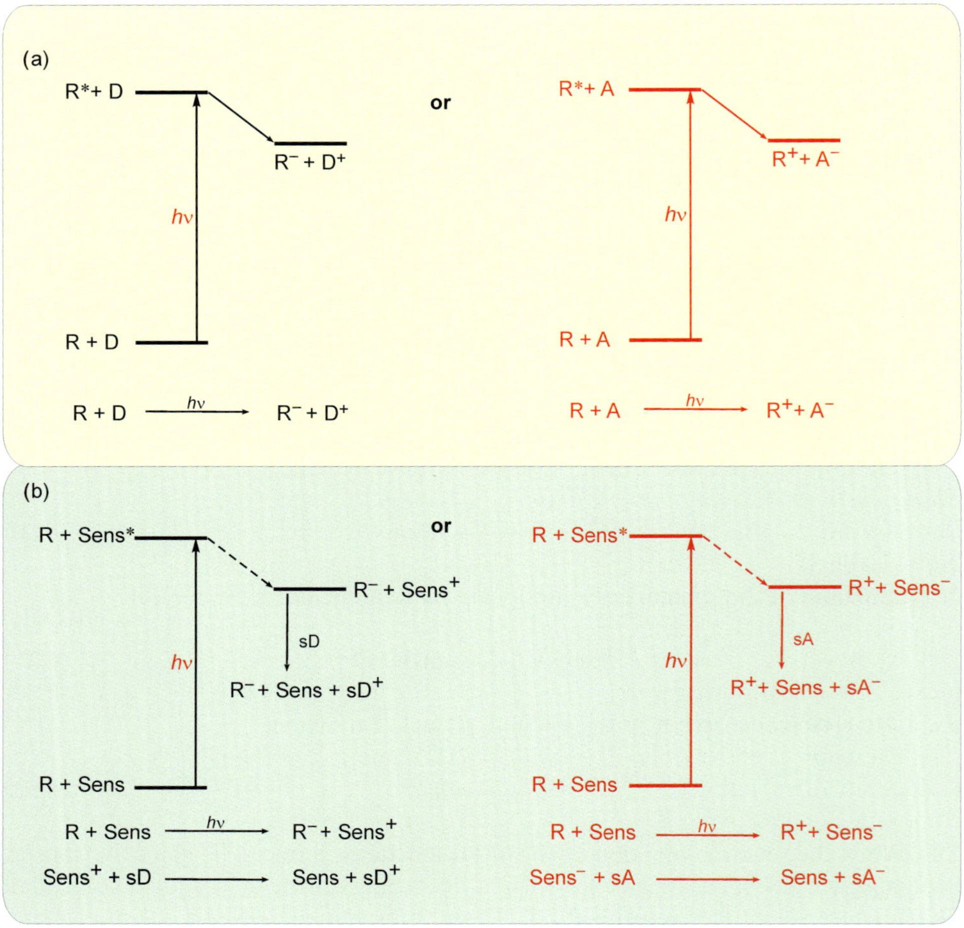

Figure 6.9 *Photoinduced electron transfers via (a) direct or (b) photosensitized processes. R, reactant; Sens, sensitizer; D and sD, donor and sacrificial donor, respectively; A and sA, acceptor and sacrificial acceptor, respectively. Reductive PET (photoinduced electron transfer) schemes are drawn in black, whereas oxidative schemes are in red*

or sacrificial electron acceptor, A:

$$\text{Sens}^{-} + \text{A} \rightarrow \text{Sens} + \text{A}^{-} \tag{6.36}$$

A sacrificial donor (acceptor) is a molecular entity that acts as the electron donor (acceptor) in a photoinduced electron transfer process and is not restored in a subsequent redox process, but is destroyed by irreversible chemical conversion [29].

The excitation by photosensitization may lead to population of excited states that are inaccessible by direct absorption, and their unsensitized and sensitized chemistry may be quite different.

The energy transfer step is assumed to be adiabatic and hence spin conserved, with sensitizers, and thus the occurrence of photosensitized reactions provides valuable information about intersystem crossing and triplet chemistry. The photosensitization can be initiated by both singlet and triplet excited states of the photosensitizers, but in the case of organic photosensitizers most of the oxidations occur by way of the triplet sensitizer. Many different substances are effective sensitizers, eg dyes, pigments, aromatic hydrocarbons, and transition metal complexes. A wide variety of organic photosensitizers may be exemplified by biacetyl, benzophenone, benzaldehyde, naphthalene, rhodamine [1, 106, 107], whereas among inorganic photosensitizers the best known are $[Ru^{II}(bpy)_3]^{2+}$ and its derivatives [108, 109], various porphyrinoid sensitizers [110, 111], and conjugated polymers [112].

The crucial meanings of photosensitization for environmental self-cleaning processes and conversion of solar energy into a useful one are discussed in Chapters 9 (section 9.4) and 21 (section 21.4) [113–124].

Photosensitized Oxidation

Photosensitized oxidations by molecular oxygen are especially important in the biological field, because of their impact on living organisms. It is known that microorganisms and other living cells can be killed in the presence of light, molecular oxygen, and sensitizing dyes.

Photosensitized oxidations involve either organic dyes or transition metal complexes as sensitizers. Most organic sensitizers operate in their triplet states, as they are relatively longer lived than the singlet states.

The triplet sensitizer may react with the O_2 triplet or an oxidizable reactant (Figure 6.10), eg with alkenes, dienes, and aromatic and other unsaturated organic compounds. For many sensitizers the reaction with oxygen is so efficient that the former process is favoured over the latter [125].

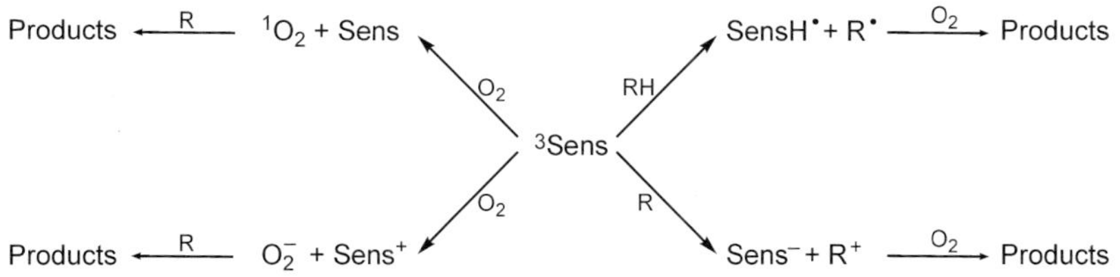

Figure 6.10 *Reactions of the triplet sensitizer: with molecular oxygen and with an oxidizable reactant (R or RH). (Adapted from Rontani [125])*

The excited O_2 states are singlets ($^1\Delta_g$, $^1\Sigma_g$), the first of which is relatively low lying (94 kJ mol^{-1}, equivalent to near-infrared radiation at 1270 nm) and thus can be easily populated by the energy transfer from the triplet states of most dyes; the second O_2 singlet ($^1\Sigma_g$), is more difficult to reach (157 kJ mol^{-1}), but is much more reactive. The first step of this photo-oxidation pathway is usually the singlet oxygen attack on the reactant molecule in its ground state:

$$R + {}^1O_2 \rightarrow RO_2 \tag{6.37}$$

The first oxidation products are often peroxides, hydroperoxides, or superoxides, which then take part in secondary oxidation steps.

One of the best photosensitizers of oxidation is chlorophyll, the role of which in vital processes cannot be overestimated (see Chapter 11, section 11.1). The highly valuable processes can, however, be accompanied by the production in chloroplasts of toxic oxygen species, eg $O_2^{-\bullet}$, $^1O_2$, H_2O_2, and $OH^\bullet$. The correct course of photosynthesis involves generation of an excited singlet state (^{1}Chl) as a result of light absorption by a chlorophyll molecule. If the excitation energy is not completely consumed by the photosynthesis reactions, a longer lived triplet state (^{3}Chl) can be populated via intersystem crossing. ^{3}Chl is not only potentially damaging itself in reaction with an oxidizable reactant of the type shown in Figure 6.10, but it may also generate toxic oxygen species in reaction with ground state oxygen via three probable pathways:

(1) energy transfer:

$$^3O_2 \xrightarrow{^3\text{Chl}} {}^1O_2 \tag{6.38}$$

(2) electron transfer:

$$^3O_2 \xrightarrow{^3\text{Chl}} O_2^{-\bullet} \xrightarrow{H^+} HOO^\bullet \xrightarrow{\text{Dismutation}} {}^1O_2 + H_2O_2 \xrightarrow{\text{Fe(III)}} OH^\bullet \tag{6.39}$$

in which the $O_2^{-\bullet}$ oxidation by strong oxidants may be an additional source of $^1O_2$

(3) formation of an active complex, which can oxidize OH^- ions to $OH^\bullet$ radicals [125]:

$$^3\text{Chl} + {}^3O_2 \rightarrow {}^3\text{Chl-}O_2 \xrightarrow{OH^-} OH^\bullet \tag{6.40}$$

The side reactions leading to $O_2^{-\bullet}$ generation are especially important when the quantity of electrons generated by light absorption exceeds the efficiency of their sink yielding CO_2; this applies especially to photosynthesis occurring under very intensive sunlight.

To avoid oxidative damage, the photosynthetic systems cooperate with carotenoids, which protect the organisms against the lethal effects of their own chlorophyll. β-Carotene was recognized as an extremely efficient quencher of singlet oxygen and inhibitor of sensitized photo-oxidation. The antioxidant mechanisms involve quenching by carotenoids (Car), both ^{3}Chl and $^1O_2$, via energy transfers occurring at very high rates (k of the order of 10^8 and $10^{10}\,\text{mol}^{-1}\,\text{s}^{-1}$, respectively):

$$^3\text{Chl} + \text{Car} \rightarrow \text{Chl} + {}^3\text{Car} \xrightarrow{\text{Radiationless}} \text{Car} \tag{6.41}$$

$$^1O_2 + \text{Car} \rightarrow {}^3O_2 + {}^3\text{Car} \xrightarrow{\text{Radiationless}} \text{Car} \tag{6.42}$$

This behaviour suggests that carotenoids not only remove the toxic singlet oxygen but can also neutralize the energy excess that is transferred from chlorophyll to molecular oxygen [1, 125].

Moreover, in the membrane environments the tocopherols are present, which remove the toxic oxygen species ($O_2^{\bullet-}$, $^1O_2$, $HOO^{\bullet}$ and $OH^{\bullet}$). The tocopherols act here as sacrificial donors, ie scavengers, which are not restored in a subsequent reduction process but are destroyed by irreversible chemical conversion [125].

The photosensitizing properties of some pollutants are of environmental relevance [126] (for details see Chapter 9, section 9.4).

6.6 Homogeneous Photocatalysis

Often the photochemical process is efficient only in the presence of a catalyst or its precursor:

$$\text{Substrate} + \text{Light} + \text{Catalyst} \rightarrow \text{Products} + \text{Catalyst} \tag{6.43}$$

The catalytic reaction involving light absorption by a substrate and catalyst species is called photocatalysis [29]. At the end of the reaction cycle, the photocatalyst is regenerated to its original state. Sometimes, however, the agent that initiates certain chemical transformations under the action of light is itself consumed in the reaction or process; then it is called a photoinitiator [127–129].

The first more comprehensive classification of the photocatalytic processes was proposed by Salomon, who divided the processes into two main classes: (1) photogenerated catalysis, which is catalytic in photons, and (2) catalyzed photolysis, which is non-catalytic in photons [130] (Figure 6.11).

Broader descriptions were proposed by many authors, eg Kutal [131, 132], Chanon and Chanon [28], Kisch [133], Hennig [134], and Serpone et al. [135–137].

The photogenerated catalysis includes catalytic reactions in which the ground states of the catalyst and the substrate are involved in the thermodynamically spontaneous (exoergic) catalytic step. Light may initiate only the reaction, which can be continued without any irradiation. Indeed, using this approach we may consider photogenerated catalysis as catalysis of a thermal reaction by a catalyst produced as a result of light absorption by the catalyst precursor. The processes are also called photogenerated (photoinduced, photoinitiated) catalytic reactions or true photocatalysis.

Alternatively, light is consumed and the reaction progress is possible only under continuous light absorption; this option, called catalyzed photolysis, includes photoassisted generation of a reactive form of substrate or photocatalyst. In the former the process is called catalyzed photochemical reaction, whereas in the latter either catalyst activation may lead to formation of catalyst or photoinitiator, which initiates chemical transformations but is consumed within a reaction cycle, or the catalyst reacts with substrate in its excited state (photosensitization); in both cases

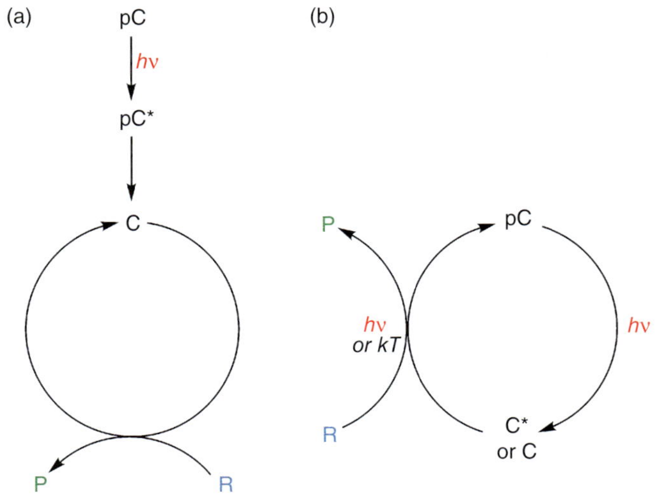

Figure 6.11 *Representational formal schemes of (a) photogenerated catalysis, which is catalytic in photons, and (b) catalyzed photolysis, which is non-catalytic in photons. pC, catalyst precursor; C, catalytic entity; R, substrate; P, product*

the reaction progress needs absorption of the next photons. Catalyzed photoreactions, photosensitized reactions, and photoassisted catalysis are rated among types of catalyzed photolysis; in these cases the nominal catalyst or substrate, or both, can be in an excited state during the catalytic step. This classification of photocatalysis is presented in Figure 6.12.

Photogenerated catalysis is a catalytic reaction that involves production of a catalyst by the absorption of light. The initiation of chemical transformations consists of the photochemical formation of substances, which (once formed) are catalytically active even without the further action of light. It means that photogenerated catalysis does not require continuous irradiation because the catalyst C is reproduced at the end of each cycle (Figures 6.13a and 16.13b) and, contrary to catalyzed photolysis, the photoinduced catalysis is catalytic in photons. The schemes resembling this presented in Figure 6.11a operate in photocatalysis of olefin hydrogenation (Figure 6.13a), *cis–trans* isomerization, or double bond migration (Figure 6.13b). In these cases the catalyst precursors are [$Fe(CO)_5$], [$Fe_3(CO)_{12}$], or [$Ru_3(CO)_{12}$], which in photochemical reactions transform into the proper catalysts [28].

Another possible alternative of photogenerated catalytic reaction is a chain reaction initiated by photochemically generated catalyst (radical) and propagated by its successor radicals (Figures 6.14 and 6.15). The photoinduced chain processes in some way resemble photocatalytic ones. Such reactions offer considerable potential for light-induced polymerizations, but also for the synthesis of fine chemicals. The initiating radical may be generated in photodissociation (Figure 6.14) or photoredox reaction (Figure 6.15) [138].

Considering the applications, it is advantageous to distinguish between two limiting cases of photocatalysis, photoinduced catalytic, and photoassisted reactions.

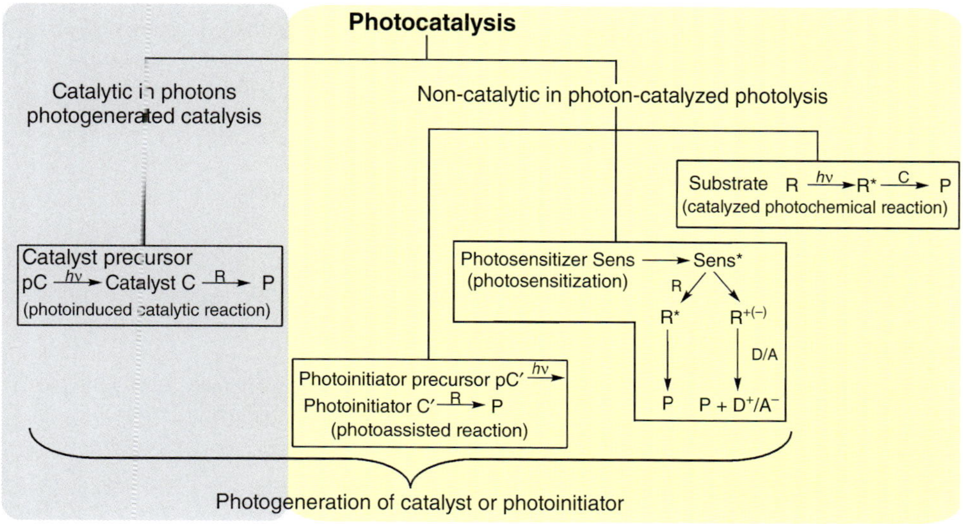

Figure 6.12 *Classification of photocatalysis and summary of various mechanism-specific labels. Assignments: C, catalytic entity; pC, catalyst precursor; R, substrate; P, product; C′, photoinitiator; pC′, photoinitiator precursor; Sens, sensitizer; D, electron donor; A, electron acceptor*

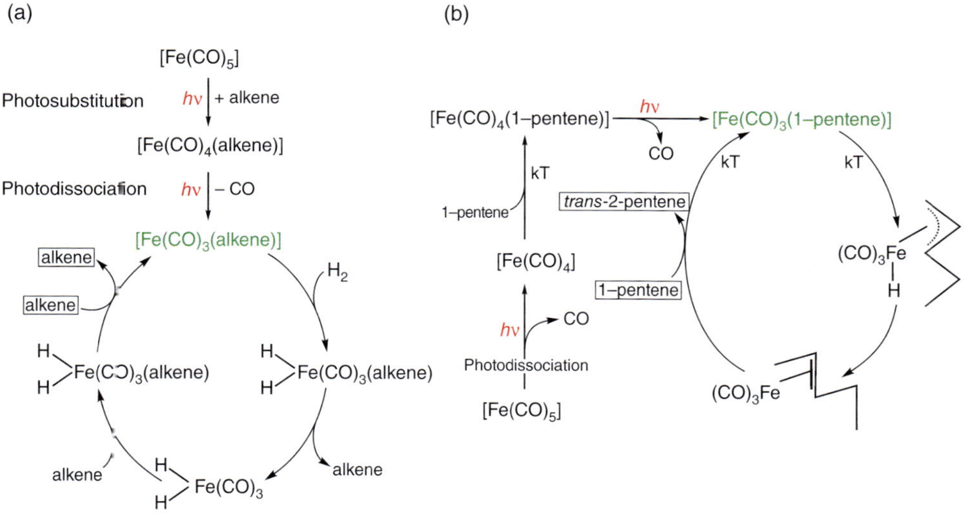

Figure 6.13 *Examples of photoinduced catalytic reactions: (a) alkene hydrogenation by the [Fe(CO)$_3$(alkene)] photocatalyst and (b) the double bond migration by the [Fe(CO)$_3$(1-pentene)] photocatalyst; both catalysts are generated in photosubstitution and photodissociation reactions of the [Fe(CO)$_5$] precursor [28]*

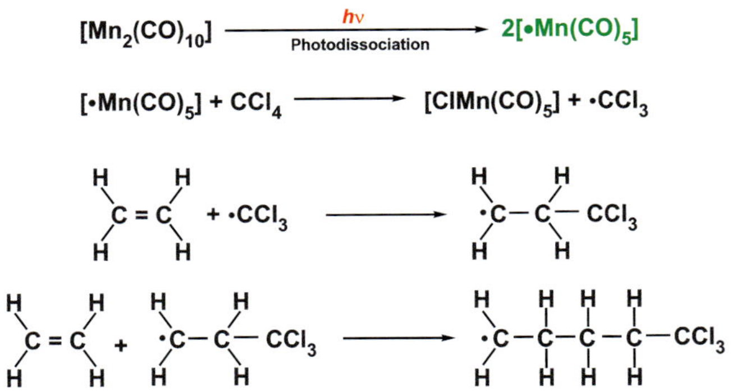

Figure 6.14 Example of polymerization chain reaction: ethylene polymerization initiated by the $[M(CO)_5]$ photoinitiator (generated from its $[Mn_2(CO)_{10}]$ precursor by photodissociation) and propagated by its successive chain carriers

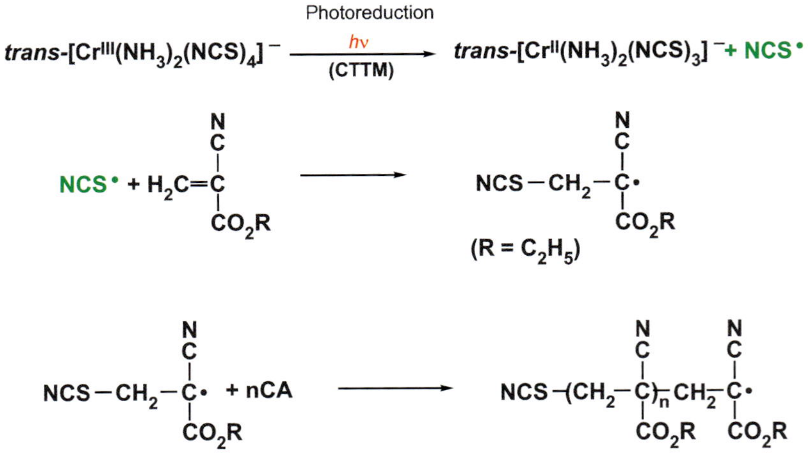

Figure 6.15 The cyanoacrylate (CA) polymerization chain reaction initiated by the $NCS^{\bullet}$ photoinitiator (generated in photoreduction of the Cr^{III} complex) and propagated by its successive chain carriers

Photoinduced catalysis means the photogeneration of a catalyst that subsequently promotes a catalyzed reaction. Photons are required to generate the catalyst only. Thus, the efficiency of such processes depends only on the activity of the catalyst produced photochemically and, in homogeneous photocatalysis, the turnover number (TON) is the useful tool. The TON is usually expressed as the number of moles of product formed per mole of catalyst and, for photoassisted catalysis, TON <1, whereas for photogenerated catalysis TON >1 and even $>>1$ [135]. Therefore, high turnovers of photochemically produced catalysts are one of the main criteria concerning efficient photocatalytic processes. Quantum yields (ratio of moles of product formed to the number of photons absorbed) >1 may occur. The same is true for photoinduced chain reactions.

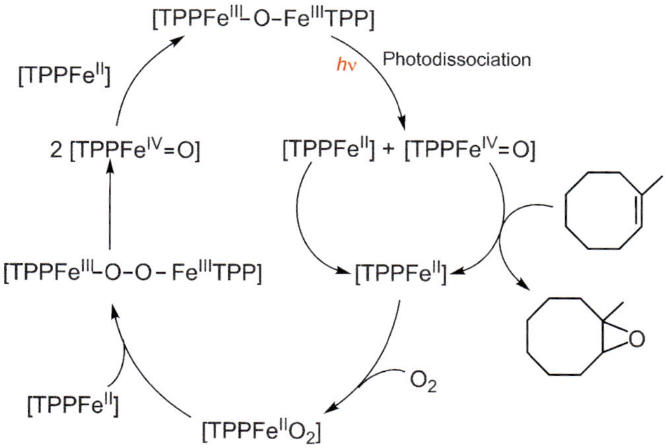

Figure 6.16 *An example of a photoassisted reaction: oxygen photoaddition initiated by [TPP-Fe^{IV}–O] complex produced by photodissociation from the [TPPFe^{III}–O–Fe^{III}TPP] precursor (TPP = triphenylphosphine)*

Photoassisted reactions include interactions between electronically excited states or short-lived intermediates and substrate molecules, leading to product formation under the concomitant regeneration of the starting complex in its electronic ground state. Product quantum yields >1 are impossible, because one photon may not initiate more than one catalytic cycle. In contrast to photoinduced catalytic processes, photoassisted reactions (Figure 6.16) can be thermodynamically uphill, thus allowing the possibility for light energy conversion.

Catalyzed photolysis refers to catalysis of a photochemical reaction, for which there is a physical pathway for decay of the system back to its ground state (Figure 6.17). When the photocatalytic process occurs through photoexcitation of the catalyst, the physical decay may occur through recombination and/or thermal photoionization of the excited states, which ultimately leads to regeneration of the original state of the catalyst. Note that catalyzed photolysis is not catalytic in photons, contrary to photogenerated catalysis.

As catalyst species may also be active in their excited states, the photosensitized reactions in some way resemble the catalyzed photolysis (Figure 6.12). The details of these reactions were given in section 6.5.

The first popular photocatalysts in homogeneous media were transition metal carbonyls (eg $M(CO)_6$, where M=Cr, Mo, W; $Fe(CO)_5$, $Fe_3(CO)_{12}$, $Ru_3(CO)_{12}$, $Mn_2(CO)_{10}$) and their derivatives [28]. Recently, other systems have been intensively studied, eg iron(III) complexes, especially in the photo-Fenton reactions [63–78, 139–142], polyoxometallates, porphyrins, metalloporphyrins, and other porphyrinoid sensitizers.

Polyoxometallates (POMs), molecular anionic metal oxides, show the ability to act as homogeneous photocatalysts in electron transfer from organic electron donors to metal ions. In the process, illumination at the LMCT band (visible and near-UV regions) renders POMs strong oxidants that are able to abstract electrons from a

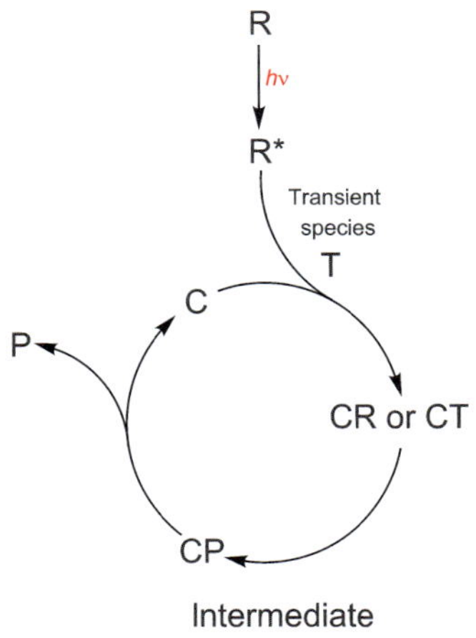

Figure 6.17 *A formal scheme of a catalyzed photochemical reaction*

great variety of organic compounds, including organic pollutants. Thus, POMs are used in a variety of photocatalytic processes; among others they are applied as photocatalysts for metal recovery in the form of fine particles. The process involves absorption of light by a POM, oxidation of an organic substrate, as sacrificial reducing reagent, and reoxidation of the reduced POMs by the metal ions, closing the photocatalytic cycle [142–154].

The photocatalyst that is active under visible light is of special importance as an essential element of solar photoenergy utilization. However, the photocatalysts based on metal oxides lack visible light activities; alternative visible light photocatalysts were then sought among visible light-absorbing metal–organic complexes. Porphyrins have been found to be of particular interest because they are the synthetic structural analogues of chlorophyll in plant photosynthesis. Porphyrins are excellent photosensitizers because of the small singlet-triplet splitting, the high quantum yield for intersystem crossing, and the long triplet state lifetimes. They readily coordinate metal ions in the central cavity, which makes the photoresponse of porphyrin stronger and broader in the visible light region.

Metalloporphyrin complexes that can absorb visible light with high molar absorption coefficients are chemically and biologically important molecules that have versatile catalytic capabilities in electron-transfer and energy-transfer reactions. Therefore, the photocatalytic reactions of a number of metalloporphyrin complexes have been investigated under visible-light irradiation. The metalloporphyrins such as iron (eg haem), magnesium (eg chlorophyll), zinc, antimony, and tin porphyrins have been widely investigated for their photoactivity in hydrogen evolution and degradation of organic compounds. Metalloporphyrins can be oxidized or reduced at three discrete sites: central metal, axial ligands, and porphyrin ring. The porphyrin ring oxidation and reduction are feasible because of the delocalized π-electron system and are influenced by the electronic interaction between the π-electron system and the central metal [110, 142, 155, 156].

An interesting alternative to the conventional one-electron photocatalysts was the discovery of two-electron photocatalysts in the form of mixed-valence coordination compounds [118]. The two-electron photocatalysis can be useful in activation of small molecules of energy consequence, especially hydrogen and oxygen (more details can be found in Chapter 21).

New prospects in photocatalysis may be in use as reaction media the supercritical fluids, which were recently found to push the homogeneous catalysis on a new trajectory [157].

References

1. Wayne CE, Wayne RP. *Photochemistry*. Oxford: Oxford University Press, 2001.
2. Herrera JM, Marvaud V, Verdaguer M, Marrot J, Kalisz M, Mathoniere C. Reversible photoinduced magnetic properties in the heptanuclear complex $[Mo^{IV}(CN)_2(CN{-}CuL)_6]^{8+}$: A photomagnetic high-spin molecule. *Angew Chem Int Ed* 2004;**43**:5468.
3. Przychodzen P, Korzeniak T, Podgajny R, Sieklucka B. Supramolecular coordination networks based on octacyanometalates: From structure to function. *Coord Chem Rev* 2006;**250**:2234–60.
4. Ohkoshi S, Tokoro H, Hashimoto K. Temperature- and photo-induced phase transition in rubidium manganese hexacyanoferrate. *Coord Chem Rev* 2005;**249**:1830–40.
5. Sato O. Photoinduced magnetization in molecular compounds. *J Photochem Photobiol C: Photochem Rev* 2004;**5**:203–223.
6. Ohkoshi S, Hashimoto K. Photo-magnetic and magneto-optical effects of functionalized metal polycyanides. *J Photochem Photobiol C: Photochem Rev* 2001;**2**:71–88.
7. Culp J, Park J, Frye F, Huha Y, Meisel MW, Talham DR. Magnetism of metal cyanide networks assembled at interfaces. *Coord Chem Rev* 2005;**249**:2642–8.
8. Einaga Y. Photo-switching magnetic materials. *J Photochem Photobiol C: Photochem Rev* 2006;**7**:69–88.
9. Lescouezec R, Toma LM, Vaissermann J, et al. Design of single chain magnets through cyanide-bearing six-coordinate complexes. *Coord Chem Rev* 2005;**249**:2691–729.
10. Dei A. Photomagnetic effects in polycyanometallate compounds: an intriguing future chemically based technology? *Angew Chem Int Ed* 2005;**44**:1160–3.
11. Balzani V, Scandola F. *Supramolecular Photochemistry*. New York: Ellis Horwood Ltd, 1991.
12. Balzani V, Bergamini G, Marchioni F, Ceroni P. Ru(II)-bipyridine complexes in supramolecular systems, devices and machines *Coord Chem Rev* 2006;**250**:1254–66.
13. Beyeler A, Belser P. Synthesis of a novel rigid molecule family for the investigation of electron and energy transfer. *Coord Chem Rev* 2002;**230**:29–39.
14. Sun SS, Lees AJ. Transition metal based supramolecular systems: synthesis, photophysics, photochemistry and their potential applications as luminescent anion chemosensors. *Coord Chem Rev* 2002;**230**:171–92.
15. Willner I, Willner B. Vectorial photoinduced electron-transfer in tailored redox-active proteins and supramolecular nanoparticle arrays. *Coord Chem Rev* 2003;**245**:139–51.
16. El-Khouly ME, Ito O, Smith PM, D'Souza F. Intermolecular and supramolecular photoinduced electron transfer processes of fullerene–porphyrin/phthalocyanine systems. *J Photochem Photobiol C: Photochem Rev* 2004;**5**:79–104.
17. D'Souza F, Ito O. Photoinduced electron transfer in supramolecular systems of fullerenes functionalized with ligands capable of binding to zinc porphyrins and zinc phthalocyanines. *Coord Chem Rev* 2005;**249**:1410–22.
18. Welter S, Salluce N, Belser P, Groeneveld M, De Cola L. Photoinduced electronic energy transfer in modular, conjugated, dinuclear Ru(II)/Os(II) complexes. *Coord Chem Rev* 2005;**249**:1360–71.

19. Kondo T, Uosaki K. Self-assembled monolayers (SAMs) with photo-functionalities. *J Photochem Photobiol C: Photochem Rev* 2007;**8**:1–17.
20. Otsuki J, Akasaka T, Araki K. Molecular switches for electron and energy transfer processes based on metal complexes. *Coord Chem Rev* 2008;**252**:32–56.
21. Koda S, Sugimoto K. Pressure effect on the absorption and photodissociation of O_2 near the dissociation threshold. *J Photochem Photobiol C: Photochem Rev* 2003;**4**:215–26.
22. Sadanaga Y, Matsumoto J, Kajii Y. Photochemical reactions in the urban air: Recent understandings of radical chemistry. *J Photochem Photobiol C: Photochem Rev* 2003;**4**:85–104.
23. Braden DA, Parrack EE, Tyler DR. Solvent cage effects. I. Effect of radical mass and size on radical cage pair recombination efficiency. II. Is geminate recombination of polar radicals sensitive to solvent polarity? *Coord Chem Rev* 2001;**211**:279–94.
24. Sykora J, Sima J. Photochemistry of coordination compounds. *Coord Chem Rev* 1990;**107**:1–225.
25. Zhao J, Li B, Onda K, Feng M, Petek H. Solvated electrons on metal oxide surfaces. *Chem Rev* 2006;**106**:4402–27.
26. Polo AS, Itokazu MK, Frin KM, Patrocinio AOT, Iha NYM. Light driven trans-to-cis isomerization of stilbene-like ligands in fac-$[Re(CO)_3(NN)(trans\text{-}L)]^+$ and luminescence of their photoproducts. *Coord Chem Rev* 2006;**250**:1669–80.
27. Kutal C, Adamson AW. Photochemical processes. In: Wilkinson G, Gillard RD, McCleverty JA, eds. *Comprehensive Coordination Chemistry*. Oxford: Pergamon Press, 1987: 385–414.
28. Channon F, Channon M. Transition metal complexes in photocatalysis. In: Serpone N, Pelizzetti E, eds. *Photocatalysis Fundamentals and Applications*. New York: Wiley- Interscience, 1989: 489–541.
29. Braslavsky SE, Houk KN, Verhoeven JW. *Glossary of Terms used in Photochemistry*. International Union of Pure and Applied Chemistry, Organic Chemistry Division, Commission on Photochemistry, 1996.
30. Billing R. Optical and photoinduced electron transfer in ion pairs of coordination compounds. *Coord Chem Rev* 1997;**159**:257–70.
31. De Cola L, Belser P. Photoinduced energy and electron transfer processes in rigid bridged dinuclear Ro/Os complexes. *Coord Chem Rev* 1998;**177**:301–46.
32. Tero-Kubota S, Katsuki A, Kobori Y. Spin-orbit coupling induced electron spin polarization in photoinduced electron transfer reactions. *J Photochem Photobiol C: Photochem Rev* 2001;**2**:17–33.
33. Endicott JF. The photophysics and photochemistry of coordination compounds. In: Solomon EI, Lever ABP, eds. *Inorganic Electronic Structure and Spectroscopy*. New York: John Wiley & Sons, Inc., 1999: 291–342.
34. Lees AJ. Quantitative photochemistry of organometallic complexes: insight to their photophysical and photoreactivity mechanisms. *Coord Chem Rev* 2001;**211**:255–78.
35. Vogler A, Kunkely H. Electronic spectra and photoreactivity of cyclopentadienyl complexes. *Coord Chem Rev* 2001;**211**:223–33.
36. Vogler A, Kunkely H. Charge transfer excitation of organometallic compounds. Spectroscopy and photochemistry. *Coord Chem Rev* 2004;**248**:273–8.
37. Vogler A, Kunkely H. Excited state properties of organometallics based on interligand interactions between aromatic ligands. *Coord Chem Rev* 2005;**249**:1511–16.
38. Zink JI. Photo-induced metal–ligand bond weakening, potential surfaces, and spectra. *Coord Chem Rev* 2001;**211**:69–96.
39. Kane-Maguire NAP. Photochemistry and photophysics of coordination compounds: Chromium. In: Balzani V, Campagna S, eds. *Photochemistry and Photophysics of Coordination Compounds I*. Berlin: Springer, 2007: 37–67.
40. Grills DC, Huang KW, Muckerman JT, Fujita E. Kinetic studies of the photoinduced formation of transition metal–dinitrogen complexes using time-resolved infrared and UV–vis spectroscopy. *Coord Chem Rev* 2006;**250**:1681–95.

41. Skibsted LH. Photoisomerization of rhodium amine complexes. The deduction of an excited state reaction mechanism. *Coord Chem Rev* 1989;**94**:151–79.
42. Bitterwolf TE. Photochemical nitrosyl linkage isomerism/metastable states. *Coord Chem Rev* 2006;**250**:1196–207.
43. Rack JJ, Winkler JR, Gray HB. Phototriggered Ru(II)-dimethylsulfoxide linkage isomerization in crystals and films. *J Am Chem Soc* 2001;**123**:2432–3.
44. Smith MK, Gibson JA, Young CG, Broomhead JA, Junk PC, Keene FR. Photoinduced ligand isomerization in dimethyl sulfoxide complexes of ruthenium(II). *Eur J Inorg Chem* 2000;1365–70.
45. Sima J. Photochemistry of azide-moiety containing inorganic compounds. *Coord Chem Rev* 2006;**250**:2325–34.
46. Vlcek A Jr, Busby M. Ultrafast ligand-to-ligand electron and energy transfer in the complexes fac-$[Re^{I}(L)(CO)_3(bpy)]^{n+}$. *Coord Chem Rev* 2006;**250**:1755–62.
47. van Slageren J, Klein A, Zalis S. Ligand-to-ligand charge transfer states and photochemical bond homolysis in metal-carbon bonded platinum complexes. *Coord Chem Rev* 2002;**230**:193–211.
48. Vogler A, Kunkely H. Photochemistry of peroxo complexes induced by LMCT, MLCT and peroxide IL/LLCT excitation. *Coord Chem Rev* 2006;**250**:1622–6.
49. Vogler A, Kunkely H. Ligand-to-ligand and intraligand charge transfer and their relation to charge transfer interactions in organic zwitterions. *Coord Chem Rev* 2007;**251**: 577–83.
50. Sutin N, Creutz C. Electron-transfer reactions of excited states. *J Chem Educ* 1983;**60**:809–14.
51. Stasicka Z, Wasielewska E. Photosubstitution and photoredox behaviour of cyanometallates: reaction models *Coord Chem Rev* 1997;**159**:271–94.
52. Szaciłowski K, Macyk W, Stochel G, Stasicka Z, Sostero S, Traverso O. Ligand and medium controlled photochemistry of iron and ruthenium mixed-ligand complexes: prospecting for versatile systems. *Coord Chem Rev* 2000;**208**:277–97.
53. Vogler A, Kunkely H. Photoreactivity of gold complexes. *Coord Chem Rev* 2001; **219–21**:489–507.
54. Szaciłowski K, Oszajca J, Barbieri A, et al. Photochemistry of the $[Fe(CN)_5N(O)SR]^{3-}$ complex. A mechanistic study. *J Photochem Photobiol A: Chem* 2001;**143**:99–108.
55. Lawrance GA, Maeder M, Neuhold Y-M, Szaciłowski K, Barbieri A, Stasicka Z. An electrospray ionization mass spectrometry study of the nitroprusside–cation–thiolate system *J Chem Soc Dalton Trans* 2002;3649–55.
56. Szaciłowski K, Wanat A, Barbieri A, et al. Reactions of the $[Fe(CN)_5NO]^{2-}$ complex with biologically relevant thiols. N *J Chem* 2002;**26**:1495–502.
57. Jaworska M, Stasicka Z. Structure and UV-Vis spectroscopy of the iron-sulfur dinuclear nitrosyl complexes $[Fe_2S_2(NO)_4]^{2-}$ and $[Fe_2(SR)_2(NO)_4]$. *New J Chem* 2005;**29**:604–612.
58. Jaworska M, Stasicka Z. Structure and UV–vis spectroscopy of Roussin black salt $[Fe_4S_3(NO)_7]^{-}$. *J Mol Struct* 2006;**785**:68–75.
59. Chmura A, Szaciłowski K, Waksmundzka-Góra A, Stasicka Z. Photochemistry of the $[Fe_4(\mu_3\text{-}S)_3(NO)_7]^{-}$ complex in the presence of S-nucleophiles: A spectroscopic study. *Nitric Oxide* 2006;**14**:247–60.
60. Chmura A, Szaciłowski K, Stasicka Z. The role of photoinduced electron transfer processes in photodegradation of the $[Fe_4(\mu_3\text{-}S)_3(NO)_7]^{-}$ cluster. *Nitric Oxide* 2006;**15**: 370–9.
61. Kocot P, Szaciłowski K, Stasicka Z. Photochemistry of the $[Fe^{III}(edta)(H_2O)]^{-}$ and $[Fe^{III}(edta)(OH)]^{2-}$ complexes in presence of environmentally relevant species. *J Photochem Photobiol A: Chem* 2007;**188**:128–34.
62. Kocot P, Karocki A, Stasicka Z. Photochemistry of the Fe(III)–EDTA complexes. A mechanistic study. *J Photochem Photobiol A: Chem* 2006;**179**:176–83.
63. Bautitz IR, Nogueira RFP. Degradation of tetracycline by photo-Fenton process – Solar irradiation and matrix effects. *J Photochem Photobiol A: Chem* 2007;**187**:33–9.

64. Farias J, Rossetti GH, Albizzati ED, Alfano OM. Solar degradation of formic acid: temperature effects on the photo-Fenton reaction. *Ind Eng Chem Res* 2007;**46**:7580–6.
65. Peralta-Hernandez JM, Meas-Vong Y, Rodriguez FJ, Chapman TW, Maldonado MI, Godinez LA. In situ electrochemical and photo-electrochemical generation of the Fenton reagent: A potentially important new water treatment technology. *Water Res* 2006;**40**:1754–62.
66. Chacon JM, Leal MT, Sanchez M, Bandala ER. Solar photocatalytic degradation of azo-dyes by photo-Fenton process. *Dyes Pigments* 2006;**69**:144–50.
67. Rodriguez SM, Galvez JB, Rubio MIM, Ibanez PF, Gernjak W, Alberola IO, Treatment of chlorinated solvents by TiO_2 photocatalysis and photo-Fenton: influence of operating conditions in a solar pilot plant. *Chemosphere* 2005;**58**:391–8.
68. Rodriguez M, Malato S, Pulgarin C, et al. Optimizing the solar photo-Fenton process in the treatment of contaminated water. Determination of intrinsic kinetic constants for scale-up. *Solar Energy* 2005;**79**:360–8.
69. Perez-Estrada LA, Malato S, Gernjak W, et al. Photo-Fenton degradation of diclofenac: identification of main intermediates and degradation pathway. *Environ Sci Technol* 2005;**39**:8300–6.
70. Parra S, Guasaquillo I, Enea O, et al. Abatement of an azo dye on structured C-nafion/Fe-ion surfaces by photo-Fenton reactions leading to carboxylate intermediates with a remarkable biodegradability increase of the treated solution. *J Phys Chem B* 2003;**107**:7026–35.
71. Gernjak W, Krutzler T, Glaser A, et al. Photo-Fenton treatment of water containing natural phenolic pollutants. *Chemosphere* 2003;**50**:71–8.
72. Malato S, Blanco J, Cáceres J, Fernández-Alba AR, Agüera A, Rodriguez A. Photocatalytic treatment of water-soluble pesticides by photo-Fenton and TiO_2 using solar energy. *Catalysis Today* 2002;**76**:209–20.
73. Kiwi J, Denisov N, Gak Y, et al. Catalytic Fe^{3+} clusters and complexes in nafion active in photo-Fenton processes. High-resolution electron microscopy and femtosecond studies. *Langmuir* 2002;**18**:9054–66.
74. Rossetti GH, Albizzati ED, Alfano OM. Decomposition of formic acid in a water solution employing the photo-Fenton reaction. *Ind Eng Chem Res* 2002;**41**:1436–44.
75. Malato S, Caceres J, Aguera A, et al. Degradation of imidacloprid in water by photo-Fenton and TiO_2 photocatalysis at a solar pilot plant: a comparative study. *Environ Sci Technol* 2001;**35**:4359–66.
76. Benitez FJ, Beltran-Heredia J, Acero JL, Rubio FJ. Chemical decomposition of 2,4,6-trichlorophenol by ozone, Fenton's reagent, and UV radiation. *Ind Eng Chem Res* 1999;**38**:1341–9.
77. Pignatello JJ, Liu D, Huston P. Evidence for an additional oxidant in the photoassisted Fenton reaction. *Environ Sci Technol* 1999;**33**:1832–9.
78. Cermenati L, Pichat P, Guillard C, Albini A. Probing the TiO_2 photocatalytic mechanisms in water purification by use of quinoline, photo-Fenton generated OH radicals and superoxide dismutase. *J Phys Chem B* 1997;**101**:2650–8.
79. Vogler A, Kunkely H. Photoreactivity of metal-to-ligand charge transfer excited states. *Coord Chem Rev* 1998;**177**:81–96.
80. Vlcek Jr. A. Highlights of the spectroscopy, photochemistry and electrochemistry of $[M(CO)_4(\alpha\text{-diimine})]$ complexes, M=Cr, Mo, W. *Coord Chem Rev* 2002;**230**:225–42.
81. Daniel C. Quantum chemistry and Dick Stufkens photochemistry. *Coord Chem Rev* 2002;**230**:65–78.
82. Vlcek Jr. A, Zalis S. Modeling of charge-transfer transitions and excited states in d^6 transition metal complexes by DFT techniques. *Coord Chem Rev* 2007;**251**:258–87.
83. McClenaghan ND, Leydet YM, Indelli MT, Campagna S. Excited-state equilibration: a process leading to long-lived metal-to-ligand charge transfer luminescence in supramolecular systems. *Coord Chem Rev* 2005;**249**:1336–50.
84. Jaworska M, Macyk W, Stasicka Z. Structure, Spectroscopy and Photochemistry of the $[M(\eta^5\text{-}C_5H_5)(CO)_2]_2$ Complexes (M=Fe, Ru). *Structure Bonding* 2004;**106**:153–72.

85. Macyk W, Herdegen A, Stochel G, Stasicka Z, Sostero S, Traverso O. Equilibrium between CO-bridged and non-bridged forms of $[(C_5H_5)Ru(CO)_2]_2$ and selective photoreactivity of the non-bridged form. *Polyhedron* 1997;**16**:3339–44.
86. Yang H, Snee PT, Kotz KT, Payne CK, Harris CB. Femtosecond infrared study of the dynamics of solvation and solvent caging. *J Am Chem Soc* 2001;**123**:4204–10.
87. Bitterwolf TE. Mechanisms and intermediates in the photochemistry of $M_2(CO)_6(\eta^5\text{-}C_5H_5)_2$, where M–Cr, Mo and W, and their ring-coupled analogs. *Coord Chem Rev* 2001;**211**:235–54.
88. Kunkely H, Stochel G, Vogler A. Photoredox reactions of $Hg(CN)_2/[Fe(CN)_6]^{4-}$ and $[HgCo_2(CN)_{10}]^{6-}$ induced by inner-sphere metal to metal charge transfer excitation. *Z Naturforsch* 1989;**44b**:145–8.
89. Balzani V, Bergamini G, Campagna S, Puntoriero F. Photochemistry and photophysics of coordination compounds: overview and general concepts. In: Balzani V, Campagna S, eds. *Photochemistry and Photophysics of Coordination Compounds I. Berlin: Verlag*, 2007: 1–36.
90. Sans MQ, Belser P. Towards new chiroptical switches. *Coord Chem Rev* 2002;**229**:59–66.
91. Balzani V, Sabbatini N, Scandola F. 'Second-sphere' photochemistry and photophysics of coordination compounds. *Chem Rev* 1986;**86**:319–37.
92. Vogler A, Kunkely H. Outer-sphere charge transfer in ion pairs with hydridic, carbanionic, sulfidic and peroxidic anions as electron donors – spectroscopy and photochemistry. *Coord Chem Rev* 2002;**229**:147–52.
93. Stevenson KL, Bell PB, Watson RE. Dinuclear excited-state complexes from UV laser excitation of aqueous ammonia, methylamine, and ethylamine copper(I) complexes. *Coord Chem Rev* 2002;**229**:133–46.
94. Lever ABP, Dodsworth ES. Electrochemistry, Charge Transfer Spectroscopy and Electronic Structure. In: Solomon EI, Lever ABP, eds. *Inorganic Electronic Structure and Spectroscopy*. New York: John Wiley & Sons, Inc., 1999: 227–291.
95. Wolcan E, Alessandrini JL, Feliz MR. On the quenching of MLCT(Re-bpy) luminescence by Cu(II) species in Re(I) polymer micelles. *J Phys Chem B* 2005;**109**:22890–8.
96. Gkika E, Troupis A, Hiskia A, Papaconstantinou E. Photocatalytic reduction of chromium and oxidation of organics by polyoxometalates. *Appl Catalysis B: Environ* 2006;**62**:28–34.
97. Mytych P, Karocki A, Stasicka Z. Mechanism of photochemical reduction of chromium(VI) by alcohols and its environmental aspects. *J Photochem Photobiol A: Chem* 2003;**160**:163–70.
98. Mytych P, Stasicka Z. Photochemical reduction of chromium(VI) by phenol and its halogen derivatives. *Appl Catalysis B: Environ* 2004;**52**:167–72.
99. Mytych P, Cieśla P, Stasicka Z. Photoredox processes in the Cr(VI)–Cr(III)–oxalate system and their environmental relevance. *Appl Catalysis B: Environ* 2005;**59**:161–70.
100. Cieśla P, Mytych P, Kocot P, Stasicka Z. Role of iron and chromium complexes in environmental self-cleaning processes. *Separation Sci Technol* 2007;**42**:1651–66.
101. Hicks C, Ye G, Levi C, et al. Excited-state acid–base chemistry of coordination complexes *Coord Chem Rev* 2001;**211**:207–22.
102. Elias B, Kirsch-De Mesmaeker A. Photo-reduction of polyazaaromatic Ru(II) complexes by biomolecules and possible applications. *Coord Chem Rev* 2006;**250**:1627–41.
103. Scandola F, Chiorboli C, Prodi A, Iengo E, Alessio E. Photophysical properties of metal-mediated assemblies of porphyrins. *Coord Chem Rev* 2006;**250**:1471–96.
104. Campagna S, Di Pietro C, Loiseau F, et al. Recent advances in luminescent polymetallic dendrimers containing the 2,3-bis(2′-pyridyl)pyrazine bridging ligand. *Coord Chem Rev* 2002;**229**:67–74.
105. Scandola F, Balzani V. Energy-transfer processes of excited states of coordination compounds. *J Chem Educ* 1983;**60**:814–23.
106. Suppan P. *Chemistry and Light*. London: The Royal Society of Chemistry, 1994

107. Carlos RM, Tfouni E, Neumann MG. The sensitization of the ligand-field state of hexaamineruthenium(II) complex ion by organic compounds. *J Photochem Photobiol A: Chem* 1997;**103**:121–6.
108. Balzani V, Juris A. Photochemistry and photophysics of Ru(II) polypyridine complexes in the Bologna group. From early studies to recent developments. *Coord Chem Rev* 2001;**211**:97–15.
109. Takahashi Y, Arakawa H, Sugihara H, et al. Highly efficient polypyridyl-ruthenium(II) photosensitizers with chelating oxygen donor ligands: β-diketonato-bis(dicarboxybipyridine)ruthenium. *Inorg Chim Acta* 2000;**310**:169–74.
110. Lang K, Mosinger J, Wagnerová DM. Photophysical properties of porphyrinoid sensitizers non-covalently bound to host molecules; models for photodynamic therapy. *Coord Chem Rev* 2004;**248**:321–50.
111. Hadjur C, Wagnieres G, Ihringer F, Monnier P, van den Bergh H. Production of the free radicals O_2^- and OH by irradiation of the photosensitizer zinc(II)phthalocyanine. *J Photochem Photobiol B: Biol* 1997;**38**:196–202.
112. Chan WK, Hui CS, Man KYK, et al. Synthesis and photosensitizing properties of conjugated polymers that contain chlorotricarbonyl bis(phenylimino)acenaphthene rhenium(I) complexes. *Coord Chem Rev* 2005;**249**:1351–9.
113. Elvington M, Brown J, Arachchige SM, Brewer KJ. Photocatalytic hydrogen production from water employing a Ru, Rh, Ru. Molecular device for photoinitiated electron collection. *J Am Chem Soc* 2007;**129**:10644–5.
114. Nowotny J, Bak T, Nowotny MK, Sheppard LR. Titanium dioxide for solar-hydrogen I. Functional properties. *Int J Hydrogen Energy* 2007;**32**:2609–29.
115. Nowotny J, Sorrell CC, Sheppard LR, Bak T. Solar-hydrogen: Environmentally safe fuel for the future. *Int J Hydrogen Energy* 2005;**30**:521–44.
116. Aroutiounian VM, Arakelyan VM, Shahnazaryan GE. Metal oxide photoelectrodes for hydrogen generation using solar radiation-driven water splitting. *Solar Energy* 2005;**78**:581–92.
117. Dempsey JL, Esswein AJ, Manke DR, Rosenthal J, Soper JD, Nocera DG. Molecular chemistry of consequence to renewable energy. *Inorg Chem* 2005;**44**:6879–92.
118. Rosenthal J, Bachman J, Dempsey JL, et al. Oxygen and hydrogen photocatalysis by two-electron mixed-valence coordination compounds. *Coord Chem Rev* 2005;**249**: 1316–26.
119. Gondal MA, Hameed A, Yamani ZH, Suwaiyan A. Production of hydrogen and oxygen by water splitting using laser induced photo-catalysis over Fe_2O_3. *Appl Catalysis* 2004;**268**:159–67.
120. Lu Y, Jiang Z, Xu S, Wu H. Efficient conversion of CO_2 to formic acid by formate dehydrogenase immobilized in a novel alginate–silica hybrid gel. *Catalysis Today* 2006;**115**:263–8.
121. Tsujisho I, Toyoda M, Amao Y. Photochemical and enzymatic synthesis of formic acid from CO_2 with chlorophyll and dehydrogenase system. *Catalysis Comm* 2006;**7**: 173–6.
122. Goldsmith JI, Hudson WR, Lowry MS, Anderson TH, Bernharol, S. Discovery and high-throughput screening of heteroleptic iridium complexes for photoinduced hydrogen production. *J Am Chem Soc* 2005;**127**:7502–10.
123. Amouyal E. Photochemical production of hydrogen and oxygen from water: A review and state of art. *Solar Energy Mater Solar Cells* 1995;**38**:249–76.
124. Hori H, Johnson FPA, Koike K, Ishitani O, Ibusuki T. Efficient photocatalytic CO_2 reduction using $[Re(bpy)(CO)_3P(OEt)_3]^+$. *J Photochem Photobiol A: Chem* 1996;**96**:171–4.
125. Rontani JF. Photodegradation of lipidic compounds during the senescence of phytoplakton. In: Boule P, ed. *Environmental Photochemistry*. Berlin: Springer-Verlag, 1999: 263–84.
126. Richard C, Grabner G. Mechanism of phototransformation of phenol and its derivatives in aqueous solution In: Boule P, ed. *Environmental Photochemistry*. Berlin: Springer Verlag, 1999: 217–40.

127. Parmon V, Emeline AV, Serpone N. Glossary of terms in photocatalysis and radiocatalysis {A Preliminary Version of IUPAC's Project # 2001–036–1}. *Int J Photoenergy* 2002;**4**:91–131.
128. Parmon VN. Photocatalysis as a phenomenon: Aspects of terminology. *Catalysis Today* 1997;**39**:137–44.
129. Stasicka Z. Photocatalysis by transition metal complexes. In: Ziółkowski J, ed. *Photocatalysis by Transition Metal Complexes*. Poznań-Wrocław, 1993: 105–23.
130. Salomon RG. Homogeneous metal-catalysis in organic photochemistry. *Tetrahedron* 1983;**39** 485–575.
131. Kutal C. Photosensitive metal-organic systems: An overview. *Adv Chem Ser* 1993;**238**: 1–28.
132. Kutal C. Photochemistry of transition metal-organic systems. *Coord Chem Rev* 1985;**64**:191–206.
133. Kisch H. What is photocatalysis? In: Serpone N, Pelizzetti E, eds. *Photocatalysis Fundamentals and Applications*. New York: Wiley-Interscience, 1989: 1–8.
134. Hennig H. Homogeneous photo catalysis by transition metal complexes. *Coord Chem Rev* 1999;**182**:101–23.
135. Serpone N, Salinaro A, Emeline A, Ryabchuk V. Turnovers and photocatalysis. A mathematical description. *J Photochem Photobiol A: Chem* 2000;**130**:83–94.
136. Serpone N, Salinaro A. Terminology, relative photonic efficiencies and quantum yields in heterogeneous photocatalysis. Part I Suggested protocol. *Pure Appl Chem* 1999;**71**:303–20.
137. Salinaro A, Emeline AV, Zhao J, Hidaka H, Ryabchuk VK, Serpone N. Terminology, relative photonic efficiencies and quantum yields in heterogeneous photocatalysis. Part II. Experimental determination of quantum yields. *Pure Appl Chem* 1999;**71**:321–35.
138. Hennig H, Rehorek D, Archer RD. Photocatalytic systems with light-sensitive coordination compounds and possibilities of their spectroscopic sensitization – an overview. *Coord Chem Rev* 1985;**61**:1–53.
139. Franch MI, Ayllon JA, Peral J, Domenech X. Enhanced photocatalytic degradation of maleic acid by Fe(III) adsorption onto the TiO_2 surface. *Catalysis Today* 2005;**101**: 245–52.
140. Cieśla P, Kocot P, Mytych P, Stasicka Z. Homogeneous photocatalysis by transition metal complexes in the environment. *J Mol Catalysis A: Chem* 2004;**224**:17–33.
141. Bajt O, Mailhot G, Bolte M. Degradation of dibutyl phthalate by homogeneous photocatalysis with Fe(III) in aqueous solution. *Appl Catalysis B: Environ* 2001;**33**:239–48.
142. Maldotti A, Molinari A, Bergamini P, Amadelli R, Battioni P, Mansuy D. Photocatalytic oxidation of cyclohexane by $(nBu_4N)W_{10}O_{32}$/Fe(III)porphyrins integrated systems. *J Mol Catalysis A: Chem* 1996;**113**:147–57.
143. Fornal E, Giannotti C. Photocatalyzed oxidation of cyclohexane with heterogenized decatungstate. *J Photochem Photobiol A: Chem* 2007;**188**:279–86.
144. Kormali P, Troupis A, Triantis T, Hiskia A, Papaconstantinou E. Photocatalysis by polyoxometallates and TiO_2: A comparative study. *Catalysis Today* 2007;**124**:149–55.
145. Tanielian C, Lykakis IN, Seghrouchni R, Cougnona F, Orfanopoulos M. Mechanism of decatungstate photocatalyzed oxygenation of aromatic alcohols. Part I. Continuous photolysis and laser flash photolysis studies. *J Mol Catalysis A: Chem* 2007;**262**: 170–5.
146. Lykakis IN, Tanielian C, Seghrouchni R, Orfanopoulos M. Mechanism of decatungstate photocatalyzed oxygenation of aromatic alcohols. Part II. Kinetic isotope effects studies. *J Mol Catalysis A: Chem* 2007;**262**:176–84.
147. Troupis A, Gkika E, Hiskia A, Papaconstantinou E. Photocatalytic reduction of metals using polyoxometallates: recovery of metals or synthesis of metal nanoparticles. *CR Chimie* 2006;**9**:851–7.
148. Kim S, Park H, Choi W. Comparative study of homogeneous and heterogeneous photocatalytic redox reactions: $PW_{12}O_{40}^{3-}$ vs TiO_2. *J Phys Chem B* 2004;**108**: 6402–11.

149. Hori H, Takano Y, Koike K, Kutsuna S, Einaga H, Ibusuki T. Photochemical decomposition of pentafluoropropionic acid to fluoride ions with a water-soluble heteropolyacid photocatalyst. *Appl Catalysis B: Environ* 2003;**46**:333–40.
150. Troupis A, Hiskia A, Papaconstantinou E. Photocatalytic reduction – recovery of silver using polyoxometalates. *Appl Catalysis B: Environ* 2003;**42**:305–15.
151. Troupis A, Hiskia A, Papaconstantinou E. Photocatalytic reduction and recovery of copper by polyoxometalates. *Environ Sci Technol* 2002;**36**:5355–62.
152. Tanielian C. Decatungstate photocatalysis. *Coord Chem Rev* 1998;**178–180**:1165–81.
153. Tanielian C, Duffy K, Jones A. Kinetic and mechanistic aspects of photocatalysis by polyoxotungstates: a laser flash photolysis, pulse radiolysis, and continuous photolysis study. *J Phys Chem B* 1997;**101**:4276–82.
154. Hill CL, Prosser-McCartha CM. Homogeneous catalysis by transition metal oxygen anion clusters. *Coord Chem Rev* 1995;**143**:407–55.
155. Shiragami T, Matsumoto J, Inoue H, Yasuda M. Antimony porphyrin complexes as visible-light driven photocatalyst. *J Photochem Photobiol C: Photochem Rev* 2005; **6**:227–48.
156. Kim W, Park J, Jin Jo HJ, Kim H-J, Choi W. Visible light photocatalysts based on homogeneous and heterogenized tin porphyrins. *J Phys Chem C* 2008;**112**:491–9.
157. Jessop PG. Homogeneous catalysis using supercritical fluids: Recent trends and systems studied. *J Supercritical Fluids* 2006;**38**:211–31.
158. Stasicka Z, ed. *Photochemical Processes in the Environment*. Kraków: Jagiellonian University Publisher, 2001, 152 pp.

7

Photochemistry and Photophysics of Supramolecular Systems and Nanoassemblies

All that glitters may not be gold, but at least it contains free electrons.
John Desmond Baernal

7.1 From Molecules through Clusters to Crystals

Photochemical behaviour of coordination compounds described in previous chapters results mainly from electronic interactions between the central metal atom or ion and ligands in the first coordination sphere. An increased size of molecular systems to clusters and nanosized crystals expands the possibility of photoinduced electron transfer between the discrete electronic states to excitation within bands. Furthermore, interactions of nanoparticles with molecules yield unique materials, combining structural versatility of molecular species with collective properties of solids.

A separated molecule or ion is characterized by its highest occupied molecular orbital (HOMO) and lowest unoccupied molecular orbital (LUMO) (Figure 7.1a). In the case of a set of molecules, every moiety is localized in an electromagnetic field of neighbours influencing the energy of orbitals (Figure 7.1b). As a result the cluster formed by several molecules or ions can be described by a number of HOMOs and LUMOs differing in energy from HOMO and LUMO of a parent moiety. Further increase in size of the molecular system to nanocrystals and microcrystals is associated with loss of the discrete levels of molecular orbitals and formation of continuum of states called bands (Figures 7.1c, d). The band formed by HOMOs is called a *valence band* (VB) whereas that originating from LUMOs is

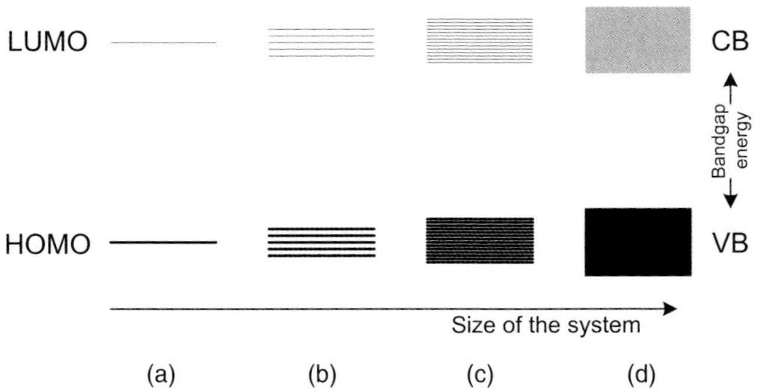

Figure 7.1 *The energy levels distribution in molecular and supramolecular systems: (a) isolated molecule, (b) cluster, (c) nanocrystal, and (d) crystal*

called a *conduction band* (CB). Depending on the bandgap energy between the edges of VB and CB, the formed crystals can show properties of a conductor, semiconductor, or insulator.

Excitation of the crystalline material, leading to the electron transfer from the VB to the CB by UV or visible light, corresponds to energy of photons in the range of approximately 1–4 eV. The materials characterized by similar bandgap energy can be classified as semiconductors or wide bandgap semiconductors.

The dependence of bandgap energy (E_g) on crystal size is given by [1]:

$$E_g = E_g^0 + \frac{\hbar^2 \pi^2}{2R^2}\left[\frac{1}{m_e} + \frac{1}{m_h}\right] - \frac{1.786e^2}{\varepsilon R} - 0.248E_{Ry}^* \tag{7.1}$$

where E_g and E_g^0 are bandgap energies of nanocrystals and microcrystals, respectively, R is a radius of the nanocrystal particle, m_e and m_h are effective masses of electron and hole, ε is a dielectric constant of the material, and E_{Ry}^* is Rydberg's energy defined as:

$$E_{Ry}^* = \frac{e^4}{2\varepsilon^2 \hbar^2 \left[\frac{1}{m_e} + \frac{1}{m_h}\right]} \tag{7.2}$$

A significant difference between electronic properties of molecular and crystalline structures reflects in other spectral features and photochemical behaviour – these topics are described later.

7.2 Metallic Nanoparticles: Metals in the Embryonic State

Metals (mostly silver and gold) in nanometric dispersions have been known since Roman times. The most beautiful example of application of these materials is the Lycurgus cup. This masterpiece of glass-blowing appears green in reflected light, but

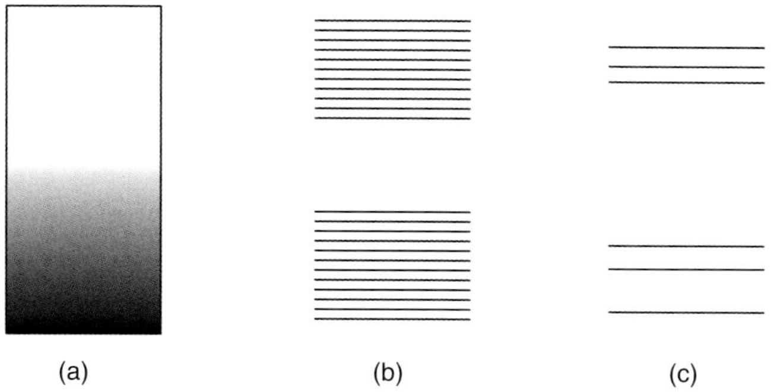

Figure 7.2 *Electronic structure of (a) bulk metal, (b) metallic nanoparticle, and (c) small molecular cluster. (Adapted from Schmid [2])*

red in transmitted light. These peculiar optical properties are characteristic of metal nanoparticles and result from their electronic structure and interaction with the surrounding dielectric medium.

If a metal particle with bulk properties is reduced to the size of a few dozen or a few hundred atoms, the density of states in the VB and the CB decreases to such an extent that the electronic properties change dramatically. These changes encompass mainly electrical, optical, and magnetic properties [2]. The most exciting issue is the question of how small the particle has to be to observe a collapse of bulk properties. This in turn implies the problem of substitution of the quasi-continuous density of states by the discrete energy level structure. In the case of very small clusters the picture is simple, eg three metal atoms form energetically well-defined bonding and antibonding molecular orbitals (Figure 7.2).

In the case of small clusters/particles another issue needs to be addressed as well. The smaller the particles, the larger the fraction of atoms that belongs to the surface. Unprotected, naked metal particles are not suited for any applications due to their extreme reactivity – the surface atoms are very reactive because they are coordinatively highly unsaturated. Their contribution to the properties of the whole particle is different from the contribution of the inner atoms, which are fully coordinated. Surface atoms can be easily distinguished even if they are ligated with protecting ligands. Therefore, it is important to distinguish the total number of atoms in the nanoparticle from the number of surface atoms, eg a dense packed cluster Au_{13} has 12 surface atoms and only one inner atom. In the cluster $[Au_{55}(PPh_3)_{12}Cl_6]$ the 'metallic' core consists of 13 atoms, whilst 42 atoms are located in the outer shell. This simple calculation indicates that the contribution of surface atoms dominates the inner part of the clusters.

Synthesis of metallic nanoparticles proceeds in many ways: they can be divided into physical and chemical. Physical methods include inert gas condensation, arc discharge, ion sputtering, and laser ablation. The main idea behind these methods is condensation of solid particles from the gas phase, the substrate for nanoparticle generation being pure metals (or their mixtures/alloys in the case of complex particle composition). Chemical methods, in turn, include various methods utilizing volatile or soluble compounds that transform into metallic particles. These methods include chemical reduction, photochemical methods, solvothermal synthesis, electrochemical reduction, and others. One of the most common methods, at least for gold and silver particles, is chemical reduction.

A variety of reducing agents is used to reduce soluble metal salts to the corresponding metal. The most common reducing agents include tetrahydroborates, alcohols, and citrate. Various additives are used to terminate the growth of the resulting particles at the desired stage. Various anions, thiols, carboxylates, and surfactants are used to control the formation of metallic nanoparticles. These additives also determine the properties of the nanoparticle surface.

Nanoparticles prepared in a solution phase tend to have a spherical shape because this shape has the smallest surface area relative to objects of other shapes (when the volume is fixed). It is worth noting that most of the gold nanospheres reported in the literature are not truly spherical. They are multiply twinned particles with more or less rounded profiles and facets on the surfaces. In a sense, these particles should be correctly called quasi-spheres. One of the most popular protocols for preparing gold nanospheres is based on the reduction of $HAuCl_4$ by citrate in water, which was first described in 1951 by Turkevitch and is now commonly referred to as the 'Turkevitch method'. In this method, citrate serves as both a reducing agent and an anionic stabilizer. It yields gold nanospheres of diameter in the range 15–150 nm, depending on the reactant ratio, with a fairly narrow size distribution through the following reaction [3]:

$$6AuCl_4^- + C_6H_8O_7 + 5H_2O \rightarrow 6CO_2 + 24Cl^- + 6Au^0 + 18H^+ \tag{7.3}$$

Other reduction methods have also been developed to achieve better control over size and uniformity of size, including the 'Schmid method' [4] and the 'Brust–Schiffrin method' [5]. These methods utilize stabilization of growing gold nanoparticles with phosphines and thiols, respectively. The latter method yields air-stable particles, which can be dissolved in numerous organic solvents. The thiol-terminated molecules can be also incorporated in order to modify the properties of the particles, eg to change their morphology, optical or electrochemical properties, affinity towards biomolecules, etc (Figure 7.3) [6]. Seed-mediated growth is another technique frequently used for producing spherical nanoparticles of different sizes. The size of gold nanospheres could be tuned from 5 nm to 40 nm by controlling the ratio between the seed and the precursor [7]. Gold nanorods can be conveniently generated also by employing the seed-mediated growth method [8].

Optical properties of metal nanoparticles embedded in dielectric media can be derived from the electrodynamic calculations within solid state theory. A simple model of electrons in metals, based on the gas kinetic theory, was presented by Drude in 1900 [9]. It assumes independent and free electrons with a common relaxation time. The theory was further corrected by Sommerfeld [10], who incorporated corrections originating from the Pauli exclusion principle (Fermi–Dirac velocity distribution). This so-called free-electron model was later modified to include minor corrections from the band structure of matter (effective mass) and termed quasi-free-electron model. Within this simple model electrons in metals are described as

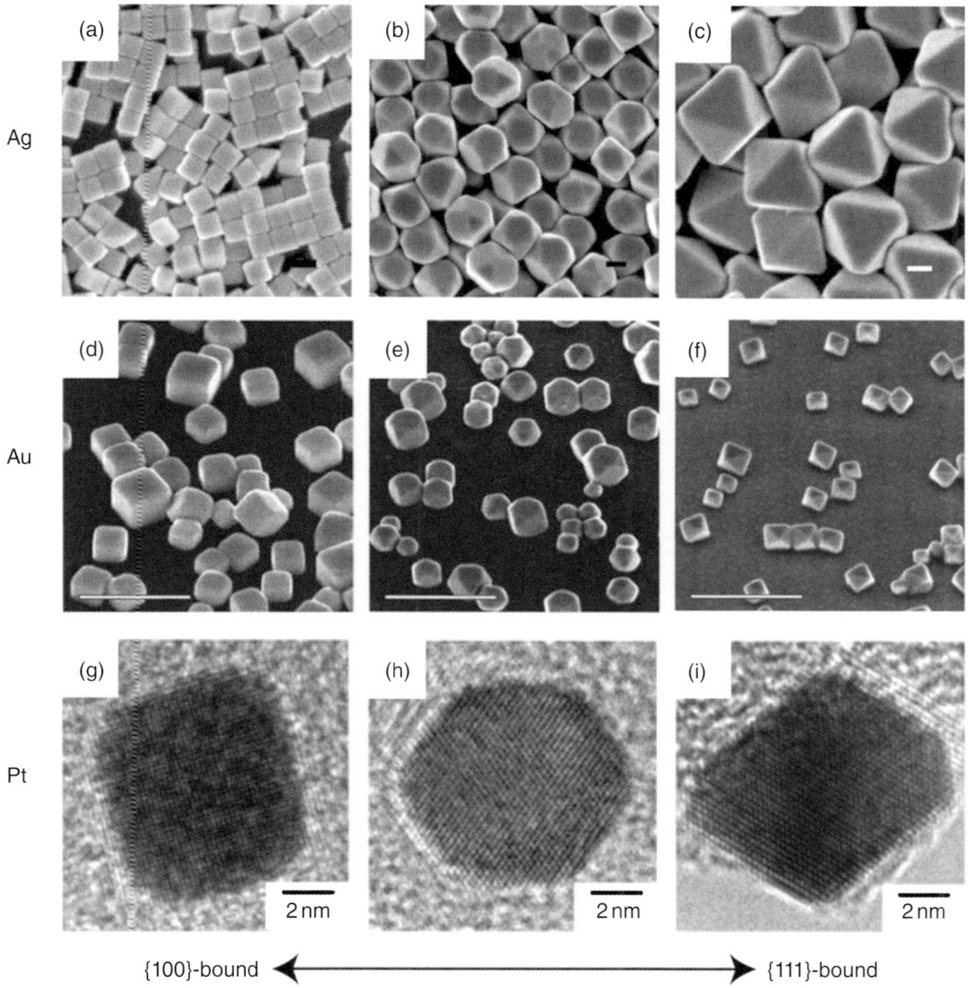

Figure 7.3 *Morphology of Ag, Au, and Pt nanoparticles obtained in the presence of ligands specifically interacting with [100] or [111] crystal planes. (Reproduced from Tao et al. [109], copyright Wiley-VCH, with permission)*

a gas of point-like charged particles. These particles move freely between independent collisions with unspecified collision centres, which occur with an average rate of γ_0. This figure is related to the electron relaxation time (τ) by:

$$\gamma_0 = \frac{1}{\tau} \tag{7.4}$$

The optical properties of metals (ie their response to alternating electromagnetic waves) is described by the dielectric function [11]:

$$\varepsilon(\omega) = \varepsilon_\infty - \frac{\omega_p^2}{\omega(\omega + i\gamma_0)} \tag{7.5}$$

If $\omega = \gamma_0$ the expression of the dielectric function simplifies to:

$$\varepsilon(\omega) \approx \varepsilon_\infty - \frac{\omega_p^2}{\omega^2} + i\frac{\gamma_0 \omega_p^2}{\omega^3} = \varepsilon_r + i\varepsilon_i \tag{7.6}$$

where:

$$\omega_p = \sqrt{\frac{ne^2}{\varepsilon_0 m^*}} \tag{7.7}$$

denotes the so-called plasma frequency, with n and m^* standing for density and effective mass of conduction band electrons, and ε_∞ embodies the contribution of the bound electrons to the dielectric function. In the case where only free electrons contribute to the dielectric function, this quantity equals 1.

The surface plasmon resonance is the coherent excitation of all the 'free' electrons within the conduction band, leading to an in-phase oscillation. In other words, the plasmon resonance of bulk metals corresponds to the point where the free electrons can no longer shield the interior of the metal from the incident radiation [12]. The surface plasmon resonance does not give rise to the most intense absorption for very small clusters but is rather strongly damped. For the larger particles of several tens of nanometres in which their size is still small compared with the wavelength of light, excitation of the surface resonance can take place with visible light. According to Mie's theory, for a spherical metallic nanoparticle with radius r ($r = \lambda$), the nanoparticle cross-section for light absorption is given by [3, 13]:

$$C_{ext} = \frac{24\pi^2 r^3 \sqrt{\varepsilon_m^3}}{\lambda} \cdot \frac{\varepsilon_i}{(\varepsilon_r + 2\varepsilon_m)^2 + \varepsilon_i^2}, \tag{7.8}$$

where ε_m is the dielectric constant of the surrounding medium and λ is the incident light wavelength. The optical extinction thus has a band maximum at the resonance condition roughly given by $\varepsilon_r = -2\varepsilon_m$, provided that the imaginary part of the dielectric constant is small or weakly dependent on the frequency. Whereas the real part determines the wavelength position of the resonance, the imaginary part determines the bandwidth [13].

For the same particles, the volume plasmon is located at very high energies (6–9 eV). The surface obviously plays a very important role for the observation of the surface plasmon resonance because it alters the boundary conditions for the polarizability of the metal and therefore shifts the resonance to optical frequencies. In this sense, the surface plasmon absorption is a small particle (or thin layer) effect but is definitely not a quantum size effect [14].

Although the analytical description of optical properties of spherical particles is rather simple (see equation 7.8), there are no analytical expressions for plasmon resonances of metallic particles of arbitrary shapes. The only exceptions are rod-shaped particles, which can be described by a modified Mie's theory. The elongated nanoparticle cross-section for light absorption is given by [3, 14]:

$$C_{ext} = \frac{2\pi V}{3\lambda}\sqrt{\varepsilon_m^3}\sum_j \frac{\frac{1}{P_j}\varepsilon_i}{\left(\varepsilon_r + \frac{1-P_j}{P_j}\varepsilon_m\right)^2 + \varepsilon_i^2} \tag{7.9}$$

where j (j = A, B, or C; A > B = C; A, length; B, C, width of nanowire) denotes the three principal axes of the nanoellipsoid, V the volume of the nanorod, and P_j the depolarization factor along a specified axis. These factors are defined as:

$$P_A = \frac{1-k^2}{k^2}\left[\frac{1}{2k}\ln\left(\frac{1+k}{1-k}\right) - 1\right] \tag{7.10}$$

$$P_B = P_C = \frac{1-P_A}{2} \tag{7.11}$$

where k is the nanorod ellipticity, which can be derived from the nanorod aspect ratio (A/B), ξ:

$$k^2 = 1 - \frac{1}{\xi^2} \tag{7.12}$$

Many properties of particle plasmons are qualitatively understood in the following semi-classical model (Figure 7.4). As the diameter of a nanoparticle is of the order of the penetration depth of electromagnetic waves in metals, the excitation light is able to penetrate the particle. The field inside the particle shifts the conduction electrons collectively with respect to the fixed positive charge of the lattice ions. The electrons build up a charge on the surface at one side of the particle. The attraction of this negative charge and the positive charge of the remaining lattice ions on the opposite side results in a restoring force. If the frequency of the excitation light field is in resonance with the eigenfrequency of this collective oscillation, even a small exciting field leads to a strong oscillation. The magnitude of the oscillation

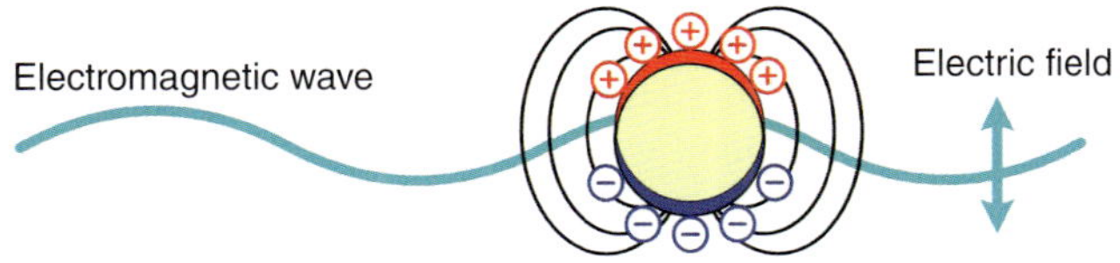

Figure 7.4 *Schematic view of plasmon excitation within metallic nanoparticle via interaction with electromagnetic field*

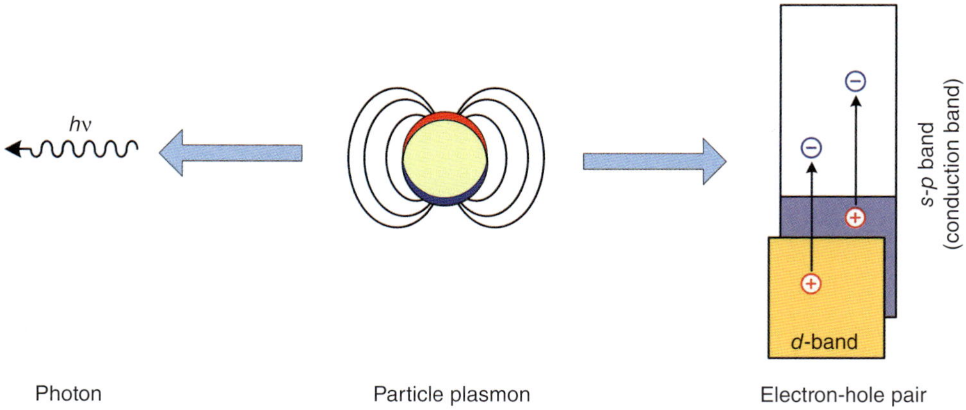

Figure 7.5 *Schematic representation of radiative and non-radiative decay of particle plasmons. (Adapted from Sönnichsen et al. [110])*

depends only on the damping involved, which can be both radiative and non-radiative. The resonance frequency is mainly determined by the strength of the restoring force. This force depends on the separation of the surface charges, ie the particle size, and the polarizability of the medium between and around the charges, which depends on the embedding medium and the polarizability of the core electrons of the metal particle. The alternating surface charges effectively form an oscillating dipole, which radiates electromagnetic waves. This simple model for particle plasmons is reminiscent of an 'optical antenna' [15].

The decay of the nanoparticle plasmons can be either radiative, ie by emission of a photon, or non-radiative (Figure 7.5). Within the Drude–Sommerfeld model the plasmon is a superposition of many independent electron oscillations. The non-radiative decay is thus due to a dephasing of the oscillation of individual electrons. In terms of the Drude–Sommerfeld model this is described by scattering events with phonons, lattice ions, other conduction or core electrons, the metal surface, impurities, etc. As a result of the Pauli exclusion principle, the electrons can be excited into empty states only in the CB, which in turn results in electron-hole pair generation. These excitations can be divided into inter- and intraband excitations by the origin of the electron either in the *d*-band or the CB (Figure 7.5) [15].

Gold nanoparticles are virtually not luminescent, but silver nanoparticles show plasmon emissions with reasonable quantum yields. Furthermore, the non-radiative decay, resulting in electron-hole pair generation, may be used for photosensitization of wide bandgap semiconductors (see Figure 7.5) [16, 17]. Similar effects may also be observed as direct photoinduced electron transfer between metal surfaces and surface-bound molecules [18].

Interparticle coupling, eg by various ligands, DNA strands, results in bathochromic shifts of the plasmon bands. Interaction of two (or more) nanoparticles results in collective resonance within the whole assembly, so the effective size of the particle is increased [13, 19, 20].

7.3 Formation and Decay of the Excited States of Semiconductors

7.3.1 Optical Excitation of Semiconductors

Out of various nanoassemblies, nano- and microcrystalline solid semiconductors deserve a particular attention due to their specific spectroscopic, photochemical, and photocatalytic properties. These properties derive from the nature of the semiconductor's excited states.

There are several optical absorption processes associated with semiconductors and their contribution depends on semiconductor properties and incident light energy. These processes include:

- fundamental absorption
- exciton absorption
- absorption due to dopants and defects
- intraband transitions
- free carrier absorption.

For most photocatalytic (electron transfer) and analytical (photoluminescence) applications only the fundamental absorption plays an important role.

At photon energies greater than the energy gap, the absorption mechanism is due to the transfer of electrons from filled VB states to the empty CB states. At energies slightly below the energy gap, the absorption mechanism is due to the excitons and transitions between dopant states and band states (eg acceptor to CB and VB to donor). Free carrier absorption due to the transitions within the energy bands results in an absorption continuum at lower energies.

In the fundamental absorption process, a photon excites an electron from the VB to the CB. Both energy and momentum must be conserved in this process. As the photon momentum is small compared with the crystal momentum, the absorption process should essentially conserve the electron momentum, ie $\hbar\mathbf{k}$. When the minimum of the CB and the maximum of the VB occur at the same value of the wave vector **k**, transitions are direct, and the material is referred to as a direct bandgap semiconductor (Figure 7.6a). If the band extrema do not occur at the same wave vector **k**, transitions are indirect, and the material is referred to as an indirect bandgap semiconductor (Figure 7.6b). For momentum conservation in such semiconductors, a participation of an extra particle, ie phonon, is required; the probability of such a process is substantially lower compared with direct transitions. Therefore, in general, fundamental absorption in indirect bandgap semiconductors is relatively weaker compared with direct bandgap materials [21].

The envelope of the semiconductor absorption spectrum is described by a simple power function:

$$\alpha(h\nu) = A(h\nu - E_g)^{n/2} \tag{7.13}$$

where $h\nu$ is a photon energy, α the absorption coefficient, A is a constant describing electrical properties of the semiconductor, and E_g the forbidden band width [21].

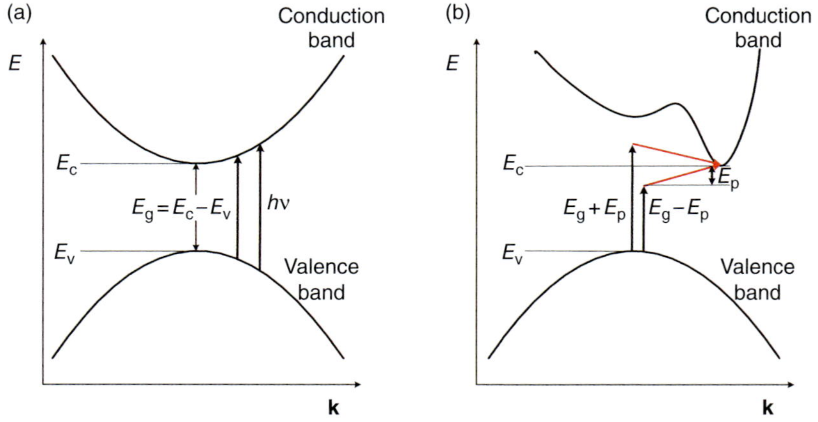

Figure 7.6 Schematic representation of fundamental absorption processes in (a) direct bandgap and (b) indirect bandgap semiconductors. Phonon emission and phonon absorption processes are marked in red. (Adapted from Yacobi [21])

The n parameter equals 1 for direct bandgap semiconductors or 4 for indirect bandgap semiconductors in the case of allowed fundamental transitions [22]. Other values of n, 2 or 3, are valid only for forbidden transitions. The proper transformation allows estimation of the bandgap energy, E_g, for both types of crystalline semiconductors. Figure 7.7 presents the procedure of E_g evaluation.

Further complication in semiconductor band shape analysis concerns the spectral region near the fundemental absorption onset. Ideal semiconductor crystal at 0 K should not absorb any photons with energies lower than E_g. Real systems, however, show pronounced absorption tails at energies lower than the bandgap energy (Figure 7.7). The absorption profile within the tail region can be very well approximated by the empirical Urbach's rule [23–26]:

$$\alpha(h\nu) = \alpha_0 \exp\left(\frac{h\nu - E_0}{E_U}\right) \tag{7.14}$$

where α_0 and E_0 are material-related parameters and E_U stands for Urbach energy. This quantity is related in turn to the lattice ionicity and structural disorders [27]:

$$E_U = \frac{E_p}{2\sigma_0}\left(X + \coth\frac{E_p}{2k_B T}\right), \tag{7.15}$$

where σ_0 is a parameter depending on the lattice ionicity, E_p the phonon energy, and X a measure of lattice disorders (see equation 7.16) [27]. This is an extension to the original model by Urbach and Martienssen, in which the structural disorders are neglected (ie $X = 0$) [23, 28]. In the extended model X is defined as the ratio of the mean square deviation of atomic positions caused by structural disorder to the zero-point uncertainty in the atomic positions [29]:

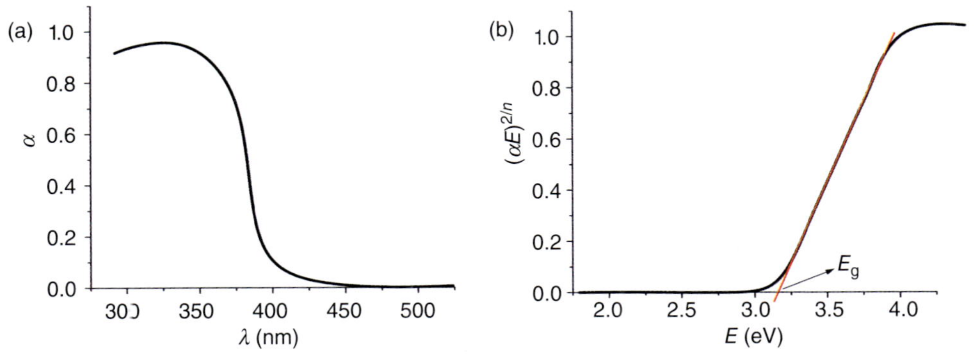

Figure 7.7 (a) Absorption spectrum of a semiconductor and (b) its transformation to $(\alpha E)^{2/n}$ vs E. The red line shows extrapolation of the linear part of the absorption band edge. Its crossing with the energy axis determines the bandgap energy E_g

$$X = \frac{\langle U_x^2 \rangle}{\langle U^2 \rangle_0} \tag{7.16}$$

In indirect bandgap semiconductor crystals both the emission and absorption of phonons are allowed to preserve the momentum (see Figure 7.6.). Therefore two contributions to the overall absorption spectrum should be considered: α_a and α_e, associated with phonon absorption and emission, respectively [21]:

$$\alpha(h\nu) = \alpha_a(h\nu) + \alpha_e(h\nu), \tag{7.17}$$

where:

$$\alpha_a(h\nu) = \frac{A(h\nu - E_g + E_p)^2}{\exp(E_p/k_B T) - 1} \tag{7.18}$$

and

$$\alpha_e(h\nu) = \frac{A(h\nu - E_g - E_p)^2}{1 - \exp(-E_p/k_B T)} \tag{7.19}$$

Denominators in both components are related to the number of phonons of E_p energy according to the Bose–Einstein statistics [21]. Therefore the approximate method for bandgap determination in the case of indirect semiconductors has an intrinsic error of the order of E_p (see Figure 7.7b).

7.3.2 Electrons and Hole Trapping

Considering the wide bandgap semiconductor crystal (there are mainly oxides, sulphides, selenides, and halides in this group of materials) the visible- or UV-light

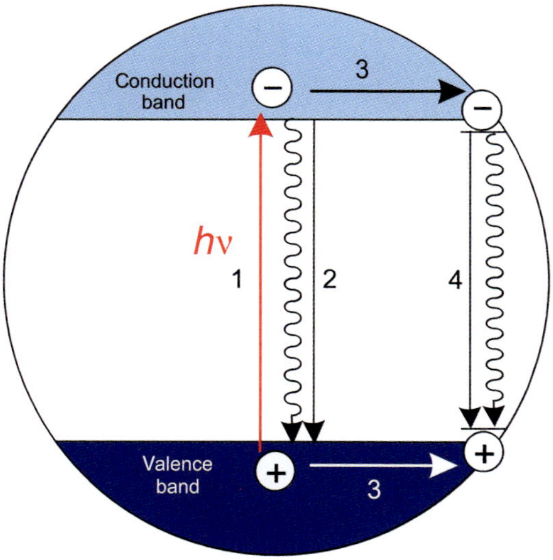

Figure 7.8 *Primary physical processes taking place after excitation of a semiconductor particle: (1) excitation, (2) radiative or non-radiative recombination, (3) charge trapping, and (4) recombination of trapped charges*

induced generation of the electron-hole pairs can be usually regarded as a ligand-to-metal charge transfer (LMCT) process. Due to a certain width of the band, electron excitation from VB to CB can be realized on irradiation within a relatively broad range of photon energy (Figure 7.8, process 1). The charge separation proceeds within a femtosecond time scale.

The fate of generated charges can be various, as demonstrated in Figure 7.8. The electron-hole pair can recombine (process 2) with heat evolution (non-radiative pathway) or light emission (radiative pathway) soon after the separation process or, later, during their transfer to the surface (processes 3 and 4). The non-radiative pathway is favoured in the case of indirect bandgap semiconductors (eg TiO_2) while the radiative pathway takes place for direct bandgap semiconductors (eg CdSe, see below).

The lifetime of separated charges increases after electron and hole trapping in certain states, eg in the case of titanium dioxide, electrons are trapped as Ti^{III} centres [30, 31] with the holes as $[>Ti^{IV}OH^{\bullet}]^{+}$ [30]. Trapping of holes proceeds in 10–100 ns, whereas this process is faster for electrons and requires a few hundred picoseconds. Charge-carrier recombination from the trapped states also proceeds in 10–100 ns.

7.3.3 Radiative vs Non-Radiative Decay

The energies of the upper edge of VB and the lower edge of CB are functions of the wave vector **k** (the momentum can be described as $\hbar\mathbf{k}$, see above). When the minimum and maximum of the CB edge energy and the VB edge energy, respectively, occur together at $\mathbf{k} = 0$, the electrons and holes can recombine conserving momentum (Figure 7.9a). Such a transition can be accompanied by photon emission.

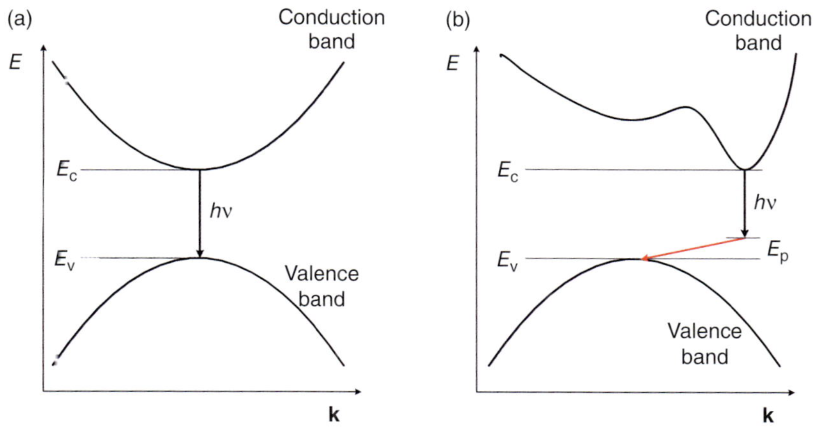

Figure 7.9 *Recombination of electrons and holes in the case of (a) direct bandgap semiconductor and (b) indirect bandgap semiconductor. The energy E is a function of the wave vector* **k**

Another situation is described in Figure 7.9b where the energy minimum and maximum occur at various places in the **k**-space. In this case charge recombination must involve a phonon or a defect in the crystal structure to ensure conservation of the momentum. Therefore photon emission accompanying the electron-hole recombination, depicted in Figure 7.9b by an inclined arrow, is much less probable (which may eventually result in only a very weak emission) and in most cases energy dissipation to the crystal lattice is observed. A good example of the indirect bandgap semiconductor is crystalline silicon – the non-radiative decay of its excited state may be accompanied by an energy transfer process to an adsorbed oxygen molecule [32]. Generated singlet oxygen may in turn participate in specific oxidation reactions.

Apart from fundamental transitions in direct-gap semiconductors, other processes may be responsible for radiative decay of the semiconductor excited states. The most common are processes associated with electron-hole annihilation involving donor and acceptor sites (Figure 7.10) [33].

These emissive processes can be easily tuned via changes in dopant redox potentials and their concentration:

$$h\nu_2 = E_g - (E_A + E_D) + \frac{e^2}{\varepsilon_0 r} \tag{7.20}$$

The emitted photon energy depends on both the relative positions of donor and acceptor levels in respect to the VBs and CBs (E_D and E_A, respectively) and the spatial separation between the donor and acceptor sites (r) [33].

The energy of the emitted photon depends also on the particle size (see equation 7.1). Lowering the particle size within the nanometre scale the maximum of emission spectrum of the quantum dot may shift within a full range of visible light (Figure 7.11).

The probability of direct electron-hole recombination even in direct semiconductors is low and various defects facilitate the light emission processes. In particular

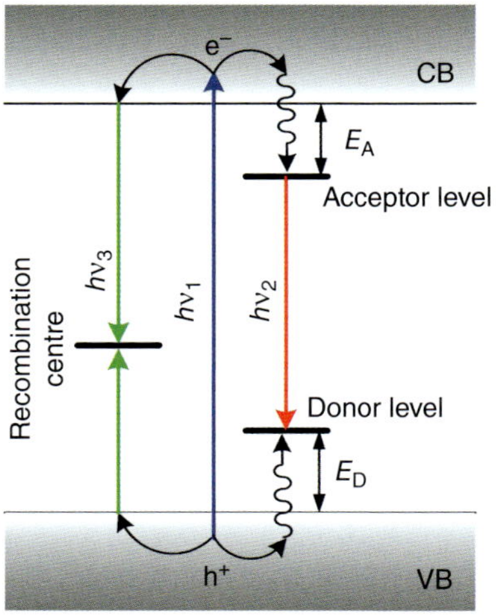

Figure 7.10 *Schematic illustration of electron and hole trapping at dopant sites and subsequent donor–acceptor-mediated photon emission. (Adapted from Ozawa and Itoh [33])*

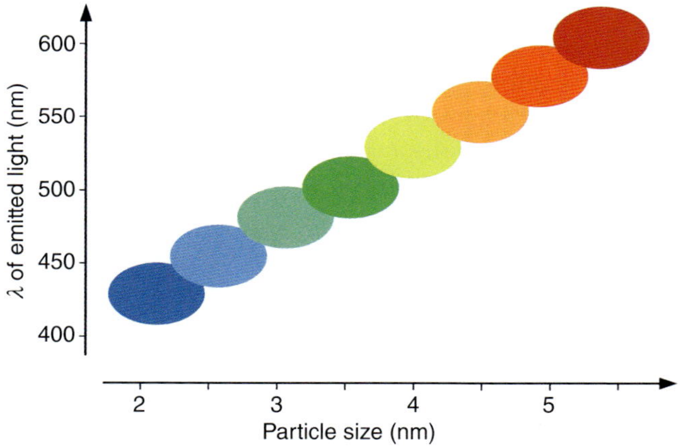

Figure 7.11 *A schematic view of λ_{em} dependence on quantum dot size in the case of cadmium selenide*

the presence of isoelectronic dopants may improve luminescent properties of semiconductors due to generation of recombination centres ($h\nu_3$ in Figure 7.10). In this case energy of emitted photons corresponds to the fundamental bandgap of the material [21].

7.3.4 Surface–Molecule Interaction: General Description

A simple quantum chemical description of photophysical processes involved in surface-modified metals and semiconductors was given recently by Galperin and

Nitzan [18, 34–36]. Properties of the surface-modified semiconductor can be derived from an interaction between isolated electronic levels of the surface molecule (especially HOMO and LUMO orbitals) and the electronic continuum of semiconductor under the influence of a radiation field. The total hamiltonian (operator describing components) of the system can be formulated as a sum of hamiltonians for all the components of the system ($\hat{\boldsymbol{H}}$) and a coupling operator including all the physical processes of interest ($\hat{\boldsymbol{W}}$) (7.21):

$$\hat{\boldsymbol{H}} = \hat{\boldsymbol{H}}_0 + \hat{\boldsymbol{W}} \tag{7.21}$$

The $\hat{\boldsymbol{H}}_0$ hamiltonian is simply a sum of the hamiltonians of individual components of the system:

$$\hat{\boldsymbol{H}}_0 = \hat{\boldsymbol{H}}_{\mathbf{M}} + \hat{\boldsymbol{H}}_{\mathbf{S}} + \hat{\boldsymbol{H}}_{\mathbf{R}} \tag{7.22}$$

where $\hat{\boldsymbol{H}}_{\mathrm{M}}$, $\hat{\boldsymbol{H}}_{\mathrm{S}}$, and $\hat{\boldsymbol{H}}_{\mathbf{R}}$, are the hamiltonians for the isolated molecule, semiconducting substrate, and radiation field, respectively. The coupling operator in turn can be formulated as:

$$\hat{\boldsymbol{W}} = \hat{\boldsymbol{W}}_{\mathbf{ET}} + \hat{\boldsymbol{W}}_{\mathbf{DC}} + \hat{\boldsymbol{W}}_{\mathbf{RM}} + \hat{\boldsymbol{W}}_{\mathbf{RS}} + \hat{\boldsymbol{W}}_{\mathbf{RMS}} \tag{7.23}$$

The coupling operator includes terms corresponding to the following processes: electron transfer coupling between each isolated molecular electronic state of the molecule and the electronic continuum of the semiconductor ($\hat{\boldsymbol{W}}_{\mathbf{ET}}$); dipole-induced dipole (energy transfer) coupling, which describes interaction of an excited molecule and the dielectric response of the substrate ($\hat{\boldsymbol{W}}_{\mathbf{DC}}$); molecule–radiation field coupling ($\hat{\boldsymbol{W}}_{\mathbf{RM}}$), describing light absorption by the isolated molecule; substrate–radiation field coupling ($\boldsymbol{W}_{\mathbf{RS}}$) depicting optical properties of the semiconductor; and molecule–substrate–radiation field coupling ($\hat{\boldsymbol{W}}_{\mathbf{RMS}}$), which describes a direct optical excitation between the substrate electronic continuum and a molecular electronic state or vice versa. Photosensitization by organic chromophores or transition metal complexes chemisorbed onto the semiconductor surface can be in most cases described simply by the combination of $\hat{\boldsymbol{W}}_{\mathbf{RM}}$ and $\hat{\boldsymbol{W}}_{\mathbf{ET}}$ terms: energy quanta are absorbed by the chromophore and electrons from its excited state are injected into the CB of the semiconductor (or holes are injected in the VB in the case of photosensitized *p*-type semiconductor). Kinetics of these processes follow the Sakata–Hashimoto–Hiramoto model [37, 38]. On the other hand the direct photoinduced electron transfer from the surface-bound molecule described by the $\hat{\boldsymbol{W}}_{\mathbf{RMS}}$ term was reported only for cyanoferrate- [39–47], catechol- [48, 49], and ascorbic acid-modified [50] titanium dioxide.

The above-mentioned theoretical background shows that, irrespective of the chemical nature of the photosensitizer and its binding mode to the semiconductor surface, one should consider two main ways of the semiconductor CB populating: direct and indirect. *Direct* processes include VB → CB excitations, photosensitization via bulk doping ($\hat{\boldsymbol{W}}_{\mathbf{RS}}$-driven processes) and photophysical processes involving the $\hat{\boldsymbol{W}}_{\mathbf{RMS}}$ term. *Indirect* processes, in turn, involve excitation of the surface and a subsequent electron transfer reaction ($\hat{\boldsymbol{W}}_{\mathbf{RM}} + \hat{\boldsymbol{W}}_{\mathbf{ET}}$).

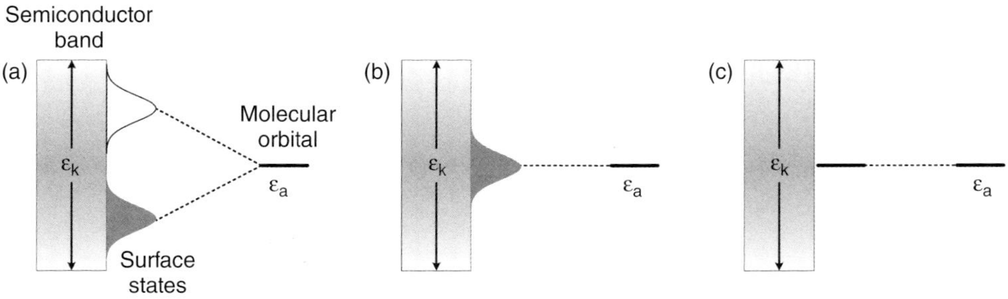

Figure 7.12 Energy diagrams for surface–molecule interactions in the case of: (a) strong electronic coupling, (b) weak electronic coupling, and (c) no electronic coupling. ε_k stands for an electronic continuum of the metal or semiconductor band, whereas ε_a represents the energy of frontier molecular orbital involved in this interaction

Various molecules can interact with metal or semiconductor surfaces. Depending on the nature of the interaction they may exert various modifications to the optical and electronic properties of semiconductor particles and especially the particle–environment interface. Covalent bonding with a π-bonding framework results in deep modification of the interface properties (Figure 7.12a). This type of bonding constitutes a good platform for a strong electronic coupling between surface species and the electronic continuum of a semiconductor (Figure 7.12a) [41, 43]. Alternatively, redox active complexes can be immobilized via ligands equipped with anchoring groups (carboxylic, phosphoric, boronic, etc). Such a solution is very often used, for example, in construction of dye-sensitized solar cells; this mode of binding supports a much weaker electronic interaction between the molecule and the semiconducting substrate (Figure 7.12b). The weakest interaction can be observed in the case of physisorption of molecules onto the semiconductor surface (Figure 7.12c) [45]. Depending on the complex–semiconductor-binding mode various mechanisms of photosensitization are observed. Furthermore, if the surface complex can be addressed independently on the bulk semiconductor, the light-induced photocurrent switching effect can be observed [51].

Interactions between semiconductor surfaces and molecular species are important not only for photosensitization of wide bandgap semiconductors. They not only modify the electronic properties of semiconductors, but also alter spectroscopic and electrochemical properties of molecules bound to the semiconductor surfaces. Especially important are interactions between redox-active species. *n*-Type semiconductors tend to bind electron donors, whereas *p*-type semiconductors bind electron acceptors. These interactions, in the case of a strong electronic coupling (see above) result in new allowed electronic transitions: molecule-to-particle charge transfer (MPCT, Figure 7.13a) and particle-to-molecule charge transfer (PMCT, Figure 7.13b) [49, 52]. The first process is usually observed in the case of TiO_2 with biomolecules possessing the enediol moiety, eg catechols, ascorbic acid [50, 53–61], whereas the latter was only scarcely observed in the case of NiO modified with electron-deficient porphyrins.

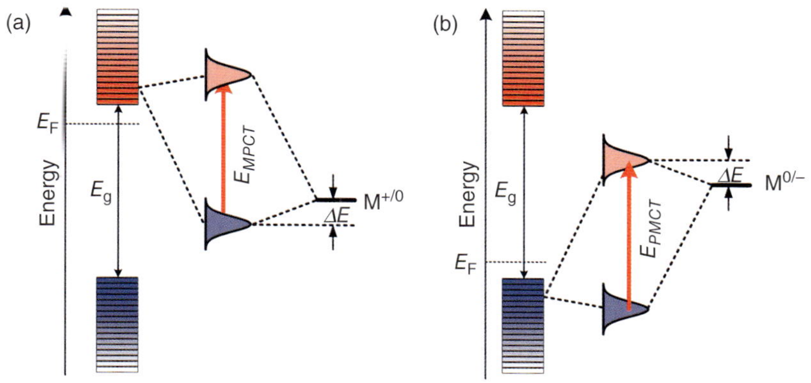

Figure 7.13 *Energy diagrams for donor–acceptor surface–molecule interactions in the case of (a) n-type semiconductor–molecular donor and (b) p-type semiconductor and molecular acceptor.* E_g *denotes the bandgap energy and* E_F *the Fermi energy. See text for other details*

Apart from spectral changes, the molecule–semiconductor interaction results in modification of redox potentials on the surface species. This results from covalent bond formation between the surface of the semiconductor crystal and the molecule [44, 46].

7.3.5 Heterogeneous Photocatalysis

Light absorption by a semiconductor may be followed by *interfacial electron transfer* (IFET) between excited material and adsorbed species. This effect opens possibilities for designing various materials for environmental, biological, and medical applications. Water and air purification, self-cleaning, and self-sterilizing surfaces are just selected examples of photoactive semiconducting material applications. The semiconductor of particular interest is TiO_2 due to its appropriate redox properties, low toxicity, good thermal and photostability, and low price.

Primary Redox Processes

The interfacial electron transfer (process 5 in Figure 7.14) competes with the recombination processes (processes 2 and 4). The photocatalytic activity of a photocatalyst depends directly on the efficiency of process 5 because most types of photocatalytic transformations are a consequence of the redox processes involving IFET. Electrons trapped at active surface sites can reduce an electron acceptor (A) adsorbed at the semiconductor (see equation 7.24). In the case of the photocatalytic oxidation processes at TiO_2 or ZnO, the adsorbed oxygen molecule plays the role of the electron acceptor. This relatively slow process requires milliseconds to proceed. Holes generated in the VB oxidize an electron donor (D) adsorbed at another surface site (see equation 7.25) within hundreds of nanoseconds. As holes generated upon UV light irradiation of TiO_2 are strong oxidants (about 2.8–3.0V at pH 0), they may directly oxidize most organic compounds. However, the main

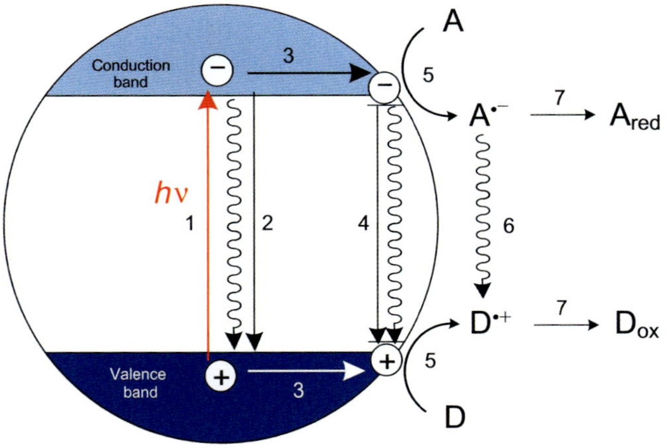

Figure 7.14 Mechanism of photocatalysis at a semiconductor particle. (Adapted from Kisch and Macyk [82])

oxidation step involves water (or hydroxyl group) oxidation, leading to formation of hydroxyl radicals. Both oxidation agents, holes and hydroxyl radicals, play a crucial role in the processes of oxidation.

$$A + e^-_{CB} \rightarrow A^{\bullet -} \tag{7.24}$$

$$D + h^+_{VB} \rightarrow D^{\bullet +} \tag{7.25}$$

The efficiency of photocatalytic reactions also depends on yields of the primary redox product ($A^{\bullet -}$ and $D^{\bullet +}$) recombination (equation 7.26, or process 6 in Figure 7.14) and back-electron transfer (equations 7.27 and 7.28):

$$A^{\bullet -} + D^{\bullet +} \rightarrow A + D \tag{7.26}$$

$$A^{\bullet -} + h^+_{VB} \rightarrow A \tag{7.27}$$

$$D^{\bullet +} + e^-_{CB} \rightarrow D \tag{7.28}$$

A fast conversion of $A^{\bullet -}$ and $D^{\bullet +}$ to secondary, more stable products (process 7 in Figure 7.14) diminishes the probability of these reactions. An increased lifetime of trapped charge carriers, IFET rate, and rates of the primary redox product conversion results in higher quantum yields of an overall photocatalyzed reaction.

A semiconductor suitable for an efficient photo-oxidation reactions of environmental relevance should fulfil several requirements. Its bandgap should allow the use of solar radiation, ie the catalyst has to absorb in the visible or near-UV light region. The redox potentials of $OH^{\bullet}/H_2O$ and $O_2/O_2^{\bullet -}$ couples should lie within this bandgap ($E^{\circ}_{OH^{\bullet}/H_2O} = 2.8\,V$ [62], $E^{\circ}_{O_2/O_2^{\bullet -}} = -0.16\,V$ [63]) in order to facilitate

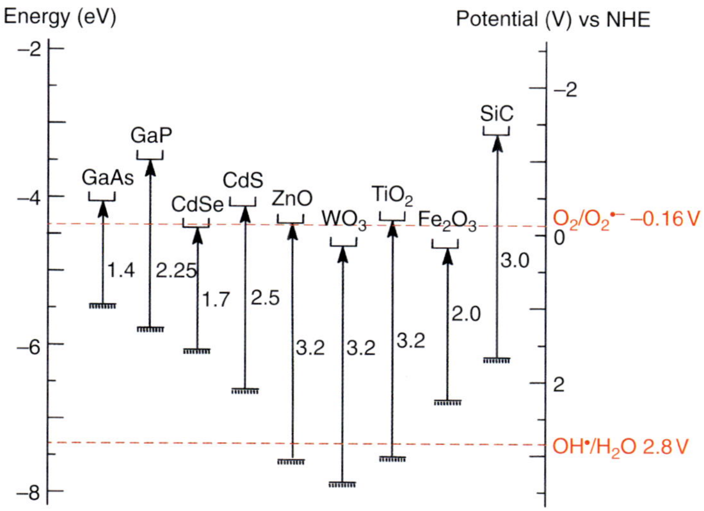

Figure 7.15 *Energy bandgaps and band edges of selected semiconductor materials in aqueous electrolytes at pH = 0. (Adapted from Linsebigler et al. [111])*

generation of superoxide anions ($O_2^{\bullet-}$), hydroxyl radicals (see equations 7.29, 7.30) and a variety of other reactive oxygen species:

$$O_2 + e^-_{CB} \rightarrow O_2^{\bullet-} \tag{7.29}$$

$$H_2O + h^+_{VB} \rightarrow OH^{\bullet} + H^+ \tag{7.30}$$

ZnO and TiO_2 fulfil these requirements (Figure 7.15) and therefore both of them can efficiently photocatalyze organic molecules degradation (and mineralization). TiO_2 shows, however, better stability and photostability (does not undergo photocorrosion) compared with ZnO. In addition this oxide is non-toxic and very cheap. Its VB and CB edges are localized at about 3.1 and –0.1 V (vs NHE at pH 0), respectively.

Mineralization (meaning a complete oxidation) of numerous organic species cannot be regarded only as an oxidation process. In many cases oxidation must be preceded by reduction steps, eg photocatalyzed transformation of CCl_4 to CO_2 and Cl^- requires first reduction of carbon(IV) to lower oxidation states, followed by its reoxidation to CO_2 [30]. In this context redox properties of an excited semiconductor play a crucial role. Photogenerated holes should support highly oxidative potential, but at the other surface sites an efficient reducer (electron) should also be available [64, 65].

Secondary Chemical Transformations

The primary redox processes leading to formation of superoxide and hydroxyl radicals (see equations 7.29 and 7.30) participate in secondary chemical transformations.

Although $OH^{\bullet}$ is responsible for the oxidation of variety of organic species present at the TiO_2 surface or in its vicinity, the fate of the superoxide radical is determined by further transformations to other reactive oxygen species, such as hydrogen peroxide, hydroxyl radical, and others (see equations 7.31– 7.36). Formation of hydroxyl radicals as a result of water oxidation with holes from the VB can be assigned as an *oxidative pathway* of $OH^{\bullet}$ generation. The *reductive pathway* of $OH^{\bullet}$ generation involves one-electron reduction of two oxygen molecules (formation of two $HO_2^{\bullet}$ radicals; see equation 7.31 and H_2O_2; see equation 7.32) followed by further step of reduction (see equation 7.35 or 7.36) [66–69]. According to this mechanism, formation of one $OH^{\bullet}$ radical on the reductive pathway requires three electrons from the conduction band (generation of two $HO_2^{\bullet}$ radicals and reduction of H_2O_2):

$$O_2^{\bullet-} + H^+ \rightarrow HO_2^{\bullet} \tag{7.31}$$

$$HO_2^{\bullet} + HO_2^{\bullet} \rightarrow H_2O_2 + O_2 \tag{7.32}$$

$$O_2^{\bullet-} + HO_2^{\bullet} \rightarrow O_2 + HO_2^- \tag{7.33}$$

$$HO_2^- + H^+ \rightarrow H_2O_2 \tag{7.34}$$

$$H_2O_2 + O_2^{\bullet-} \rightarrow OH^{\bullet} + OH^- + O_2 \tag{7.35}$$

$$H_2O_2 + e_{CB}^- \rightarrow OH^{\bullet} + OH^- \tag{7.36}$$

Another form of reactive oxygen that can be formed on TiO_2 irradiation is singlet oxygen. The mechanisms of $^1O_2$ formation that are usually discussed are:

- energy transfer from the excited semiconductor to O_2 molecule [70] (see below)
- $HO_2^{\bullet}$ radicals recombination with concomitant H_2O_2 generation [71] (see equation 7.32)
- $O_2^{\bullet-}/h_{VB}^+$ recombination [72, 73].

The most likely singlet oxygen is formed according to the first mechanism, ie in the photosensitization process [70].

Indirect Photocatalysis

The term direct ‘photocatalysis’ can be defined as the photocatalytic process in which a semiconductor particle absorbs light. This situation has been described above. In addition to this type of heterogeneous photocatalysis an *indirect photocatalysis* can also be considered when light is absorbed not by a semiconductor particle but by donor molecules (D) adsorbed at the photocatalyst surface. Depending on the relative redox potentials of the dye excited state and the lower CB edge, an electron from the excited state of the molecule (D*) may be injected into CB of the semiconductor (Figure 7.16). The oxidized molecule undergoes further transformations leading to self-degradation. The degradation process is enhanced by reactive

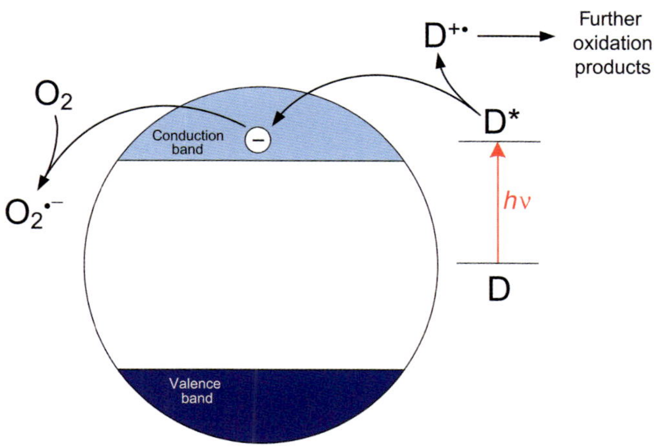

Figure 7.16 *The mechanism of indirect photocatalysis*

oxygen species generated as a consequence of oxygen reduction by electrons from the conduction band. This type of the indirect photocatalysis was successfully applied for several coloured organic pollutants degradation produced mainly by textile industry [74–80].

Photosensitization

A relatively high bandgap energy of TiO_2 enables two main redox processes facilitating formation of various reactive oxygen species (see equations 7.29 and 7.30), but the electron-hole pair generation requires UV light absorption. The efficiency of photocatalysis at TiO_2 surface may be increased after photosensitization towards visible light. There are several methods suitable for broadening the photoactivity spectral range. A brief overview of possible photosensitization methods is presented here.

Dye Sensitization. This method of TiO_2 photosensitization is based on the interaction of adsorbed (or chemisorbed) species with the semiconductor matrix as described in section 7.3.4. The photosensitization may be achieved by adsorption of sensitizer molecules (S) absorbing visible light (Figure 7.17a) or by a surface charge transfer complex formed on binding an organic or inorganic moiety to the TiO_2 surface (Figure 7.17b). The redox potential of the excited state ($S^{\bullet+}/S^*$ pair) should be lower than the CB edge of the semiconductor. The electron injection from the excited state S^* to the CB of the semiconductor particle takes place with concomitant formation of $S^{\bullet+}$ transient species. Regeneration of the photosensitizer S is assured in the presence of an electron acceptor D. The efficiency of this process strongly influences the photostability of the sensitized photocatalyst. Electrons from the conduction band reduce adsorbed oxygen thus generating reactive oxygen species according to equations 7.29 and 7.31– 7.36. Therefore the semiconductor ensures an efficient charge separation and plays the role of a support mediating electron transfer from the donor to the acceptor. The effect of TiO_2 photosensitization

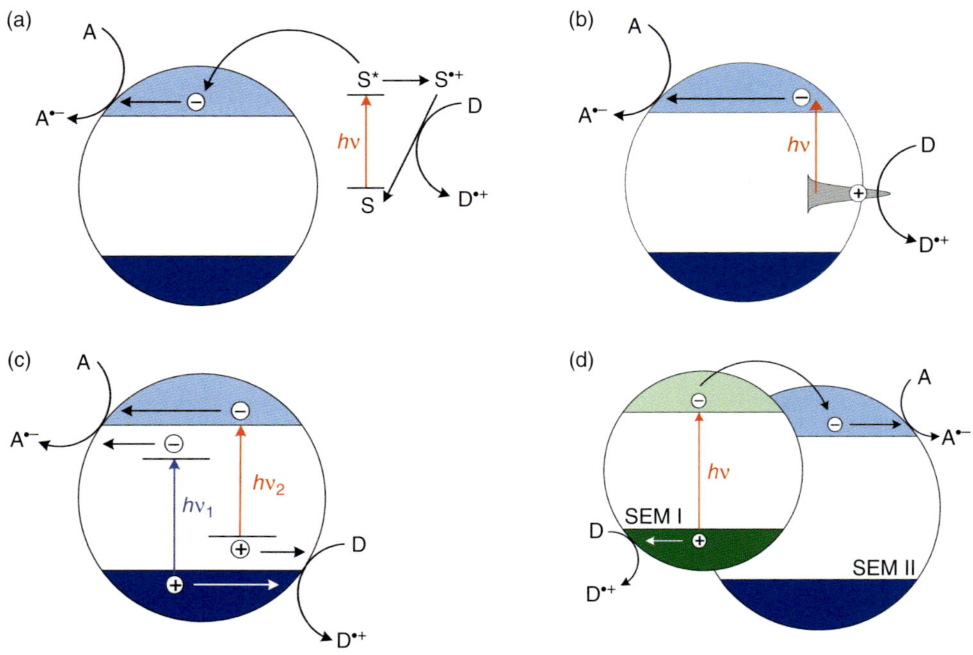

Figure 7.17 *Modes of TiO_2 photosensitization: (a) photosensitization with organic or inorganic chromophores chemisorbed onto titanium dioxide surface; (b) formation of surface complexes exhibiting metal-to-band charge transfer transitions (MBCT); (c) bulk doping resulting in formation of acceptor or donor levels; and (d) formation of composite semiconductors. 'A' denotes the electron acceptor, 'D' the electron donor*

presented in Figure 7.17a was observed in the case of platinum(IV) chloride complexes chemisorbed at the TiO_2 surface [81–86] or with several $[Ru(bpy)_3]^{2+}$ derivatives used as a dye in photovoltaic cells [87, 88].

A significantly different photosensitization mechanism, shown in Figure 7.17b, takes place when a charge transfer surface complex (eg MPCT) is formed on TiO_2 surface modification. The excitation is associated with a direct electron transfer from the surface moiety to the CB or from the VB to the surface moiety. Details of this type of photosensitisation have been described in section 7.3.4. Figure 7.17b is valid for TiO_2 with chemisorbed cyanide complexes [39–47] of Fe^{II}, Ru^{II}, Os^{II}, Mo^{IV}, W^{IV}, and Re^{III}, or with surface-bound catechols [48, 49].

Semiconductor Doping. An additional energy levels within the bandgap of the semiconductor can be provided by introducing acceptor or donor levels (see Figure 7.17c). Electron excitation from the VB to acceptor level ($h\nu_1$) or from the donor level to CB ($h\nu_2$) requires lower photon energy compared with the direct semiconductor excitation. The photogenerated holes oxidize the electron donor D and primary oxidation products while the electron acceptor A is reduced. The photosensitization described by $h\nu_2$ was achieved with several metal ions, eg Fe^{3+} [89, 90], Co^{2+} [90], Li^+, Zn^{2+} [91], Mn^{4+}, Ni^{2+} [92], Cr^{3+} [90, 92, 93], and others [94, 95].

Nowadays non-metal doping attracts more and more attention. Among others nitrogen- [96–98] and carbon-doped titania [99, 100] show interesting visible light activity.

Composite Semiconductors. Other visible light active systems involve hybrid catalysts consisting of two semiconductors characterized by various bandgaps and redox properties, as depicted in Figure 7.17d. Lower photon energy (visible light) irradiation leads to a charge separation within the semiconductor of lower bandgap energy (SEM I). Electron may move to the conduction band of the second semiconductor (SEM II) where it can be accessible for the reduction of the electron acceptor. The oxidation of D takes place at the surface of the first semiconductor. Such a composite system ensures not only activity upon visible light irradiation, but also an efficient charge separation, as well as reduction and oxidation process separation, suppression of the recombination, and increased yield of the catalytic reactions. The isopropanol oxidation at semiconductor–zeolite composites [101] or chlorophenol oxidation at CdS/TiO_2 [102, 103] may serve as examples of this approach. Also Fe_2O_3/TiO_2 colloids were reported to be active towards dichloroacetic acid photodegradation upon visible light [104].

Interfacial Energy Transfer

Although the charge transfer processes govern the photocatalytic and photoelectrochemical properties of semiconductors, the energy transfer may also play an important role in photocatalysis. Oxygen molecule in the excited state, namely singlet oxygen, may be generated at the surface of excited semiconductors as a result of either energy transfer or – as sometimes postulated – redox reactions [71, 72, 105]:

$$2\,HOO^{\bullet} \rightarrow H_2O_2 + {}^1O_2 \tag{7.37}$$

$$O_2^{\bullet-} + h_{VB}^{+} \rightarrow {}^1O_2 \tag{7.38}$$

The energy difference between the excited (${}^1O_2$) and ground state (${}^3O_2$) of oxygen equals 0.98 or 1.63 eV for ${}^1\Delta$ and ${}^1\Sigma$ states, respectively [106]. An efficient energy transfer process requires matching of the excited state energies of the photosensitizer (semiconductor particle) and O_2 [32, 106]. Thus the photosensitization of oxygen molecule can be realized at semiconductors characterized by a bandgap energy close to 1 eV rather than at broad bandgap semiconductors. In fact, relatively efficient singlet oxygen formation was observed in the presence of nanocrystalline silicon (particle size 2–10 nm) [32]. Silicon is an indirect bandgap semiconductor facilitating energy transfer rather than emission processes.

Despite a huge energy mismatch between the TiO_2 bandgap energy and oxygen excited states, singlet oxygen generation has been reported at the surface of irradiated TiO_2 [70–73, 107]. Recent studies have shown that relative efficiencies of the electron and energy transfer processes may be controlled to some extent by a simple surface modification [108]. Exchange of the surface hydroxyl groups by inorganic ions, complexes, and organic groups can effectively suppress an interfacial electron

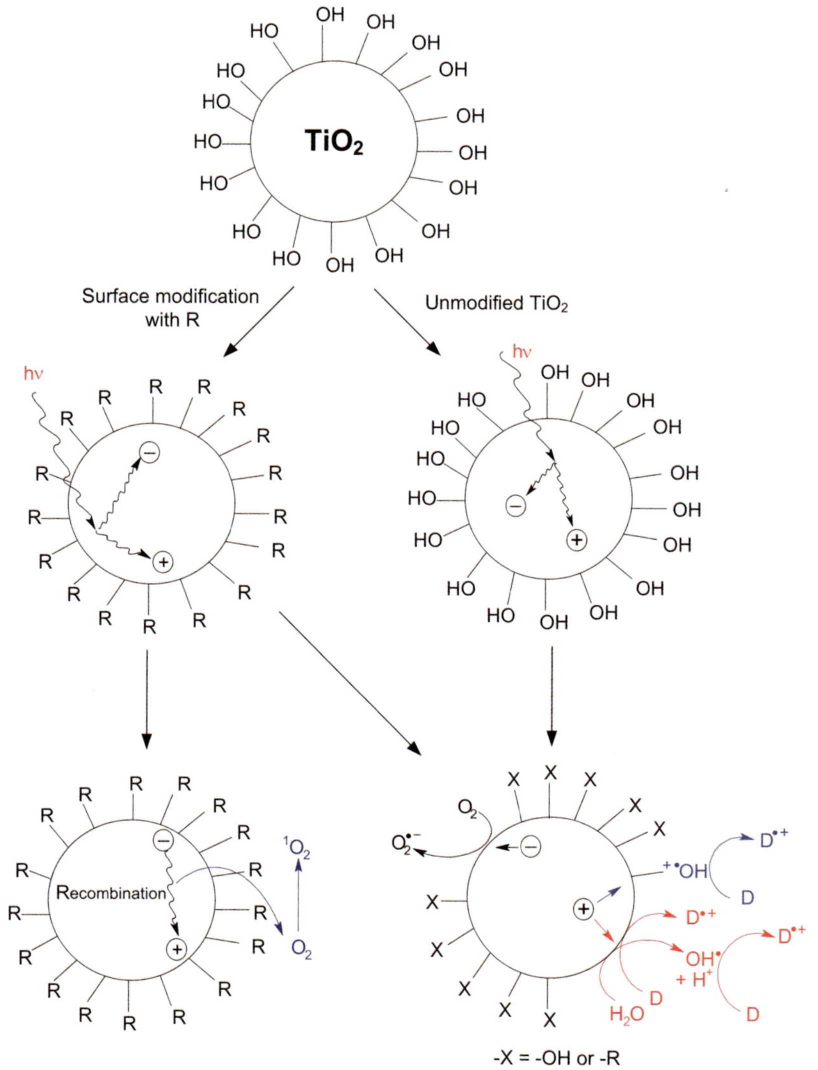

Figure 7.18 *Primary processes at neat TiO_2 (right path) and at TiO_2 with substituted surface –OH groups (left path). (Adapted from Jańczyk et al. [108])*

transfer and improve the energy transfer processes. Lower efficiencies of interfacial electron transfer processes – such as direct hole oxidation, reduction with electrons from the conduction band, photocurrent generation – were observed, for example, in the case of fluorinated TiO_2 (F-TiO_2) or silylated TiO_2 compared with neat TiO_2. Therefore two alternative mechanisms of photocatalysis at titanium dioxide may be considered: one involves the electron transfer whereas the other the energy transfer (Figure 7.18). As the result of the first mechanism, hydroxyl radicals, superoxide radicals, and other reactive oxygen species are generated, whereas the second may be responsible for singlet oxygen formation. Surface modification of the TiO_2 in some cases enhances the energy transfer pathway and diminishes the efficiency of interfacial electron transfer. At the surface of unmodified TiO_2 the electron transfer processes prevail.

References

1. Wang Y, Herron N. Nanometer-sized semiconductor clusters materials synthesis, quantum size effects, and photophysical properties. *J Phys Chem* 1991;**95**:525.
2. Schmid G. Large clusters and colloids. Metals in the embryonic state. *Chem Rev* 1992;**92**:1709–27.
3. Hu M, Chen J, Li Z-Y, et al. Gold nanostructures: engineering their plasmonic properties for biomedical applications. *Chem Soc Rev* 2006;**35**:1084–94.
4. Schmid G, Pfeil R, Boese R, et al. $Au_{55}[P(C_6H_5)_3]_{12}Cl_6$ – A gold cluster of an exceptional size. *Ber Dtsch Chem Ges* 1981;**114**:3634–42.
5. Brust M, Walker M, Bethell D, Schiffrin DJ, Whyman R. Synthesis of thiol-derivatised gold nanoparticles in a two-phase liquid–liquid system. *J Chem Soc Chem Commun* 1994:801–2.
6. Daniel M-C, Astruc D. Gold nanoparticles: assembly, supramolecular chemistry, quantum size-related properties, and applications toward biology, catalysis, and nanotechnology. *Chem Rev* 2004;**104**:293–346.
7. Jana NR, Gearheart L, Murphy CJ. Evidence for seed-mediated nucleation in the chemical reduction of gold salts to gold nanoparticles. *Chem Mater* 2001;**13**:2313–23.
8. Pérez-Juste J, Pastoriza-Santos I, Liz-Marzán LM, Mulvaney P. Gold nanorods: Synthesis, characterization and applications. *Coord Chem Rev* 2005;**249**:1870–901.
9. Drude P. Zur Elektronentheorie der Metalle. *Ann Phys* 1900;**312**:687–92.
10. Sommerfeld A, Bethe H. *Elektronentheorie der Metalle*. Heidelberg: Springer Verlag, 1933.
11. Del Fatti N, Vallée F, Hamanaka Y, Nakamura A. Electron dynamics and surface plasmon resonance nonlinearities in metal nanoparticles. *Chem Phys* 2000;**251**:215–26.
12. Adair JH, Li T, Kido T, et al. Recent developments in the preparation and properties of nanometer-size spherical and platelet-shaped particles and composite particles. *Mater Sci Eng* 1998;**R23**:139–242.
13. Jain PK, Huang X, El-Sayed IH, El-Sayed MA. Review of some interesting surface plasmon resonance-enhanced properties of noble metal nanoparticles and their applications to biosystems. *Plasmonics* 2007;**2**:107–18.
14. Link S, El-Sayed MA. Shape and size dependence of radiative, non-radiative and photothermal properties of gold nanocrystals. *Int Revs Phys Chem* 2000;**19**:409–53.
15. Sönnichsen C. Plasmons in metal nanostructures. PhD dissertation, Munich, 2001.
16. Furube A, Du L, Hara K, Katoh R, Tachiya M. Ultrafast plasmon-induced electron transfer from gold nanodots into TiO_2 nanoparticles. *J Am Chem Soc* 2007;**129**:14852–3.
17. Tian Y, Tatsuma T. Mechanisms and applications of plasmon-induced charge separation at TiO_2 films loaded with gold nanoparticles. *J Am Chem Soc* 2005;**127**:7632–7.
18. Lindstrom CD, Zhu X-Y. Photoinduced electron transfer at molecule-metal interface. *Chem Rev* 2006;**106**:4281–300.
19. Ghosh SK, Pal T. Interparticle coupling effect on the surface plasmon resonance of gold nanoparticles: from theory to applications. *Chem Rev* 2007;**107**:4797–862.
20. Noguez C. Surface plasmons on metal nanoparticles: the influence of shape and physical environment. *J Phys Chem C* 2007;**111**:3806–39.
21. Yacobi BG. *Semiconductor Materials. An introduction to basic principles*. New York: Kluwer Academic Publishers, 2003.

22. Pankove JI. *Optical Processes in Semiconductors*. Englewood Cliffs, NJ: Prentice-Hall, 1975.
23. Urbach F. The long-wavelength edge of photographic sensitivity and of the electronic absorption of solids. *Phys Rev* 1953;**92**:1324.
24. Skettrup T. Urbach's rule derived from thermal fluctuations in the band-gap energy. *Phys Rev B* 1978;**18**:2622–31.
25. Mahan GD. Phonon-broadened optical spectra: Urbach's Rule. *Phys Rev* 1966;**145**: 602–8.
26. Monroe D, Kastner MA. Exactly exponential band in a glassy semiconductor. *Phys Rev B* 1986;**33**:8881–4.
27. Cody GD, Tiedje T, Abeles B, Brooks B, Goldstein Y. Disorder and the optical-absorption edge of hydrogenated amorphous silicon. *Phys Rev Lett* 1981;**47**:1480–3.
28. Martienssen W. Über die Excitonenbanden der Alkalihalogenidkristalle. *J Phys Chem Solids* 1957;**2**:257–67.
29. Rakhshani AE. Study of Urbach tail, bandgap energy and grain-boundary characteristics in CdS by modulated photocurrent spectroscopy. *J Phys: Condens Mater* 2000;**12**:4391–400.
30. Hoffmann MR, Martin ST, Choi W, Bahnemann DW. Environmental applications of semiconductor photocatalysis. *Chem Rev* 1995;**95**:69–96.
31. Kiselev VF, Krylov OV. *Adsorption and Catalysis On Transition Metals and Their Oxides*. Berlin: Springer-Verlag, 1989.
32. Kovalev D, Fujii M. Silicon nanocrystals: photosensitised for oxygen molecules. *Adv Mater* 2005;**17**:2531–44.
33. Ozawa L, Itoh M. Cathode ray tube phosphors. *Chem Rev* 2003;**103**:3835–55.
34. Galperin M, Nitzan A. Current induced light emission and light induced current in molecular tunelling junctions. *Phys Rev Lett* 2005;**95**:206802.
35. Galperin M, Nitzan A, Ratner MA. Molecular transport junctions: current from electronic excitations in the leads. *Phys Rev Lett* 2006;**96**:166803.
36. Galperin M, Nitzan A. Optical properties of current carrying molecular wires. *J Chem Phys* 2006;**124**:234709.
37. Sakata T, Hashimoto K, Hiramoto M. New Aspects of electron transfer on semiconductor surface: dye-sensitization system. *J Phys Chem* 1990;**94**:3040–5.
38. Kitao O. Photoinduced electron transfer in dye-sensitized solar cells: modified Sakata-Hashimoto-Hiramoto Model (MSHH). *J Phys Chem C* 2007;**111**:15889–902.
39. Vrachnou E, Grätzel M, McEvoy AJ. Efficient visible light photoresponse following surface complexation of titanium dioxide with transition metal cyanides. *J Electroanal Chem* 1989;**258**:193–205.
40. Vrachnou E, Vlachopoulos N, Grätzel M. Efficient visible light sensitization of TiO_2 by surface complexation with $Fe(CN)_6^{4-}$. *J Chem Soc Chem Commun* 1987:868–9.
41. Gosh HN, Ashbury JB, Weng Y, Lian T. Interfacial electron transfer between $[Fe(CN)_6]^{4-}$ and TiO_2 nanoparticles: direct electron injection and nonexponential recombination. *J Phys Chem B* 1998;**102**:10208–15.
42. Khoudiakov M, Parise AR, Brunschwig BS. Interfacial electron transfer in $[Fe(CN)_6]^{4-}$ – sensitized TiO_2 nanoparticles: a study of direct charge injection by electroabsorption spectroscopy. *J Am Chem Soc* 2003;**125**:4637–42.
43. De Angelis F, Tilocca A, Selloni A. Time-dependent DFT study of $[Fe(CN)_6]^{4-}$ sensitization of TiO_2 nanoparticles. *J Am Chem Soc* 2004;**126**:15024–5.
44. Hebda M, Stochel G, Szaciłowski K, Macyk W. Optoelectronic switches based on wide bandgap semiconductors. *J Phys Chem B* 2006;**110**:15275–83.
45. Macyk W, Stochel G, Szaciłowski K. Photosensitization and photocurrent switching effect in nanocrystalline titanium dioxide functionalized with iron(II) complexes: A comparative study. *Chem Eur J* 2007;**13**:5676–87.
46. Szaciłowski K, Macyk W, Hebda M, Stochel G. Redox-controlled photosensitization of nanocrystalline titanium dioxide. *ChemPhysChem* 2006;**7**:2384–91.

47. Harris JA, Trotter K, Brunschwig BS. Interfacial electron transfer in metal cyanide-sensitized TiO_2 nanoparticles. *J Phys Chem B* 2007;**111**:6695–702.
48. Rego LGC, Batista VS. Quantum dynamics simulations of interfacial electron transfer in sensitized TiO_2 semiconductors. *J Am Chem Soc* 2003;**125**:7989–97.
49. Creutz C, Brunschwig BS, Sutin N. Interfacial charge transfer absorption: Semiclassical treatment. *J Phys Chem B* 2005;**109**:10251–60.
50. Rajh T, Nedeljkovic JM, Chen LX, Poluektov O, Thurnauer MC. Improving optical and charge separation properties of nanocrystalline TiO_2 by surface modification with vitamin C. *J Phys Chem B* 1999;**103**:3515–19.
51. Szacilowski K, Macyk W. Photoelectrochemical photocurrent switching effect: A new platform for molecular logic devices. *Chimia* 2007;**61**:831–4.
52. Creutz C, Brunschwig BS, Sutin N. Interfacial charge transfer absorption: 3. Application to semiconductor-molecule assemblies. *J Phys Chem B* 2006;**110**:25181–90.
53. Creutz C, Chou MH. Binding of catechols to mononuclear titanium(IV) and to 1- and 5-nm TiO_2 nanoparticles. *Inorg Chem* 2008;**47**:3509–14.
54. de la Garza L, Saponjic ZV, Dimitrijevic NM, Thurnauer MC, Rajh T. Surface states of titanium dioxide nanoparticles modified with enediol ligands. *J Phys Chem B* 2006;**110**:680–6.
55. Wu H-P, Cheng T-L, Tseng W-L. Phosphate-Modified TiO_2 nanoparticles for selective detection of dopamine, levodopa, adrenaline, and catechol based on fluorescence quenching. *Langmuir* 2007;**23**:7880–5.
56. Araujo PZ, Morando PJ, Blesa MA. Interaction of catechol and gallic acid with titanium dioxide in aqueous suspensions. 1. Equilibrium studies. *Langmuir* 2005;**21**:3470–4.
57. Xagas APB, Hugot-Le Goff A, Spyrellis N, Loizos Z, Falaras P. Surface modification and photosensitisation of TiO_2 nanocrystalline films with ascorbic acid. *J Photochem Photobiol A: Chem* 2000;**132**:115–20.
58. Lana-Villarreal T, Rodes A, Perez JM, Gómez R. A Spectroscopic and electrochemical approach to the study of the interactions and photoinduced electron transfer between catechol and anatase nanoparticles in aqueous solution. *J Am Chem Soc* 2005;**127**: 12601–11.
59. Redfern PC, Zapol P, Curtiss LA, Rajh T, Thurnauer MC. Computational studies of catechol and water interactions with titanium oxide nanoparticles. *J Phys Chem B* 2003;**107**:11419–27.
60. Rajh T, Chen LX, Lukas K, Liu T, Thurnauer MC, Tiede DM. Surface restructuring of nanoparticles: An efficient route for ligand-metal oxide crosstalk. *J Phys Chem B* 2002;**106**:10543–52.
61. Niederberger M, Garnweitner G, Krumeich F, Nesper R, Cölfen H, Antonietti M. Tailoring the surface and solubility properties of nanocrystalline titania by a nonaqueous in situ functionalization process. *Chem Mater* 2004;**16**:1202–8.
62. Latimer W. *Oxidation Potentials*, 2nd edn. Englewood Cliffs, NJ: Prentice-Hall, 1952.
63. Stanbury DM. Reductions potentials involving inorganic free radicals in aqueous solution. *Adv Inorg Chem* 1989;**33**:69–138.
64. Serpone N, Pelizzetti E. *Photocatalysis. Fundamentals and applications*. New York: Wiley, 1989.
65. Maurino V, Minero C, Pelizzetti E, Piccinini P, Serpone N, Hidaka H. The fate of organic nitrogen under photocatalytic conditions: degradation of nitrophenols and aminophenols on irradiated TiO_2. *J Photochem Photobiol A: Chem* 1997;**109**: 171–6.
66. Cundall RB, Rudham R, Salim MS. Photocatalytic oxidation of propan-2-ol in the liquid phase by rutile. *J Chem Soc Faraday Trans 1* 1976;**72**:1642–51.
67. Harvey PR, Rudham R, Ward S. Photocatalytic oxidation of liquid propan-2-ol by titanium dioxide. *J Chem Soc Faraday Trans 1* 1983;**79**:1381–90.
68. Fujihira M, Satoh Y, Osa T. Heterogeneous photocatalytic reactions on semiconductor materials. III. Effect of pH and Cu^{2+} ions on the photo-Fenton type reaction. *Bull Chem Soc Jpn* 1982;**55**:666–71.

69. Okamoto K, Yamamoto Y, Tanaka H, Itaya A. Kinetics of heterogeneous photocatalytic decomposition of phenol over anatase TiO_2 powder. *Bull Chem Soc Jpn* 1985;**58**: 2023–8.
70. Konaka R, Kasahara E, Dunlap WC, Yamamoto Y, Chien KC, Inoue M. Irradiation of titanium dioxide generates both singlet oxygen and superoxide anion. *Free Rad Biol Med* 1999;**27**:294–300.
71. Sugimoto H, Sawyer DT, Kanofsky JR. Transient formation of hydrogen tetraoxide [HOOOOH] from the dimerization of perhydroxyl (HOO): Concerted homolytic dissociation to singlet oxygen ($^1O_2$) and hydrogen peroxide. *J Am Chem Soc* 1988;**110**: 8707–8.
72. Mayeda EA, Bard AJ. The production of singlet oxygen in electrogenerated radical ion electron transfer reaction. *J Am Chem Soc* 1973;**95**:6223–26.
73. Pappas SP, Fischer RM. Photo-chemistry of pigments: Studies on the mechanism of chalking. *J Paint Technol* 1974;**46**:65–72.
74. Hidaka H, Nohara K, Zhao J, Pelizzetti E, Serpone N. Photodegradation of surfactants. XIV. Formation of NH_4^+ and NO_3^- ions for the photocatalyzed mineralization of nitrogen-containing cationic, non-ionic and amphoteric surfactants. *J Photochem Photobiol A: Chem* 1995;**91**:145–52.
75. Liu G, Wu T, Zhao J, Hidaka H, Serpone N. Photoassisted degradation of dye pollutants. 8. Irreversible degradation of alizarin red under visible light radiation in air-equilibrated aqueous TiO_2 dispersions. *Environ Sci Technol* 1999;**33**:2081–7.
76. Wu T, Liu G, Zhao J, Hidaka H, Serpone N. Photoassisted degradation of dye pollutants. V. Self-photosensitized oxidative transformation of rhodamine B under visible light irradiation in aqueous TiO_2 dispersions. *J Phys Chem B* 1998;**102**:5845–51.
77. Wu T, Lin T, Zhao J, Hidaka H, Serpone N. TiO_2-assisted photodegradation of dyes. 9. Photooxidation of a squarylium cyanine dye in aqueous dispersions under visible light irradiation. *Environ Sci Technol* 1999;**33**:1379–87.
78. Wu T, Liu G, Zhao J, Hidaka H, Serpone N. Evidence for H_2O_2 generation during the TiO_2-assisted photodegradation of dyes in aqueous dispersions under visible light illumination. *J Phys Chem B* 1999;**103**:4862–7.
79. Zhang F, Zhao J, Shen T, Hidaka H, Pelizzetti E, Serpone N. TiO_2-assisted photodegradation of dye pollutants. II. Adsorption and degradation kinetics of eosin in TiO_2 dispersions under visible light irradiation. *Appl Catal B: Environ* 1998;**15**:147–56.
80. Zhao J, Wu T, Wu K, Oikawa K, Hidaka H, Serpone N. Photoassisted degradation of dye pollutants. 3. Degradation of the cationic dye rhodamine B in aqueous anionic surfactant/TiO_2 dispersions under visible light irradiation: evidence for the need of substrate adsorption on TiO_2 particles. *Environ Sci Technol* 1998;**32**:2394–400.
81. Kisch H, Burgeth G, Macyk W. Visible light photocatalysis by a titania transition metal complex. *Adv Inorg Chem* 2004;**56**:241.
82. Kisch H, Macyk W. Visible-light photocatalysis by modified titania. *ChemPhysChem* 2002;**3**:399–400.
83. Burgeth G, Kisch H. Photocatalytic and photoelectrochemical properties of titania-chloroplatinate(IV). *Coord Chem Rev* 2002;**230**:40–7.
84. Macyk W, Burgeth G, Kisch H. Photoelectrochemical properties of platinum(IV)-chloride surface modified TiO_2. *Photochem Photobiol Sci* 2003;**2**:322–8.
85. Macyk W, Kisch H. Mechanism of titanium dioxide photosensitization by platinum(IV)chloride surface complexes. *J Inf Rec* 2000;**25**:435–8.
86. Macyk W, Kisch H. Photosensitization of crystalline and amorphous titanium dioxide by Pt^{IV}-chloride surface complexes. *Chem Eur J* 2001;**7**:1862.
87. Hagfeldt A, Grätzel M. Molecular photovoltaics. *Acc Chem Res* 2000;**33**:269–77.
88. Moser JE, Bonnote P, Grätzel M. Molecular photovoltaics. *Coord Chem Rev* 1998;**171**:245–50.
89. Ranjit KT, Viswanathan B. Synthesis, characterization and photocatalytic properties of iron-doped TiO_2 catalysts. *J Photochem Photobiol A: Chem* 1997;**108**:79–84.
90. Ranjit KT, Viswanathan B. Photocatalytic reduction of nitrite and nitrate ions over doped TiO_2 catalyst. *J Photochem Photobiol A: Chem* 1997;**107**:215–20.

91. Brezová V, Blazková A, Karpinský L, et al. Phenol decomposition using M^{n+}/TiO_2 photocatalysts supported by the sol-gel technique on glass fibres. *J Photochem Photobiol A: Chem* 1997;**109**:177–83.
92. Blasse G, de Korte PHM, Mackor A. The colouration of titanates by transition metal ions in view of solar energy applications. *J Inorg Nucl Chem* 1981;**43**:1499–503.
93. Borgarello E, Kiwi J, Grätzel M, Pelizzetti E, Visca M. Visible light induced water cleavage in colloidal solutions of chromium-doped titanium dioxide particles. *J Am Chem Soc* 1982;**104**:2996–3002.
94. Kruczynski L, Gesser HD. Factors affecting hydrogen evolution from hydrated $Ti(OH)_3$: a model system for photoassisted hydrogen production on titania. *Inorg Chim Acta* 1983;**72**:161–3.
95. Yamashita H, Honda M, Harada M, et al. Preparation of titanium oxide photocatalysts anchored on porous silica glass by a metal ion-implantation method and their photocatalytic reactivities for the degradation of 2-propanol diluted in water. *J Phys Chem B* 1998;**102**:10707–11.
96. Asahi R, Morikawa T, Ohwaki T, Aoki K, Taga Y. Visible-light photocatalysis in nitrogen-doped titanium oxides. *Science* 2001;**293**:269–71.
97. Sakthivel S, Kisch H. Photocatalytic and photoelectrochemical properties of nitrogen-doped titanium dioxide. *ChemPhysChem* 2003;**4**:487–90.
98. Morikawa T, Asahi R, Ohwaki T, Aoki K, Taga Y. Band-gap narrowing of titanium dioxide by nitrogen doping. *Jpn J Appl Phys* 2001;**40**:L561–3.
99. Lettmann C, Hildenbrand K, Kisch H, Macyk W, Maier WF. Visible light photodegradation of 4-chlorophenol with a coke-containing titanium dioxide photocatalysts. *Appl Catal B: Environ* 2001;**32**:215–27.
100. Sakthivel S, Kisch H. Daylight photocatalysis by carbon-modified titanium dioxide. *Angew Chem Int Ed* 2003;**42**:4908–911.
101. Green KJ, Rudham R. Photocatalytic oxidation of propan-2-ol by semiconductor-zeolite composites. *J Chem Soc Faraday Trans* 1993;**89**:1867–70.
102. Serpone N, Maruthamuthu P, Pichat P, Pelizzetti E, Hidaka H. Exploiting the interparticle electron transfer process in the photocatalysed oxidation of phenol, 2-chlorophenol and pentachlorophenol: chemical evidence for electron and hole transfer between coupled semiconductors. *J Photochem Photobiol A: Chem* 1995;**85**:247–55.
103. Kang MG, Han H-E, Kim K-J. Enhanced photodecomposition of 4-chlorophenol in aqueous solution by deposition of CdS on TiO_2. *J Photochem Photobiol A: Chem* 1999;**125**:119–25.
104. Bahnemann D, Bockelmann D, Goslich R, Hilgendorff M. Photocatalytic detoxification of polluted aquifers: novel catalysts and solar applications. In: Helz GR, Zepp RG, Crosby DG, eds. *Aquatic and Surface Photochemistry*. Boca Raton, FL: Lewis, 1994: 349–67.
105. Pappas SP, Fischer RM. Photo-chemistry of pigments: Studies on the mechanism of chalking. *J Paint Tech* 1974;**46**:65–72.
106. Schweitzer C, Schmidt R. Physical mechanisms of generation and deactivation of singlet oxygen. *Chem Rev* 2003;**103**:1685–757.
107. Nosaka Y, Daimon T, Nosaka AY, Murakami Y. Singlet oxygen formation in photocatalytic TiO_2 aqueous suspension. *Phys Chem* 2004;**6**:2917–18.
108. Jańczyk A, Krakowska E, Stochel G, Macyk W. Singlet oxygen photogeneration at surface modified titanium dioxide. *J Am Chem Soc* 2006;**128**:15574–5.
109. Tao AR, Habas S, Yang P. Shape control of colloidal metal nanocrystals. *Small* 2008;**4**:310–25.
110. Sönnichsen C, Franzl T, Wilk T, von Plessen G, Feldman J. Drastic reduction of plasmon damping in gold nanorods. *Phys Rev Lett* 2002;**88**:077402.
111. Linsebigler AL, Lu G, Yates JT Jr. Photocatalysis on TiO_2 surfaces: principles, mechanisms, and selected results. *Chem Rev* 1995;**95**:735–58.

Part III

Natural Photoprocesses Involving Inorganic Compounds

8

From Interstellar Space to Planetary Atmospheres

Time is in fact the hero of the plot. The time with which we have to deal is of the order of two billion years ... . Given so much time the 'impossible' becomes possible, the possible probable, the probable virtually certain. One has only to wait, time itself performs the miracles.

George Wald

Most of the (bio)chemical processes found within all the living organisms are well understood at the molecular level, whereas the origin of life remains one of the most vexing issues in chemistry, biology, and philosophy [1]. At different stages in the development of civilization and science the question has been answered differently. Nowadays scientists are almost sure that life on Earth emerged approximately 3.7–4 billion years ago, soon after the Earth's crust formed and heavy bombardment with meteorites declined. Numerous theories tackle the origin of life, but there is no direct evidence supporting any of them [2], and no fossil evidence of primordial life forms. It is, however, generally accepted that both metabolism and replication are factors that feature the most in differentiating animate from inanimate matter [3]. Another generally accepted hypothesis states that liquid water with dissolved simple inorganic and organic compounds is an indispensable prerequisite for life to appear and evolve on a planet. Solar light also seems to be an essential factor.

The origin of fluid water and all the necessary organic compounds, the building blocks on which primitive life was based, is, however, uncertain. The protostellar cloud and solar nebula definitely contained not only inorganic particles and dust, but also highly volatile materials such as hydrogen, water, methane, or ammonia. The just-formed protoplanets (including proto-Earth) could also have contained large quantities of this volatile material. Solar radiation and heavy bombardment with meteorites resulted, however, in rapid release of most of the volatile material to interplanetary space. Only the giant planets (Jupiter, Saturn, Uranus, and Neptune) could retain most of their light components. Therefore, it is supposed that all the water and most of the organic compounds present on Hadean Earth must have been of extraterrestrial origin [4]. Comets can be recognized as the main source of water and other volatile compounds of carbon, nitrogen, oxygen, sulphur, and phosphorus. There are no reliable data on the chemical composition of comets from the Oort cloud, but investigations on the Halley comet showed that it contains 23% rock particles, 41% water, and 36% volatile 'CHON' organic compounds [5].

Even now there is a large influx of extraterrestrial material in the form of meteorites, micrometeorites, and interplanetary dust. Independent investigations indicate that Earth captures 50–100 tons of interplanetary matter daily. In the past the influx of cosmic dust was approximately three orders of magnitude stronger. Eighty per cent of all the micrometeorites are carbonaceous particles, containing, on average, 2% of carbon in the form of various organic compounds. It is assumed that over 300 million years of bombardment this process supplied Earth with 10^{20} g organic carbon, which exceeds the amount of carbon engaged in the whole biomass on Earth (10^{18} g) [4].

The fundamental source of most organic compounds found in space is various photochemical reactions involving simple inorganic (H_2, O_2, CO, N_2, NH_3) and organic (CH_4) molecules. These reactions can proceed in the gas phase (within interstellar and interplanetary space, in planetary atmospheres), in icy phases (comets and ice particles), and on the surface of inorganic particles (cosmic dust and planetary surfaces). On planetary surfaces various forms of energy could have been relevant to the formation of complex organic molecules. Also in this case the most important seems to be light, although thermal energy could be relevant only at the geothermal vents and at the sites of meteorite impact. Although the role of the photochemical processes in the origin of life has not been unambiguously proved, the supply of complex organic molecules was essential for the emergence of living entities.

8.1 Homogeneous Systems: From Interstellar Space to Planetary Atmospheres and Primitive Soup Models

The main components of the Universe are hydrogen (74%) and helium (26%). In the interstellar gas these primitive elements contain small admixtures of heavier elements formed in nucleosynthesis processes in stars (from lithium to iron) and in supernova explosions (heavier than iron). The most abundant heavier elements are: oxygen (0.06%), carbon (0.03%), neon (0.02%), and nitrogen (0.008%). Approximately 5% of the mass of the galaxy is contained in molecular nebulae. Diffuse clouds contain up to 100 atoms and molecules per cubic centimetre, at the average temperature of 40–100 K. The highest density of matter is observed within dense molecular nebulae (10^4–10^8 atoms/cm^3); apart from gas they also contain dust particles (Figure 8.1) [6]. At very low temperatures (10–30 K) and upon constant irradiation with stellar radiation a large family of multiatomic molecules is formed (Table 8.1) [7].

Figure 8.1 Interstallar gas and dust. The pillars of the Eagle nebula as observed by Hubble Space Telescope using Wide Field and Planetary Camera 2. (Picture by courtesy of NASA and European Space Agency)

At extremely low pressures in molecular nebulae, intermolecular collisions are scarce and highly unsaturated molecules are stable. In hot cores of large molecular clouds associated with heavy protostars, complex (photo)chemical processes yield heavier alcohols, nitriles, and isonitriles. Gaseous outflows of heavy Wolf–Rayet stars contain high concentrations of helium, carbon, and oxygen whereas nitrogen and hydrogen are absent. In these gas phases fullerenes and graphitic particles are formed in large quantities, contributing to the complex chemistry of interstellar space [3].

A much more complex chemistry and much larger diversity of compounds are observed in planetary atmospheres, especially in the outer solar system [9–13]. The main components of giant planets' atmospheres are hydrogen and helium. Photochemical reactions of these two gases occur in thermospheres and contribute to the formation of ionospheres due to production of charged species [9]. The H_2 molecule undergoes photodissociation on irradiation with light of wavelengths shorter than 110.9 nm (equation 8.1), while more energetic radiation (80.4 nm) induces photoionization of hydrogen (equations 8.2–8.3). Photoionization of helium at $\lambda < 50.4$ nm also contributes to H_2^+ formation (equations 8.4–8.5) [9]:

$$H_2 \xrightarrow{h\nu} H + H \tag{8.1}$$

$$H_2 \xrightarrow{h\nu} H_2^+ + e^- \tag{8.2}$$

Table 8.1 *The most abundant interstellar and circumstellar ions and molecules [7]*

Two atoms	Three atoms	Four atoms	Five atoms	Six atoms	Seven atoms	Eight or more atoms
H_2	C_2H	C_2H_2	C_4H	C_2H_4	C_6H	CH_3COOH
C_2	CH_2	C_3H	C_3H_2	C_4H_2	C_5HN	$HCOOCH_3$
CH	HCN	NH_3	HCOOH	CH_3OH	CH_2CHCN	CH_3C_3N
CN	HNC	HNCO	CH_2CO	CH_3CN	CH_3C_2H	C_7H
CO	HCO	$HOCO^+$	HC_3N	CH_3NC	CH_3CHO	CH_3C_4H
CS	HCO^+	$HCNH^+$	CH_2CN	CH_3SH	CH_3NH_2	CH_3OCH_3
OH	HOC^+	HNCS	NH_2CN	NH_2CHO	$(CH_2)_2O$	CH_3CH_2CN
NH	N_2H^+	C_3N	CH_2NH	C_3H_2O		C_7HN
NO	NH_2	C_3O	CH_4	C_5H		CH_3C_4CN
NS	H_2O	H_2CS	SiH_4	C_3HNH^+		C_9HN
SiC	HCS^+	C_3S	C_4Si	C_5N		C_6H_2
SiO	H_2S	HCCN	C_5			$C_{11}HN$
SiS	OCS	H_3O^+	HCCNC			
SiN	N_2O	H_2CN	HNCCC			
SO	SO_2	H_2CO	HCO_2H^+			
HCl	SiC_2	SiC_3				
CP	C_2S					
SO^+	C_2O					
NaCl	C_3					
AlCl	MgNC					
KCl	NaCN					
AlF	HNO					
PN	H_3^+					
CO^+						
SiH						
SiF						

$$H_2 \xrightarrow{h\nu} H^+ + H + e^- \tag{8.3}$$

$$He \xrightarrow{h\nu} He^+ + e^- \tag{8.4}$$

$$H_2 + He^+ \rightarrow H_2^+ + He \tag{8.5}$$

Subsequent thermal reactions generate higher quantities of atomic hydrogen (equations 8.6–8.7), the recombination of which at thermospheric pressure is extremely improbable [9]:

$$H_2^+ + H_2 \rightarrow H_3^+ + H \tag{8.6}$$

$$H_3^+ + e^- \rightarrow 3H \tag{8.7}$$

These reactions supply atomic hydrogen to lower parts of atmospheres and thus initiate numerous reactions of carbon, oxygen, and nitrogen species. The most important are the reactions with methane (equations 8.8–8.10), which finally yield ethane and thus contribute to formation of higher hydrocarbons (equation 8.11) [9]:

$$H^+ + CH_4 \rightarrow CH_3^+ + H_2 \tag{8.8}$$

$$H^+ + CH_4 \rightarrow CH_4^+ + H \tag{8.9}$$

$$CH_4^+ + H_2 \rightarrow CH_5^+ + H \tag{8.10}$$

$$CH_3^+ + CH_4 \rightarrow C_2H_5^+ + H_2 \tag{8.11}$$

Methane itself is also photochemically active below 145 nm [9]:

$$CH_4 \xrightarrow{h\nu} CH_3 + H \tag{8.12}$$

$$CH_4 \xrightarrow{h\nu} {}^{1\text{ or }3}CH_2 + 2H \tag{8.13}$$

$$CH_4 \xrightarrow{h\nu} {}^{1}CH_2 + H_2 \tag{8.14}$$

$$CH_4 \xrightarrow{h\nu} CH + H + H_2 \tag{8.15}$$

Subsequent thermal and photochemical reactions lead to formation of ethane (equation 8.16), ethene (equation 8.17) and ethine (equation 8.18) [9]:

$$2CH_3 \rightarrow C_2H_6 \tag{8.16}$$

$$CH + CH_4 \rightarrow C_2H_4 + H \tag{8.17}$$

$$C_2H_4 \xrightarrow{h\nu} C_2H_2 + 2H \tag{8.18}$$

Further photochemical reaction involving C_1 and C_2 species yield higher monomeric and polymeric hydrocarbons [11, 12].

Although photochemical reactions of hydrogen and hydrocarbons proceed easily and with high quantum yields, nitrogen photochemistry is not that straightforward. The triple bond within the N_2 molecule is extremely difficult to break ($E \geq 9.7\,\text{eV}$). Furthermore, there are no optically allowed excitation paths into repulsive electronic excited states, and dissociation can occur only via indirect paths. Solar radiation below 100 nm can excite predissociating electronic states and constitutes a minor source of N atoms. Dissociative ionization of N_2 by either electron impact or solar extreme UV (10–121 nm) radiation produces one N atom and one N^+ ion [9]:

$$N_2 \xrightarrow{h\nu,\, e^-} N + N^+ \tag{8.19}$$

Thus, generated nitrogen atoms can react with methane yielding the NH radicals, whereas reactions with carbene (CH_2) and methyl radicals yield hydrogen cyanide.

Photodissociation of HCN yields CN radicals, which can further react with methane (equations 8.20–8.21). Reaction of CN radicals with ethane and ethine yields corresponding unsaturated nitriles (equations 8.22–8.23) [14].

$$HCN \xrightarrow{h\nu} H + CN \tag{8.20}$$

$$CN + CH_3 \rightarrow HCN + CH_2 \tag{8.21}$$

$$CN + C_2H_2 \rightarrow HCCCN + H \tag{8.22}$$

$$CN + C_2H_4 \rightarrow CH_2CHCN + H \tag{8.23}$$

$$CN + HCN \rightarrow (CN)_2 + H \tag{8.24}$$

It is also suggested that these reactions yield acetonitrile (CH_3CN), but the atmospheres of Titan (the largest moon of Saturn) and giant planets contain only traces of this compound. Its formation may require more reducing conditions, ie higher hydrogen content.

Contrary to molecular nitrogen, ammonia undergoes photochemical reactions very easily, and the same occurs with phosphine, which is present in the atmosphere of Saturn. Ammonia is easily dissociated upon irradiation with UV light below 230 nm (equation 8.25). Subsequent thermal and photochemical reactions of the NH_2 radical are considered as the main source of nitrogen in primordial atmospheres (equations 8.26–8.28) [9].

$$NH_3 \xrightarrow{h\nu} NH_2 + H \tag{8.25}$$

$$2NH_2 \rightarrow N_2H_4 \tag{8.26}$$

$$N_2H_4 \xrightarrow{h\nu} N_2H_3 + H \tag{8.27}$$

$$2N_2H_3 \rightarrow N_2H_4 + N_2 + H_2 \tag{8.28}$$

The main source of oxygen in prebiotic planetary atmospheres is ablation of cometary material. The primary source of various oxygen species in all atmospheric photochemical reactions is water. Water molecules are easily photolyzed by the solar Lyman α line (121.6 nm) and on irradiation within the water UV absorption bands (140–190 nm). The main products are hydroxyl radicals, oxygen atoms in their ground (^3P) and excited (^1D) states, and hydrogen atoms [14].

$$H_2O \xrightarrow{h\nu} H + OH \tag{8.29}$$

$$H_2O \xrightarrow{h\nu} 2\,H + O \tag{8.30}$$

$$H_2O \xrightarrow{h\nu} H_2 + O^* \tag{8.31}$$

Both OH radicals and oxygen atoms are responsible for formation of carbon monoxide equations (8.32–8.34), carbon dioxide (see equation 8.35), and formaldehyde (see equations 8.37–8.38):

$$O + {}^3CH_2 \rightarrow CO + 2H \tag{8.32}$$

$$OH + {}^3CH_2 \rightarrow CO + H_2 + H \tag{8.33}$$

$$OH + CH_3 \rightarrow CO + 2H_2 \tag{8.34}$$

Further reaction of carbon monoxide with hydroxyl radical yields carbon dioxide (equation 8.35), whereas reaction of carbon monoxide with carbine yields ketene (equation 8.36) [14]. Atomic hydrogen, in turn, converts carbon monoxide to formaldehyde (equations 8.37–8.38), which in principle may be a substrate for prebiotic synthesis of carbohydrates:

$$OH + CO \rightarrow CO_2 + H \tag{8.35}$$

$$CO + {}^3CH_2 \rightarrow CH_2CO \tag{8.36}$$

$$CO + H \rightarrow HCO \tag{8.37}$$

$$2HCO \rightarrow CH_2O + CO \tag{8.38}$$

Recombination of various radical species formed during photolysis of water and methane yield acetaldehyde (equation 8.39), methanol (equation 8.40), and other organic species [10, 13]:

$$CH_3 + HCO \rightarrow CH_3CHO \tag{8.39}$$

$$CH_3 + OH \rightarrow CH_3OH \tag{8.40}$$

At low temperatures these products tend to condense, especially in the presence of dust particles, forming an organic haze, which is observed in the upper parts of Titan and the giant planets (Figure 8.2). These dust particles result from ablation of mineral meteorite material as well as condensation of hydrocarbons and hydrogen cyanide polymers (so-called tholins, after the Greek *θυλερός* = muddy or dirty) [15]. After condensation both solid particles and droplets of organic compounds are transported towards the surfaces of planets. In the case of Earth this organic supply, together with ammonia, carbon dioxide, and water, may have formed a liquid phase usually called *primitive soup* – a complex aqueous solution, which, according to some theories, was the birthplace of life [2, 16–18]. According to the trace analysis of gas inclusions in the oldest rock minerals, the primitive atmosphere of Earth should resemble the present atmosphere of Titan – it might have contained mainly nitrogen with small admixtures of ammonia, methane, carbon monoxide, and water.

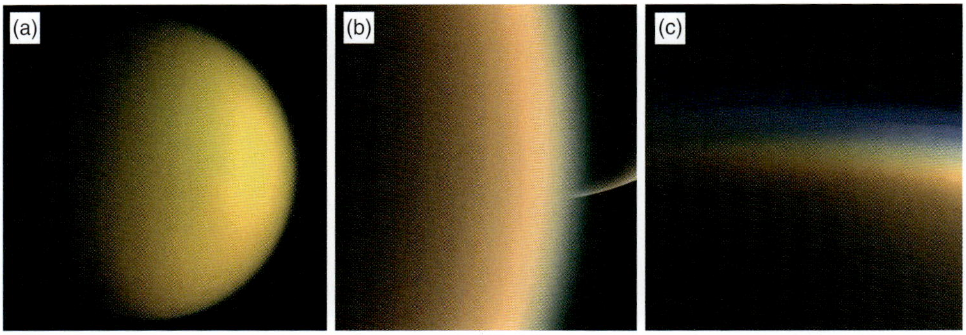

Figure 8.2 The natural colour images of Titan taken with the Cassini spacecraft: (a) general view on 16 April 2005, at average distance of 170,000 km, (b) the photochemical smog around Titan seen from 26,000 km (south pole of Saturn can be seen at the background), and (c) the Titan's upper atmosphere as seen from the distance of 9,500 km. The haze preferentially scatters blue and UV wavelengths of light, making its complex layered structure more easily visible at the shorter wavelengths used in this image. (Pictures by courtesy of NASA)

Therefore, gas phase photochemistry of inorganic and simple organic gases may have contributed to the prebiotic synthesis of organic materials, which subsequently may have been building blocks for animate matter [19].

An alternative theory was presented by Miller and Urey in 1953. They were successful with electric discharge-induced synthesis of a complex mixture of amino acids (containing glycine, aspartic acid, alanine, and aminobutyric acid) from ammonia, hydrogen, water, and methane (Figure 8.3) [20–22]. These conditions simulated the hypothetical primitive Earth's atmosphere. As solar radiation is a more reliable energy source than electric discharges in the atmosphere, the hypothesis of photochemical synthesis of prebiotic organic molecules seems to be fully justified. This assumption was recently tested experimentally by Serpone et al. [23]. Irradiation of aqueous solutions of methanol and ammonia (a simplified model of primitive soup) resulted in formation of the whole family of complex organic molecules. At ambient temperature two products were formed: hexamethylenetetramine and 1-methylisourea [23]. When irradiation was carried out at 12 K the main products were CO, CO_2, CH_4, HCHO, and the HCO radical, all of which vanished on warming the ice to 200 K, yielding formamide ($HCONH_2$), acetamide (CH_3CONH_2), some amines, and ethanol. Further warming to near-ambient temperatures (300 K) produced hexamethylenetetramine, ethanol, $HCONH_2$, and CH_3CONH_2, as well as polyoxomethylenes, $(-CH_2-O-)_n$, resulting from the polymerization of HCHO [23]. This gives clear proof that prebiotic photosynthesis of various organic precursors was possible under the hypothetical primitive Earth conditions.

An aqueous solution containing ammonia, hydrogen cyanide, and simple organic compounds such as formaldehyde and cyanoethine, in contact with rocks, constituted substrates and the environment for the prebiotic synthesis of more complex organic molecules: purine and pyrimidine bases, amino acids, and sugars – the building blocks of all the organisms.

Figure 8.3 *A photograph of (a) original set-up for the Miller–Urey experiment and (b) a reproduction of this experiment at NASA laboratories. (Partially reproduced from Lazcano and Bada [22] with kind permission from Springer Science + Business Media. © Kluwer Academic Publishers 2003. Picture (b) by courtesy of NASA)*

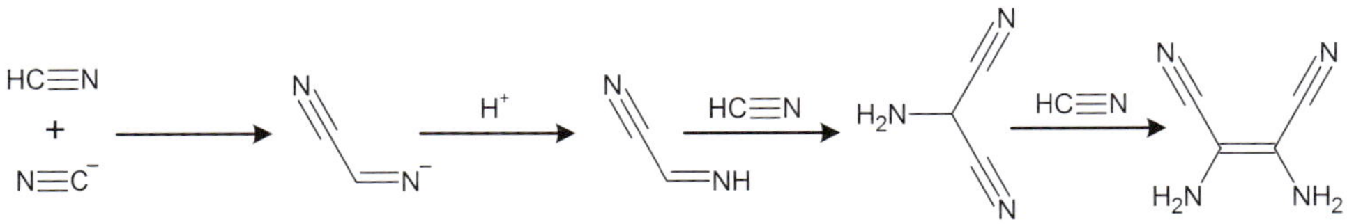

Figure 8.4 *Mechanism of tetramerization of hydrogen cyanide leading to the formation of a stable tetramer: diaminomaleonitrile. (Adapted from Orgel and Lohrman [24])*

The synthesis of various purine bases from hydrogen cyanide as the only carbon-containing substrate was documented by Orgel and coworkers [24]. The only stable precursor found in polymerization of hydrogen cyanide in concentrated (1–11 mol/l) aqueous solutions is its tetramer, diaminomaleonitrile. It is easily formed in the reaction of hydrogen cyanide with ammonium cyanide (Figure 8.4). At the same time, in concentrated solutions of ammonium cyanide in the presence of ammonia, formamidine is formed:

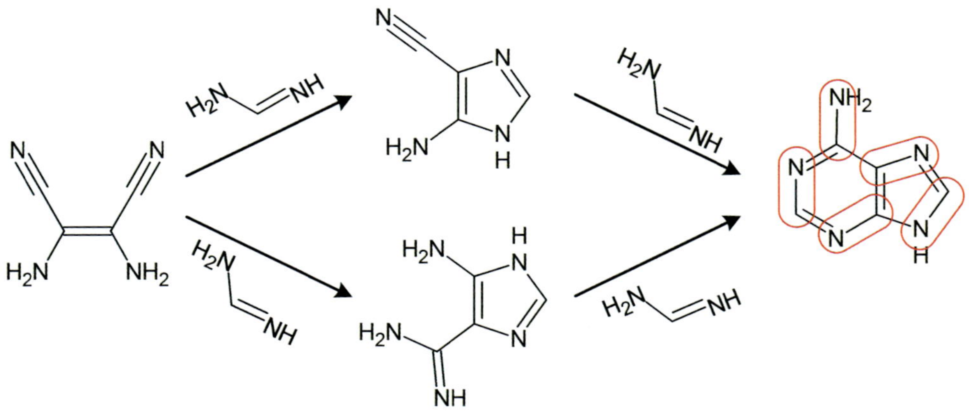

Figure 8.5 Thermal synthesis of adenine from hydrogen cyanide and ammonia. Red ovals indicate individual CN moieties in adenine molecule. (Adapted from Orgel and Lohrman [24])

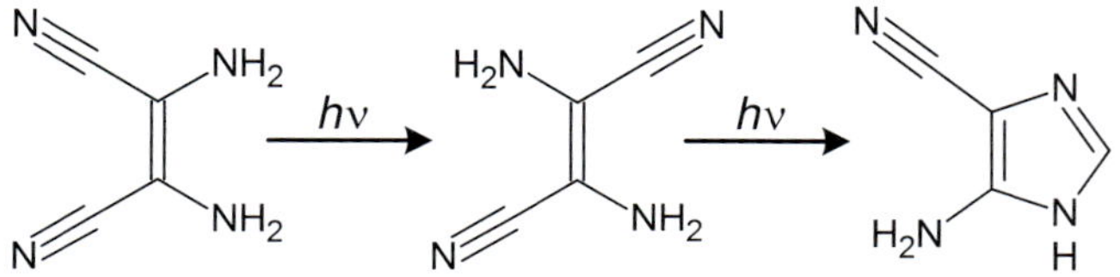

Figure 8.6 Photoisomerization and photocyclization of diaminomaleonitrile. (Adapted from Orgel and Lohrman [24])

$$HCN + NH_3 \rightarrow HN{=}CH{-}NH_2 \tag{8.41}$$

Formamidine can in turn react with diaminomaleonitrile to yield adenine, which can be considered as hydrogen cyanide pentamer, via two different intermediates (Figure 8.5). This reaction can proceed only at high concentrations of ammonium cyanide, because in dilute solutions the equilibrium concentration of formamidine is very low. This problem can be solved by involvement of photochemical isomerization of HCN tetramer, which in the absence of oxygen proceeds almost quantitatively on irradiation with 350 nm light (Figure 8.6) [25]. This reaction can proceed even at very diluted NH_4CN solutions at low temperatures (–10 to –22 °C): freezing results in separation of a highly concentrated liquid phase from ice. Thus, the 4-aminoimidazole-5-carbonitrile formed is a suitable precursor for synthesis of the whole family of purine bases (Figure 8.7). All the carbon atoms of the purine ring are, thus, provided by HCN molecules via a complex step-by-step condensation process. In particular, oligomers of HCN, such as the HCN trimer, aminomaleonitrile, and the HCN tetramer, diaminomaleonitrile, were found to be intermediates in this transformation [26]. The same group of compounds can be obtained on acidic and alkaline hydrolysis of the HCN polymer produced in a frozen ammonium

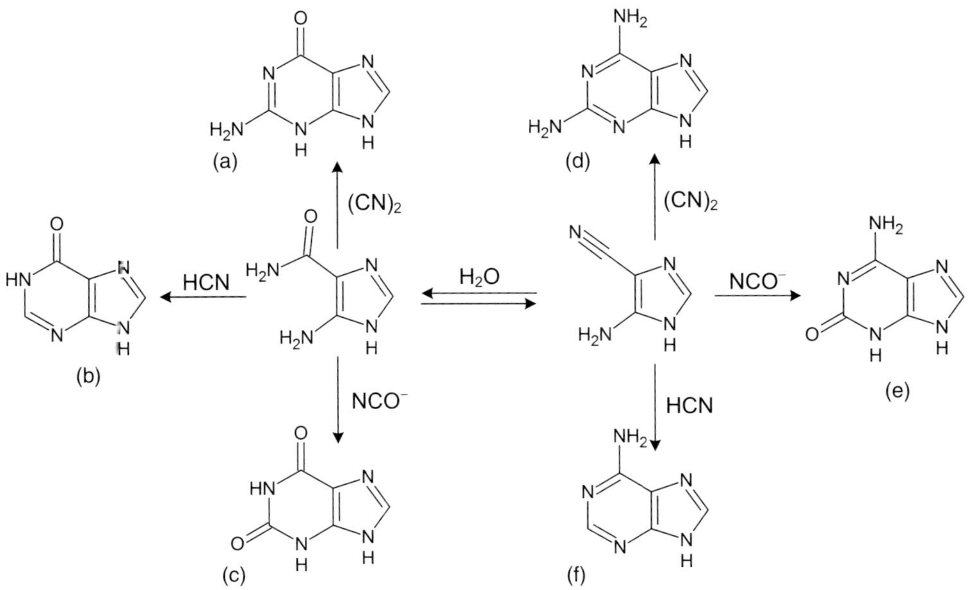

Figure 8.7 Formation pathways of various purine derivatives: (a) guanine, (b) hypoxanthine, (c) xanthine, (d) diaminopurine, (e) isoguanine, and (f) adenine

cyanide solution at –78 °C within 27 years. Purine derivatives can also be obtained from the products of HCN hydrolysis: formamide and ammonium formate [26].

More complex, but still feasible, is the synthesis of pyrimidine bases from simple prebiotic substrates, although the reported yields of these reactions are relatively low. In this context, two main prebiotic precursors have been identified: cyanoethine and a primary product of its hydrolysis, cyanoacetaldehyde. These compounds contain a preformed C–C bond which is incorporated in the C5–C6 position of the pyrimidine ring. In 1968 Ferris and co-workers reported that the reaction of cyanoethine with cyanate at 30 °C yields cytosine and, after its hydrolysis, uracil in acceptable yield [27]. *trans*-Cyanovinylurea was recovered as a key intermediate for this transformation. However, this reaction requires relatively high concentrations of cyanate (>0.1 mol/l), unlikely to occur in aqueous media due to its rapid degradation to carbon dioxide and ammonia. Cyanoethine also reacts with cyanate and yields cytosine and uracil at elevated temperatures. In this reaction urea or guanidine (also considered as prebiotic organic compounds) can easily replace cyanate (Figure 8.8) [26].

Sugars play a dual role in all living organisms: they are the principal source of energy for cellular processes and are indispensable components of DNA and RNA information carriers. The possibility of prebiotic synthesis of various sugars was demonstrated by Butlerow in 1861 [28], who showed that a simple prebiotic molecule – formaldehyde – can be used for the synthesis of a mixture of various sugars in a $Ca(OH)_2$-catalyzed series of aldol condensations. The reaction starts

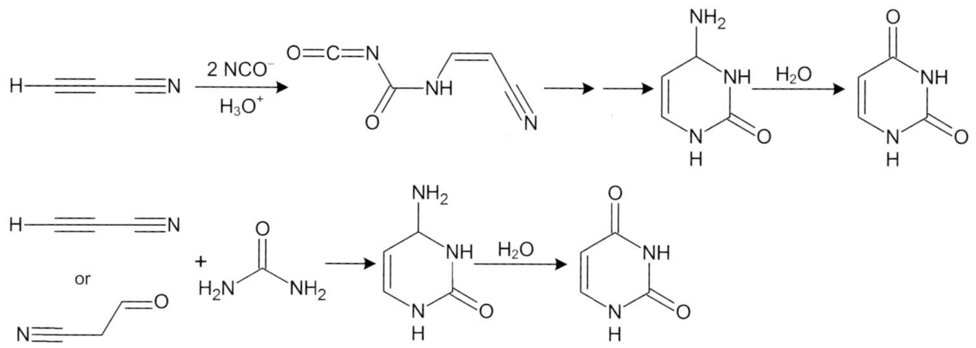

Figure 8.8 Synthesis of cytosine and uracil from simple organic molecules

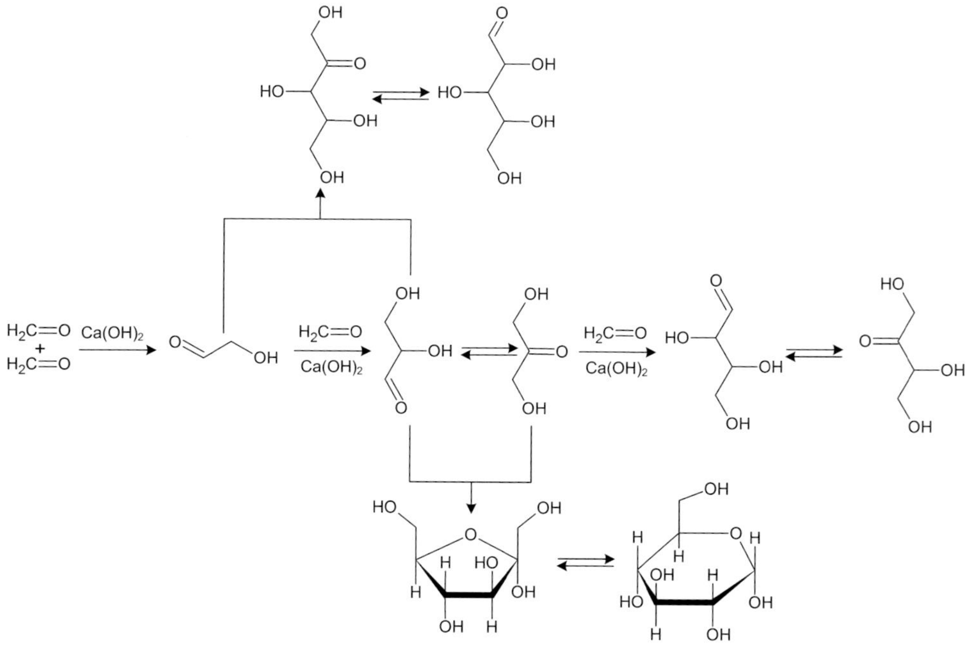

Figure 8.9 Aldol condensation of formaldehyde as a source of carbohydrates in the prebiotic primitive soup

with condensation of two formaldehyde molecules, yielding glycoaldehyde, which may then undergo a series of subsequent aldol condensations (Figure 8.9). In this way carbohydrates up to seven carbon atoms long can be easily prepared.

The principal building blocks of all the living entities, amino acids, can also be formed in the primitive soup model [29]. As HCN and aldehydes were produced directly from methane and ammonia, both photochemically and during electric discharge, the Strecker reaction was proposed very early as a likely pathway for the

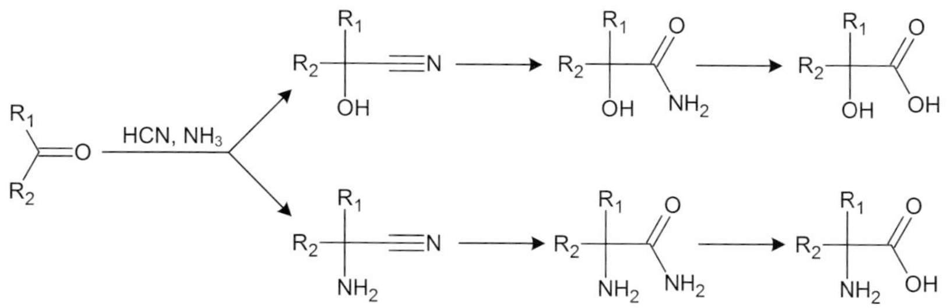

Figure 8.10 *Reaction pathways of amino acid and hydroxyacid formation from carbonyl compounds, hydrogen cyanide, and ammonia*

prebiotic synthesis of amino acids [29]. This reaction, discovered in 1850, is the most anciently known abiotic synthesis of α-amino acids; it originally consisted of the formation of an α-aminonitrile from a carbonyl compound (either aldehyde or ketone), ammonia, and hydrogen cyanide in a moderately alkaline aqueous solution followed by aminonitrile hydrolysis in a strong acid. The mixture of a carbonyl compound, ammonia, and hydrogen cyanide equilibrates into aminonitrile and cyanohydrine, the product ratio being pH dependent (Figure 8.10) [29]. Hydrolysis of aminonitriles yields amino acid amides and finally amino acids, whereas cyanohydrins yield α-hydroxyacids.

8.2 Heterogeneous Photochemistry in Ice Phases

The collision of two atoms in vacuum very rarely results in the formation of a chemical bond because they carry a great deal of kinetic energy and there is no third body to absorb any excess energy. Therefore the only way to form chemical bonds must proceed on the surface of dust particles. These grains must be considered the thermodynamic entities and can convert collision energy into vibrational energy of multi-atom body [30]. The first step of any chemical reaction in space must involve binding of energetic atoms on to the surface of interstellar dust particles.

Gaseous components of molecular clouds tend to condensate on the surface of solid dust particles at low temperatures. These dust particles are produced in the atmospheres of cool giant stars by condensation of refractory materials such as olivine and other silicates, whereas condensation of supernova ejecta results in metallic particles.

Refractory dust particles not only serve as condensation nuclei for ices (see above), but are also necessary for energy dissipation during gas phase reaction between energetic atoms and molecules. Furthermore, solid inorganic particles (mostly oxides and sulphides) may play a role of photocatalysts in transformations of organic materials in both space and the primitive soup [5, 31].

The ice mantles formed on the surface of dust particles can be divided into two main categories according to the chemical composition, which in turn result from

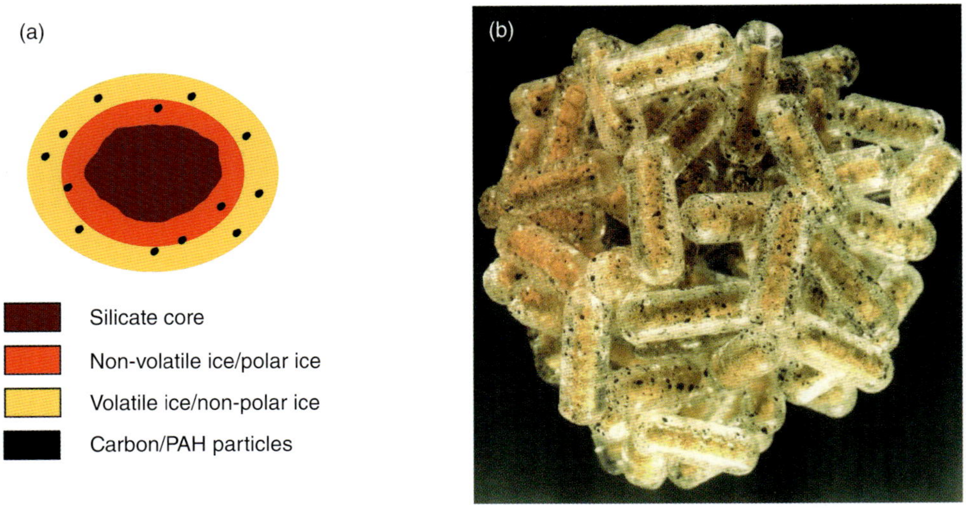

Figure 8.11 *(a) Possible structure of single ice particle and (b) a model of a piece of a comet consisting of an aggregate of ice particles [6, 33, 34]. (Reprinted from Greenberg [34], with permission from Elsevier)*

the condensation temperature. Hydrogen-rich ices (so-called polar ices) are dominated by water molecules, with significant admixture of CO, CO_2, CH_4, NH_3, and CH_3OH, and trace amounts of formaldehyde and formic acid (Figure 8.11). This material is formed in hydrogen-rich areas of molecular nebulae. Polar ice is volatile at temperatures higher than 90 K. Ice formed in hydrogen-poor areas is made from more volatile gases (CO, O_2, and N_2), and tend to evaporate at termperatures >20 K. Therefore polar ice particles can survive for a long time in close proximity of stars [6].

Experiments show that energetic processing of carbon-containing ices, by UV, energetic ions, or even, in some cases, simple warm-up leads to the formation of complex organic species. Sources of energy such as energetic ions and stellar photons could be ubiquitous in the dense environments near young stellar objects or on the outskirts of dense nebulae [32]. Therefore interstellar ice particles are possibly the photochemical reaction pots where complex organic molecules are produced upon stellar illumination. Model studies indicate the formation of complex mixtures of organic compounds during cryogenic photolysis, followed by thermal processing of water ice containing methanol, carbon monoxide, and ammonia (Table 8.2).

Observations of dense molecular nebulae in star-forming regions indicate the presence of complex organic molecules such as acetic acid, glycine, propionitrile, and acetone, together with water, methane, ammonia, and carbon monoxide. The suggested source of these molecules is photochemical processes occurring in the icy mantles of dust particles induced by photons of energy in the range 4–13 eV.

The primary photochemical steps should include dissociation of various hydrogen compounds with formation of hydrogen atoms and the following radicals: OH,

Table 8.2 *Organic molecules formed during UV photolysis followed by thermal processing of water-based ices [32]*

Ice composition	Products
$H_2O + CO + NH_3$	$HOCH_2CH_2OH$
	$HOCH_2COOH$
	$HOCH_2CONH_2$
	$(CH_2)_6N_4$
	H_2NCONH_2
	$HOCH_2CH(OH)CH_2OH$
	$HOCH_2CH(OH)COOH$
	$HOCH_2CH(OH)CONH_2$
$H_2O + CH_3OH + NH_3$	$(CH_2)_6N_4$
	CH_3CH_2OH
	CH_3CONH_2
	$HCONH_2$

CH_3, NH_2 (see equations 8.12, 8.25, and 8.29). Once formed, the radicals remain trapped in the ice phase. Heat produced on collisions causes the radical to become mobile and react with other radicals and molecules within the ice particle, thus producing complex organic compounds:

$$CO + OH \rightarrow COOH \tag{8.42}$$

$$CH_3 + COOH \rightarrow CH_3COOH \tag{8.43}$$

$$NH_2 + CH_3COOH \rightarrow NH_2CH_2COOH + H \tag{8.44}$$

In the case of high radical concentration within an ice particle, heating may lead to liberation of large quantities of chemical energy, which in turn leads to particle explosion. In this way large quantities of organic material can be released to ambient space [33]. Cosmic dust particles may undergo many cycles of condensation, photochemical processing, heating, and collisions, which may result in the rather complex structure of the icy mantle deposited onto the silicate core [34].

Although information on the chemical composition and reactivity of interstellar ices can be obtained only from remote observations and laboratory simulations, cometary ices and dusts are subject to direct studies, eg within Vega and Giotto (comet Halley), Stardust (comet Wild 2) and Rosetta (comet Churyumov–Gerasimenko) missions (Figure 8.12).

The Stardust mission succeeded in bringing solid particles from the Wild 2 comet to the Earth. Comets are the intriguing celestial bodies that are believed the last witnesses of the formation of the solar system. They are considered to be aggregates of primordial interstellar dust particles, the last reminders of protosolar nebula. Even simple spectroscopic observation indicates that comets are rich in simple organic species, which may undergo further photochemical transformations during their life within the solar system [35, 36]. Hundreds of organic molecules

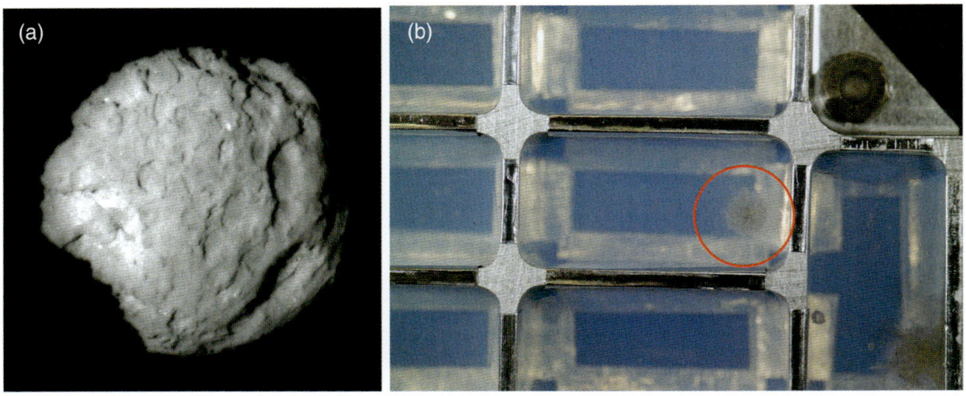

Figure 8.12 *(a) Photograph of Wild 2 comet taken by the Stardust spaceship and (b) a close-up view of a cometary particle captured into aerogel and brought back to Earth in the Stardust Sample Return Canister. (Pictures by courtesy of NASA)*

have been detected in cometary ice including alcohols, aldehydes, esters, polyaromatic hydrocarbons and others [35]. High concentration of simple organic compounds and proximity to the Sun (ie much higher photon flux) make comets much better 'synthetic laboratories' compared with ice mantles of interstellar dust particles. Furthermore, comets are considered the main source of water in the solar system. Therefore one can assume that prebiotic Earth was 'fertilized' with organic materials, which subsequently became the building blocks of animate matter.

References

1. Hiscox JA. An overview of the origin of life: the case for biological prospecting on Mars. *Earth Moon Planets* 2001;**87**:191–212.
2. Orgel LE. The origin of life – a review of facts and speculations. *Trends Biochem Sci* 1998;**23**:491–5.
3. Lifson S. On the crucial stages in the origin of animate matter. *J Mol Evol* 1997;**44**:1–8.
4. Brack A. Life in the solar system. *Adv Space Res* 1999;**24**:417–33.
5. McClendon JH. The origin of life. *Earth Sci Rev* 1999;**47**:71–93.
6. Ehrenfreund P, Schutte WA. ISO observations of interstellar ices: implications for the pristinity of comets. *Adv Space Res* 2000;**25**:2177–88.
7. Irvine W. The composition of interstellar molecular clouds. *Space Sci Rev* 1999;**90**:203–18.
8. Tielens AGGM, Charnley SB. Circumstellar and interstellar synthesis of organic molecules. *Origins Life Evol Biosphere* 1997;**27**:23–51.
9. Strobel D. Photochemistry in outer solar system atmospheres. *Space Sci Rev* 2005;**116**:155–70.
10. Majeed T, Waite JH Jr, Bougher SW, et al. The ionospheres-thermospheres of the giant planets. *Adv Space Res* 2004;**33**:197–211.
11. Gladstone GR, Allen M, Yung YL. Hydrocarbon photochemistry in the upper atmosphere of Jupiter. *Icarus* 1996;**119**:1–52.

12. Moses JI, Bézard B, Lelouch E, Gladstone GR, Feuchtgruber H, Allen M. Photochemistry of Saturn's atmosphere. 1. Hydrocarbon chemistry and comparisons with ISO observations. *Icarus* 2000;**143**:244–98.
13. Moses JI, Lelouch E, Bézard B, Gladstone GR, Feuchtgruber H, Allen M. Photochemistry of Saturn's atmosphere. 2. Effects and influx of external oxygen. *Icarus* 2000;**145**:166–202.
14. Hébrard E, Dobrijevic M, Bénilan Y, Raulin F. Photochemical kinetics uncertainties in modeling Titan's atmosphere: a review. *J Photochem Photobiol C: Photochem Rev* 2006;**7**:211–30.
15. Sagan C, Khare BN. Tholins: organic chemistry of interstellar grains and gas. *Nature* 1979;**277**:102–7.
16. Edwards MR. From a soup or a seed? Pyritic metabolic complexes in the origin of life. *Trends Ecol Evol* 1998;**13**:178–81.
17. Sutherland JD, Whitfield JN. Prebiotic chemistry: a bioorganic perspective. *Tetrahedron* 1997;**53**:11493–527.
18. Maurel M-C, Décout J-L. Origins of life: molecular foundations and new approaches. *Tetrahedron* 1999;**55**:3141–82.
19. Trainer MG, Pavlov AA, DeWitt HL, et al. Organic haze on Titan and the early Earth. *Proc Natl Acad Sci USA* 2006;**103**:18035–42.
20. Miller SL, Urey HC. Organic compound synthesis on the primitive Earth. *Science* 1959;**130**:245–51.
21. Miller SL. A production of amino acids under possible primitive Earth conditions. *Science* 1953;**117**:528–9.
22. Lazcano A, Bada JL. The 1953 Stalney L. Miller experiment: fifty years of prebiotic organic chemistry. *Origins Life Evol Biosphere* 2003;**33**:235–42.
23. Dondi D, Merli D, Pretali L, Fagnoni M, Albini A, Serpone N. Prebiotic chemistry: chemical evolution of organics on the primitive Earth under simulated prebiotic conditions. *Photochem Photobiol Sci* 2007;**6**:1210–17.
24. Orgel LE, Lohrman R. Prebiotic chemistry and nucleic acid replication. *Acc Chem Res* 1974;**7**:368–77.
25. Ferris JP, Orgel LE. An unusual photochemical rearrangement in the synthesis of adenine from hydrogen cyanide. *J Am Chem Soc* 1966;**88**:1074.
26. Saladino R, Crestini C, Costanzo G, DiMauro E. On the prebiotic synthesis of nucleobases, nucleotides, oligonucleotides, pre-RNA and pre-DNA molecules. *Top Curr Chem* 2005;**259**:29–68.
27. Ferris JP, Sanchez R, Orgel LE. Studies in prebiotic synthesis III. Synthesis of pyrimidines from cyanoacetylene and cyanate. *J Mol Biol* 1968;**33**:693–704.
28. Butlerow A. Bildung einer zuckerartigen Substanz durch Synthese. *Liebigs Ann Chem* 1861;**120**:295–8.
29. Pascal R, Boiteau L, Commeyras A. From the prebiotic synthesis of α-amino acids towards a primitive translation apparatus for the synthesis of peptides. *Top Curr Chem* 2005;**259**:69–122.
30. Williams DA, Herbst E. It's a dusty Universe: surface science in space. *Surf Sci* 2002;**500**:823–37.
31. Rajski SR, Barton JK. How different DNA-binding proteins affect long-range oxidative damage to DNA? *Biochemistry* 2001;**40**:5556–64.
32. Schutte WA. Production of organic molecules in interstellar ices. *Adv Space Res* 2002;**3**:1409–17.
33. Sorrell WH. Interstellar grains as aminoacid factories and the origin of life. *Astrophys Space Sci* 1997;**253**:27–41.
34. Greenberg JM. Cosmic dust and our origins. *Surf Sci* 2002;**500**:793–822.
35. Cottin H, Gazeau MC, Raulin F. Cometary organic chemistry: a review from observations, numerical and experimental simulations. *Planet Space Sci* 1999;**47**:1141–62.
36. Donn B. Comets: chemistry and chemical evolution. *J Mol Evol* 1982;**18**:157–60.

9

Solar Radiation and Terrestrial Environment

Nichts Süßres gibt es, als der Sonne Licht zu schaun.
[Nothing is sweeter than watching the sunlight.]
Friedrich Schiller

9.1 Solar Radiation

Sunlight supplies energy to the whole terrestrial environment: atmosphere, hydrosphere, lithosphere, and biosphere. The sun can be considered as a spherical light source of diameter 1.4×10^6 km, located 1.8×10^8 km from the Earth's surface, which emits radiation of about 2.82×10^{23} kJ s^{-1}. The total energy flux incident at the top of the Earth's atmosphere is 1.8×10^{14} kJ s^{-1}; the sea level receives only about 45% of this energy [1]. Nevertheless, about 1370 J falls on the average square metre of land and sea area within 1 s (so-called solar constant).

Extraterrestrial global irradiance, ie the solar irradiance just above the Earth's atmosphere, is called AM(0) sunlight (AM = air mass). The solar irradiance traversing the atmosphere (terrestrial global irradiance) is called AM(1) sunlight, when the sun is perpendicular to the Earth's surface (solar zenith angle, $Z = 0°$) [2]. AM(1) sunlight is much weaker than AM(0), because passing over the atmosphere solar radiation is absorbed and scattered by atmospheric gases and particles. The effect depends strongly on light wavelengths and solar zenith angles, which are shown in Figure 9.1 [3, 4].

Although the solar spectrum is very broad, extending from cosmic rays to radio waves (see Figure 2.1), the photochemical reactions are driven only by visible and ultraviolet (UV) radiation (actinic light – Figure 9.2).

The most energetic radiation (far UV, X-rays, and γ-radiation) induces ionization:

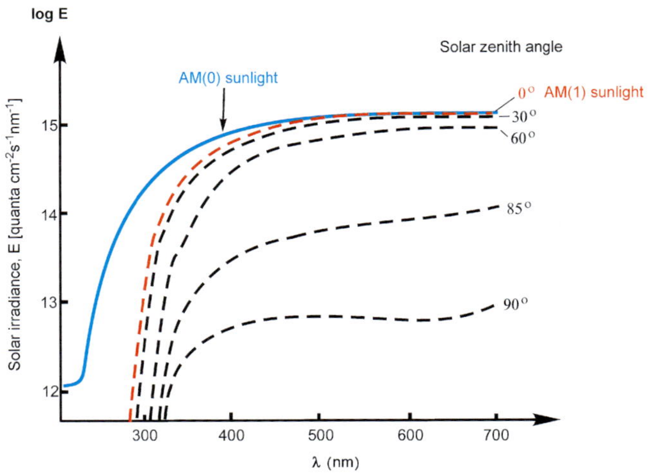

Figure 9.1 *Solar irradiance reaching the top of the Earth's atmosphere (AM(0) sunlight) and the sea level for different solar angles of 0°, 30°, 60°, 85°, and 90°*

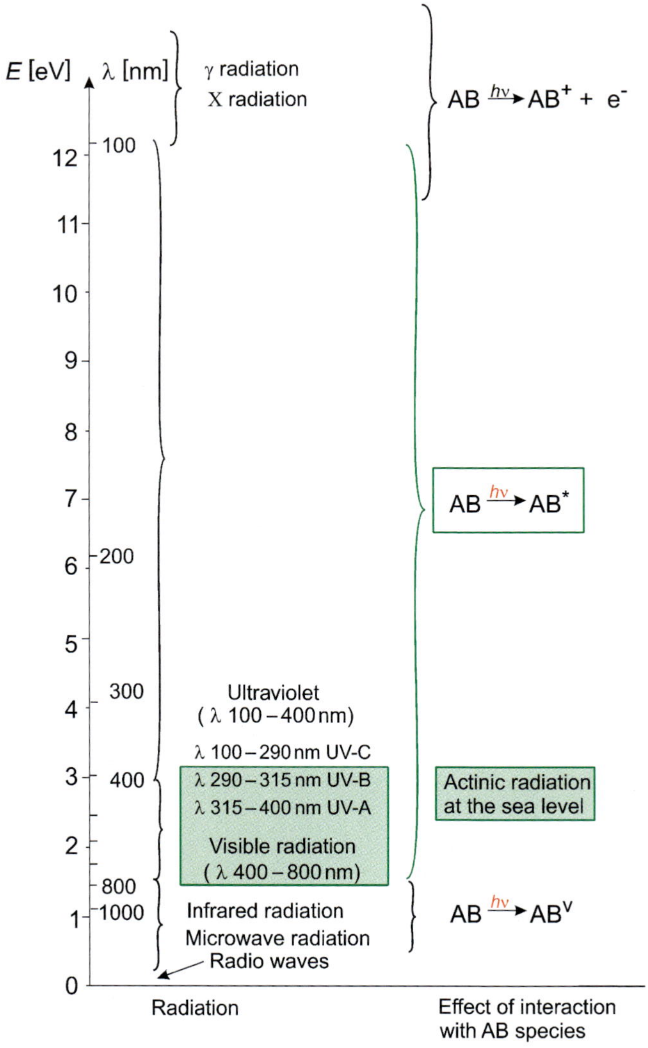

Figure 9.2 *Solar radiation and the effects of its interaction with matter*

$$AB \xrightarrow{h\nu} AB^{+} + e^{-} \tag{9.1}$$

whereas infrared (IR) radiation of wavelengths longer than 800 nm ($h\nu < 150$ kJ/einstein) excites the AB molecules mostly to their higher vibrational states:

$$AB \xrightarrow{h\nu<150\,\text{kJ/einstein}} AB^{v} \tag{9.2}$$

Photochemically active (actinic) solar radiation includes visible light (formally limited to 400–800 nm, 300–150 kJ/einstein) and UV radiation, the contribution of which depends on the altitude: at sea level the high-energy limit is 414 kJ/einstein ($\lambda = 290$ nm); the higher the altitude above sea level, the larger the participation of more energetic UV light.

UV radiation is highly noxious to living beings and, with respect to this, the UV radiation has been divided into three parts: most noxious, UV-C (100–280 nm, 1200–429 kJ/einstein), medium harm, UV-B (280–315 nm, 429–381 kJ/einstein), and least injurious, UV-A (315–400 nm, 381–300 kJ/einstein) [4].

9.2 Atmospheric Photochemistry

Atmospheric gases are dissipated in the atmosphere roughly according to their density: those of smallest density (H_2, He) occur only in the thermosphere and mesosphere; heavier gases (N_2 and O_2) can be found in all parts, from the troposphere to the mesosphere; the heaviest (CO_2) are situated near to ground level. Gases emitted from the Earth's surface concentrate mostly in the troposphere from where they migrate slowly to the upper parts of the atmosphere. An exception to this rule is ozone, which is generated and decomposed photochemically at two different altitudes: in the troposphere and at the border of the stratosphere and mesosphere (Figure 9.3).

Some of the atmospheric gases (eg N_2, H_2O, and CO_2) cannot undergo photochemical reactions because they do not meet the radiation, which could be absorbed by them and initiate the photochemical reaction, eg N_2 molecule absorbs radiation of $\lambda < 100$ nm, which is absorbed by gases occurring in the upper part of the atmosphere and does not reach the area of N_2 occurrence (Figure 9.3).

Mainly hydrogen is responsible for trapping the most energetic UV radiation, which is present in the thermosphere in small but noticeable amounts. H_2 absorbs the far-UV radiation ($\lambda_{max} = 110.9$ nm) and undergoes photodissociation:

$$H_2 \xrightarrow{h\nu\,(\text{UV-C})} 2H \tag{9.3}$$

followed by a successive recombination:

$$2H \rightarrow H_2 + Q \tag{9.4}$$

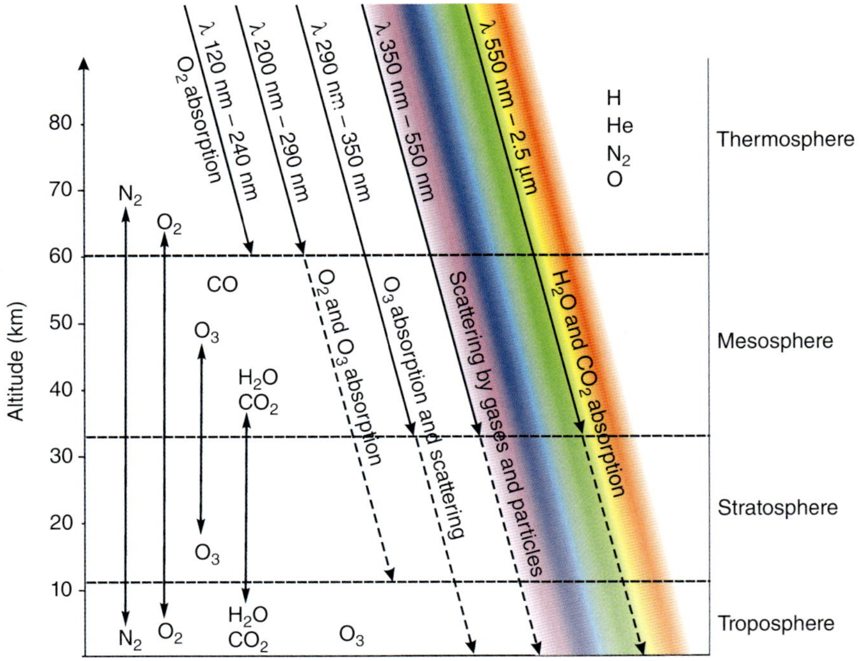

Figure 9.3 *Attenuation of solar radiation by atmospheric gases [4]*

Thus, these processes remove the most energetic UV-C radiation from the atmosphere and convert it into thermal energy.

The most important atmospheric photoreagent is, however, molecular oxygen: it absorbs UV-C radiation within 100–240 nm, with a maximum at 160 nm, and undergoes photodissociation to the O atoms in their ground states $O(^3P)$:

$$O_2 \xrightarrow{h\nu\,(\lambda<240\,\text{nm})} 2O \tag{9.5}$$

The formation of odd oxygen species under pressurized conditions is possible even in the wavelength region that is somewhat longer than the dissociation threshold at 242.4 nm; this is important as the initial reaction in the laser-induced oxidation of hydrocarbons in the hydrocarbon/O_2/supercritical CO_2 mixtures at 248 nm [5].

The photodissociation is followed by recombination:

$$2O \rightarrow O_2 + Q \tag{9.6}$$

and/or addition of an O atom to the O_2 molecule:

$$O + O_2 \rightarrow O_3 \tag{9.7}$$

The last reaction is responsible for generation of the ozone layer at 20–40 km above sea level. In the troposphere ozone is also synthesized by reaction between O_2

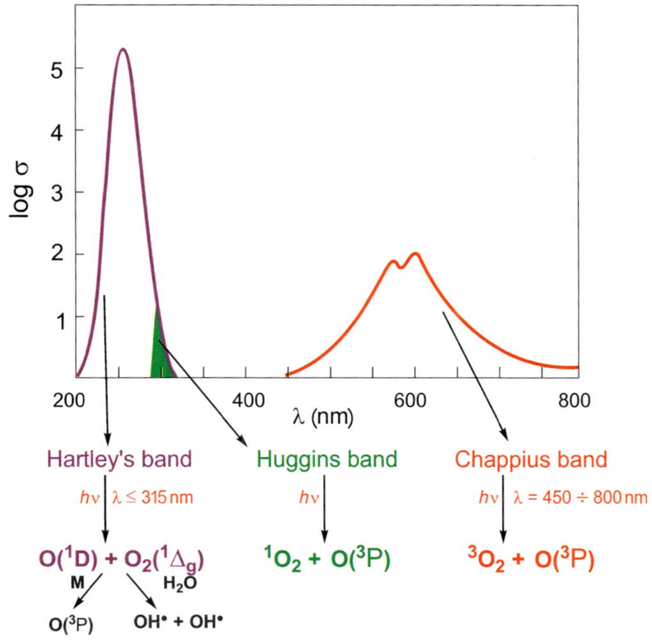

Figure 9.4 *Electronic spectrum of ozone and its photodissociation products generated upon exciting within different bands [6]*

molecules and O atoms in their ground states, but $O(^3P)$ atoms originate from the NO_2, not O_2, photodissociation (see below).

In both parts of the atmosphere, ozone decays mainly by photodissociation to molecular and atomic oxygen, but, depending on the exciting energy, the products are generated in their ground or excited states (Figure 9.4). Among the products, especially hazardous is atomic oxygen in its excited state, $O(^1D)$, which readily reacts with H_2O, yielding $OH^•$ radicals.

Besides the photochemical dissociation, ozone decays in oxidation–reduction reactions with different species. The stratospheric O_3 reacts rapidly with nitric oxide and products of photodissociation of halogenated hydrocarbons (Figure 9.5).

Another gas that undergoes photodissociation in the stratosphere is N_2O, which is emitted to the troposphere, where its lifetime is about 120–150 years; it then migrates slowly to the upper parts of the atmosphere until it meets UV radiation energetic enough to induce its photodissociation ($\lambda < 250\,\text{nm}$). This can yield molecular nitrogen and atomic oxygen or nitric oxide and atomic nitrogen:

$$N_2O \xrightarrow{h\nu} N_2 + O \qquad (9.8)$$

$$N_2O \xrightarrow{h\nu} NO + N \qquad (9.9)$$

Nitric oxide is also generated by reaction of N_2O with atomic oxygen in its excited state:

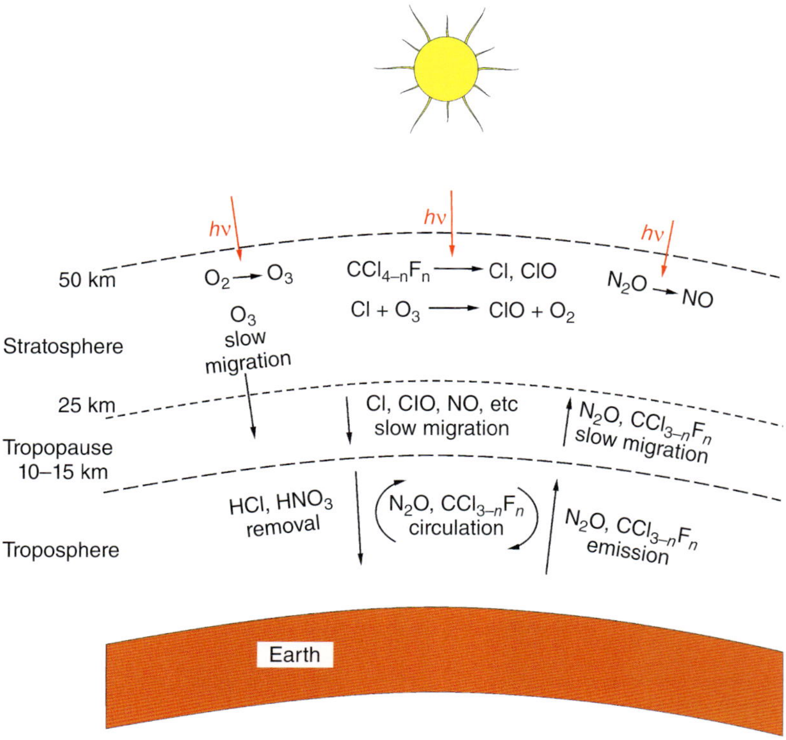

Figure 9.5 *Photochemical reactivity and thermal behaviour of more important gases in different parts of the atmosphere. (Adapted from Finlayson-Pitts and Pitts [6])*

$$N_2O + O^* \rightarrow 2NO \tag{9.10}$$

Nitric oxide reacts readily with ozone

$$NO + O_3 \rightarrow NO_2 + O_2 \tag{9.11}$$

and thereby its formation in the stratosphere causes damage to the ozone layer.

Nitrogen dioxide also undergoes photodissociation (see below) or thermal reaction with water vapour and molecular oxygen, yielding nitric acid, which falls slowly to ground level. Similar behaviour is experienced with hydrogen chloride, which is the endproduct of photochemical and thermal reactions of halogenated hydrocarbons (Figure 9.5).

Nitrogen dioxide is the most important photochemically active component of the troposphere. It absorbs not only in UV but also in the visible range (Figure 9.6), so it is able to undergo photochemical reactions within the whole atmosphere.

Excitation by the actinic radiation in the troposphere ($290 < \lambda < 420$ nm, Figure 9.6b) induces dissociation to nitric oxide and atomic oxygen in their ground states:

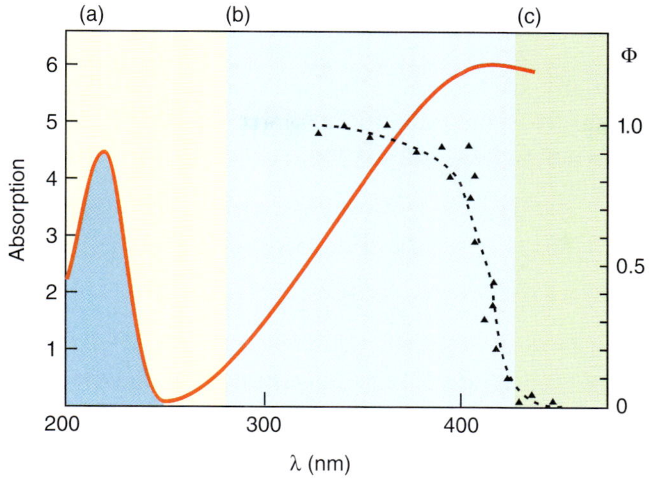

Figure 9.6 *Electronic spectrum of nitrogen dioxide and ranges of its different photochemical behaviour; Φ, quantum yield of NO_2 photodissociation between 300 and 420 nm [6]*

$$NO_2 \xrightarrow{h\nu\,(290-420\,nm)} NO + O(^3P) \tag{9.12}$$

$O(^3P)$ reacts readily with molecular oxygen producing tropospheric ozone:

$$O(^3P) + O_2 \rightarrow O_3 \tag{9.13}$$

The other spectral ranges are also photochemically active: radiation of $\lambda < 290\,nm$ (Figure 9 6a) generates the same products as $\lambda > 290\,nm$, but both are in their excited states, whereas radiation of $\lambda > 430\,nm$ (Figure 9.6c) is effective in energy transfer and formation of the singlet oxygen:

$$NO_2 + O_2 \xrightarrow{h\nu\,(\lambda > 430\,nm)} NO_2 + O_2(^1\Delta_g) \tag{9.14}$$

The tropospheric nitric oxide is produced from molecular nitrogen or its compounds during high temperature combustion processes; thus it is mostly of anthropogenic origin. Nitrogen dioxide is produced by further oxidation of NO; from these compounds other nitrogenous species are formed.

During the reaction of nitrogen dioxide with ozone the $NO_3^{\bullet}$ radical is generated:

$$NO_2 + O_3 \rightarrow NO_3^{\bullet} + O_2 \tag{9.15}$$

whereas during the reaction with water vapour nitric oxides (NO_x) form nitrogen acids:

$$NO + NO_2 + H_2O \rightarrow 2HNO_2 \tag{9.16}$$

$$2NO_2 + H_2O \rightarrow HNO_2 + HNO_3 \tag{9.17}$$

All these compounds are active participants in environmental photochemistry. Nitrate radicals absorb solar radiation at approximately 600–700 nm and in the excited state dissociate to NO_x and oxygen. There are two possible sets of the photodissociation products:

$$NO_3^{\bullet} \xrightarrow{h\nu} NO + O_2 \tag{9.18}$$

$$NO_3^{\bullet} \xrightarrow{h\nu} NO_2 + O \tag{9.19}$$

As a strong oxidant ($E = 2.3$–2.7 V vs SCE [saturated calomel electrode]), the nitrate radical is also very active in thermal reactions, eg it induces oxidation of the aliphatic hydrocarbons:

$$NO_3^{\bullet} + RH \rightarrow HNO_3 + R^{\bullet} \tag{9.20}$$

During the reaction of this radical with nitrogen dioxide, dinitrogen pentoxide is produced:

$$NO_3^{\bullet} + NO_2 \rightarrow N_2O_5 \tag{9.21}$$

which may be a source of nitric acid in the troposphere. N_2O_5 absorbs the UV solar radiation and undergoes different photochemical reactions, from which the most important is generation of nitrate radicals:

$$N_2O_5 \xrightarrow{h\nu} NO_2 + NO_3^{\bullet} \tag{9.22}$$

Both nitrogen acids and their esters undergo photodissociation under sunlight ($\lambda < 400$ nm), regenerating NO_x and yielding hydroxyl radical or its derivatives:

$$HONO \xrightarrow{h\nu} OH^{\bullet} + NO \tag{9.23}$$

$$RONO \xrightarrow{h\nu} RO^{\bullet} + NO \tag{9.24}$$

$$HONO_2 \xrightarrow{h\nu} OH^{\bullet} + NO_2 \tag{9.25}$$

$$RONO_2 \xrightarrow{h\nu} RO^{\bullet} + NO_2 \tag{9.26}$$

The hydroxyl radical, $OH^{\bullet}$, is one of the strongest oxidants ($E = 1.77$–2.85 V vs SCE, depending on pH), and is generated in relatively high concentrations so its formation and reactivity are of crucial importance for removal of environmental pollutants.

One of the most important effects is oxidation of hydrocarbons, which occurs in the natural troposphere by $OH^{\bullet}$ attack [6], eg:

$$CH_4 + OH^\bullet \rightarrow CH_3^\bullet + H_2O \quad (9.27)$$

yielding a hydrocarbon radical, which is oxidized by molecular oxygen:

$$CH_3^\bullet + O_2 \rightarrow CH_3O_2^\bullet \quad (9.28)$$

The fate of the peroxy radical depends on the ambient concentration of nitric oxide. In the presence of sufficient NO the $CH_3O_2^\bullet$ radical is converted to methanal:

$$CH_3O_2^\bullet + NO \rightarrow CH_3O^\bullet + NO_2 \quad (9.29)$$

$$CH_3O^\bullet + O_2 \rightarrow HCHO + HO_2^\bullet \quad (9.30)$$

The $HO_2^\bullet$ radical oxidizes another NO molecule to NO_2, regenerating a hydroxyl radical:

$$HO_2^\bullet + NO \rightarrow NO_2 + OH^\bullet \quad (9.31)$$

Although two NO molecules are oxidized during the oxidation of CH_4 to HCHO they are regenerated in the NO_2 photodissociation:

$$NO_2 \xrightarrow{h\nu} NO + O \quad (9.32)$$

The methane oxidation to methanal is thus realized in the catalytic cycle in which atmospheric O_2 is the oxidant and the $OH^\bullet$ radicals are the catalyst, and which is coupled to photoassisted dissociation of nitrogen dioxide (Figure 9.7). The latter process yields two ozone molecules per photocatalytic cycle.

If the NO concentration is too low, the $CH_3O_2^\bullet$ and $HO_2^\bullet$ radicals react with each other, yielding finally acetic acid and molecular oxygen [6]:

$$CH_3O_2^\bullet + HO_2^\bullet \rightarrow CH_3COOH + O_2 \quad (9.33)$$

CH_3COOH can, however, be included in the cycle by thermal transformation into $CH_3O_2^\bullet$ radicals (in reaction with $OH^\bullet$) or by photodissociation to $CH_3O^\bullet$ and $OH^\bullet$ radicals.

Heavier hydrocarbons behave similarly, although rate constants of their $OH^\bullet$ oxidation are the higher the more carbon atoms there are in the molecule (eg rate constant for pentane is of the order of 10^3 higher than that for methane); the oxidation is also much faster in the case of unsaturated or aromatic organic hydrocarbons and their derivatives.

Products of these reactions, aldehydes and ketones, undergo photodissociations under tropospheric sunlight. Aldehydes absorb actinic UV-A radiation and methanal absorption extends out to approximately 370 nm, whereas the heavier aldehydes absorb only to approximately 345 nm. Methanal has two photodissociation paths:

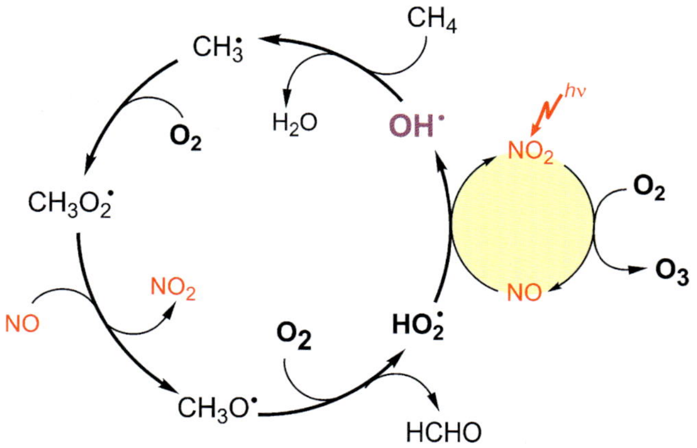

Figure 9.7 *Main pathways of methane oxidation to methanal induced by OH• radicals and assisted by NO₂ photodissociation*

$$HCHO \xrightarrow{h\nu} H_2 + CO \tag{9.34}$$

$$HCHO \xrightarrow{h\nu} H^\bullet + HCO^\bullet \tag{9.35}$$

The $H^\bullet$ and $HCO^\bullet$ radicals react with each other and with atmospheric O_2, resulting in the final production of CO and H_2 (Figure 9.8).

Heavier aldehydes behave similarly, eg ethanal dissociates to methane, carbon oxide, and two different sets of free radicals. The reaction mode depends on the energy of the absorbed radiation:

$$CH_3CHO \xrightarrow{h\nu\,(\lambda<290\,nm)} CH_4 + CO \tag{9.36}$$

$$CH_3CHO \xrightarrow{h\nu\,(\lambda<300\,nm)} H^\bullet + CH_3CO^\bullet \tag{9.37}$$

$$CH_3CHO \xrightarrow{h\nu\,(\lambda<320\,nm)} CH_3^\bullet + CHO^\bullet \tag{9.38}$$

The possible photodissociation modes of aldehydes may be thus generally written as:

$$RCHO \xrightarrow{h\nu} RH + CO \tag{9.39}$$

$$RCHO \xrightarrow{h\nu} R'^\bullet + R''CO^\bullet \tag{9.40}$$

$$RCHO \xrightarrow{h\nu} R^\bullet + CHO \tag{9.41}$$

Ketones absorb the actinic UV out to approximately 330 nm and then undergo photodissociation to radicals, eg acetone yields $CH_3^\bullet$ and $CH_3CO^\bullet$:

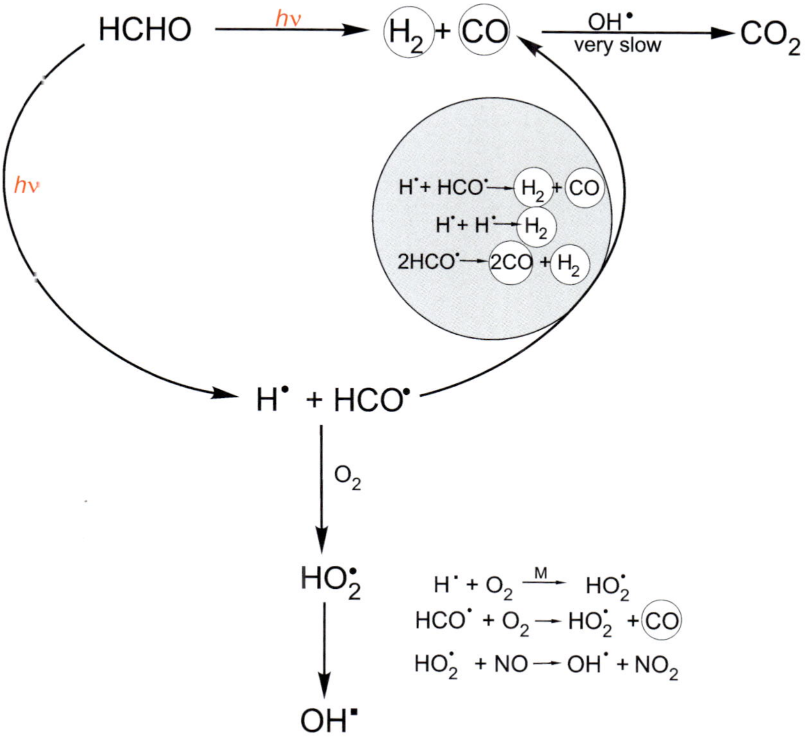

Figure 9.8 *Photodissociation of methanal and successive thermal reactions*

$$CH_3COCH_3 \xrightarrow{h\nu\,(220<\lambda<340\,nm)} CH_3^{\bullet} + CH_3CO^{\bullet} \tag{9.42}$$

For asymmetric ketones, such as methylethyl ketone, there are two possible bond cleavages, producing different sets of free radicals [6]:

$$CH_3COC_2H_5 \xrightarrow{h\nu} C_2H_5^{\bullet} + CH_3CO^{\bullet} \tag{9.43}$$

$$CH_3COC_2H_5 \xrightarrow{h\nu} CH_3^{\bullet} + C_2H_5CO^{\bullet} \tag{9.44}$$

Rupture of the bond forming the most stable radicals usually predominates.

Among the species active in the environment the most important are nitrogen oxides (NO_x) and hydroxo radicals $HO_x^{\bullet}$ ($OH^{\bullet}$ and $HO_2^{\bullet}$), which are the key species for photochemical ozone production in urban air. The sophisticated role of NO_x consists of alternating formation and destruction of ozone, depending on the environmental conditions. Briefly, NO is rapidly oxidized to NO_2 by O_3, $HO_2^{\bullet}$, and $RO_2^{\bullet}$ after being emitted to the atmosphere. However, NO_2 under sunlight is also rapidly photodissociated back to NO and atomic oxygen, which in consequence produces ozone. At typical tropospheric O_3 levels, the resulting chain between NO and NO_2 occurs with a time constant of a few minutes [7].

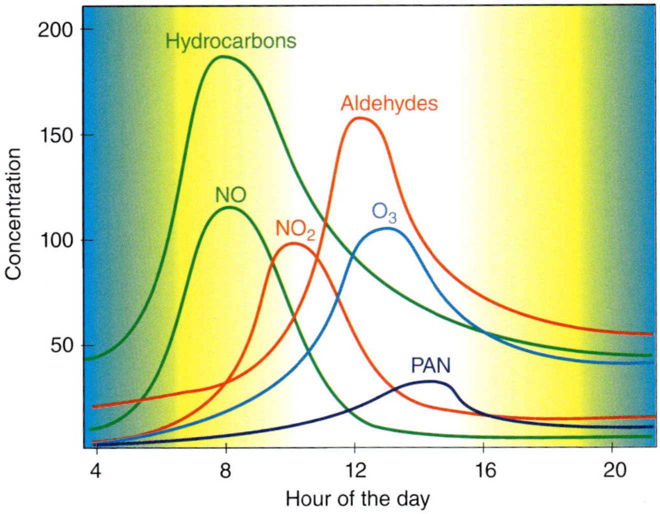

Figure 9.9 *Typical profiles of primary (green lines), secondary (red lines) and tertiary pollutants (blue lines) of the photochemical smog. PAN, peroxyacetyl nitrate,* $CH_3CO_3NO_2$

These species are also determinants of photochemical smog generation. The main sources of the smog are nitric oxide and hydrocarbons, emitted with combustion gases to the urban troposphere. The emission is especially intensive at the start of a week day, when concentrations of NO and hydrocarbons increase steeply (Figure 9.9). The production competes with NO oxidation to NO_2 and with hydrocarbon oxidation to aldehydes, induced by $HO_x^{\bullet}$ and photoassisted by NO_x (see Figure 9.7). Increased sunlight intensity enhances the photochemical reactions of aldehydes and NO_2, and thus ozone production. The latter is decomposed in photochemical (see Figure 9.4) and thermal reactions with NO, different radicals, and organic pollutants. Besides the major products, a lot of minor species are generated: some are very aggressive and dangerous and even life threatening; an especially noxious lacrimator is peroxyacetyl nitrate (PAN).

The reactions observed in the photochemical smog, especially those concerning decomposition and oxidation of volatile organic substances, are accelerated by atmospheric aerosols, eg oxidation of some halogenated hydrocarbons, isoprene, monoterpenes, and aromatic hydrocarbons is enhanced by the surfaces of metal oxides, desert sand, volcanic ash, and sea salt [8].

9.3 Photochemistry in the Hydrosphere and Soil

Most of the Earth's surface is covered by water, which forms oceans, seas, rivers, estuaries, lakes, ponds, streams, pools, and puddles. The water basins contain various chemicals of both natural and anthropogenic origin. In the aquatic medium, more condensed than the atmosphere, the bimolecular reactions are more frequent. Some of the dissolved substances undergo transformation in sunlight via direct or indirect photochemical reactions. Direct photoreactions concern those reactions in which the same species absorbs sunlight and undergoes photochemical reaction, whereas indirect photoreactions refer to those reactions in which one species undergoes photoreaction in result of light absorption by another species (photosensitizer).

The environmental photoreaction rates depend on both sunlight and spectral and photochemical properties of the relevant substances, ie their molar absorption coefficients and quantum yields. Sunlight parameters depend on the solar irradiance at the water surface, radiative transfer from air to water, and transmission of sunlight in the water body [9]. However, the attenuation of sunlight is drastically greater on surface and marine waters than it is in the atmosphere. The absorption and scattering of the light by compounds and particles present in natural water give rise to a depth-dependent profile in the spectral composition and irradiance of sunlight, which, in turn, stimulates depth-dependent differences in reactions, chemical composition, and specifications of substances in the water column.

The maximum depth for the light level corresponding to 0.1% of incident light for the wavelength of interest involved in a given photochemical reaction in surface or marine water is called the 'photochemical zone'. Typical values of absorption coefficients for surface and marine waters at 400 nm correspond to a depth of 6–30 m, whereas for shorter wavelengths the depth is much smaller. Thereby, photochemical reactions in water basins are confined to their top layers [10].

There are three typical classes of compounds, for which photochemistry plays a crucial role in surface and marine waters: nitrate ions, humic substances, and transition metal ions.

9.3.1 Nitrate Photochemistry

Nitrate and nitrite ions are usually present in natural waters, including cloud water. Their concentrations depend strongly on the local biological activity, but they are usually lower than 10^{-3} mol dm^{-3} and 2 µmol dm^{-3}, for NO_3^- and NO_2^- ions, respectively [11].

Both ions absorb terrestrial sunlight, although their absorption coefficients are very low: $\varepsilon_{max} = 7.4\,dm^3\,mol^{-1}cm^{-1}$ at λ_{max} 300 nm (NO_3^-) and $\varepsilon_{max} = 24\,dm^3\,mol^{-1}cm^{-1}$ at λ_{max} 354 nm (NO_2^-) (Figure 9.10) [6, 11].

Upon excitation both ions undergo photodissociation to the relevant oxide and oxygen anion radical:

$$NO_2^- \xrightarrow{h\nu\,(\lambda<410\,nm)} {}^*NO_2^- \rightarrow NO + O^{\bullet-} \quad (9.45)$$

$$NO_3^- \xrightarrow{h\nu\,(\lambda<350\,nm)} {}^*NO_3^- \rightarrow NO_2 + O^{\bullet-} \quad (9.46)$$

The radical immediately reacts with water molecule, generating $OH^{\bullet}$ radicals:

$$O^{\bullet-} + H_2O \rightarrow OH^{\bullet} + OH^- \quad (9.47)$$

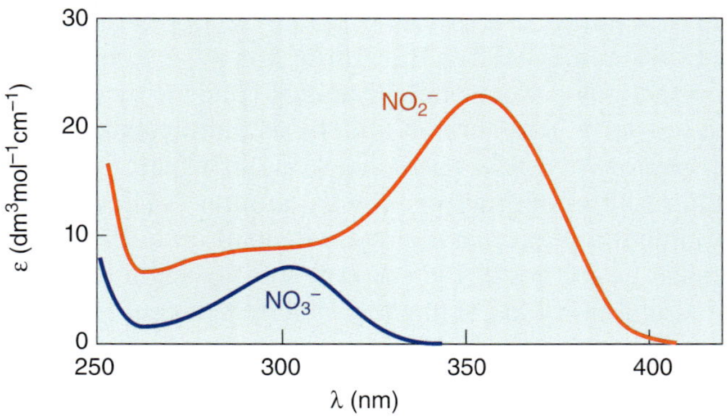

Figure 9.10 *Electronic spectra of NO_2^- and NO_3^- ions*

Photolysis of nitrate ion may yield NO_2^- and O atoms as well:

$$^*NO_3^- \rightarrow NO_2^- + O \tag{9.48}$$

Quantum yields (Φ) of these photodissociations are fairly low: for NO_2^- Φ amounts of 0.06–0.07 within 300–340 nm, whereas for NO_3^- about 0.01 for both dissociations, with more efficient paths yielding the $OH^•$ radicals [11].

Due to photogeneration of nitrogen oxides, nitrate and nitrite ions induce nitrations or nitrosations of organic compounds. The most important effect is, however, photogeneration of $OH^•$ radicals [12], which induce oxidation of most organic substances, although nitration and nitrosation are favoured by increasing concentration of NO_2^- ions, because the ions behave as both a source and a quencher of $OH^•$ radicals [11]. Recently, nitrate and nitrite ions were found to induce photodegradation of monuron, a pesticide, in aqueous solution under solar light irradiation [13].

9.3.2 Role of Humic Substances

Among the great variety of photoreactions occurring in surface waters, many result from light absorption by humic substances (HSs) which are widespread in different soils and aquatic media. HSs are degradation products from plants, animals, and microbiological materials, and therefore they are heterogeneous mixtures of a variety of organic compounds, consisting of aromatic, aliphatic, phenolic, and quinolic functional groups with varying molecular sizes and properties. Humic substances are known to form very stable complexes with iron and other transition metals.

Degradation of many organic pollutants under the action of sunlight was found to proceed with involvement of humic substances as photocatalysts or photosensitizers (see sections 6.5 and 6.6 in Chapter 6). HSs absorb solar radiation in the range 300–500 nm and excitation by laser pulse of λ 355 nm was found to induce two different processes:

$$HS^* \rightarrow HS^{\bullet +} + e^-_{aq} \tag{9.49}$$

$$^1HS^* \rightarrow {}^3HS^* \tag{9.50}$$

The primary quantum yields ranged between 5×10^{-3} and 8×10^{-3} [11]. In oxygenated media solvated electrons (e^-_{aq}) are mainly trapped by molecular oxygen, generating hydrogen peroxyl radicals and superoxide anions; this disproportionation is a source of hydrogen peroxide and thereby hydroxyl radicals.

Excited triplet states of HSs are the main species responsible for photoinduced degradation of aquatic pollutants. There are two main pathways of the degradation: photosensitized oxygenations:

$$^3HS^* + {}^3O_2 \rightarrow {}^1HS + {}^1O_2 \tag{9.51}$$

or a direct reaction between $^3HS^*$ and an organic reactant:

$$^3HS^* + {}^1R \rightarrow {}^1HS + {}^3R \tag{9.52}$$

which may proceed not only via energy transfer (equation 9.52) but also by electron transfer or hydrogen atom transfer reactions [11]. Among others the HSs are known to induce photodecomposition of some herbicides [14, 15] and pesticides [16, 17].

The HSs and other aquatic and soil organic compounds interact with other components of the environmental system. Various kinds of clay organic intercalation compounds have been described, which are able to initiate unique photochemical reactivities characteristic of their specific molecular arrangements; these differ considerably from solution photochemistry [18].

9.3.3 Photocatalysis by Fe^{III}/Fe^{II} Complexes

Especially important for cleaning of aquatic systems are iron(III) complexes with HSs. These thermally very stable compounds undergo photoredox reactions, creating ferrous ion and free $HS^{\bullet}$ radicals. In a neutral or alkaline medium, the Fe^{II} species is reoxidized to Fe^{III} by molecular oxygen, whereas in acidic medium a more efficient oxidant (H_2O_2) is needed (Fenton reaction) [11, 19].

Transition metal ions, and especially iron(III) species present in aquatic compartments, are involved in photochemical cycles, which play an important role in the degradation of various pollutants. These are often relevant to coordination of the Fe^{III} centre as ligands and/or are susceptible to oxidation by the excited Fe^{III} centre. Typically Fe^{III} complexes absorb sunlight undergoing excitation to their ligand-to-metal-charge transfer (LMCT) states (see section 3.2 in Chapter 3), which results in the photoinduced electron transfer to the Fe centre from the inner (ligand) or external electron donor. The photochemically generated Fe^{II} species is then oxidized by molecular oxygen or other oxygen oxidants (eg H_2O_2), closing the photocatalytic cycle (Figure 9.11). As a result some of the environmental pollutants (L or D) are oxidized by molecular oxygen in photoassisted processes catalyzed by Fe

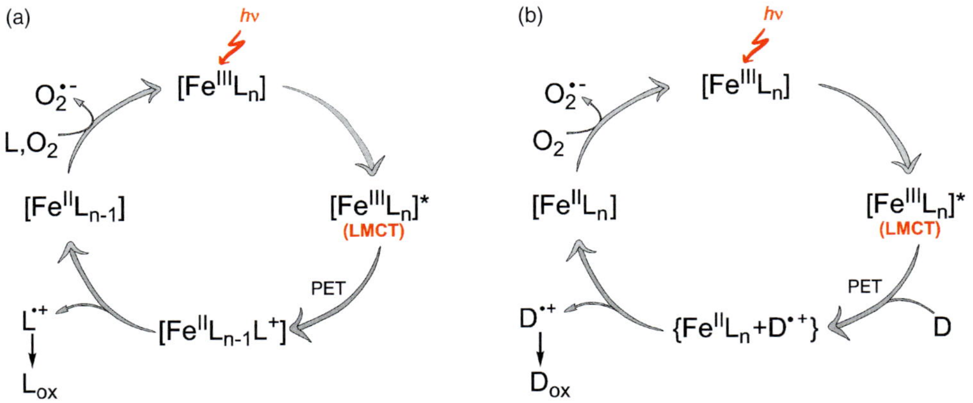

Figure 9.11 *Two alternative pathways of redox cycles photocatalyzed by Fe^{III} complexes: (a) inner-sphere photoinduced electron transfer (PET) with ligand (L) electron donor and (b) outer-sphere PET with an external electron donor (D). (Adapted from Cieśla et al. [20])*

complexes. Excitation of the aqua Fe^{III} complex leads to the generation of hydroxyl radicals, because the complex readily undergoes hydrolysis and at $pH \geq 3$ the hydroxo complex dominates [4]:

$$[Fe(H_2O)_6]^{3+} + H_2O \rightleftharpoons [Fe(H_2O)_5OH]^{2+} + H_3O^+ \tag{9.53}$$

$$[Fe(H_2O)_5OH]^{2+} + H_2O \xrightarrow{h\nu\,(LMCT)} [Fe(H_2O)_6]^{2+} + OH^{\bullet} \tag{9.54}$$

In the case of Fe^{III} complexes with chelating ligands, generation of ligand radicals is observed. Thus, in the case of the Fe^{III} complexes with aminopolycarboxylic acids, eg $[Fe^{III}edta]^-$, photoreduction of $Fe^{III} \rightarrow Fe^{II}$ is accompanied by decarboxylation and formation of the respective organic radical [11, 20–22].

Recent studies have shown that LMCT excited states of the $[Fe^{III}edta]^-$ complexes can be quenched not only by molecular oxygen and other electron acceptors, eg chromate(VI), but also by electron donors, such as free EDTA, sulphate(IV), nitrate(III), or alcohols; the suggested mechanism is presented in Figure 9.12 [22, 23].

Regeneration of Fe^{III} forms from Fe^{II} products is rather a simple task: in alkaline media it is readily done by molecular oxygen; Fe^{II} complexes with chelating ligands behave similarly in acidic media, eg oxidation of $[Fe^{II}L]$ complexes (L = EDTA, diethylenetriaminepentaacetate, and *N*-(hydroxyethyl)ethylenediaminetriacetate) by O_2 was reported to proceed in three steps:

$$[Fe^{II}edta]^{2-} + O_2 \rightleftharpoons [Fe^{II}edta\text{-}O_2]^{2-} \tag{9.55}$$

$$[Fe^{II}edta\text{-}O_2]^{2-} \rightarrow [Fe^{III}edta]^- + O_2^- \tag{9.56}$$

$$[Fe^{II}edta]^{2-} + O_2^- + 2H^+ \rightarrow [Fe^{III}edta]^- + H_2O_2 \tag{9.57}$$

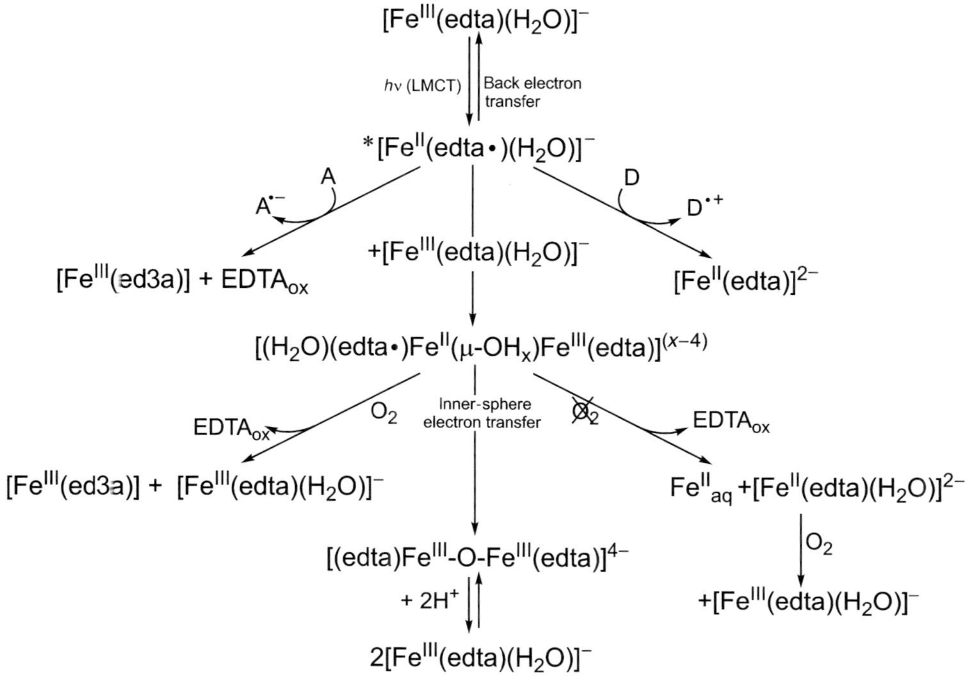

Figure 9.12 *Mechanism of photochemical behaviour of the $[Fe(edta)H_2O]^-$ complex in the presence of some environmentally relevant electron donors and acceptors; the reactivity scheme for $[Fe(edta)(OH)]^{2-}$ is analogous. D, electron donor; A, electron acceptor; ed3a, ethylenediaminetriacetate*

The processes are fast enough at pH > 2 and for the $[Fe^{II}edta]^{2-}$ complex the maximum rate is reached at pH of about 3 [24–26].

The recent flash photolysis study has revealed, that the $*[Fe^{III}edta]^-$ excited state is quenched by electron transfer processes by both electron donors and electron acceptors (see Figure 9.12). The latter cause oxidation of the EDTA ligand and the acceptor reduction, whereas the Fe^{III} centre does not change its oxidation state. One of the acceptors is molecular oxygen, which results in direct EDTA oxidation at the expense of O_2 reduction [21, 22]:

$$*[Fe^{III}edta]^- \xrightarrow{O_2} [Fe^{III}ed3a] + HCHO + CO_2 + OH^- \tag{9.58}$$

There is a more serious problem for the $[Fe(H_2O)_6]^{2+}$ complex in acidic medium, which needs more efficient oxidants than O_2. In such a case H_2O_2 is sufficient for regenerating Fe^{III} with simultaneous formation of hydroxyl radicals (Fenton reaction):

$$[Fe(H_2O)_6]^{2+} + H_2O_2 \rightarrow [Fe(H_2O)_6]^{3+} + OH^{\bullet} + OH^- \tag{9.59}$$

When Fe^{III}/Fe^{II} complexes are irradiated with H_2O_2, Fe^{III} undergoes photoreduction at the expense of an electron donor, whereas Fe^{II} is reoxidized by H_2O_2; as a result the $OH^{\bullet}$ radicals are generated in the photo-Fenton cycle at the expense of H_2O_2 (sacrificial acceptor). The $OH^{\bullet}$ radical is one of the strongest environmental oxidants responsible for decomposition or even complete mineralization of most environmental pollutants. The mechanism and applications of the photo-Fenton reactions, especially using solar energy, are the subject of many studies (for more information see Chapter 21) [27–56].

Fe^{III} complexes were reported as effective photocatalysts for oxidation of many different organic pollutants, eg: alcohols and their derivatives [20, 29]; organic acids, such as formic [50, 53, 56], oxalic [37], citric [57], and maleic [58]; EDTA [11, 20–23], phenol and its derivatives [35, 36, 45, 59, 60], other aromatic pollutants [38, 43, 51, 61–64], non-biodegradable azo dyes [40, 41, 48, 55, 59, 65], herbicides [54, 66–70], pesticides [32, 46, 71, 72], insecticides [44], pharmaceuticals and wastewater from medical laboratories [39, 47, 73], chlorinated solvents [33, 74], municipal wastewater [75], and many others [20]. The photo-Fenton process was explored as photochemical pre-treatment to improve its biodegradability, especially of biorecalcitrant wastewater from the textile industry [76, 77]; the method was also proposed for water disinfection [78, 79].

It should be noted, however, that not always do sunlight and Fe^{III} compounds result in the abatement of the pollution: they were reported to induce photochemical reduction of Hg^{II} species and thus to be responsible for the production of dissolved gaseous mercury in freshwater and its subsequent toxicity [80].

9.3.4 Photocatalysis by Cu^{II}/Cu^{I} Complexes

The effects of other transition metal ions are qualitatively similar, although their extent is decidedly smaller. Complexes of copper, chromium, and manganese with central atoms in their higher oxidation states are fairly good candidates to play the role of an environmental photocatalyst [20].

Cu^{II} complexes undergo photoreduction to the Cu^{I} species and may be the photocatalyst in photo-oxidation cycles of organic environmental matter, quite similar to the Fe^{III} species [20, 81] (see Figure 9.11). Dissolved copper compounds are important to transformation reactions, because they react with hydroperoxyl ($HO_2^{\bullet}$) and superoxide ($O_2^{\bullet-}$) radicals much faster than other species present in the solution. Oxidation of Cu^{I} and Fe^{II} by H_2O_2 is a source of the $OH^{\bullet}$ radicals in oceans comparable with nitrite photolysis, whereas photochemistry of Cu^{II} chlorocomplexes provides $Cl_2^{\bullet-}$ radicals [81]:

$$[Cu^{II}Cl_x]^{2-x} \xrightarrow{h\nu\,(LMCT)} [Cu^{I}Cl_{x-1}]^{2-x} + Cl^{\bullet} \tag{9.60}$$

$$Cl^{\bullet} + Cl^{-} \rightarrow Cl_2^{\bullet-} \tag{9.61}$$

Photoredox reactions of the Cu^{II} chelates result in inner-sphere electron transfer:

$$[Cu^{II}(LL)_2]^{2+} \xrightarrow{h\nu\,(LMCT)} [Cu^{I}LL]^{+} + LL^{\bullet+} \tag{9.62}$$

which lead to photodecarboxylation and oxidation of carboxylates (eg oxalates, malonates, acetates), aminocarboxylates (eg glycine, glutamine, alanine, nitrilotriacetic acid [NTA], iminodiacetic acid [IDA], EDTA), phenols, polyamines, etc [20, 60, 81–85]. For example, photoreduction of Cu^{II} acetate by sunlight:

$$[Cu^{II}(OOCCH_3)solv]^+ \xrightarrow{h\nu\,(LMCT)} [Cu^{I}solv]^+ + CH_3^{\bullet} + CO_2 \qquad (9.63)$$

was reported to yield Cu^{I}, carbon dioxide and $CH_3^{\bullet}$ radicals [81].

9.3.5 Photocatalysis by Chromium Compounds

The two common oxidation states of Cr present in the environment, ie Cr^{III} and Cr^{VI}, are drastically different in charge, physicochemical properties as well as chemical and biochemical reactivity. The immobile Cr^{III} form is one of the trace elements essential for the proper functioning of living organisms, whereas the mobile Cr^{VI} form exerts toxic effects on biological systems. In the environment both forms are transformed into each other thermally or photochemically. The reactions depend on pH, oxygen concentration, and presence of appropriate electron donors, acceptors, and mediators acting as reagents, ligands, or catalysts. The optimal conditions needed to convert Cr^{III} to Cr^{VI} are completely different from those required for the opposite conversion, $Cr^{VI} \rightarrow Cr^{III}$; this makes a full Cr cycle rather difficult [86, 87].

Both chromium forms are photochemically active, and the metal-centred (MC) photochemistry of the Cr^{III} complexes has been known for a long time [88]. Photosubstitution or photoisomerization is, however, of minor environmental relevance; to facilitate or accelerate the mutual conversion of Cr^{III} and Cr^{VI} only photoredox behaviours are essential.

The photoredox reactivity was reported for Cr^{III} complexes with some relatively good electron donors as ligands, such as CN^-, NCS^-, EDTA, and $C_2O_4^{2-}$ [87, 89–92]. The LMCT excitation within the UV range was needed to induce photoreduction to Cr^{II} of $[Cr(CN)_{6-x}(OH)_x]^{3-}$ [89], $[CrA_2(NCS)_4]^-$ ($A = NH_3$, aniline, *n*-propylamine, morpholine) and $[Cr(A\text{-}A)(NCS)_4]^-$ (A-A = en, tn) complexes [90, 93]. Cr^{III} complexes with EDTA, and especially with oxalate, were found to undergo photoreduction on excitation by solar light [87, 91, 92].

The LMCT excitation of Cr^{III} complex with EDTA [91]:

$$[Cr^{III}(edta)OH]^{2-} \xrightarrow{h\nu\,(LMCT)} {}^*[Cr^{II}(edta^{\bullet})OH]^{2-} \qquad (9.64)$$

is followed by back electron transfer:

$${}^*[Cr^{II}(edta^{\bullet})OH]^{2-} \rightarrow [Cr^{III}(edta)OH]^{2-} \qquad (9.65)$$

and/or photoinduced electron transfer accompanied by dissociation of the oxidized ligand. In the presence of molecular oxygen $^*[Cr^{II}(edta^{\bullet})OH]^{2-}$ is quenched, yielding the substituted Cr^{III} complex or even the Cr^{VI} species. The latter Cr^{II} oxidation:

$$Cr^{II}_{aq} + O_2 \xrightarrow{OH^-} CrO_4^{2-} \qquad (9.66)$$

is possible under specific conditions, ie in alkaline medium and at a relatively large excess of O_2 over the Cr^{II} concentration [87, 89–91, 93].

Although chromate(VI) is photochemically inactive in all of its forms in neat aqueous solution, the photochemical oxidation of alcohols by chromate(VI) has been known for more than 80 years and interpreted in terms of photochemical reactivity of the chromate(VI) esters [94]. Recent studies have shown, however, that LMCT excitation of Cr^{VI} species is quenched not only by inner-sphere but also by outer-sphere electron transfer [23, 87, 92, 94, 95]. Moreover, inner-sphere electron transfer in chromate(VI) esters was found to involve two electrons, yielding a Cr^{IV} species and appropriate aldehyde or ketone:

$$^{*}R_1R_2CHOCrO_3^- + 5H^+ + 3H_2O \rightarrow R_1R_2CO + [Cr(H_2O)_6]^{4+} \tag{9.67}$$

At higher pH, ie at more loose interaction between chromate(VI) and alcohol, the outer-sphere transfer of one electron generates a Cr^V species and the alcohol radical [94]:

$$^{*}HCrO_4^- + R_1R_2CHOH \rightarrow H_2Cr^VO_4^- + R_1R_2CHO^{\bullet} \tag{9.68}$$

The parallel outer-sphere PET mechanism was found in systems containing chromate(VI) plus an electron donor different from alcohol, such as oxalate or phenol and its halogen derivatives [92, 95].

The photochemical interconversion of Cr^{VI} and Cr^{III} has also been analyzed with regard to the possible existence of complete photocatalytic cycles functioning in nature at day-and-night intervals. Under conditions similar to natural ones, ie in aqueous solution of pH only slightly above 7, Cr^{III} and Cr^{VI} species were irradiated by discrete portions of light from the same spectral range as sunlight. In the model system consisting of $[Cr(C_2O_4)_3]^{3-}$, CrO_4^{2-}, and $C_2O_4^{2-}$, the Cr^{VI} concentration oscillated according to the light and dark sequences, suggesting cyclic conversions of Cr^{III} to Cr^{VI} and Cr^{VI} to Cr^{III} (Figure 9.13) [23, 92].

Each $Cr^{VI} \rightarrow Cr^{III}$ conversion starts from the LMCT excitation of the Cr^{VI} species and its quenching by photoinduced electron transfer from the available donor, eg oxalate:

$$^{*}HCrO_4^- + C_2O_4^{2-} \rightarrow HCr^VO_4^{2-} + C_2O_4^{\bullet -} \tag{9.69}$$

Both products undergo further thermal reactions; $C_2O_4^{\bullet -}$ radical is oxidized by molecular oxygen:

$$C_2O_4^{\bullet -} + O_2 \rightarrow 2CO_2 + O_2^- \tag{9.70}$$

whereas Cr^V species is submitted to a dismutation process, producing Cr^{III} and reproducing Cr^{VI}:

$$3Cr^V \rightarrow Cr^{III} + 2Cr^{VI} \tag{9.71}$$

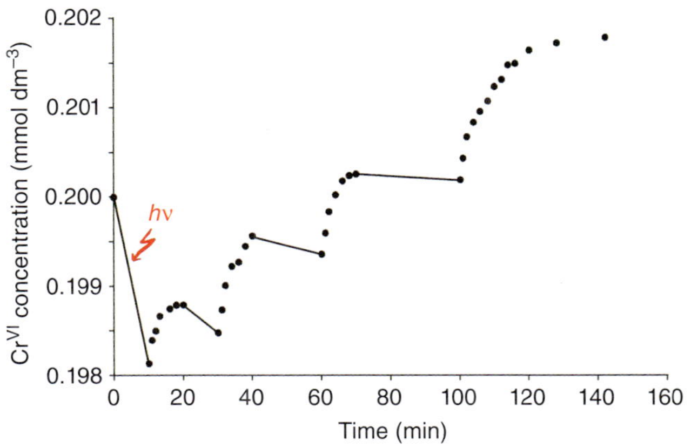

Figure 9.13 *Irradiation (solid lines) and post-irradiation (dotted lines) changes in* Cr^{VI} *concentration in the system* $[Cr(C_2O_4)_3]^{3-} + Cr^{VI} + C_2O_4^{2-}$

Preceding reactions can be used to interpret the decrease in the Cr^{VI} concentration during irradiation (equation 9.69) and its partial recover in the dark sequences (equation 9.71), as shown in Figure 9.13. After a few cycles, however, the Cr^{VI} level is higher than initial one and this result can be interpreted only by $Cr^{III} \rightarrow Cr^{VI}$ photoconversion (see above).

The $Cr^{III} \rightarrow Cr^{VI}$ conversion is also initiated by the LMCT excitation:

$$[Cr^{III}(C_2O_4)_3]^{3-} \xrightarrow{h\nu\,(LMCT)} {}^*[Cr^{II}(C_2O_4)_2(C_2O_4^{\bullet -})]^{3-} \qquad (9.72)$$

followed by the back electron transfer:

$$^*[Cr^{II}(C_2O_4)_2(C_2O_4^{\bullet -})]^{3-} \rightarrow [Cr^{III}(C_2O_4)_3]^{3-} \qquad (9.73)$$

and/or photoinduced electron transfer accompanied by dissociation of the oxidized ligand:

$$^*[Cr^{II}(C_2O_4)_2(C_2O_4^{\bullet -})]^{3-} \rightarrow [Cr^{II}(C_2O_4)_2]^{2-} + C_2O_4^{\bullet -} \qquad (9.74)$$

In the aerated atmosphere $^*[Cr^{II}(C_2O_4)_2(C_2O_4^{\bullet -})]^{3-}$ is quenched by molecular oxygen yielding the substituted Cr^{III} complex:

$$^*[Cr^{II}(C_2O_4)_2(C_2O_4^{\bullet -})]^{3-} + O_2 + 2H_2O \rightarrow [Cr^{III}(C_2O_4)_2(H_2O)_2]^{-} + C_2O_4^{\bullet -} + O_2^{-} \qquad (9.75)$$

or the Cr^{VI} species:

$$^*[Cr^{II}(C_2O_4)_2(C_2O_4^{\bullet -})]^{3-} + O_2 + 4OH^{-} \rightarrow CrO_4^{2-} + C_2O_4^{\bullet -} + 2\,C_2O_4^{2-} + 2H_2O \qquad (9.76)$$

whereas the $C_2O_4^{\bullet -}$ radical is oxidized to CO_2 (9.70) [87, 89–91, 93].

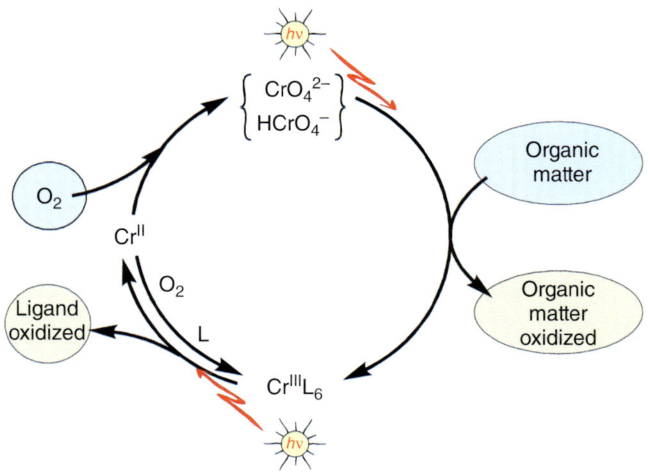

Figure 9.14 A model of environmental Cr^{III}/Cr^{VI} cycling based on the sunlight photoconversions between the chromium forms, accompanied by oxidation of internal or external electron donors by molecular oxygen

Each of these stages is thus accompanied by oxidation of the electron donor, which is connected with the Cr^{III} centre as a ligand or an external donor (typical for Cr^{VI} species). Both these steps are integral to each other in photocatalytic cycles, in which Cr^{III} and Cr^{VI} compounds play the role of photocatalysts and lead to removal of electron donors by their oxidation at the expense of molecular oxygen (Figure 9.14); the cycling leads finally to degradation or even complete mineralization of the environmental organic matter.

The chromium photocatalytic cycle differs from the iron or copper cycle, because it needs excitations of two different Cr forms in one cycle. This results in a greater diversification and larger mass of degraded organic pollutants. Beside monometallic cycles, the mixed metal systems can function in nature. The cooperation between the photocatalytic cycles of two transition metals can not only lead to variegation of the degraded pollutants, but also change the cycle efficiency, which might be either higher or lower in comparison with that of the single cycle [20, 96].

The photocatalytic cycles may be treated as an 'anti-photosynthesis' because they consume atmospheric oxygen and disperse energy accumulated by photosynthesis. These processes are of vital importance, because, theoretically, most atmospheric oxygen could be consumed, which would be a threat to life on Earth. Fortunately, in nature photosynthesis dominates and photocatalysis by transition metals plays only a minor, albeit useful, role of cleaning water basins of excess organic pollutants.

9.4 Photochemical Self-Cleaning in the Environment

Photochemical and photocatalytic processes are responsible for various self-cleaning behaviours in atmosphere, soils, or water basins.

The main atmospheric cleaner is ozone, photodissociation of which:

$$O_3 \xrightarrow{h\nu\,(\lambda<340\,\mathrm{nm})} O_2 + O(^1D) \tag{9.77}$$

is responsible for production of $OH^\bullet$ and $HO_2^\bullet$ radicals:

$$O(^1D) + H_2O \rightarrow 2OH^\bullet \tag{9.78}$$

$$OH^\bullet + O_3 \rightarrow HO_2^\bullet + O_2 \tag{9.79}$$

$$HO_2^\bullet + O_3 \rightarrow OH^\bullet + 2O_2 \tag{9.80}$$

as well as for production of atmospheric peroxides [3]:

$$HO_2^\bullet + HO_2^\bullet \rightarrow H_2O_2 + O_2 \tag{9.81}$$

Ozone, hydroxyl, and hydroperoxyl radicals are active in oxidation of hydrocarbons, aldehydes and ketones, whereas hydrogen peroxide is a source of hydroxyl radicals in the Fenton or photo-Fenton reactions (see section in 9.3.3 in Chapter 9 and Chapter 21).

Photocatalysis and photosensitization processes are common in natural waters and are of special importance because they can stimulate a transformation of molecules that resist direct photolysis, such as transparent species or chromophores, the reactive states of which are inefficiently populated by sunlight absorption.

Photocatalysis by transition metals shows an effective self-cleaning ability of soils and waters by the degradation and complete mineralization of the dissolved organic matter. Moreover, the oxidation processes are driven by sunlight and atmospheric oxygen and do not pollute the environment by risky side-products (see section 9.3, Chapter 9).

This activity is particularly useful for degradation of strongly hazardous substances or recalcitrant pollutants that are difficult to remove in chemical or biochemical processes. In this respect any pathway leading to abatement of chromate(VI) pollution arouses interest. One such pathway seems to be created by cooperation between iron and chromium photocatalytic cycles, which were reported as effective in conversion of chromate(VI) into Cr^{III} species [20–23, 97]. A synergistic photoreduction of Cr^{VI} and Cu^{II} mediated by TiO_2 [98], or photocatalytic reduction of Cr^{VI} and oxidation of organic matter by environmental polyoxometallates as photocatalysts [99], may constitute alternative possibilities.

The efficiency of the Cr(VI) photoreduction is, however, difficult to evaluation because intermediate species of Cr^V and Cr^{IV} undergo dismutation reactions, resulting in formation of Cr^{III} and partial regeneration of Cr^{VI}:

$$3Cr^V \rightarrow Cr^{III} + 2Cr^{VI} \tag{9.82}$$

Moreover, Cr^{III} complexes undergo photoreduction followed by oxidation to Cr^{VI} (equation 9.66). Both processes depend strongly on the pH and composition of the solution, component concentrations, and other parameters [91, 92, 94, 95].

Not only transition metal ions and complexes are active in the environmental self-cleaning, but also naturally occurring NO_2^- and NO_3^- ions have recently been reported to contribute to the photodegradation of pesticides [13].

Environmental photosensitization processes may be stimulated not only by natural but also by anthropogenic pollutants, such as phenols and dihydroxybenzenes, which are efficient photosensitizers. Among others, phenol was found to sensitize the dehalogenation of halogenophenols [99]. This behaviour shows that the simultaneous presence of adequate sensitizer and quencher in polluted environment can lead to the photosensitized decomposition of the latter and thereby to abatement of the pollution.

The self-cleaning mechanisms of photocatalysis and photosensitization apply also to organic matter included in microorganisms, so it can also lead to self-disinfection. This is especially expected near a semiconductor (particularly TiO_2) and/or iron deposits, which are responsible for solar photogeneration of reactive oxygen species capable of water disinfection [78, 79, 100, 101].

References

1. Parmon VM, Zamaraev KI. Photocatalysis in energy production. In: Serpone N, Pelizzetti E, eds. *Photocatalysis: Fundamentals and applications*. New York: John Wiley & Sons, 1989:565–602.
2. Braslavsky SE, Houk KN, Verhoeven JW. *Glossary of Terms used in Photochemistry*. International Union of Pure and Applied Chemistry, Organic Chemistry Division, Commission on Photochemistry, 1996.
3. Madronich S, Flocke S. The role of solar radiation in atmospheric chemistry In: Boule P, ed. *Environmental Photochemistry*. Berlin: Springer, 1999: 1–26.
4. Stasicka Z. *Photochemical Processes in the Environment*. Kraków: Jagiellonian University, 2001, 152 pp.
5. Koda S, Sugimoto K. Pressure effect on the absorption and photodissociation of O_2 near the dissociation threshold. *J Photochem Photobiol C: Photochem Rev* 2003;**4**:215–26.
6. Finlayson-Pitts BJ, Pitts JJN. *Atmospheric Chemistry: Fundamentals and experimental techniques*. New York: Wiley-Interscience, 1986.
7. Sadanaga Y, Matsumoto J, Kajii Y. Photochemical reactions in the urban air: Recent understandings of radical chemistry. *J Photochem Photobiol C: Photochem Rev* 2003;**4**:85–104.
8. Isidorov V, Klokova E, Povarov V, Kalkova S. Photocatalysis on atmospheric aerosols: Experimental studies and modeling. *Catalysis Today* 1997;**39**:233–42.
9. Leifer A. *The Kinetics of Environmental Aquatic Photochemistry*. ACS professional reference book. Washington DC: American Chemical Society, 1988.
10. Faust BC. Aquatic Photochemical Reactions in Atmospheric, Surface and Marine Waters: Influences of Oxidant Formation and Pollutant Degradation. In: Boule P, ed. *Environmental Photochemistry*. Berlin: Springer, 1999: 101–122.
11. Boule P, Bolte M, Richard C. Transformations photoinduced in aquatic media by NO_3^-, Fe(III) and humic substances. In: Boule P, ed. *Environmental Photochemistry*. Berlin: Springer, 1999: 181–216.
12. Vione D, Falletti G, Maurino V, et al. Sources and sinks of hydroxyl radicals upon irradiation of natural water samples. *Environ Sci Technol* 2006;**40**:3775–81.

13. Nelieu S, Shankar MV, Kerhoas L, Einhorn J. Phototransformation of monuron induced by nitrate and nitrite ions in water: Contribution of photonitration. *J Photochem Photobiol A: Chem* 2008;**193**:1–9.
14. Sakkas VA, Lambropoulou DA, Albanis TA. Photochemical degradation study of irgarol 1051 in natural waters: influence of humic and fulvic substances on the reaction. *J Photochem Photobiol A: Chem* 2002;**147**:135–41.
15. Kamiya M, Kameyama K. Photochemical effects of humic substances on degradation of organophosphorus pesticides. *Chemosphere* 1998;**36**:2337–44.
16. Chan KH, Chu W. Effect of humic acid on the photolysis of the pesticide atrazine in a surfactant-aided soil-washing system in acidic condition. *Water Research* 2005;**39**: 2154–66.
17. Zhao X, Quan X, Zhao H, Chen S, Zhao Y, Chen J. Different effects of humic substances on photodegradation of *p,p*-DDT on soil surfaces in the presence of TiO_2 under UV and visible light. *J Photochem Photobiol A: Chem* 2004;**167**:177–83.
18. Shichi T, Takagi K. Clay minerals as photochemical reaction fields. *J Photochem Photobiol C: Photochem Rev* 2000;**1**:113–30.
19. Smejkalova D, Piccolo A. Rates of oxidative coupling of humic phenolic monomers catalyzed by a biomimetic iron-porphyrin. *Environ Sci Technol* 2006;**40**:1644–9.
20. Ciesla P, Kocot P, Mytych P, Stasicka Z. Homogeneous photocatalysis by transition metal complexes in the environment. *J Mol Catalysis A: Chem* 2004;**224**:17–33.
21. Kocot P, Karocki A, Stasicka Z. Photochemistry of the Fe^{III}–EDTA complexes. A mechanistic study. *J Photochem Photobiol A: Chem* 2006;**179**:176–83.
22. Kocot P, Szaciłowski K, Stasicka Z. Photochemistry of the $[Fe^{III}(edta)(H_2O)]^-$ and $[Fe^{III}(edta)(OH)]^{2-}$ complexes in presence of environmentally relevant species. *J Photochem Photobiol A: Chem* 2007;**188**:128–34.
23. Cieśla P, Mytych P, Kocot P, Stasicka Z. Role of iron and chromium complexes in environmental self-cleaning processes. *Separ Sci Technol* 2007;**42**:1651–66.
24. Zang V, van Eldik R. Kinetics and mechanism of the autoxidation of iron(II) induced through chelation by ethylenediaminetetraacetate and related ligands. *Inorg Chem* 1990;**29**:1705–11.
25. Sebig S, van Eldik R. Kinetics of $[Fe^{II}(edta)]$ oxidation by molecular oxygen revisited. new evidence for a multistep mechanism. *Inorg Chem* 1997;**36**:4115–20.
26. Laine DF, McAllister SD, Cheng IF. Electrochemical characterization of oxygen reduction by Fe^{II}[ethylenediaminetetraacetate]. *J Electroanal Chem* 2007;**608**:111–16.
27. Dunford HB. Oxidations of iron(II)/(III) by hydrogen peroxide: from aquo to enzyme. *Coord Chem Rev* 2002;**233/234**:311–18.
28. Bozzi A, Yuranova T, Mielczarski E, et al. Superior biodegradability mediated by immobilized Fe-fabrics of waste waters compared to Fenton homogeneous reactions. *Appl Catalysis B: Environ* 2003;**42**:289–303.
29. Masarwa A, Rachmilovich-Calis S, Meyerstein N, Meyerstein D. Oxidation of organic substrates in aerated aqueous solutions by the Fenton reagent. *Coord Chem Rev* 2005;**249**:1937–43.
30. Lee J, Kim J, Choi W. Oxidation on zerovalent iron promoted by polyoxometalate as an electron shuttle. *Environ Sci Technol* 2007;**41**:3335–40.
31. Andreozzi R, Caprio V, Insola A, Marotta R. Advanced oxidation processes (AOP) for water purification and recovery. *Catalysis Today* 1999;**53**:51–9.
32. Malato S, Blanco J, Cáceres J, Fernández-Alba AR, Agüera A, Rodrıguez A. Photocatalytic treatment of water-soluble pesticides by photo-Fenton and TiO_2 using solar energy. *Catalysis Today* 2002;**76**:209–20.
33. Rodrıguez SM, Galvez JB, Rubio MIM, Ibanez PF, Gernjak W, Alberola IO. Treatment of chlorinated solvents by TiO_2 photocatalysis and photo-Fenton: influence of operating conditions in a solar pilot plant. *Chemosphere* 2005;**58**:391–8.
34 Peralta-Hernandez JM, Meas-Vong Y, Rodrıguez FJ, Chapman TW, Maldonado MI, Godınez LA. In situ electrochemical and photo-electrochemical generation of the

Fenton reagent: A potentially important new water treatment technology. *Water Res* 2006;**40**:1754–62.
35. Gernjak W, Krutzler T, Glaser A, et al. Photo-Fenton treatment of water containing natural phenolic pollutants. *Chemosphere* 2003;**50**:71–8.
36. Rodrıguez M, Malato S, Pulgarin C, et al. Optimizing the solar photo-Fenton process in the treatment of contaminated water. Determination of intrinsic kinetic constants for scale-up. *Solar Energy* 2005;**79**:360–8.
37. Quici N, Morgada ME, Piperata G, Babay P, Gettar RT, Litter MI. Oxalic acid destruction at high concentrations by combined heterogeneous photocatalysis and photo-Fenton processes. *Catalysis Today* 2005;**101**:253–60.
38. Katsumata H, Kaneco S, Suzuki T, Ohta K, Yobiko Y. Degradation of polychlorinated dibenzo-*p*-dioxins in aqueous solution by Fe(II)/H_2O_2/UV system. *Chemosphere* 2006;**63**:592–9.
39. Bautitz IR, Nogueira RFP. Degradation of tetracycline by photo-Fenton process – Solar irradiation and matrix effects. *J Photochem Photobiol A: Chem* 2007;**187**:33–9.
40. Gumy D, Fernandez-Ibanez P, Malato S, Pulgarin C, Enea O, Kiwi J. Supported Fe/C and Fe/Nafion/C catalysts for the photo-Fenton degradation of Orange II under solar irradiation. *Catalysis Today* 2005;**101**:375–82.
41. Feng J, Hu X, Yue PL, Zhu HY, Lu GQ. Degradation of azo-dye orange II by a photoassisted Fenton reaction using a novel composite of iron oxide and silicate nanoparticles as a catalyst. *Ind Eng Chem Res* 2003;**42**:2058–66.
42. Parra S, Henao L, Mielczarski E, et al. Synthesis, testing, and characterization of a novel nafion membrane with superior performance in photoassisted immobilized Fenton catalysis. *Langmuir* 2004;**20**:5621–9.
43. Sun J-H, Sun S-P, Fan M-H, Guo H-Q, Lee Y-F, Sun R-X. Oxidative decomposition of *p*-nitroaniline in water by solar photo-Fenton advanced oxidation process. *J Hazardous Mater* 2008;**153**:187–93.
44. Evgenidou E, Konstantinou I, Fytianos K, Poulios I. Oxidation of two organophosphorus insecticides by the photo-assisted Fenton reaction. *Water Res* 2007; **41**:2015–27.
45. Benitez FJ, Beltran-Heredia J, Acero JL, Rubio FJ. Chemical decomposition of 2,4,6-trichlorophenol by ozone, Fenton's reagent, and UV radiation. *Ind Eng Chem Res* 1999;**38**:1341–9.
46. Malato S, Cáceres J, Agüera A, et al. Degradation of imidacloprid in water by photo-Fenton and TiO_2 photocatalysis at a solar pilot plant: a comparative study. *Environ Sci Technol* 2001;**35**:4359–66.
47. Perez-Estrada L, Malato S, Gernjak W, et al. Photo-Fenton degradation of diclofenac: identification of main intermediates and degradation pathway. *Environ Sci Technol* 2005;**39**:8300–6.
48. Parra S, Guasaquillo I, Enea O, et al. Abatement of an azo dye on structured C-Nafion/Fe-Ion surfaces by photo-Fenton reactions leading to carboxylate intermediates with a remarkable biodegradability increase of the treated solution. *J Phys Chem B* 2003;**107**:7026–35.
49. Pignatello JJ, Liu D, Huston P. Evidence for an additional oxidant in the photoassisted Fenton reaction. *Environ Sci Technol* 1999;**33**:1832–9.
50. Farias J, Rossetti GH, Albizzati ED, Alfano OM. Solar degradation of formic acid: temperature effects on the photo-Fenton reaction. *Ind Eng Chem Res* 2007;**46**:7580–6.
51. Cermenati L, Pichat P, Guillard C, Albini A. Probing the TiO_2 photocatalytic mechanisms in water purification by use of quinoline, photo-Fenton generated OH radicals and superoxide dismutase. *J Phys Chem B* 1997;**101**:2650–8.
52. Kiwi J, Denisov N, Gak Y, et al. Catalytic Fe^{3+} clusters and complexes in Nafion active in photo-Fenton processes. high-resolution electron microscopy and femtosecond studies. *Langmuir* 2002;**18**:9054–66.
53. Rossetti GH, Albizzati ED, Alfano OM. Decomposition of formic acid in a water solution employing the photo-Fenton reaction. *Ind Eng Chem Res* 2002;**41**:1436–44.

54. Katsumata H, Kaneco S, Suzuki T, Ohta K, Yobiko Y. Photo-Fenton degradation of alachlor in the presence of citrate solution. *J Photochem Photobiol A: Chem* 2006;**180**:38–45.
55. Chacon JM, Leal MT, Sanchez M, Bandala ER. Solar photocatalytic degradation of azo-dyes by photo-Fenton process. *Dyes Pigments* 2006;**69**:144–50.
56. Rossetti GH, Albizzati ED, Alfano OM. Modeling of a flat-plate solar reactor. Degradation of formic acid by the photo-Fenton reaction. *Solar Energy* 2004;**77**:461–70.
57. Quici N, Morgada ME, Gettar RT, Bolte M, Litter MI. Photocatalytic degradation of citric acid under different conditions: TiO_2 heterogeneous photocatalysis against homogeneous photolytic processes promoted by Fe(III) and H_2O_2. *Appl Catal B: Environ* 2007;**71**:117–24.
58. Franch MI, Ayllon JA, Peral J, Domenech X. Enhanced photocatalytic degradation of maleic acid by Fe(III) adsorption onto the TiO_2 surface. *Catalysis Today* 2005;**101**: 245–52.
59. Ji H, Song W, Chen C, Yuan H, Ma W, Zhao J. Anchored oxygen-donor coordination to iron for photodegradation of organic pollutants. *Environ Sci Technol* 2007;**41**:5103–7.
60. Sykora J, Pado M, Tatarko M, Izakovic M. Homogeneous photo-oxidation of phenols: influence of metals. *J Photochem Photobiol A: Chem* 1997;**110**:167–75.
61. Andreozzi R, Marotta R. Removal of benzoic acid in aqueous solution by Fe(III) homogeneous photocatalysis. *Water Res* 2004;**38**:1225–36.
62. Andreozzi R, Canterino M, Marotta R. Fe(III) homogeneous photocatalysis for the removal of 1,2-dichlorobenzene in aqueous solution by means UV lamp and solar light. *Water Res* 2006;**40**:3785–92.
63. Lam SW, Chiang K, Lim TM, Amal R, Low GK-C. The role of ferric ion in the photochemical and photocatalytic oxidation of resorcinol. *J Catal* 2005;**234**:292–9.
64. Martyanov IN, Savinov EN. Mineralization of organic compounds in photochemical and photocatalytic systems. Comparative analysis for the example of methylviologen photooxidation. *Catalysis Today* 1997;**39**:197–205.
65. Domınguez JR, Beltran J, Rodrıguez O. Vis and UV photocatalytic detoxification methods (using TiO_2, TiO_2/H_2O_2, TiO_2/O_3, $TiO_2/S_2O_8^{2-}$, O_3, H_2O_2, $S_2O_8^{2-}$, Fe^{3+}/H_2O_2 and $Fe^{3+}/H_2O_2/C_2O_4^{2-}$) for dyes treatment. *Catalysis Today* 2005;**101**:389–95.
66. Poulain L, Mailhot G, Wong-Wah-Chung P, Bolte M. Photodegradation of chlortoluron sensitised by iron(III) aquacomplexes. *J Photochem Photobiol A: Chem* 2003;**159**: 81–8.
67. Meštánková H, Krýsa J, Jirkovský J, Mailhot G, Bolte M. The influence of Fe(III) speciation on supported TiO_2 efficiency: example of monuron photocatalytic degradation. *Appl Catal B: Environ* 2005;**58**:185–191.
68. Kwan CY, Chu W, Lam WS. The role of oxalate in the kinetics of 2,4-D oxidation over ferrous ion-supported catalysts. *J Mol Catal A: Chem* 2007;**274**:50–7.
69. Mestankova H, Mailhot G, Jirkovsky J, Krysab J, Bolte M. Mechanistic approach of the combined (iron–TiO_2) photocatalytic system for the degradation of pollutants in aqueous solution: an attempt of rationalisation. *Appl Catal B: Environ* 2005;**57**:257–65.
70. Chen Y, Wu F, Lin Y, Deng N, Bazhin N, Glebov E. Photodegradation of glyphosate in the ferrioxalate system. *J Hazardous Mater* 2007;**148**:360–5.
71. Catastini C, Sarakha M, Mailhot G, Bolte M. Iron(III) aquacomplexes as effective photocatalysts for the degradation of pesticides in homogeneous aqueous solutions. *Sci Total Environ* 2002;**298**:219–28.
72. Oller I, Gernjak W, Maldonado MI, Perez-Estrada LA, Sanchez-Perez JA, Malato S. Solar photocatalytic degradation of some hazardous water-soluble pesticides at pilot-plant scale. *J Hazardous Mater B* 2006;**138**:507–17.
73. Sioi M, Bolosis A, Kostopoulou E, Poulios I. Photocatalytic treatment of colored wastewater from medical laboratories: Photocatalytic oxidation of hematoxylin. *J Photochem Photobiol A: Chem* 2006;**184**:18–25.
74. Yang P, Lu C, Hua N, Du Y. Titanium dioxide nanoparticles co-doped with Fe^{3+} and Eu^{3+} ions for photocatalysis. *Mater Lett* 2002;**57**:794–801.

75. Antoniadis A, Takavakoglou V, Zalidis G, Poulios I. Development and evaluation of an alternative method for municipal wastewater treatment using homogeneous photocatalysis and constructed wetlands. *Catalysis Today* 2007;**124**:260–5.
76. Dantas TLP, Mendonca VP, Jose HJ, Rodrigues AE, Moreira RFPM. Treatment of textile wastewater by heterogeneous Fenton process using a new composite Fe_2O_3/carbon. *Chem Eng J* 2006;**118**:77–82.
77. Rodrıguez M, Sarria V, Esplugas S, Pulgarin C. Photo-Fenton treatment of a biorecalcitrant wastewater generated in textile activities: biodegradability of the photo-treated solution. *J Photochem Photobiol A* 2002;**151**:129–36.
78. Rincon A-G, Pulgarin C. Fe^{3+} and TiO_2 solar-light-assisted inactivation of *E. coli* at field scale. Implications in solar disinfection at low temperature of large quantities of water. *Catalysis Today* 2007;**122**:128–36.
79. Rincon A-G, Pulgarin C. Comparative evaluation of Fe^{3+} and TiO_2 photoassisted processes in solar photocatalytic disinfection of water. *Appl Catal B: Environ* 2006;**63**:222–31.
80. Zhang H, Lindberg SE. Sunlight and iron(III)-induced photochemical production of dissolved gaseous mercury in freshwater. *Environ Sci Technol* 2001;**35**:928–35.
81. Sykora J. Photochemistry of copper complexes and their environmental aspects. *Coord Chem Rev* 1997;**159**:95–108.
82. Madden TH, Datye AK. Oxidation of metal-EDTA complexes by TiO_2 photocatalysis. *Environ Sci Technol* 1997;**31**:3475–81.
83. Yang J-K, Davis AP. Photocatalytic Oxidation of Cu(II)-EDTA with illuminated TiO_2: mechanisms. *Environ Sci Technol* 2000;**34**:3796–801.
84. Yang J-K, Davis AP. Competitive photocatalytic oxidation of Cu(II)-EDTA and Cd(II)-EDTA with illuminated TiO_2. *Environ Sci Technol* 2001;**35**:3566–70.
85. Rehorek D, Jarzynowski T, Stasicka Z. Über die Photoreduktion von Carboxylato (cuproin)kupfer(II) Chelaten in Alkoholen. *J prakt Chem* 1980;**322**:411–16.
86. Kotaś J, Stasicka Z. Chromium occurrence in the environment and methods of its speciation. *Environ Pollution* 2000;**107**:263–83.
87. Mytych P, Cieśla P, Stasicka Z. Photoredox reactions of environmental chromium. *Int J Photoenergy* 2001;**3**:1–6.
88. Kane-Maguire NAP. Photochemistry and photophysics of coordination compounds: chromium. In: Balzani V, Campagna S, eds. *Photochemistry and Photophysics of Coordination Compounds*. Berlin: Springer, 2007: 37–67.
89. Marchaj A, Stasicka Z, Rehorek D. Photochemical production of chromate(VI) from some chromium(III) complexes. *Polyhedron* 1983;**2**:1281–6.
90. Mainusch B, Karocki A, Guldi DM, Stasicka Z, Wasgestian F. Charge transfer photochemistry of Reinecke's salt and some of its analogues containing organic amines. *Inorg Chim Acta* 1997;**255**:87–93.
91. Cieśla P, Karocki A, Stasicka Z. Photoredox behaviour of the Cr–EDTA complex and its environmental aspects. *J Photochem Photobiol A: Chem* 2004;**162**:537–44.
92. Mytych P, Cieśla P, Stasicka Z. Photoredox processes in the Cr(VI)–Cr(III)–oxalate system and their environmental relevance. *Appl Catal B: Environ* 2005;**59**:161–70.
93. Mainusch B, Wasgestian F, Stasicka Z, Karocki A. Photochemie von Thiocyanatochromaten bei Einstrahlung in die Charge-Transfer-Bande. *J Inf Rec Mat* 1994;**21**:687–9.
94. Mytych P, Karocki A, Stasicka Z. Mechanism of photochemical reduction of chromium(VI) by alcohols and its environmental aspects. *J Photochem Photobiol A: Chem* 2003;**160**: 163–70.
95. Mytych P, Stasicka Z. Photochemical reduction of chromium(VI) by phenol and its halogen derivatives. *Appl Catal B: Environ* 2004;**52**:167–72.
96. Szaciłowski K, Macyk W, Stochel G, Stasicka Z, Sostero S, Traverso O. Ligand and medium controlled photochemistry of iron and ruthenium mixed-ligand complexes: prospecting for versatile systems. *Coord Chem Rev* 2000;**208**:277–97.
97. Liu Y, Deng L, Chen Y, Wu F, Denga N. Simultaneous photocatalytic reduction of Cr(VI) and oxidation of bisphenol A induced by Fe(III)–OH complexes in water. *J Hazard Mater B* 2007;**139**:399–402.

98. Goeringer S, Chenthamarakshan CR, Rajeshwar K. Synergistic photocatalysis mediated by TiO_2: mutual rate enhancement in the photoreduction of Cr(VI) and Cu(II) in aqueous media. *Electrochem Commun* 2001;**3**:290–292.
99. Richard C, Grabner G. Mechanism of phototransformation of phenol and its derivatives in aqueous solution. In: Boule P, ed. *Environmental Photochemistry*. Berlin: Springer Verlag, 1999: 217–40.
100. Li Q, Xie R, Li YW, Mintz EA, Shang JK. Enhanced visible-light-induced photocatalytic disinfection of *E. coli* by carbon-sensitized nitrogen-doped titanium oxide. *Environ Sci Technol* 2007;**41**:5050–6.
101. Sichel C, Blanco J, Malato S, Fernandez-Ibanez P. Effects of experimental conditions on E. coli survival during solar photocatalytic water disinfection. *J Photochem Photobiol A: Chem* 2007;**189**:239–46.

10

Heterogeneous (Photo)Catalysis and Biogenesis on Earth

An ocean traveler has even more vividly the impression that the ocean is made of waves than that it is made of water.
Arthur S. Eddington

10.1 (Photo)catalysis of Chalcogenide Semiconductors

Water basins on early Earth must have contained large quantities of suspended solid particles, resulting from mechanical erosion of rocks as well as from meteorites and comets. Therefore the 'primitive' or 'primordial' soup can also be referred to as the 'clay soup'. Aqueous suspensions of clay minerals are known to absorb various organic molecules from solutions [1]. Inorganic particles serve as mechanical support or restraint for organic molecules, which can in turn influence their photochemical reactivity [1–4]. Layered mineral particles could also serve as primitive genetic information carriers [5]. Furthermore, numerous inorganic materials exhibit semiconducting properties, so photocatalytic processes may contribute to transformation of organic materials [6, 7].

Semiconductors are considered to be catalytic particles that contributed to the development of primitive metabolism. According to Wähtershäuser life could have developed on the surface of iron sulphide minerals (eg mackinawite or pyrrhotite [FeS] or pyrite [FeS_2]) [8]. The chemoautotrophy theory is based on the reaction between iron sulphide and hydrogen sulphide, which acts as a reducing agent, whereas iron sulphide provides adsorption sites for substrates and acts as a catalyst (equation 10.1):

$$FeS + H_2S \rightarrow FeS_2 + 2[H] \qquad (10.1)$$

Figure 10.1 *Hydrothermal vents, so-called 'black smokers', carry high concentration of sulphur compounds with traces of methane and hydrogen together with suspended sulphide particles. (Picture by courtesy of Professor Verena Tunnicliffe, University of Victoria, Canada)*

$$CH_3SH + CO \rightarrow CH_3COSH \tag{10.2}$$

$$CH_3COSH + H_2O \xrightarrow{(Fe,Co,Ni)S} CH_3COOH + H_2S \tag{10.3}$$

The FeS–H_2S system is capable of reducing alkenes to alkanes; furthermore it catalyzes the transformation of thioacetic acid to acetic acid (equations 10.1–10.3) as well as more complex reactions [9]. This reaction cannot, however, result in direct reduction of carbon dioxide. Alternatively, CO_2 is first converted to COS and CS_2, which are subsequently reduced by the pyrrhotite/pyrite system. Numerous complex organic compounds can be found in these reaction mixtures, including methyl mercaptan (CH_3SH) and dimethylsulphide (CH_3SCH_3). The former is a good starting material for subsequent FeS-catalyzed transformations (equations 10.1–10.3). Thioacetic acid is a molecular precursor of the acetyl-CoA cofactor involved in citrate cycle. In this context the hydrothermal vents located at the bottom of the oceans (so-called black smokers, Figure 10.1) seem to be the most suited place for prebiotic synthesis of organic compounds. Furthermore, synthesis of hydrocarbons according to the Fischer–Tropsch mechanism should be considered under conditions within hydrothermal vents [10]. At these places a supercritical (360 °C) aqueous solution containing dissolved gases and minerals meets cold water (2–4 °C). Acidic hydrothermal fluid mixes with cold alkaline medium, creating a very large thermal and chemical gradient. This results in precipitation of numerous minerals (sulphides, chalcopyrite, anhydrite) in the form of 'chimneys' and 'beehives', which are considered as natural reaction kettles for synthesis of complex organic materials [10, 11]. The vicinity of the hydrothermal vents is usually colonized by strange animals such as giant tube worms (eg *Riftia pachyptila*), which are marine invertebrates in the phylum Annelida.

An extension of Wähtershäuser's original idea was presented by Edwards [12]. This alternative photoautotrophic scenario accounts for significant semiconductor properties of pyrites and other transition metal sulphides. The high quantum efficiencies of charge separation at the surface of pyrites on visible irradiation may be due to the presence of the surface redox-active iron–sulphur clusters acting as charge traps. Furthermore, these centres should be considered in this context as functional analogues of the cores of ferredoxins and other iron–sulphur proteins. In a photoautotrophic mechanism these surface iron–sulphur clusters might be seen as forming a redox system analogous to the FeS_2–FeS–H_2S system, 'powered' by photochemically generated charge carriers. In a cyclic process photoelectrons generated in the pyrite surface might initially convert clusters in the oxidized state to the reduced state. In the next step the clusters would then be reoxidized in an exergonic reaction analogous to the driving reaction of Wähtershäuser's model. The electrons liberated in this step would be used to reduce external cluster ligands, eg CO_2 or CO, to biochemically relevant species such as formate and acetate. These products should initially remain coordinated to the surface clusters; furthermore, they can undergo further photoinduced redox reactions. The holes in the pyrite could later be neutralized by electrons from aqueous redox reactions or can contribute to oxidation of the underlying pyrite body (ie photocorrosion). In the presence of sulphides or organic sulphur compounds photocorrosion would, however, be of minor importance. To start the next cycle the clusters in the oxidized state would again be reduced photochemically. In contrast with the model proposed by Wähtershäuser, there would thus be no formation of a pyrite waste product, which might potentially disrupt the delicate assemblies of chemical species [12].

The photocatalytic processes that could have been involved in the synthesis of prebiotic molecules are not limited to iron sulphides. Photocatalytic processes occurring on the surface of other sulphide and oxide materials should be considered as well.

10.2 Photocatalytic Nitrogen Fixation

In 1977 Schrauzer and Guth reported photocatalyzed reduction of N_2 to ammonia upon irradiation of the rutile–TiO_2 doped with 0.2% Fe_2O_3 [13]. Photoreduction was observed only in the presence of humid nitrogen, and water vapour was oxidized to molecular oxygen. In general a higher content of iron oxide resulted in inactive materials. The process was also studied in detail by other groups, although the nature of reducing agent remains unclear because in many cases no oxygen formation was detected [14–20]. Typical ammonia concentrations reached in this process were in the range 1–10 $\mu mol\, l^{-1}$. Although the iron-doped material shows visible light absorption, the photoreduction occurs only on UV-light irradiation.

Kisch et al. reported photoreduction of N_2 at iron titanate films with a Ti:Fe molar ratio 1:1 [21, 22]. The postulated new semiconducting phase $Fe_2Ti_2O_7$

catalyzed nitrogen photoreduction in the presence of ethanol even upon visible light irradiation. The following mechanism points at main steps of the process (SEM is a semiconductor particle which upon excitation offers reactive changes – e_r^- and h_r^+:

$$SEM + h\nu \rightarrow SEM(e_r^-, h_r^+) \quad (10.4)$$

$$SEM(e_r^-, h_r^+) + H_2O \rightarrow SEM(h_r^+) + H_{ad} + OH^- \quad (10.5)$$

$$SEM(h_r^+) + CH_3CH_2OH \rightarrow SEM + CH_3^{\bullet}CHOH + H^+ \quad (10.6)$$

$$SEM + CH_3^{\bullet}CHOH \rightarrow SEM(e_r^-) + CH_3CHO + H^+ \quad (10.7)$$

$$SEM(e_r^-) + H_2O \rightarrow SEM + H_{ad} + OH^- \quad (10.8)$$

$$N_2 + 2H_{ad} \rightarrow N_2H_2 \quad (10.9)$$

$$SEM(e_r^-, h_r^+) \rightarrow SEM + heat \quad (10.10)$$

Ammonia and hydrazine formation was also observed when a neat semiconductor or semiconductor (TiO_2, WO_3) doped with metal ions other than Fe^{III} (Cr, Mn, Co) were used as photocatalyst [23, 24]. In the presence of oxygen nitric oxides and nitrates can also be formed.

10.3 Photocatalytic Carbon Dioxide Reduction

Reduction of CO_2 photocatalyzed by semiconducting materials may lead to formation of formaldehyde, formic acid, methanol, methane, and oxalate among other products [7]. The earliest report on photocatalytic reduction of CO_2 with water at TiO_2 was published by Inoue et al. [25]. Almost at the same time the photoreduction of CO_2 to methane at $SrTiO_3$ was reported by Hemminger et al. [26]. The same process can also be performed on AgCl/zeolites [27].

Metal(0) deposited at the TiO_2 surface may enhance yields and selectivity of the reduction products [28–30]. In the presence of copper(0)-loaded TiO_2 formaldehyde and methanol were detected as main products [31]. Mercury- and platinum-loaded TiO_2 favoured formation of carbon and formaldehyde [28, 29]. Methane was obtained with a high selectivity in the case of palladium-doped TiO_2, whereas in the presence of Rh, Au, and Ru metals acetic acid was also produced [30].

Carbon dioxide photoreduction was also observed with other heterogeneous photocatalysts such as SiC, CdS, GaP, ZnO, WO_3, $BaTiO_3$ [32], $SrTiO_3$ [33], and GaP [34, 35]. Low yields of the process result from the reverse oxidation reactions.

Heterogeneous systems similar to those described above, capable of CO_2 and N_2 photofixation and catalysing other transformations of small molecules, may be responsible for genesis of life in interstellar space and on the young Earth. This hypothesis was recently discussed by Serpone et al. [32] and Greenberg [36]. Also photosynthesis of urea from inorganic compounds of carbon and nitrogen catalyzed by TiO_2 was reported [37].

10.4 'Fossils' of Prebiotic Catalysts: Metal Clusters in Active Centres of Metalloenzymes

Although there is no direct evidence for involvement of semiconductor particles in the prebiotic synthesis of organic matter and the origin of life, we can search for some hints in modern organisms. All living organisms contain numerous proteins with embedded metal ions, which in many cases form cluster assemblies. These cluster motifs closely resemble the structure of semiconducting minerals such as mackinawite [38], greigite [39], smythite [40], and sphalerite [41] (Figure 10.2).

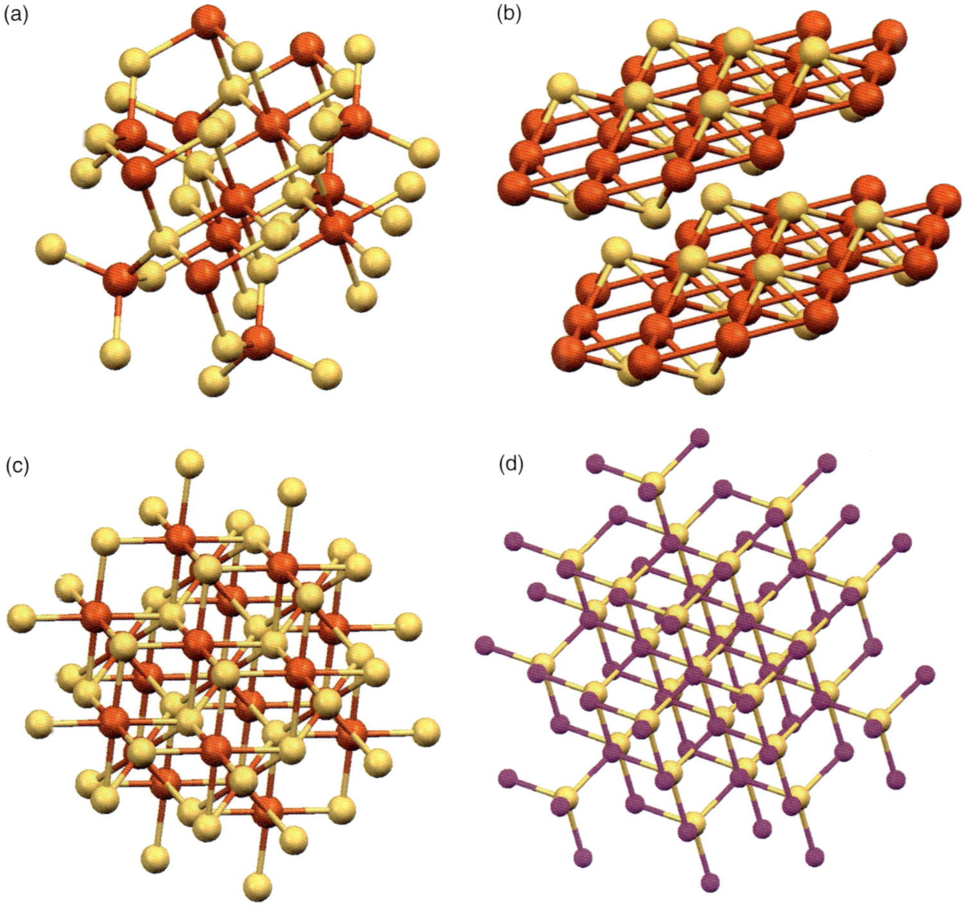

Figure 10.2 Crystal structures of (a, b, c) natural iron and (d) zinc sulphides: (a) greigite, (b) mackinawite, (c) smythite, and (d) sphalerite

Numerous transition metals ions form cluster complexes with chalcogenide anions [42–52]. Iron and sulphur are unique elements in the sense that no two other elements can generate such a large diversity of cluster structures. This is the consequence of two stable oxidation states of iron ions and strong Fe–S bonds of significantly covalent character [53]. Moreover, numerous structures are stable in several oxidation states, so these clusters serve as electron reservoirs in biological systems [51]. This is why iron–sulphur proteins usually catalyze redox reactions.

The first metal–sulphide clusters were recognized in photosynthetic centres of all plants, nitrogen-fixing bacteria, and mitochondrial proteins of mammalian origin. It soon turned out that all these proteins contain closely related iron–sulphur clusters (Figure 10.3). More recent studies indicated that iron–sulphur proteins are ubiquitous in all living matter, including extremophilic Archaebacteria.

Iron–sulphur complexes and clusters are important prosthetic groups for many metalloenzymes. They constitute the most ancient, ubiquitous, and structurally and functionally diverse class of enzymes. Their primary function is mediation of one-electron redox processes, substrate binding, and maintenance of protein structure. The iron–sulphur proteins may be regarded as the interface between the inorganic and the biological world [61, 62]. As a result of the close structural relationship to natural iron sulphides they are supposed to be descendants of the first catalytic systems on Earth. Iron–sulphur cluster species could have been of prime importance to biogenesis, especially near high-temperature volcanic vents, because iron and sulphur compounds are still the source of energy for chemoautotrophic organisms living under extreme conditions, eg Archaebacteria, which often live in extreme conditions that were once considered inhospitable to life [63, 64]. Enzymes containing Fe–S cores usually catalyze reactions involving components of the Earth's primordial atmosphere, as N_2, CO, and H_2 [61].

Apart from simple monomeric complexes such as $[Fe(SR)_4]^-$, numerous oligomeric species containing from 2 to 18 iron ions are formed in solutions containing iron ions, sulphide, and aliphatic or aromatic thiols, which in most cases are identical with the iron–sulphur cofactors found in enzymes [45, 51, 52, 65–67]. Structures containing infinite polymeric chains are also reported [68]. Despite the structural diversity of iron–sulphur clusters all of them are built from very few simple building blocks. The main constitutional unit for most of the Fe–S clusters is the $[Fe_2S_2]$ rhomb [43, 69, 70]. Edge sharing, Fe-vertex sharing, and S-vertex sharing lead to various complex structures. Also cubic core $[Fe_4S_4]$ may be treated as a building block in some more complex structures of higher nuclearity (see Figure 10.3f, g) [71, 72]. In natural systems the protein-bound Fe–S units are usually coordinated by cysteine ligands; in some cases histidine and proline ligands are also found. Synthetic analogues of Fe–S units may contain a large variety of terminal ligands, including thiolates, phosphines, amines, cyclopentadienes, carbonyls, and nitrosyls [45]. Regardless of the cluster structure, its charge, and nature of the terminal ligand, the Fe–S bond lengths, and Fe–S–Fe and S–Fe–S bond angles are very similar. The Fe–S distance falls within 0.215 nm and 0.235 nm, the acute Fe–S–Fe angle within 74–76° and the obtuse S–Fe–S angle within 103–115° [45, 69].

Iron–sulphur clusters constitute integral parts of several natural structures occurring in a large family of biologically relevant metalloproteins. The [Fe–S] units

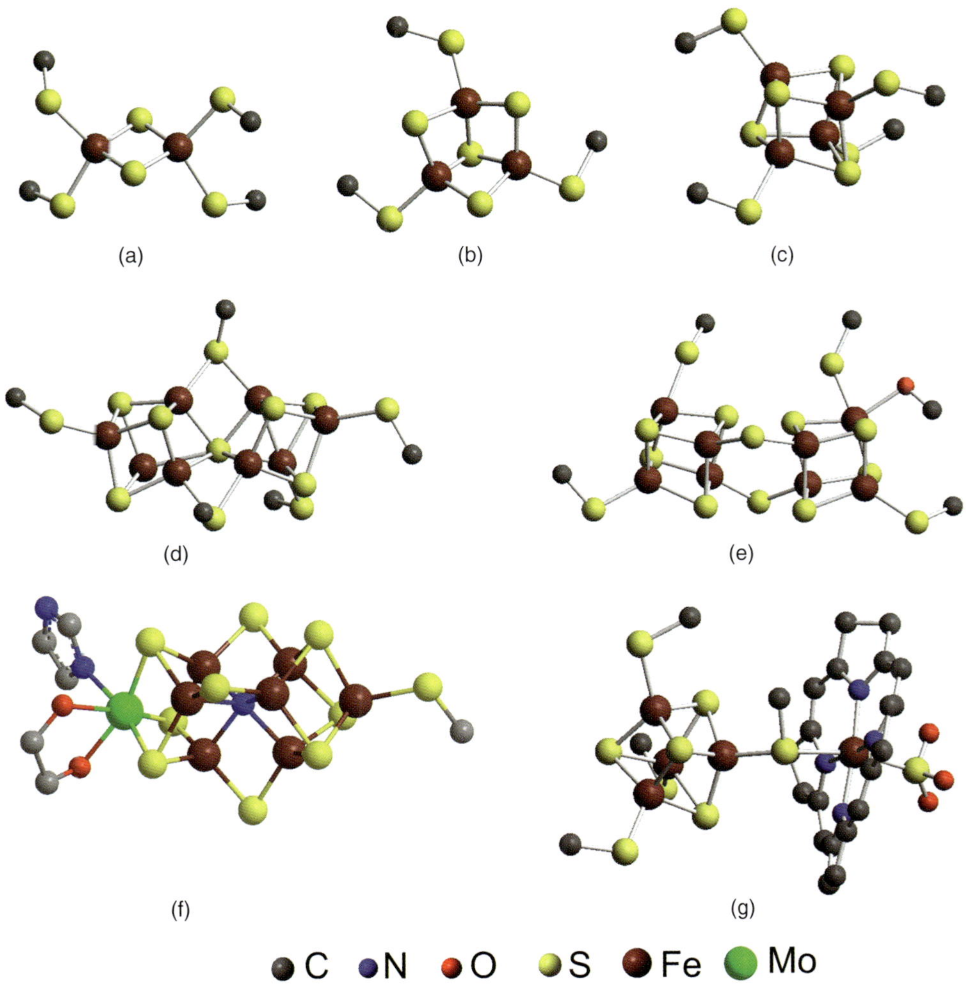

Figure 10.3 *Crystallographic structures of the most important iron–sulphur centres in proteins: (a) dimeric centre of ferrodoxin from Spinacia oleracea [54], (b) trimeric centre of ferredoxin from Bacillus schlegelii [55], (c) cubane cluster of nitrogenase reductase from Azotobacter vinelandii [56], (d) nitrogenase octameric cluster from Azotobacter vinelandii [57], (e) nitrogenase octameric cluster from Clostridium pasteurianum [58], (f) MoFe cluster of nitrogenase from Azotobacter vinelandii [59], and (g) active centre of sulphite reductase from Escherichia coli [60]*

form the active sites of enzymes, which play a crucial role in the processes of living organisms, eg electron transfer in photosynthesis and cellular respiration, nitrogen fixation, and many various catalytic reactions; they can also operate as biosensors of oxidants and iron ions. There are four principal iron–sulphur structures observed in biological systems: $[Fe(SR)_4]^n$, $[Fe_2S_2(SR)_2]^n$, $[Fe_3S_4(SR)_2]^n$, and $[Fe_4S_4(SR)_4]^n$ (see Figure 10.3) [73]. Centres containing up to eight iron ions are supposed to be derived from smaller clusters via sulphide or thiolate bridges (see above). Structures of these iron centres are almost preserved in all iron–sulphur proteins, whilst their redox potentials vary from –600 mV to +500 mV, a range larger than for any other simple redox cofactor. At the same time all the bond lengths and angles within the FeS moiety are almost unchanged.

These iron–sulphur cores isolated from various enzymes are identical with structural fragments found in natural iron sulphides. In the structure of natural iron sulphide mineral, greigite, two different structural motives can be found: tetrahedral iron centres with four sulphide ligands [FeS_4] and iron–sulphur cubes [Fe_4S_4] (see Figure 10.2a). Another modification of iron sulphide, mackinawite, contains infinite two-dimensional layers consisting of [Fe_2S_2] units (Figure 10.2b). Smythite, in turn, contains [Fe_3S_4] units (see Figure 10.2c). All these structural motifs are crucial elements of iron–sulphur cores of numerous enzymes. Therefore iron–sulphur proteins can be regarded as 'molecular fossils' of the most primitive enzymatic centres: surfaces of natural iron sulphide crystals can be considered as an interface between the inanimate and animate world.

Another important couple, which presumably originates from the prebiotic era, is the zinc–sulphur system. Zinc is an abundant element in physiological media and easily forms complexes with amino acids' oxygen, nitrogen and sulphur atoms. In most of the protein–zinc complexes the metal ion is coordinated with sulphur atoms: approximately 3% of all the human genes encode zinc–sulphur proteins [74]. There are two main families of zinc–sulphur proteins: zinc fingers and metallothioneins. The former family usually contains one zinc ion coordinated to cysteine and histidine side chains, whereas the latter contains numerous zinc cations assembled in cluster moieties, containing up to four cations in each cluster. Metallothioneins may also contain cadmium–sulphur and copper–sulphur clusters [46]. All metallothionein clusters are based on tetrahedral coordination motif which can be found also in zinc (wurzite, sphalerite) and cadmium (grenockite, hawleyite) sulphides. Metallothioneins are responsible for control of intracellular zinc concentration, protection against disulfide shock and may be also involved in some sulphur-centred reactions.

References

1. Shichi T, Takagi K. Clay minerals as photochemical reaction fields. *J Photochem Photobiol C: Photochem Rev* 2000;**1**:113–30.
2. Corma A, Garcia H. Zeolite-base photocatalysts. *Chem Commun* 2004:1443–59.
3. Schulz-Ekloff G, Wöhrle D, van Duffel B, Schoonheydt RA. Chromophores in porous silicas and minerals: preparation and optical properties. *Microporous Mesoporous Mater* 2002;**51**:91–138.
4. Takagi S, Eguchi M, Tryk DA, Inoue H. Porphyrin photochemistry in inorganic/organic hybrid materials: clays, layered semiconductors, nanotubes and mesoporous materials. *J Photochem Photobiol C: Photochem Rev* 2006;**7**:102–26.
5. Cairns-Smith AG. Chemistry and the missing era of evolution. *Chem Eur J* 2008;**14**: 3830–9.
6. Kisch H. Semiconductor photocatalysis for organic synthesis. In: Neckers DC, von Bünau G, Jenks WS, eds. *Advances in Photochemistry*. Chichester: John Wiley & Sons, Inc., 2001: 93–143.

7. Linsebigler AL, Lu G, Yates JT Jr. Photocatalysis on TiO_2 surfaces: Principles, mechanisms, and selected results. *Chem Rev* 1995;**95**:735–58.
8. Wähtershäuser G. The origin of life and its methodological challenge. *J Theor Biol* 1997;**187**:483–94.
9. Lazcano A, Miller SL. On the origin of metabolic pathways. *J Mol Evol* 1999;**49**:424–31.
10. McCollon TM, Seewald JS. Abiotic synthesis of organic compounds in deep-sea hydrothermal environments. *Chem Rev* 2007;**107**:382–401.
11. Rickard D, Luther GW III. Chemistry of iron sulfides. *Chem Rev* 2007;**107**:514–62.
12. Edwards MR. Metabolite channeling in the origin of life. *J Theor Biol* 1996;**179**:313–22.
13. Schrauzer GN, Guth TD. *J Am Chem Soc* 1977;**99**:7189–93.
14. Soria J, Conesa JC, Augugliaro V, Palmisano L, Schiavello M, Sclafani A. Dinitrogen photoreduction to ammonia over titanium dioxide powders doped with ferric ions. *J Phys Chem* 1991;**95**:274–82.
15. Sclafani A, Palmisano L, Schiavello M. N_2 photoreduction and phenol and nitrophenol isomers photooxidation as examples of heterogeneous photocatalytic reactions. *Res Chem Intermed* 1992;**18**:211–26.
16. Taqui-Khan MM, Chatterjee D, Bala M. Photocatalytic reduction of N_2 to NH_3 sensitized by the [Ru^{III}-ethylenediaminetetraacetate-2,2'-bipyridyl]-complex in a Pt-TiO_2 semiconductor particulate system. *J Photochem Photobiol A: Chem* 1992;**67**:349–52.
17. Augugliaro V, Soria J. Concerning 'An opinion on the heterogeneous photoreduction of N_2 with H_2O': First letter. *Angew Chem Int Ed Eng* 1993;**105**:579.
18. Palmisano L, Schiavello M, Sclafani A. Concerning 'An opinion on the heterogeneous photoreduction of N_2 with H_2O': Second letter. *Angew Chem Int Ed Eng* 1993;**105**:580.
19. Rao NN, Dub S, Manjubala P, Natarajan P. Photocatalytic reduction of nitrogen over (Fe, Ru or Os)/TiO_2 catalysts. *Appl Catal B: Environ* 1994;**5**:33–42.
20. Litter MI, Navio JA. Photocatalytic properties of iron-doped titania semiconductors. *J Photochem Photobiol A: Chem* 1996;**98**:171.
21. Rusina O, Eremenko A, Frank G, Strunk HP, Kisch H. Nitrogen photofixation at nanostructured iron titanate films. *Angew Chem Int Ed* 2001;**40**:3993.
22. Rusina O, Linnik O, Eremenko A, Kisch H. Nitrogen photofixation on nanostructured iron titanate films. *Chem Eur J* 2003;**9**:561–5.
23. Hoshino K, Inui M, Kitamura T, Kokado H. Fixation of dinitrogen to a mesoscale solid salt using a titanium oxide/conducting polymer systems. *Angew Chem* 2000; **112**:2558–61.
24. Schiavello M. Basic concepts in photocatalysis. In: Schiavello M, ed. *Photocatalysis and Environment Trends and Applications*. Dordrecht: Kluwer, 1988: 351–6.
25. Inoue I, Fujishima A, Konishi S, Honda K. Photoelectrocatalytic reduction of carbon dioxide in aqueous suspensions of semiconductor powders. *Nature* 1979;**277**:637.
26. Hemminger JC, Carr R, Somorjai GA. The photoassisted reaction of gaseous water and carbon dioxide adsorbed on $SrTiO_3(111)$ face to form methane. *Chem Phys Lett* 1978;**57**:100–4.
27. Saladin F, Kamber I, Pfanner K, Calzaferri G. Photochemical water oxidation to oxygen at the solid/gas interface of AgCl on zeolite A. *J Photochem Photobiol A: Chem* 1997;**109**:47–52.
28. Tennakone K. Photoreduction of carbonic acid by mercury coated *n*-titanium dioxide. *Sol Energy Mater* 1984;**10**:235–8.
29. Rophael WM, Malati MA. The photocatalysed reduction of aqueous sodium carbonate to carbon using platinised titania. *J Chem Soc Chem Commun* 1987:1418–20.
30. Ishitani O, Inoue C, Suzuki Y, Ibusuki T. Photocatalytic reduction of carbon dioxide to methane and acetic acid by an aqueous suspension of metal-deposited TiO_2. *J Photochem Photobiol A: Chem* 1993;**72**:269–71.
31 Hirano K, Inoue K, Yatsu T. Photocatalysed reduction of CO_2 in aqueous TiO_2 suspension mixed with copper powder. *J Photochem Photobiol A: Chem* 1992;**64**:255–8.

32. Emeline AV, Otroshchenko VA, Ryabchuk VK, Serpone N. Abiogenesis and photostimulated heterogeneous reactions in the interstellar medium and on primitive earth. Relevance to the genesis of life. *J Photochem Photobiol C: Photochem Rev* 2003;**3**:203–24.
33. Aurian-Blajeni B, Halmann M, Manasses J. Photoreduction of carbon dioxide and water into formaldehyde and methanol on semiconductor materials. *Sol Energy* 1980;**25**:165–70.
34. Ito K, Yoshida M, Ohta S, Iida T. On the reduction products of carbon dioxide at a *p*-type gallium phosphide photocathode in aqueous electrolytes. *Bull Chem Soc Jpn* 1984;**57**:583–4.
35. Ikeda S, Yamamoto A, Noda H, Maeda M, Ito K. Influence of surface treatment of the *p*-GaP photocathode on the photoelectrochemical reduction of carbon dioxide. *Bull Chem Soc Jpn* 1993;**66**:2473–7.
36. Greenberg JM. Cosmic dust and our origins. *Surf Sci* 2002;**500**:793–822.
37. Kuwabata S, Yamauchi H, Yoneyama H. Urea photosynthesis from inorganic carbon and nitrogen compounds using TiO_2 as photocatalyst. *Langmuir* 1998;**14**:1899.
38. Lennie AR, Redfern SAT, Schofield PF, Vaughan DJ. Synthesis and Rietveld crystal structure refinement of mackinawite, tetragonal FeS. *Miner Mag* 1995;**59**:677–83.
39. Skinner BJ, Erd RC, Grimaldi FS. Greigite, the thio-spinel of iron; a new mineral. *Am Mineralogist* 1964;**49**:543–55.
40. Erd RC, Evans HT, Richter DH. Smythite, a new iron sulfide, and associated pyrrhotite from Indiana. *Am Mineralogist* 1957;**42**:309–33.
41. Skinner BJ. Unit-cell edges of natural and synthetic sphalerites. *Am Mineralogist* 1961;**46**:1399–411.
42. Degroot MW, Corrigan JF. High nuclearity clusters: metal–chalcogenide polynuclear complexes. In: McCleverty JA, Meyer TJ, eds. *Comprehensive Coordination Chemistry II*. Amsterdam: Elsevier, 2003: 57–123.
43. Niu Y, Zheng H, Hou H, Xin X. Heterothiometallic polymeric clusters. *Coord Chem Rev* 2004;**248**:169–83.
44. Sokolov MN, Fedin VP. Chalcogenide clusters of vanadium, niobium and tantalum. *Coord Chem Rev* 2004;**248**:925–44.
45. Ogino H, Inomata S, Tobita H. Abiological iron-sulfur clusters. *Chem Rev* 1998; **98**:2093–121.
46. Henkel G, Krebs B. Metallothioneins: zinc, cadmium, mercury and copper thilates and selenolates mimicking protein active site features – structural aspects and biological implications. *Chem Rev* 2004;**104**:801–24.
47. Kalinina IV, Fedin VP. Cubane chalcogenide complexes of group IV–VI metals: Synthesis, structure, and properties. *Russ J Coord Chem* 2003;**29**:597–615.
48. Kelly PF. Oxygen, sulfur, selenium and tellurium. *Annu Rep Prog Chem Sect A* 1999;**95**:67–78.
49. Dey S, Jain VK. Platinum group metal chalcogenides. *Platinum Metals Rev* 2004; **48**:16–29.
50. Shibahara T. Syntheses of sulfur-bridged molybdenum and tungsten coordination compounds. *Coord Chem Rev* 1993;**123**:73–147.
51. Broderick JB. Iron-sulfur clusters in enzyme catalysis. In: McCleverty JA, Meyer TJ, eds. *Comprehensive Coordination Chemistry II*. Amsterdam: Elsevier, 2003: 739–58.
52. Rao PV, Holm RH. Synthetic analogues of the active sites of iron-sulfur proteins. *Chem Rev* 2004;**104**:527–59.
53. Holm RH. Electron transfer: iron-sulfur clusters. In: McCleverty JA, Meyer TJ, eds. *Comprehensive Coordination Chemistry II*. Amsterdam: Elsevier, 2003: 61–90.
54. Binda C, Coda A, Aliverti A, Zanetti G, Mattevi A. Structure of the mutant E92K of [2Fe-2S] ferrodoxin I from *Spinacia oleracea* at 1.7 Å resolution. *Acta Crystallogr D* 1998;**54**:1353–8.
55. Aono S, Bentrop D, Bertini I, Donaire A, Luchinat C. Solution structure of the oxidized Fe_7S_8 ferredoxin from the thermophilic bacterium *Bacillus schlegelii* by ^1H NMR spectroscopy. *Biochemistry* 1998;**37**:9812–26.

56. Strop P, Tatahara PM, Chiu H-J, Angove HC. Crystal structure of the all-ferrous $[4Fe\text{-}4S]^0$ form of the nitrogenase iron protein from Azotobacter vinelandii. *Biochemistry* 2001;**40**:651–6.
57. Schmid B, Ribbe MW, Einsle O, et al. Structure of a Cofactor-Deficient Nitrogenase MoFe Protein. *Science* 2002;**296**:352–6.
58. Kim J, Rees DC. Structural models for the metal centers in the nitrogenase molybdenum-iron protein. *Science* 1992;**257**:1677–82.
59. Einsle O, Tezcan FA, Andrade SL, et al. Nitrogenase MoFe-protein at 1.16 Å resolution: A central ligand in the FeMo-cofactor. *Science* 2002;**297**:1696–700.
60. Crane BR, Siegel LM, Getzoff ED. Probing the catalytic mechanism of sulfite reductase by X-ray crystallography: Structures of the *Escherichia coli* hemoprotein in complex with substrates, inhibitors, intermediates, and products. *Biochemistry* 1997;**36**:12120–37.
61. Rees DC, Howard JB. The interface between the biological and inorganic worlds: iron-sulfur clusters. *Science* 2003;**300**:929–31.
62. Beinert H, Holm RH, Münck E. Iron-sulfur clusters, nature's modular, multipurpose structures. *Science* 1997;**277**:653–9.
63. Huber C, Wächtershäuser G. Peptides by activation of amino acids with CO on (Ni,Fe)S surfaces: implications for the origin of life. *Science* 1998;**281**:670–2.
64. Johnson MK. Iron-sulfur clusters: new roles for old compounds. *Curr Opinion Chem Biol* 1998;**2**:173–81.
65. Nordlander E, Whalen AM. Iron 1992. *Coord Chem Rev* 1995;**142**:43–99.
66. Ward MD. Iron. *Coord Chem Rev* 1992;**115**:1–116.
67. Nelson SM. Iron(III) and higher oxidation states. In: Wilkinson G, Gillard RD, McClevery JA, eds. *Comprehensive Coordination Chemistry*. Oxford: Pergamon Press, 1987: 234–47.
68. Bronger W. Darstellung, Kristallstruktur und magnetische Eigenschaften von Alkalithioferraten(III). *Z Anorg Allg Chem* 1968;**359**:225–33.
69. You JF, Snyder BS, Papaefthymiou GC, Holm RH. On the molecular/solid state boundary. A cyclic iron-sulfur cluster of nuclearity of eighteen: synthesis, structure and properties. *J Am Chem Soc* 1990;**112**:1067–76.
70. Long GR, Holm RH. Protein-bound iron-sulfur clusters: application of a structural database. *Inorg Chim Acta* 1995;**229**:229–39.
71. Goh C, Segal BM, Huang J, Long GR, Holm RH. Polycubane clusters: synthesis of $[Fe_4S_4(PR_3)_4]^{1+,0}$ (R = Bu^t, Cy, Pr^i) and $(Fe_4S_4)^0$ core aggregation upon loss of phosphine. *J Am Chem Soc* 1996;**118**:11844–53.
72. Holm RH. Site-differentiated and sulfide-bridged cubane clusters in chemistry and biology. *Pure Appl Chem* 1998;**70**:931–8.
73. Holm RH, Kennepohl P, Solomon EI. Structural and functional aspects of metal sites in biology. *Chem Rev* 1996;**96**:2239–314.
74. Maret W. Zinc and sulfur: A critical biological partnership. *Biochemistry* 2004;**43**: 3301–9.

11

Foundation and Evolution of Photosynthesis

All sunshine makes a desert.
Arab proverb

The surface crust and atmosphere of the Earth are exposed to two sources of energy: geothermal energy and solar radiation [1, 2] (Figure 11.1). Sunlight supplies energy to the whole terrestrial environment: atmosphere, hydrosphere, lithosphere, and biosphere. Living organisms require a continual input of free energy from the environment. Photosynthetic organisms (phototrophs) obtain this energy by trapping sunlight, whereas chemotrophs obtain energy through the oxidation of foodstuffs generated by phototrophs.

A living process can be seen as a network of material, energy, and information flow in a structured cell or set of cells in contact with the environment. The energy harvesting and consumption in biosystems are related to inorganic elements because magnesium is involved in light harvesting, iron and copper in electron transfer, protons, sodium, and potassium cations as well as chloride anion in electrolytic currents, and phosphorous (as phosphate and pyrophosphate) in energy storage. Therefore the energy network must be seen not only in the context of genome and proteome but mostly in the context of the metallome (inorganic species) involved in almost all aspects of bioenergetics [1, 2].

Earth probably exists for over 4 billion years. Organic molecules were formed haphazardly for a long period both through the action of light and on catalytic surfaces. Organization of chemistry in space and time evolved and is called life once it could reproduce. For a long period only simple cells, prokaryotes, developed. At some stage from about 2.7 billion years an evolution of molecular oxygen in the atmosphere developed due to prokaryote metabolism and then photosynthesis [1–4] (Figure 11.2).

Photosynthesis is a biological process in which the energy of light radiation is converted into the energy of chemical bonds [3, 5–8]. The ability to convert light

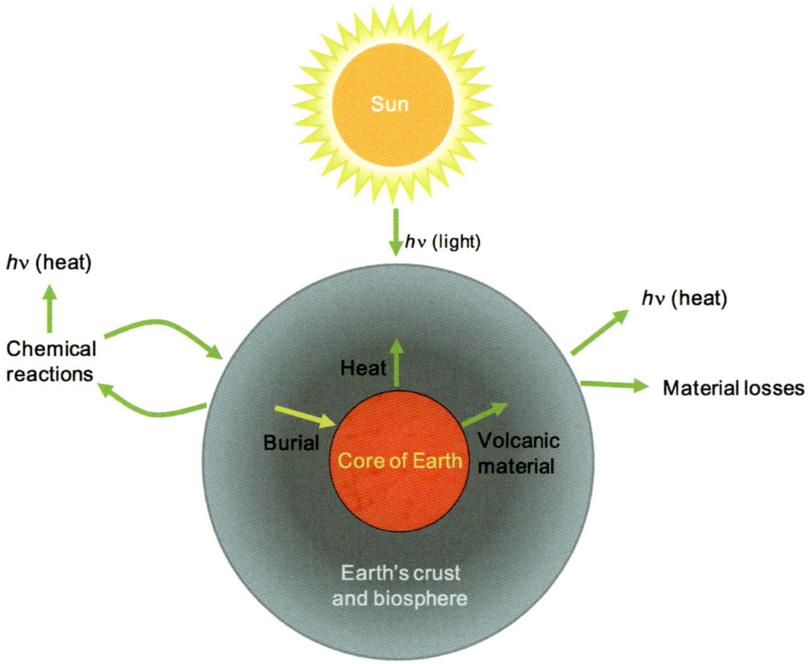

Figure 11.1 *The energy and material transfer to and from Earth's crust and biosphere*

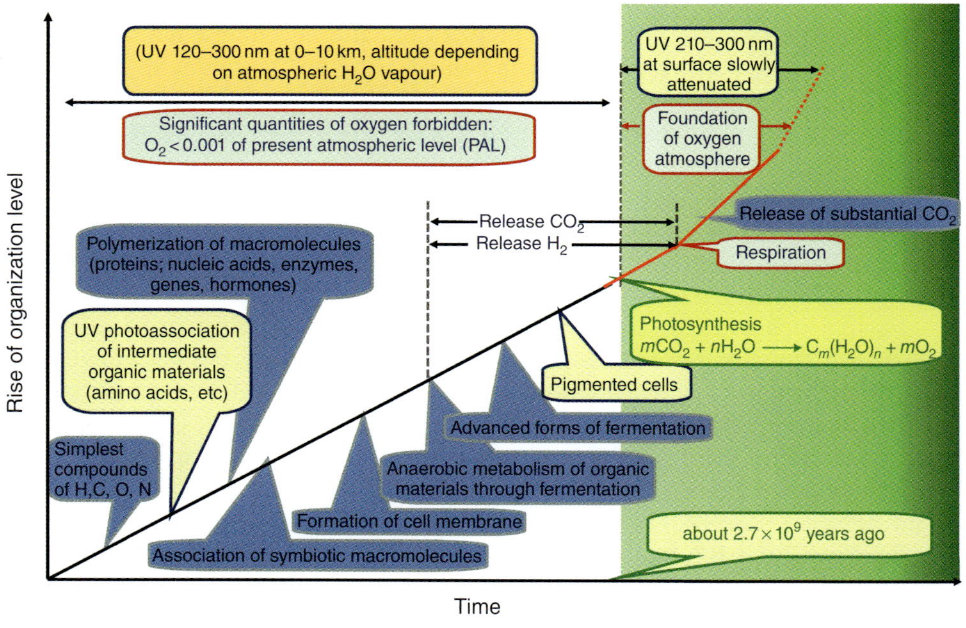

Figure 11.2 *Photosynthesis and oxygen atmosphere evolution on Earth. (Adapted from Frausto Da Silva et al. [1])*

energy into chemical energy is a tremendous evolutionary advantage. Photosynthesis did not evolve immediately at the origin of life (Figure 11.2). As no photosynthetic organisms have been discovered in the domain of *Archea*, it is assumed that photosynthesis evolved in the domain of bacteria. The foundations on which photosynthetic structures evolved were presumably electron transfer chains consisting of such components as the ubiquinone–cytochrome *c* oxidoreductase, cytochrome *bf* family or iron–sulphur clusters [1–3, 9].

Geological evidence suggests that oxygenic photosynthesis became important more than 2 billion years ago (Figure 11.2). Anaerobic photosynthetic systems arose much earlier in the 3.5-billion-year history of life on the Earth. Among anaerobic groups of photosynthetic prokaryotes green and purple sulphur bacteria use H_2, H_2S, and S as electron donors whereas green non-sulphur and purple non-sulphur bacteria usually use organic molecules:

$$CO_2 + 2H_2S \rightarrow (CH_2O) + 2S + H_2O \qquad (11.1)$$

One of the most important milestones in the Earth's history occurred when a photosynthetic organism figured out how to use water as a reductant in photosynthesis. This marked the onset of aerobic photosynthesis, and resulted in the transformation of the atmosphere from its primordial anaerobic state to the present aerobic state. The biosphere was changed dramatically by this reaction. Among the many transformations was the evolution of aerobic respiration, which enabled high-energy requiring organisms such as humans to exist [4]. The basic equation of aerobic photosynthesis is simple but the mechanism involved is complex:

$$CO_2 + 2H_2O \rightarrow (CH_2O) + O_2 \qquad (11.2)$$

From the point of view of organic synthesis, the overall process consists of the formation of carbohydrates $(CH_2O)_n$ by the reduction of carbon dioxide. The essence of the process is the use of photochemical energy to split water and concomitantly to reduce CO_2. Many proteins and small molecules are involved in photosynthetic machinery. Inorganic species are in the centre of photosynthesis as pigments in light harvesting, substrates, products, catalysts, and electron transfer mediators.

Oxygen released as a by-product of photosynthesis continues to sustain respiring organisms. The conversion of solar energy into biomass by photosynthesis is the source of most of the energy available to life on Earth. This includes not only the on-going production of high-energy organic molecules by plants that support the food chain, but also the excess production of biomass over the Earth's geological history that has been buried in the form of coal, oil, and gas. Most of our current energy comes from burning these fossil fuels. It is estimated that the total mass of organic material produced by green plants during the biological history of the Earth represents 1% of the planet's mass, and that photosynthesis annually fixes the equivalent of ten times humankind's energy consumption.

11.1 Photosynthetic Structures

Primary processes of photosynthesis take place in pigment–protein complexes of thylakoid membranes of plant or algae chloroplasts, cyanobacteria, and photosynthetic bacteria. Green sulphur bacteria contain reaction centres of type I with extra-membrane light-harvesting antennae complexes, chlorosomes; purple bacteria contain reaction centres of type II with two major types of membrane light-harvesting complexes LH1 and LH2. The photosynthetic apparatus of cyanobacteria, red algae, green algae, and higher plants consists of photosystem I (PSI), photosystem II (PSII,) and various membrane or extra-membrane light-harvesting antennae complexes [10].

The most primitive photosystem is that of green bacteria which deposit elemental sulphur. It resembles photosystem I of the chloroplasts of plants today. Here the transfer of charge generates reducing equivalents via NAD inside the cell and oxidizing equivalents as sulphur from sulphide outside the cell. Simultaneously a proton gradient is created across the membrane (acid outside) and this gradient is an energy store to be used subsequently to make ATP. In this photosystem magnesium ions in bacteriochlorophylls and iron in cytochrome *b/c* are involved. (Figure 11.3a) [1, 11].

The cyclic photosynthesis occurred in the purple bacteria system (Figure 11.3b). Light drives electrons and hence the separation of charge across the membrane to give an electrical potential ψ. Membrane-bound potential ψ is initially formed by action of light on a reaction centre of two bacteriochlorophyll molecules, to give an excited biradical state $BChl^+BChl^-$. The electron passes from $BChl^-$ to another single bacteriochlorophyll molecule and then in succession to single bacteriopheophytin and quinone molecules, before H^+ moves into the membrane of the bacteria from water, leaving negative charge as OH^- in the inner aqueous phase of the bacterial cell. Meanwhile, the initial positive charge of $BChl^+$ has been moved from this site in the opposite direction by the input of electrons from a cytochrome molecule, which in turn gets electrons from QH_2 leaving positive charge as H^+ in the outer aqueous phase. This system is more like PSII of modern chloroplasts (Figure 11.3b). The cyclic nature of the system follows from the flow of quinone Q, between an oxidized form (Q) and a reduced form (QH_2) [1].

A schematic view of bacterial photosynthetic reaction centre with a chain of electron-carrying metal prosthetic groups is shown in Figure 11.4. The metal ions involved are magnesium (bacteriochlorophylls) and iron (haem and non-haem iron species).

Today, in chloroplasts the electrons of photosystem II go to Q_A/Q_B but not just in a cyclic fashion, as these quinones are coupled to non-cyclic electrons from H_2O giving O_2.

The invention of aerobic photosynthesis, the light-driven oxidation of water to oxygen, stands as one of the pivotal evolutionary innovations in the history of life on Earth. The process is carried out only at the oxygen-evolving complex (OEC) of PSII in plants and algae, as well as in cyanobacteria. Despite the biological uniqueness of water oxidation to O_2, several of the core proteins of PSII have homologues in the so-called type I and type II anaerobic photosynthetic reaction

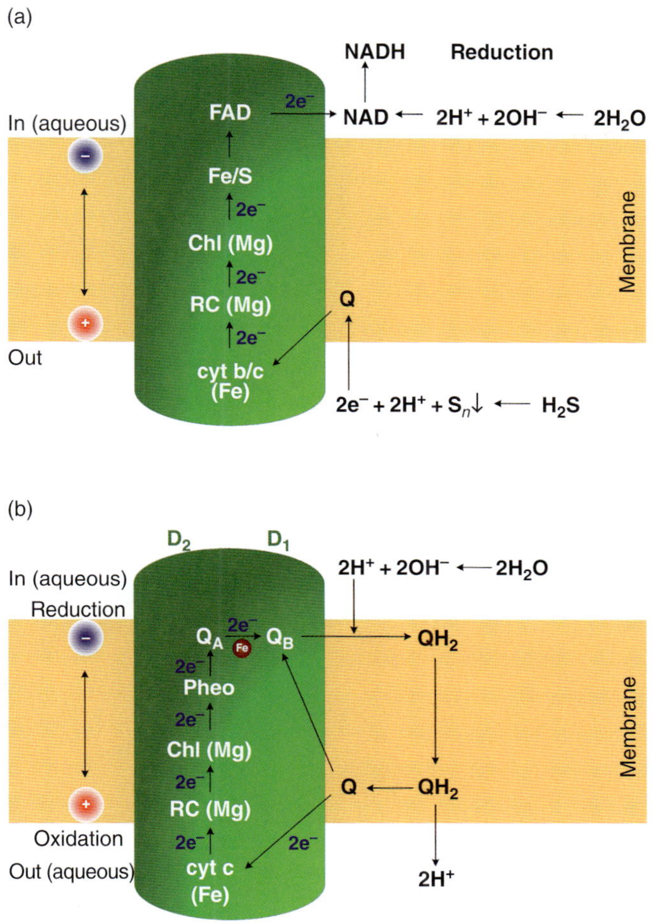

Figure 11.3 *Schematic view of the simple anaerobic photosynthetic systems in (a) green bacteria and (b) purple bacteria*

centres found in six groups of bacteria. The key difference between aerobic and anaerobic photosynthetic machinery is the presence and key role of manganese cluster at the site of water oxidation in PSII, enabling H_2O to serve as a stepwise electron donor. This is distinct from bacterial anaerobic reaction centres, which rely on redox active periplasmic proteins as electron donors. Ultimately, these electrons are derived from a variety of (exogenous) reduced compounds such as hydrogen sulphide, ferrous iron, organic carbon, or nitrite. However, no anaerobic reaction centres have been described that are able to generate a redox potential strong enough to oxidize water, a very weak electron donor. The OEC, and its unique chemistry, is foremost among the conserved components of the aerobic photosynthetic machinery [11].

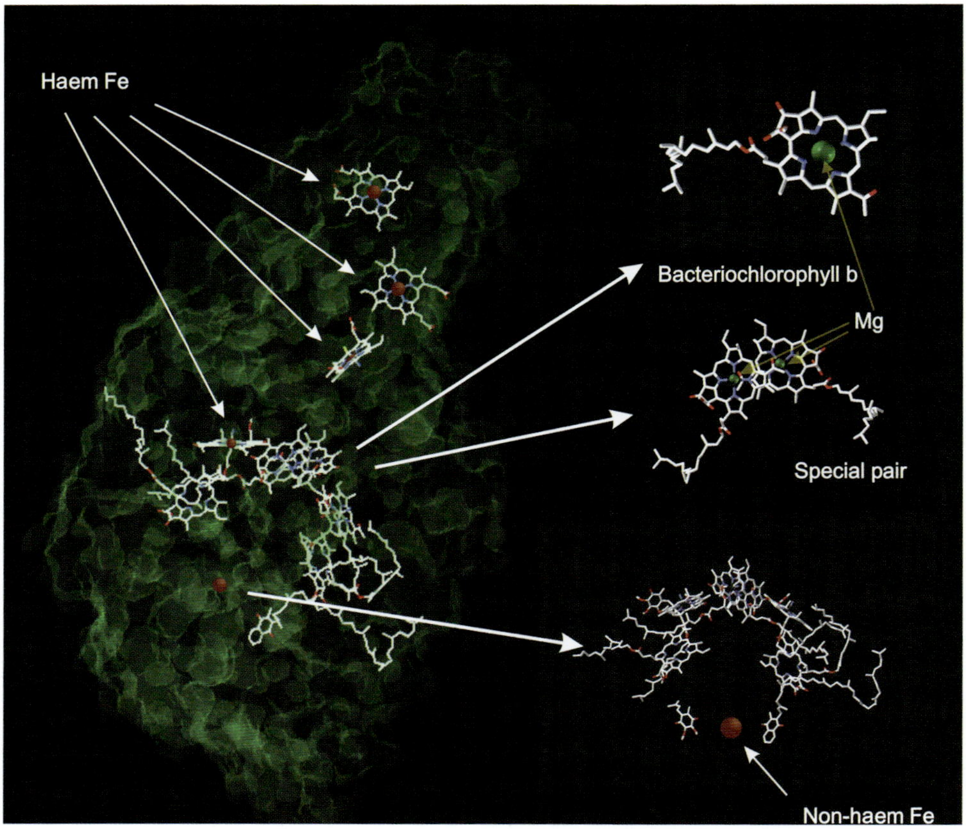

Figure 11.4 *View of the bacterial photosynthetic reaction centre showing inorganic cofactors. (Data from Protein Data Bank (DOI 10.2210/pdb2I5N/pdb))*

11.2 Aerobic Photosynthesis

Aerobic photosynthesis is carried out by cyanobacteria, green algae, and higher plants, and is the fundamental biological process by which solar energy converts atmospheric CO_2 and water into biomass and the molecular oxygen that is essential for all human and animal life on the Earth.

Photosynthesis in green plants takes place in organelles called chloroplasts. Chloroplasts have an outer and an inner membrane, with an intervening intermembrane space. The inner membrane surrounds a space called stroma, which is the site of the carbon chemistry of photosynthesis. In the stroma there are membrane structures called thylakoids. In the thylakoid membranes are localized the energy-transforming machinery, light-harvesting proteins, reaction centre, electron transport chains, and ATP synthase (Figure 11.5).

Photosynthesis proceeds in two parts: the light and the dark reactions. The light reactions transform light energy into ATP and biosynthetic reducing energy,

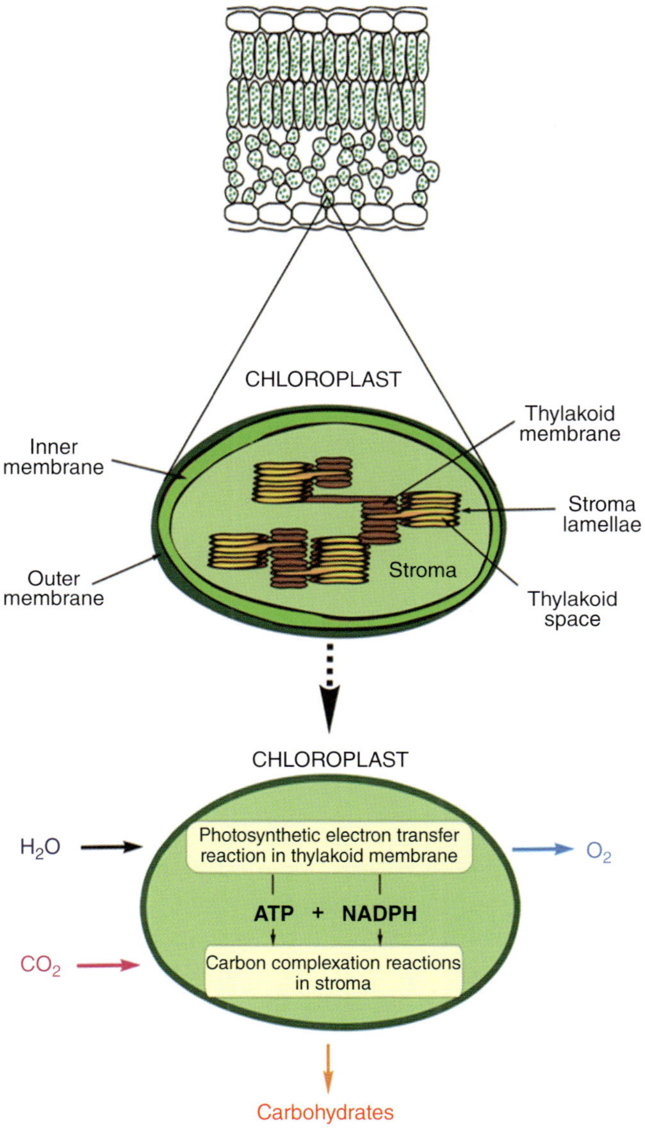

Figure 11 5 *Chloroplast and information about light and dark reactions of photosynthesis*

NADPH. The dark reactions use the ATP and NADPH produced by light reactions to reduce carbon atoms from CO_2 to a more reduced state as carbohydrates (Figure 11.5).

The light-driven, energy- and electron-transfer processes trigger reactions that finally lead to the oxidation of water, the reduction of $NADP^+$, and the build-up of a proton gradient across the photosynthetic membrane to produce ATP. These processes are catalyzed by two membrane-embedded pigment–protein complexes, called

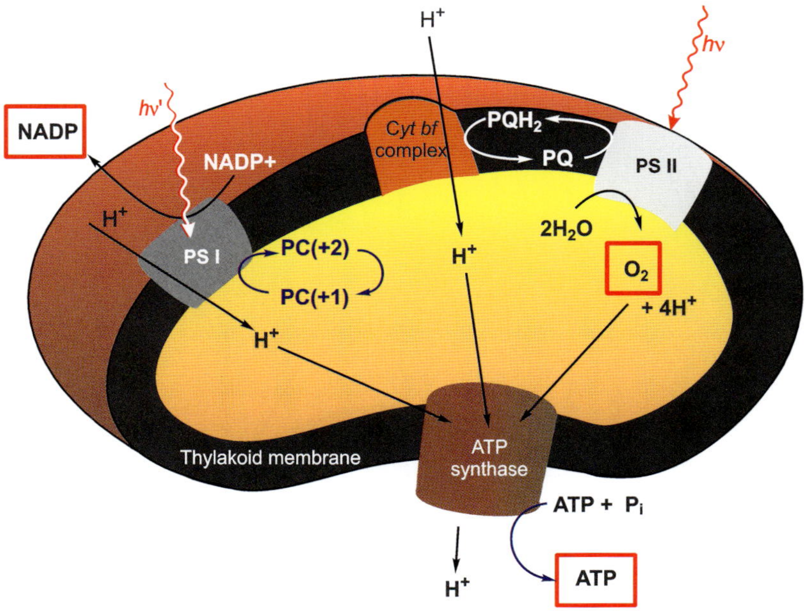

Figure 11.6 *Schematic view of photosynthetic centre in chloroplasts (photosystems I and II, cytochrome* bf *complex) and light reactions of photosynthesis*

PSI and PSII. PSII contains the site of water cleavage, and utilizes the electrons extracted from water to reduce plastoquinone to plastoquinol. The latter diffuses through the membrane until it is reoxidized by another membrane protein, the cytochrome *bf* complex, which transfers the electrons to a water-soluble electron carrier (plastocyanin or cytochrome c_6). This carrier in turn is oxidized by PSI, which delivers the electrons via ferredoxin to the enzymes that produce NADPH (Figure 11.6) [12].

11.2.1 Photosystem II (PSII)

PSII is an enzyme found in the thylakoid membranes of aerobic photosynthetic organisms. The protein complex forms a dimer with a molecular mass of around 650 kDa. Although PSII is a large protein complex, much of its bulk is involved in harvesting and transferring photonic energy. Most of the redox active cofactors are bound to the central proteins D1 and D2. The actual oxidation of water to oxygen is carried out at the metal cluster referred to as the OEC. Biochemical and biophysical experiments have established that four manganese atoms and one calcium atom constitute the inorganic part of the OEC. Beyond these five atoms, a number of other components have been suggested for OEC, among them chloride and bicarbonate ions [13].

In the past 5 years, there has been dramatic progress in the structural characterization of PSII by X-ray crystallography. Combined with a wealth of experimental data from biochemical investigations, site-directed mutagenesis, spectroscopic studies, computational modelling, and inorganic model chemistry, our understanding of the structure and catalytic chemistry of the Mn_4Ca cluster is becoming increasingly clear. However, with the current resolution of 0.30–0.35 nm, there are still uncertainties about the amino acid residues that coordinate to the metal ions [14, 15].

11.2.2 Photosystem I (PSI)

The final step of the light part of photosynthesis in chloroplasts is catalyzed by transmembrane complex called PSI. PSI is a pigment–protein complex consisting of many subunits, which binds more than a hundred cofactors (including chlorophylls, carotenoids, lipids, iron–sulphur clusters, phylloquinones). It combines the functions of light-harvesting and transmembrane charge separation in one complex. Through this combination, each photon absorbed by the antenna pigments is funnelled to the reaction centre (RC). The RC is made up of two symmetry-related branches of redox-active cofactors. Here, the excitation energy is converted into an initial charge-separated state. The subsequent fast transport of an electron over a chain of successive electron acceptors stabilizes the charge separation, which is a prerequisite for the following slower redox reactions with the soluble electron carriers [12, 16].

11.3 Light Harvesting Antennae (LHC)

During the course of evolution, both PSI and PSII have acquired extrinsic or intrinsic, peripheral light-harvesting antennas (LHCs) (Figure 11.7). The extension of the light-harvesting capacity has been necessary because sunlight is a relatively dilute energy source. Even on a cloudless day the radiation flux of solar energy would result in absorption by each chlorophyll (Chl) molecule of only approximately 10 photons per second. Moreover the density of chlorophyll *a* molecules in the reaction centre is not very high. The light-harvesting antennae allow photosynthetic organisms to be more efficient. On the other hand, selective pressure during the evolution of the antenna complexes under changing environmental conditions was directed towards efficient delivery of the excitation energy to the RC before its dissipation. Undelivered or excess energy is harmful for photosystems because it is converted into the long-lived Chl triplet states capable of generating reactive singlet oxygen. For PSI, nature solved this dilemma by acquiring antennae with significant structural diversity. The remarkably different structural forms of the various antennae complexes of PSI, including distinct pigment types and organizational principles, strongly suggest that they have evolved independently in response to both the general need to increase the absorption cross-section and the specific

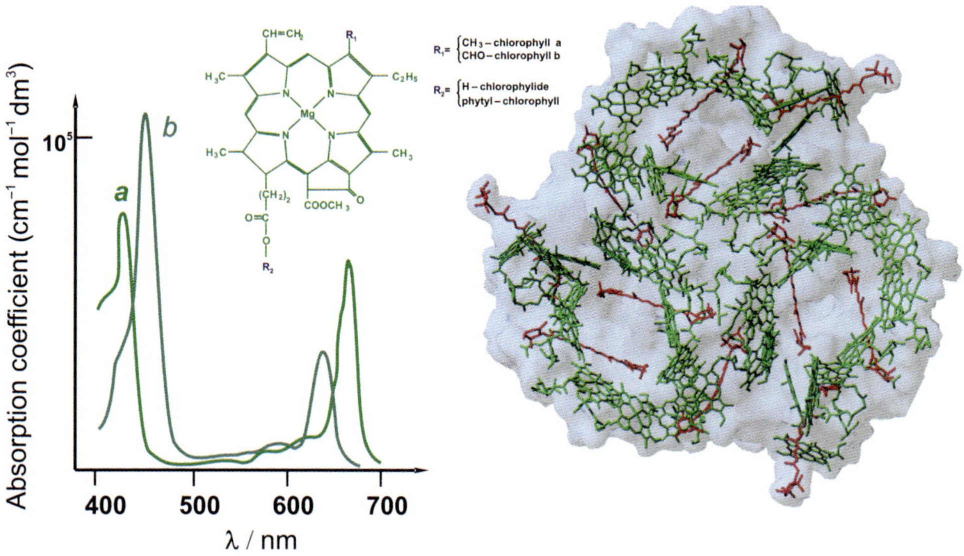

Figure 11.7 Structures and absorption spectra of chlorophylls a *and* b*; schematic view of light harvesting complex. (Data from Protein Data Bank (DOI 10.2210/pdb2BHW/pdb))*

environmental conditions such as light quality, its intensity, nutrient availability [3, 13, 16].

LHCs that surround the reaction centres in chloroplasts of green plant contain additional molecules of chlorophyll *a*, as well as carotenoids and chlorophyll *b* molecules, which absorb light in the centre of the visible spectrum. These accessory pigments increase the efficiency of light capture by absorbing light and transferring the energy to reaction centres through resonance energy transfer. The excited state of the special pair of chlorophyll molecules is lower in energy than that of single chlorophyll molecules, allowing the RC to trap the energy transferred from other molecules.

The excitation energy for the functioning of the primary electron donor in the PSI RC is delivered via a core antenna made up of chlorophyll molecules bound to the protein. In contrast to most other photosynthetic systems, where the antennae and RCs are located on distinct complexes, PSI has a combined system in which the light-harvesting Chls are associated with the same protein that binds the redox cofactors of the electron transfer in the RC [16].

Supramolecular complexes of PSI acquire remarkably different structural forms of the peripheral LHCs, including distinct pigment types and organizational principles. The PSI core antenna, being the main functional unit of supercomplexes, provides an increased functional connectivity in the chlorophyll antenna network due to dense pigment packing, resulting in a fast spread of the excitation energy among the neighbours. Functional connectivity within the network as well as the spectral overlap of antenna pigments allows equilibration of the excitation energy in the depth of the whole membrane within picoseconds and loss-free delivery of the excitation to primary donor P700 within 20–40 ps [16].

11.3.1 Chlorophyll

Chlorophyll is the dominant molecule for light harvesting in biology. Moreover, the special pair of chlorophyll molecules in the reaction centre not only harvests light energy but also provides the first step of energy transduction. At this centre the photon generates a biradical responsible for electron flow, which results in the first membrane potential [1].

Chlorophyll molecules (Figure 11.7) are tetrapyrroles with a central magnesium ion that absorb light quite efficiently because they are polyenes. The requirement for magnesium rather than another metal ion in this molecule is perhaps due to the fact that an element of such low atomic number does not enhance fluorescence and does not itself have redox properties. Magnesium as a non-redox metal does not interfere with the photochemical and free radical reactions of RC chlorins and can assist them. Moreover, the presence of magnesium in the chlorin ring gives the ring an approximate tetragonal symmetry and therefore enhances the molecular extinction coefficient to around 650–700 nm relative to the lower symmetry metal-free pheophytin molecule [1].

If the involvement of a pair of chlorophyll molecules as the reaction initiator signifies the fact that a large distance of charge separation is necessary for redox reactions to compete significantly with charge return, perhaps the role of Mg is to control the degree of association between the two molecules [17].

11.3.2 Bacteriochlorophyll

Bacteriochlorophylls (found in bacteria) are similar to chlorophylls, except for the reduction of an additional pyrrole ring and other minor differences that shift their absorption maxima to near infrared, up to wavelengths of about 900–1000 nm.

11.4 Electron Transfer Pathways in PSII and PSI

Oxygen is produced during the light period of photosynthesis in the thylakoid membranes of green plant chloroplasts and the internal membranes of bacteria. PSII oxidoreductase oxidizes water, producing molecular oxygen on one side of the membrane:

$$2H_2O \rightarrow O_2 + 4H^+_{lumen} + 4e^- \quad (11.3)$$

and at the other side it reduces plastoquinone to plastoquinol:

$$PQ + 2H^{+}_{stroma} + 2\,e^{-} \rightarrow PQH_2 \qquad (11.4)$$

The first of these processes is a four-electron oxidation of two water molecules, producing four protons and one oxygen molecule, whereas the second is a two-electron, two-proton process. As chlorophyll photochemistry is a univalent charge-separating process characterized by high quantum yield, the enzyme must interface the one-electron processes with the multielectron chemistry occurring at each side of the enzyme. The enzyme is often conceptualized as being made of two parts: (1) a photochemical charge separating device that produces a reductant and a strong oxidant on opposite sides of the membrane and (2) a catalytic device in which the oxidizing power is accumulated in the form of increased valence of a cluster of manganese ions in the OEC [18].

The sequence of electron transfer (ET) processes in these two parts is as follows. Absorption of a photon by the specialized pair of chlorophyll *a*, P680, triggers a chain of ET reactions. The excited singlet state of P680 decays to the oxidized state $P680^{+}$ by ET to a nearby pheophytin (Pheo), within approximately 2 ps after photoexcitation of P680. The charge-separated state is stabilized by ET to a primary quinone electron acceptor (Q_A), which functions as a one-electron carrier. Subsequently, a secondary plastoquinone electron acceptor (PQ) functions as a two-electron carrier and forms plastohydroquinone (PQH_2) upon two-electron reduction and protonation with protons from the stromal side of the membrane. PQH_2 has a low affinity for the PQ site and is released as neutral dihydroquinol into the stroma. The photo-oxidized chlorophyll *a* species $P680^{+}$ is reduced by a redox active tyrosine, which is in turn reduced by the oxidation of water at the OEC, releasing protons to the lumen. The combined redox reactions (equations 11.3 and 11.4) effectively transfer protons from the stroma to the lumen, establishing a transmembrane pH (free-energy) gradient that is essential for ATP biosynthesis (Figure 11.8) [14, 18–20].

The final step of the light part of photosynthesis in chloroplasts is catalyzed by a transmembrane complex called PSI. The core of the PSI system is a pair of similar subunits. A special pair of chlorophyll molecules lies at the centre of the structure and absorbs light maximally at 700 nm. This centre, called P700, initiates photoinduced charge separation. The electron passes from P700 down a pathway through chlorophyll at site A_0 and quinone at site A_1 to a set of 4Fe–4S clusters. The next step is the transfer of the electron to ferrodoxin (Fd), a soluble protein containing a 2Fe–2S cluster coordinated to four cysteine residues. Ferrodoxin transfers electrons to $NADP^{+}$. Ferrodoxin–$NADP^{+}$ reductase converts $NADP^{+}$ to form NADPH. In the meantime $P700^{+}$ captures an electron from reduced plastocyanin to return to P700 so that P700 can be excited again [3, 12].

The cooperation between PSI and PSII creates a flow of electrons from H_2O to $NADP^{+}$ (Figure 11.8) [3, 21]. Inorganic cofactors play a dominant role in this ET chain. In PSII starting from special pair of chlorophylls electrons are transferred to pheophytin (Figure 11.9), then through cytochrome *bf* complex (Figure 11.10) to plastocyanine (Figure 11.11), and in PSI Fe–S clusters take part in ET (Figure 11.12).

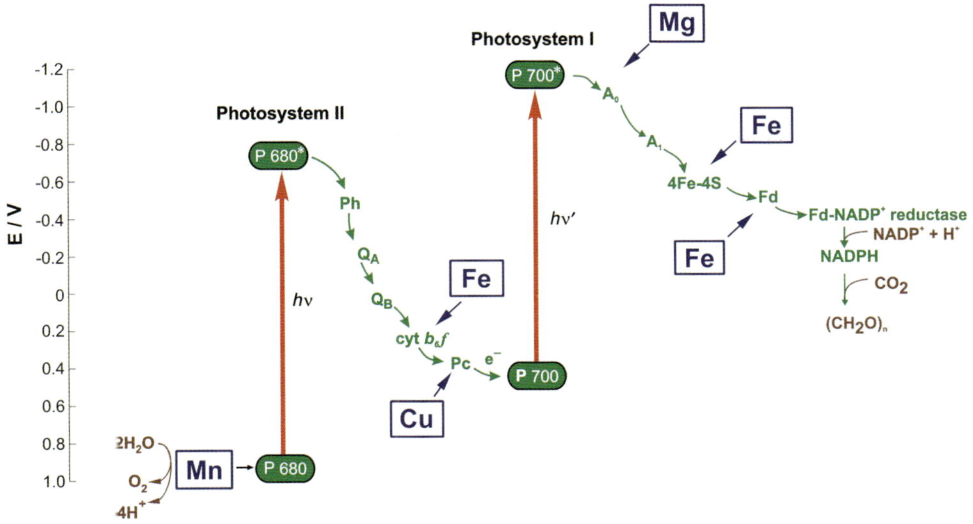

Figure 11.8 *Scheme of electron transfer flow in photosystems PSII and PSI (from H_2O to $NADP^+$) shown on a potential scale*

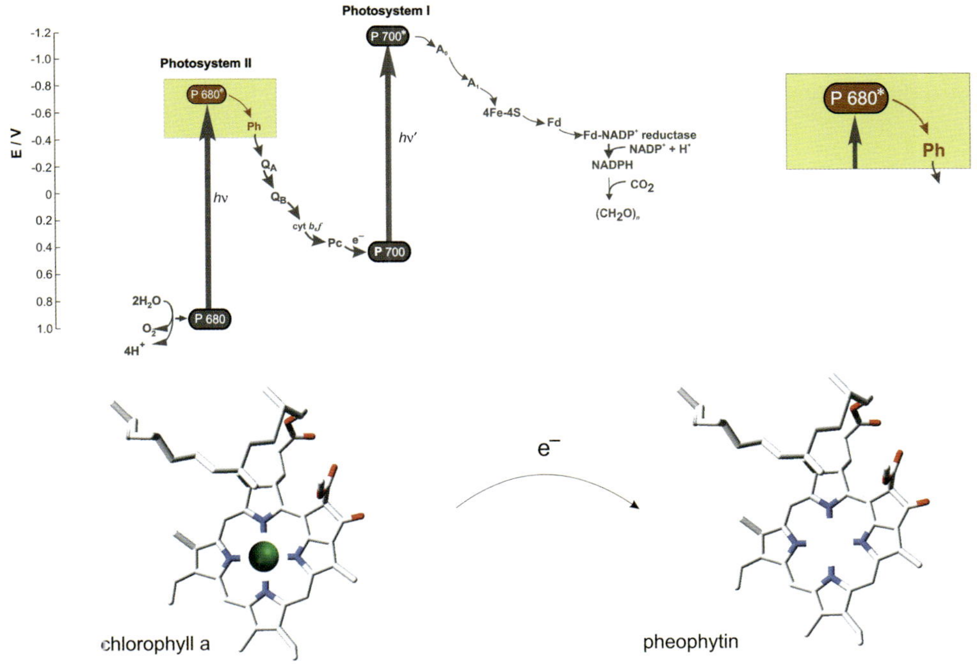

Figure 11.9 *First part of electron transfer pathway in photosystem II (PSII) (from excited special pair of chlorophylls, P680* to pheopytin) involving metal compound (Mg)*

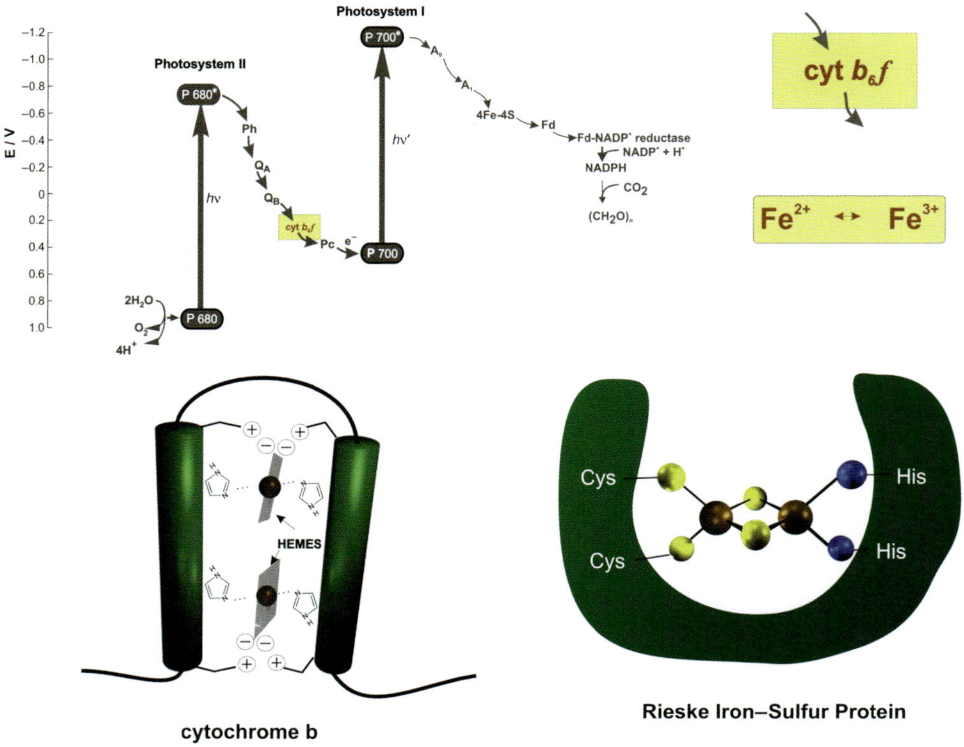

Figure 11.10 *Second part of electron transfer pathway in photosystem II (PSII) (cytochrome* bf *complex) involving metal compounds (Fe)*

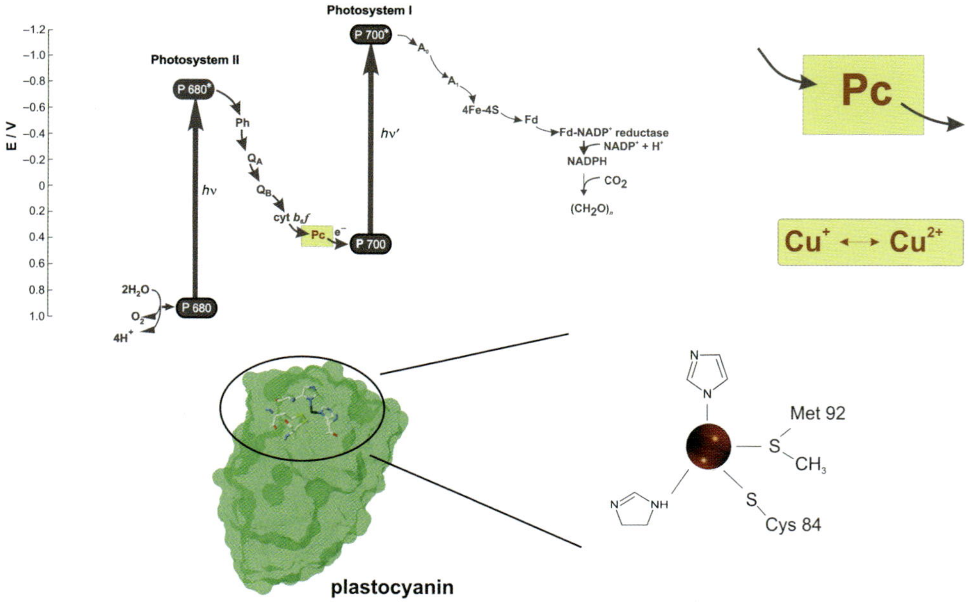

Figure 11.11 *Third part of electron transfer pathway in photosystem II (PSII) (plastocyanine) involving metal compound (Cu)*

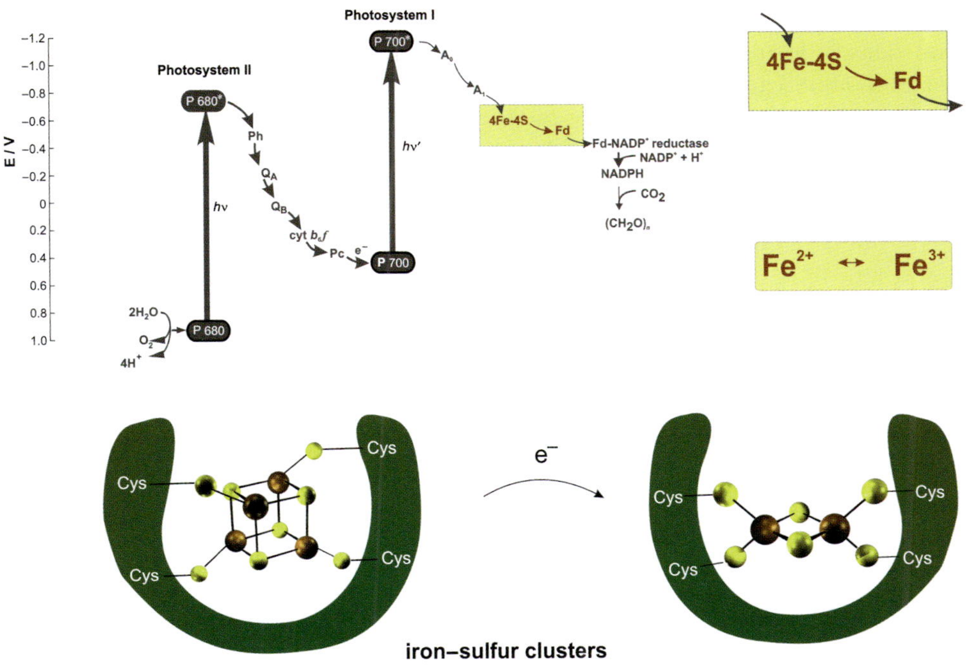

Figure 11.12 *Part of electron transfer pathway in photosystem I (PSI) (from Fe–S clusters to ferrodoxin) involving metal compound (Fe)*

11.5 Oxygen-Evolving Complex (OEC)

As a sharp contrast to the large number of enzymes utilizing O_2 as an oxidant, there is only a single enzyme capable of forming O_2 from water. This is the OEC in PSII. A pentanuclear Mn_4Ca complex bound to amino acid residues in the interior of PSII constitutes the heart of the OEC [14].

X-ray crystallography has provided some details of the location, structure, and protein environment of the Mn_4Ca^{2+} cluster. However, because of the low resolution of the crystal structures reported to date, and the possibility of radiation damage at the catalytic centre, the precise position of each metal ion remains a matter of debate. To some extent these problems have been overcome by applying spectroscopic techniques [22].

By site-directed mutagenesis and crystallographic structure analysis, it has been found that the Mn_4Ca cluster is coordinated by a relatively small number of carboxylate, imidazole, and, perhaps, phenolate ligands. There are differences in the interpretation of the Mn coordination geometry within the 0.30–0.35 nm resolution limit and there is still no final consensus on the precise coordination of the Mn_4Ca cluster, eg the substrate-binding sites have not been identified and the Mn coordination (6 vs 5 coordinate) is an open question. In addition, structural changes during the redox state transition most probably play a role [23]. The results obtained by

X-ray absorption spectroscopy (XAS), together with the compactness of the electron density assignable to the Mn complex and the relatively small number of potential amino acid ligands, suggest several μ-oxo bridges between Mn ions [24].

One of the ongoing mysteries of the OEC is how PSII could have evolved the ability to use water as an electron donor. It would seem that a number of changes, such as formation of a very high-potential oxidant in the charge-separation reaction and incorporation of a four-electron water-oxidation catalyst, would have had to occur simultaneously in order to convert an anaerobic photosynthetic reaction centre into PSII.

Photosynthetic water oxidation in PSII involves the binding of two water molecules, the removal of four electrons and four protons from the Mn complex/substrate (water) entity, O–O bond formation, and dioxygen release.

The whole process starts from absorption of photon in the antenna system which leads to a charge separation and, through a series of ET steps, to the creation of a neutral tyrosyl radical. This radical in turn extracts electrons from the OEC. For each photon that is absorbed, the tyrosyl radical extracts one electron from the OEC, taking it through a series of oxidation states, called S_0 to S_n, finally releasing O_2 when returning from S_4 back to S_0. A simplified reaction scheme is shown in Figure 11.13 [25].

There is a consensus that Mn-centred oxidation occurs during the S_0 to S_1 and S_1 to S_2 transitions. Within the context of localized oxidation, two sets of oxidation states have been proposed. The formal oxidation states of the native S_1 state have been assigned to Mn(III,III,IV,IV) or Mn(III,III,III,III) and S_2 to Mn(III,IV,IV,IV) or Mn(III,III,III,IV). Uncertainty remains about the S_0 and S_3 states. The uncertainty concerning the S_2 to S_3 transition has led to two different types of proposed

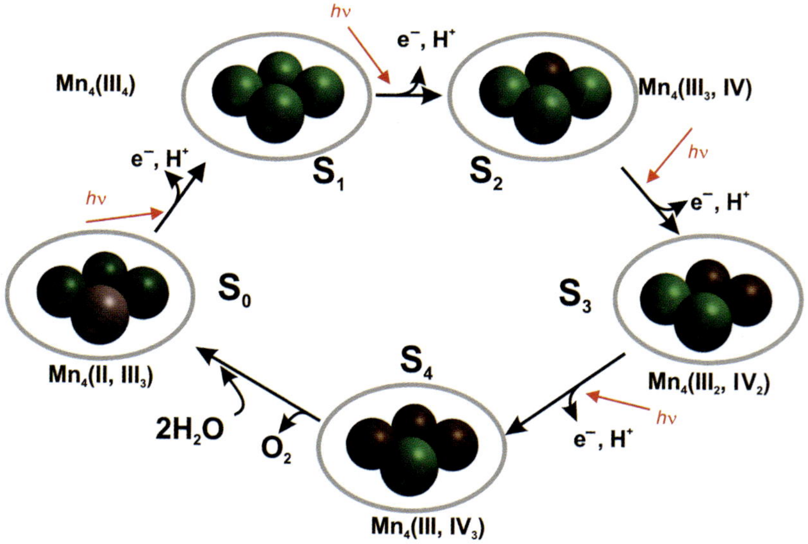

Figure 11.13 One of the suggested forms of S-state cycle describing oxygen evolution from manganese centre

O_2 evolution mechanisms, with one type incorporating the oxidation of ligand or substrate in the S_3 state, and the other invoking Mn oxidation during the S_2 to S_3 transition [24].

Despite these ambiguities, the new structural models mark a new era in which mechanistic hypotheses can be developed and tested in the light of insights derived from structural information. In addition, the new models serve as valuable guides for experiments designed to identify the specific Mn ion(s) that undergo oxidation during each step in the S state cycle, to identify the amino acid residues that facilitate the deprotonation and oxidation of the Mn_4Ca cluster during the individual steps of the S state cycle, and to identify the amino acid residues that participate in proton transfer pathways leading from the Mn_4Ca cluster to the thylakoid lumen [26].

Among many proposals for the molecular mechanism of water oxidation at the manganese cluster the critical point is how the O–O bond is formed. One type of mechanisms involves the formation of an O–O bond between two μ-oxo bridges upon oxidation of the complex; another involves the formation of a peroxo bond between two adjacent Mn–oxo species. These species would be formed when Mn-bound water becomes deprotonated as the manganese oxidation state increases. In a third type of mechanism, a high μ-valent Mn–oxo species forms a bond between the electron-poor oxygen and an electron-rich oxygen of unbound water [27, 28].

11.5.1 Inorganic Species in OEC

Inorganic species are at the centre of the OEC. Moreover, inorganic model chemistry has played an important and synergistic role in studies of the OEC.

Manganese

Manganese appears to be unique in catalyzing O_2 evolution in all aerobic phototrophs. No other metal ion has been found to date that can replace it. Manganese was probably selected via evolution for this function because of its ability to exist in multiple oxidation states (Mn^{2+}, Mn^{3+}, Mn^{4+}, Mn^{5+}) and to form strong bonds with oxygen ligands [29].

Calcium

Ca^{2+} is an essential ion in the photoassembly reaction in all PSII OECs examined to date. The experimental data indicated that Ca^{2+} is involved in accelerating the final step in the S-state catalytic cycle, involving the O–O bond formation reaction [29].

Oxide Bridges/Proton Release

To build the Mn_4Ca cluster from its free metal ions, it is necessary to deprotonate four or more aqua ligands attached to the initially bound Mn^{2+} [29].

Bicarbonate Ion

Bicarbonate ion is important not only as a substrate for carbon fixation in the Calvin cycle but also for its role in PSII ET efficiency and photoassembly of the OEC cluster in PSII [29].

Chloride Ion
The role of chloride ion in photoactivation is not clearly revealed, as its influence occurs only after the rate-limiting step and thus affects only O_2 evolution activity of the fully assembled cluster.

To elucidate the details of the OEC reaction mechanism, bioinorganic chemists have made many new complexes to mimic the proposed biological mechanisms. Work on small molecules has clarified many issues relevant to the OEC such as the necessity of multiple metals and the importance of proton transfer. Functional model systems are still very much less active than the OEC, but current mechanistic models show activity very analogous to mechanisms proposed for the OEC [13, 28, 30–34].

References

1. Frausto Da Silva JJR, Williams RJP. *The Biological Chemistry of the Elements. The inorganic chemistry of life*. Oxford: Oxford University Press, 2001.
2. Williams RJP. My past and a future role for inorganic biochemistry. *J Inorg Biochem* 2006;**100**:1908–24.
3. Berg JM, Tymoczko JL, Stryer L. *Biochemistry*. New York: WH Freeman & Co., 2007.
4. Catling DC, Claire MW. How Earth's atmosphere evolved to an oxic state: A status report. *Earth Planet Sci Lett* 2005;**237**:1–20.
5. Renger G. *Primary Processes of Photosynthesis – Part 1*. Cambridge: Royal Society of Chemistry, 2008.
6. Renger G. *Primary Processes of Photosynthesis – Part 2*. Cambridge: Royal Society of Chemistry, 2008.
7. Blankenship R. *Molecular Mechanisms of Photosynthesis*. St Louis, MO: Washington University 2002.
8. Collings AF, Critchley C, eds. *Artificial Photosynthesis: From basic biology to industrial application*. Weinheim: Wiley-VCH 2005.
9. Szaciłowski K, Chmura A, Stasicka Z. Interplay between iron complexes, nitric oxide and sulfur ligands: Structure, (photo)reactivity and biological importance. *Coord Chem Rev* 2005;**249**:2408–36.
10. Vacha F, Bumba L, Kaftan D, Vacha D. Microscopy and single molecule detection in photosynthesis. *Micron* 2005;**36**:483–502.
11. Raymond J, Blankenship RE. The origin of the oxygen-evolving complex. *Coord Chem Rev* 2008;**252**:377–83.
12. Golbeck JH, ed. *Advances in Photosynthesis and Respiration*. Dordrecht: Springer, 2006.
13. Cady C, Crabtree RH, Brudvig GW. Functional models for the oxygen-evolving complex of photosystem II. *Coord Chem Rev* 2008;**252**:444–55.
14. McEvoy JP, Brudvig GW. Water-splitting chemistry of photosystem II. *Chem Rev* 2006;**106**:4455–83.
15. Miller RA, Presley A, Francis MB. Self-assembling light-harvesting systems from synthetically modified tobacco mosaic virus coat proteins. *J Am Chem Soc* 2007;**129**:3104–9.
16. Melkozernov AN, Barber J, Blankenship RE. Light harvesting in photosystem I supercomplexes. *Biochemistry* 2006;**45**:331–46.
17. Harrowfiels JM. Biological coordination chemistry, a confluence of chemistry and biochemistry. *Compt Rend Chimie* 2005;**8**:199–210.

18. Sproviero E, Gascon J, McEvoy JP, Brudvig GW, Batista VS. Computational studies of the O_2-evolving complexes of photosystem II and biomimetic oxomanganese complexes. *Coord Chem Rev* 2008;**252**:395–415.
19. Murray J, Barber J. Structural characteristics of channels and pathway in photosystem II including the identification of an oxygen channel. *J Struct Biol* 2007;**159**:228–37.
20. Herrero C, Lassalle-Kaiser B, Leibl W, Rutherford AW, Aukauloo A. Artificial systems related to light driven electron transfer processes in PSII. *Coord Chem Rev* 2008;**252**:456–68.
21. Iverson TM. Evolution and unique bioenergetic mechanism in oxygenic photosynthesis. *Curr Opinion Chem Biol* 2006;**10**:91–100.
22. Barber J, Murray JW. Revealing the structure of the Mn-cluster of photosystem II by X-ray crystallography. *Coord Chem Rev* 2008;233–43.
23. Hilier W, Wydrzynski T. ^{18}O-water exchange in photosystem II: Substrate binding and intermediates of the water splitting cycle. *Coord Chem Rev* 2008;**252**:306–17.
24. Sauer K, Yano J, Yachandra VK. X-ray spectroscopy of the photosynthetic oxygen-evolving comlpex. *Coord Chem Rev* 2008;**252**:318–35.
25. Dau H, Haumann M. Time-resolved X-ray spectroscopy leads to an extension of the classical S-state cycle model of photosynthetic oxygen evolution. *Photosynth Res* 2007;**92**:327–43.
26. Debus RJ. Protein ligation of the photosynthetic oxygen-evolving center. *Coord Chem Rev* 2008;**252**:244–58.
27. Hammarström L, Sun L, Aakermark B, Styring S. A biomimetic approach to artificial photosynthesis: Ru(II)-polypyridine photosensitizers linked tyrosine and manganese electron donors. *Spectrochim Acta* 2001;**37**:2145–60.
28. Sun L, Raymond MK, Magnuson A, et al. Towards an artificial model for photosystem II: a manganese(II,II) dimer covalently linked to ruthenium(II) tris-bipyridine via a tyrosine derivative. *J Inorg Biochem* 2000;**78**:15–22.
29. Dasgupta J, Ananyev G, Dismukes GC. Photoassembly of the water-oxidizing complex in photosystem II. *Coord Chem Rev* 2008;**252**:347–60.
30. Han G, Li J, Chen G, et al. Reconstruction of the water-oxidizing complex in manganese-depleted Photosystem II using synthetic manganese complexes. *J Photochem Photobiol B: Biol* 2005;**81**:114–20.
31. Huang P, Magnuson A, Lomoth R, et al. Photoinduced oxidation of a dinuclear Mn_2(II,II) complex to the Mn_2(III,IV) state by inner and intramolecular electron transfer to Ru(III) tris-bipyridine. *J Inorg Biochem* 2002;**91**:159–72.
32. Magnuson A, Liebisch P, Hogblom J, et al. Bridging-type changes facilitate successive oxidation steps at about 1V in two binuclear manganese complexes – implications for photosynthetic water-oxidation. *J Inorg Biochem* 2006;**100**:1234–3.
33. Lomoth R, Magnuson A, Sjodin M, Huang P, Styring S, Hammarstrom L. Mimicking the electron donor side of Photosystem II in artificial photosynthesis. *Photosynth Res* 2006;**87**:25–40.
34. Mullins CS, Pecoraro V. Reflections on small molecule manganese models that seek to mimic photosynthetic water oxidation chemistry. *Coord Chem Rev* 2008;**252**:416–43.

Part IV

Photochemistry and Photophysics in Bioinspired Systems: Studies and Modelling

12
Photoenzymes

Trying to determine the structure of a protein by UV spectroscopy was like trying to determine the structure of a piano by listening to the sound it made while being dropped down a flight of stairs.
Francis Crick

Natural photoenzymes that are able to convert optical signals into chemical action are very rare. Therefore it is an interesting challenge for protein engineering to modify naturally occurring enzymes and/or proteins to control their activity by light [1, 2]. The proper design of a photoactive enzyme/protein would allow not only control of the triggering of the desired activity but also modulation of the redox state of the active site of the enzyme by photoinduced electron transfer. Some biomimetic systems have been constructed on the basis of natural enzymes with artificial photosensitive cofactors [3–11]. These systems allow precise photochemical control of enzyme activity, and have already found applications as biosensors [6, 11] or can lead to new features, such as artificial photosynthetic centres or enzyme-like catalysts [8, 12, 13]. A great deal of attention has also been turned into new design of bio-inspired catalysts acting as photocatalytic artificial enzymes – artificial photoenzymes.

12.1 Natural Photoenzymes

Several natural photoenzymes with activity controlled by light have been reported to date. Among them are DNA and (6–4) photoproduct photolyases, which are highly efficient light-driven DNA-repair enzymes [14, 15], protochlorophyllide reductase, which is an important enzyme in the chlorophyll biosynthesis pathway [16, 17], nitrile hydratase, which hydrates aliphatic and aromatic nitriles to the corresponding amides [18–20], and xanthine oxidase, which takes part in the metabolic pathway of uric acid formation [21, 22].

DNA damage caused by UV-B light includes the formation of cyclobutane pyrimidine dimers (T<>T, Figure 12.1a) and pyrimidine (6–4) pyrimidone photoproducts ((6–4)PPs, Figure 12.1b). The T<>T dimers can be removed by a photoinduced repair mechanism containing the enzyme DNA photolyase [14, 23–25] whereas (6–4)PPs are repaired by the (6–4) photoproduct photolyase [14]. The photolyases contain two types of cofactors: flavine adenine dinucleotide (FAD) as a redox-active cofactor and the light-harvesting one, either 5,10-methenyltetrahydrofolylpolyglutamate (MTHF) or 8-hydroxy-5-deazaflavin (8-HDF). T<>T repair starts from light independent binding of the substrate to DNA photolyase. This is followed by UV-A violet and blue light (320–500 nm) absorption by MTHF or 8-HDF cofactor, and the excitation energy is transferred to $FADH^-$. The excited flavin ($*FADH^-$) transfers one electron to the pyrimidine dimer, generating a flavin/substrate radical pair. The pyrimidine dimer radical ($T<>T^{\bullet -}$) undergoes bond rearrangement, leading to scission of the carbon–carbon bond of pyrimidine dimers and back electron transfer to the flavin radical ($FADH^{\bullet}$). As a result $FADH^-$ and repaired dinucleotide (TT) are formed. The catalytic cycle is schematically presented in Figure 12.1c [14, 25]. The repair mechanism of (6–4)PPs mediated by (6–4) photoproduct photolyase is less well understood and tentatively it has been proposed that the catalytic cycle is similar to the one suggested for the repair of T<>T dimers [14].

Another enzymatic system with activity controlled by light is protochlorophyllide oxidoreductase (POR) [16, 17]. This enzyme catalyzes the reduction of the

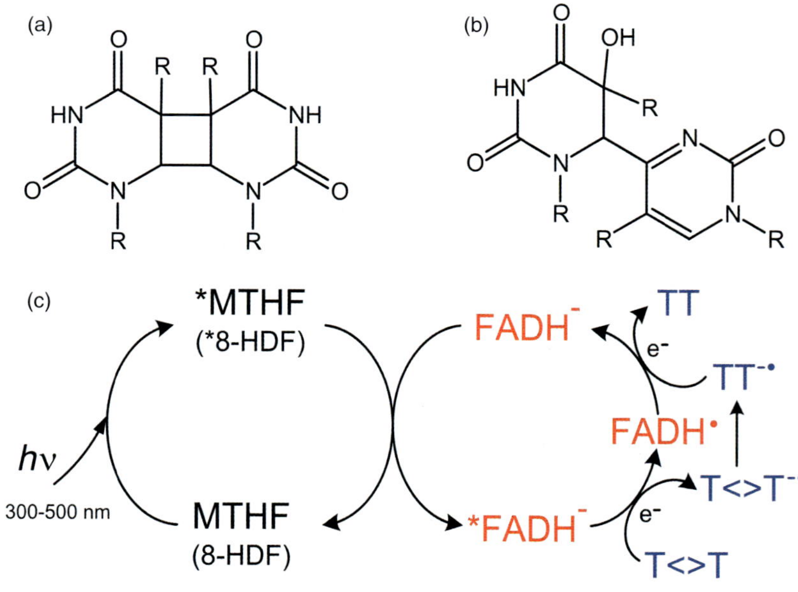

Figure 12.1 Structures of (a) cyclobutane pyrimidine dimer (T<>T), (b) pyrimidine (6–4) pyrimidone photoproduct ((6–4)pp). (c) The scheme of the catalytic cycle for the repairing of T<>T dimers by DNA photolyase [14]

C17–C18 double bond of the D-ring of protochlorophylide (Pchlide), producing the chlorophylide (Chlide). In other words, it converts the light-yellow protochlorophyllide to green chlorophylide. It has been postulated that the irradiation of the Pchlide–FOR–NADPH complex results in charge separation across the C17–C18 double bond which makes it feasible to transfer the hydride ion from the NADPH cofactor to the C17 position of Pchlide. This process creates a negatively charged, non-fluorescent intermediate A696 and a positively charged cofactor [16]. The subsequent steps of the catalytic cycle leading to formation of chlorophylide proceed as dark reactions.

Nitrile hydratases are hydrolytic metalloenzymes containing non-haem iron or non-corrinoid cobalt at their catalytic centre [26]. These are responsible for the sequential metabolism of nitrile compounds in some bacteria and fungi, which are capable of utilizing aliphatic nitriles as the sole source of nitrogen and carbon, and converting them into corresponding amides. These enzymes have been efficiently used for the industrial production of acrylamide from acrylonitrile and for removal of nitriles from wastewater [27, 28]. The active centre of Fe-type nitrile hydratases contains low-spin Fe^{III} centre. The first coordination sphere consists of two amide nitrogen atoms from the protein skeleton and three sulphur atoms from cysteine. Two of the three cysteinate ligands are a modified, active form of the enzyme contains one sulphenic (cys-SOH) and one sulphinic (cys-SO_2H) form of cysteine. The sixth coordination place is occupied by a water molecule or hydroxide anion (in the active form of the enzyme) or nitric oxide (in the inactive form) [19, 29]. Light is an essential factor regulating activity of nitrile hydratase. In the dark, active enzyme is deactivated by direct binding of nitric oxide to ferric iron in the non-haem iron centre, whereas irradiation of this inactive form leads to cleavage of the Fe–NO bond and NO release, which results in restoration of the enzyme's activity [19].

Recently a new photoenzyme system, namely xanthine oxidase (XOD), has been reported by Hwang et al. [21, 22]. XOD is a homodimer, which contains in each unit a molybdenum(VI), two Fe–S clusters and one FAD moiety. The enzyme oxidizes the xanthine to uric acid through activation of the water molecule. During this reaction the Mo^{VI} ion is reduced to Mo^{IV}, which in turns transfers two electrons via the Fe–S cluster to the FAD cofactor, reducing it to $FADH_2$ [21]. Some diseases lead to accumulation of excess uric acid in blood plasma and, in order to decrease uric acid formation, the allopurinol, as an alternative substrate for the XOD, is applied for treatment. It has been noticed that the oxidized product of this drug, ie alloxanthine, is a potent inhibitor of XOD and it can bind to the active site of the enzyme only if it is in the reduced form (XOD-Mo^{IV}). The enzyme activity can be restored upon irradiation by light in the wavelength range 340–430 nm, which causes the release of the inhibitor (alloxanthine). It has been postulated that, on photoexcitation, the alloxanthine–XOD–Mo^{IV} complex transfers an electron via a Fe–S cluster to the FAD cofactor and generates Mo^{V} and $FADH^{\bullet}$ intermediates. The formation of a Mo^{V} centre in photochemical reaction results in dissociation of alloxanthine from the active site of XOD and transfer of the second electron from Mo^{V} to the Fe–S cluster or $FADH^{\bullet}$ in the dark reaction. In the final step the active XOD–Mo^{VI} enzyme is formed [21, 22]. Thus the activity of the XOD can be regulated by light.

12.2 Modified Natural Proteins/Enzymes

One of the approaches towards light-controlled and light-powered enzymes encompasses modification of naturally occurring proteins with xenobiotic photoactive cofactors. The modification of proteins (myoglobin, haemoglobin) with a photosensitizer can lead to construction of light-activated biocatalysts – photoenzymes [12, 13] – or light-regulated oxygen carriers [30]. Hamachi et al. have shown that the binding of molecular oxygen can be regulated by light [30]. The modified myoglobin containing a covalently attached ruthenium polypyridine complex can be photoswitched from a resting to an active state (changing the redox state of the metal centre in the haem group) by visible light irradiation. The detailed mechanism of photoinduced electron transfer from the ruthenium complex to the metal centre in the haem group of proteins is described in section 13.1. Willner et al. obtained the photoactive biocatalysts by reconstruction of apo-myoglobin or apo-α and β subunits of haemoglobin with Co^{II}-protoporphyrin IX. Furthermore, in the last step, reconstituted proteins were modified by the attachment of an eosin chromophore through the cysteine residue [12]. Upon photoexcitation of the eosin chromophore the excited triplet state transfers an electron to the Co^{II}–protoporphyrin IX thus reducing metal centre to Co(I). In the presence of a sacrificial electron donor (eg EDTA) the oxidized chromophore is reduced whereas Co(I) porphyrin, the active intermediate for the hydrogenation process, is accumulated. Thus, the catalytic properties of the designed photoenzyme are triggered by steady-state illumination, as revealed by hydrogenation of acetylene to ethylene (Figure 12.2). In addition, the semisynthetic photoenzyme was coupled with the oxidative biocatalyst (lactate dehydrogenase, LDH) through a diffusional electron mediator (eg *N*-methylferrocene caproic acid), resulting in a cyclic photosynthetic system (Figure 12.2). Interestingly, a cyclic photoinduced oxidation of lactic acid by acetylene, with formation

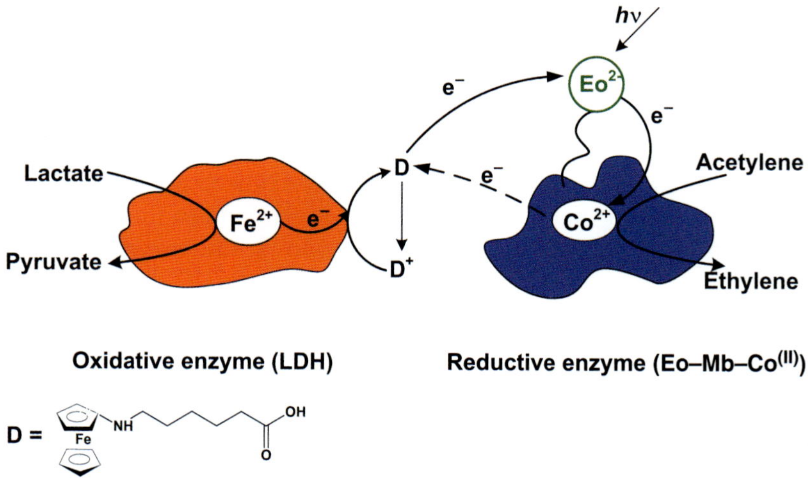

Figure 12.2 *Cyclic photosynthetic system composed of a reductive photoenzyme Eo–Mb–Co^{II} (myoglobin reconstructed with Co^{II}-protoporphyrin IX and with attached eosin chromophore to cysteine residue) and an oxidative biocatalyst lactate dehydrogenase (LDH) [12]*

of pyruvic acid and ethylene in the constructed system, mimics the functions of photosynthetic the bacterium *Chloroflexus*.

Niemeyer et al. have reported the design of quantum dot/enzyme nanohybrids that are capable of catalyzing an organic transformation upon optical excitation of semiconductor quantum dots (QDs) [31]. The hybrid device was composed of semiconductor CdS nanoparticles and cytochrome $p450_{BS\beta}$ enzyme. It has been proposed that irradiation of QDs leads to formation of excitons (h^+–e^- pairs) that on dissociation generate superoxide and hydroxyl radicals in interfacial electron transfer process (see Chapter 7). These radicals in turn activate the enzyme adsorbed at the QD surface. The activated enzyme is able to catalyze mono-oxygenation of fatty acids, but has a lower activity than the native enzyme [31].

The use of photoisomerizable units for the design of photoenzymes has been extensively developed over last two decades [9, 13]. The attachment of photoisomerizable molecules (eg diazonium salts, carboxyazobenzene, spiropyran, and many others) to enzymes allows control of their biocatalytic activity by light. The light-triggered isomerization of the anchored molecule can induce distortion of the protein structure, which results in a changed enzyme activity. Several systems involving photoswitchable activation and deactivation of enzymes by a covalently linked photoisomerizable unit have been designed and are reviewed in detail by Willner et al. [9, 13, 32].

The light-induced activation and deactivation of enzymes (particularly redox enzymes) using photoisomerizable molecules coupled to electrodes has led to the construction of biomaterials applied for amperometric transduction and amplification of recorded optical signals. In one of the approaches, the electrode was modified by a photoisomerizable monolayer to act as a photoactive command surface, particularly for association and dissociation of the redox protein to and from the electrode interface [33]. This method was applied for activation of the cytochrome *c* (cyt *c*), which acts as a cofactor by mediation of electron transfer in many redox enzymes [13, 33]. The gold electrode was functionalized by mixed, photoisomerizable, protonated, nitromerocyanine units and mercaptopyridine molecules; thus the electrical contact with positively charged cyt *c* was blocked. However, the photoisomerization of a monolayer to the nitrospiropyran/mercaptopyridine assemblies resulted in formation of neutral monolayer interface and the electric contact of cyt *c* was feasible. The association of cyt *c* to pyridine sites on the electrode switched on the redox functions of this haemoprotein. The electric response of cyt *c* could be reversibly switched off by promoting the back photoisomerization of nitrospiropyran to nitromerocyanine; thus by switching on or off the electric contact of cyt *c* it is possible to control activation and deactivation of the enzyme (Figure 12.3). As cyt *c* acts as an electron transfer-mediating cofactor to different redox proteins the photoswitchable system can be applied for the activation of an enzyme cascade. It has been shown that this system can mediate electron transfer to cytochrome oxidase and activate the bioelectrocatalyzed reduction of O_2 to H_2O (Figure 12.3). The same system can be coupled to an oxidative bioelectrocatalyzed transformation. Cytochrome *c* acts as an electron mediator for LDH promoting oxidation of lactate to pyruvate. Moreover, the tailored system represents an optobioelectronic assembly mimicking the vision process [13, 33].

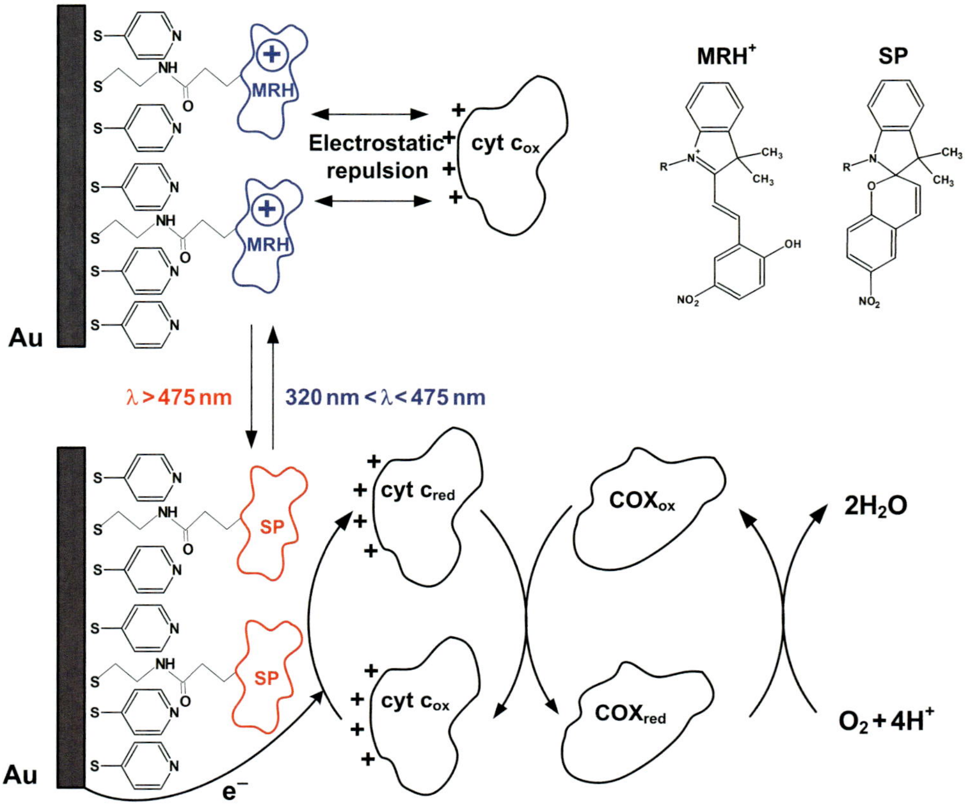

Figure 12.3 Coupling of photoinduced electrical interaction of cytochrome c (cyt c) with the photoisomerizable monolayer modified Au electrode to the biocatalyzed reduction of O_2 by cytochrome oxidase (COX) [33]

Another approach to photoregulating the biocatalyst was applied for construction of a photoactive glucose sensor based on a chemical modification of glucose oxidase by a photoisomerizable component, and immobilization of the photoswitchable biocatalyst onto the electrode surface [6, 34]. The native enzyme was transformed to the apoenzyme by FAD cofactor extraction and subsequently the enzyme was reconstituted with synthetic merocyanine–FAD cofactor. The enzyme was covalently anchored to the gold surface with a thiolate-containing linker. The modified enzyme oxidizes glucose to gluconic acid and reduces ferrocenium to ferrocenecarboxylate. The modified cofactor efficiently controls the conformation of the enzyme and hence its activity. Photoisomerization of the merocyanine to the spiropyran form inhibits glucose oxidation and ferrocenium reduction. If the cofactor contains the spiro form, the contact between the electron mediator (ferrocene) and the protein redox centre (FAD) is blocked, whereas photoisomerization to the merocyanine form activates the contact between the reagents (Figure 12.4). The system may work as an electrochemical glucose sensor or a transducer converting optical signals into electrical ones [6, 34].

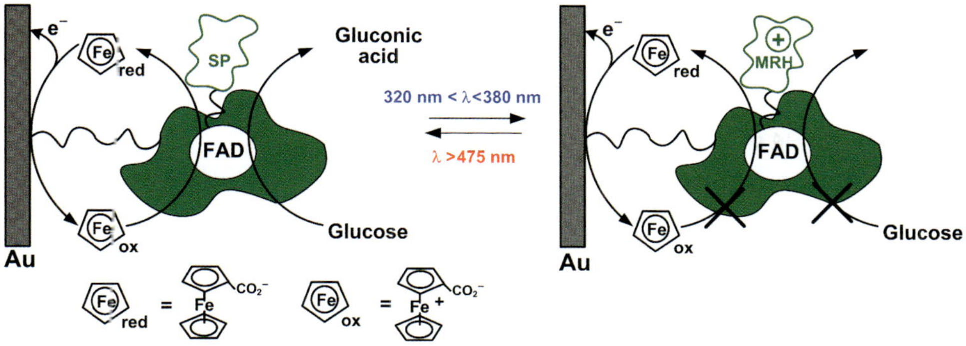

Figure 12.4 *Photoinduced bioelectrocatalyzed oxidation of glucose to gluconic acid by glucose oxidase (GOD) reconstituted with a nitrospiropyran-modified FAD cofactor (Sp-FAD) assembled as a monolayer on the Au electrode. The Sp-FAD reveals reversible photoisomerizable properties yielding nitromerocyanine-FAD isomer (MRH⁺-FAD) [34]*

12.3 Artificial Photoenzymes

The model systems for enzymes activated by light can be called 'artificial photoenzymes'. This expression comprises two important requirements, which must be fulfilled by system: its activity should be triggered by light and it should behave like an 'artificial enzyme'. The term 'artificial enzyme' is very often used inappropriately and applied to any catalysts. The detailed explanation of this term, together with critical analysis of enzyme model systems, can be found in an excellent review by Kirby [35]. In general, enzymes are more than just highly developed catalysts:

- they are highly specific species able to distinguish between similar parts of the substrate molecule (regiospecific), and between optical isomers of the substrate (stereospecific) by controlling the second coordination sphere
- they are subject to a variety of regulations
- their reaction rates show substrate saturation, which is not the case with catalysts.

Usually mimetics of enzymes are not as sophisticated and their catalytic efficiency is taken into account as a main objective. The important feature of enzyme mimetics is an initial binding interaction between substrate and the catalyst, thus giving rise to Michaelis–Menten kinetics [35]. Therefore for a model system to be called an 'artificial enzyme' it should combine more than one of the key features identified for enzymes [35]. For some systems it is difficult to decide if they can be considered as 'artificial photoenzymes' or if they are just photocatalysts (described in Chapters 6, 7, 10, and 21). It is not the purpose of this book to judge whether model systems can be called 'artificial photoenzymes' or should be considered as just chemical photocatalysts. To get some feeling which type of model systems have been investigated we describe a few model systems that have claimed to have enzyme-like activity.

Most bioinspired light-activated model systems of enzymes are based on metalloporphyrin complexes as visible light-driven simple models for photoenzymes. They have been designed for catalyzing hydrogen or oxygen evolution from water, reduction of carbon dioxide, as well as for use in many oxidation reactions by molecular oxygen. There are a large number of these photocatalysts that try to mimic oxidoreductases and hydrogenases. Their diversity arises from application the different metal ions of the synthetic models (among others, Fe, Zn, Sn, Co, Sb, Al, Cu, and Ru), as well as various porphyrin ligands. They have been extensively investigated and the original works are cited by Shiragami et al. [36] and Maldotti et al. [37]. Among a number of metalloporphyrin complexes, the high-valent metalloporphyrins with Sb^V ions were found to be good photocatalysts [36, 38]; they have been also studied as models for light – activated artificial enzymes [39, 40]. Knör has reported the construction of the artificial photoenzyme mimicking the oxidoreductase enzyme, which is able to catalyze the hydrogen atom transfer process [40]. The synthetic photoenzyme consists of an antimony metal centre with a terminal oxo group ($Sb^V = O$), axial OH^- ligand, and macrocyclic ligand *meso*-tetraphenylporphyrin (tpp). The designed system efficiently catalyzes the oxidation of primary alcohols to aldehydes when activated by light, as shown in Figure 12.5. The X-ray crystal structure of antimony(V) octaethylporphyrin derivative has revealed that antimony(V) porphyrins are capable of binding alcohols in their secondary coordi-

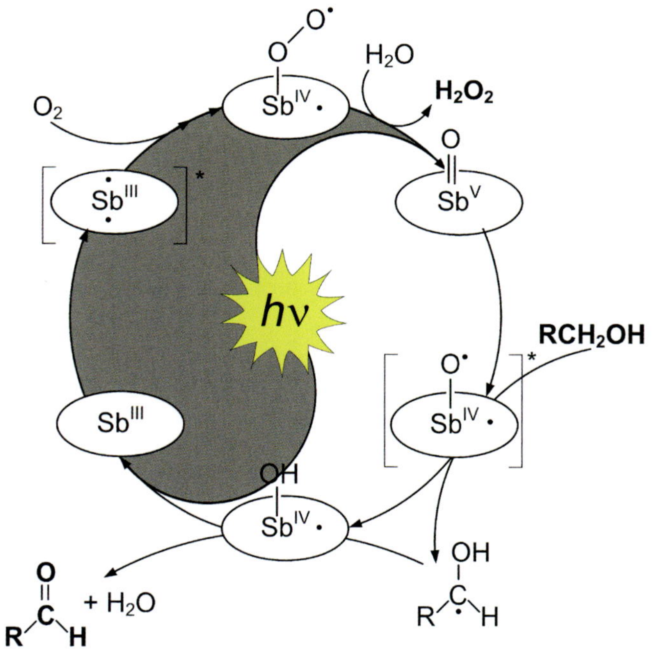

Figure 12.5 *Proposed mechanism for the cyclic oxidation of primary alcohols by molecular oxygen catalyzed by a photodriven model of native oxidoreductase – the antimony(V) oxo species*

nation sphere. Moreover, the active site shows substrate selectivity, which allows for avoidance of a competitive catalytic transformation of the aldehyde reaction product. It has been shown that the photoassisted substrate conversion is completely switched off in the dark while turned on by light. The efficiency of the photochemically triggered substrate conversion is comparable to some native enzymes such as galactose oxidase or cytochrome P450; it is, however, significantly lower than alcohol oxidase. The activity of the artificial photoenzyme corresponds to an average turnover number of ≥4000 cycles of each catalysts molecule before degradation.

Another type of potential photocatalysts mimicking the oxygenase enzymes is represented by di-iron(III) bisporphyrins, belonging to the group of so-called Pacman porphyrins [41]. The Pacman porphyrins are built from two cofacial iron(III) porphyrins connected by spacers such as anthracene, biphenylene, xanthene, diphenyl furan, or others. The architecture of the molecule can be tuned by either using different spacers or employing appropriate substituents along the periphery of the porphyrins, resulting in metal–metal distance ranging from 0.35 nm to >0.85 nm [41, 42]. Control of the pocket size in the Pacman motif makes this system sensitive to the size of the attacking substrate and can sterically direct substrate to the electronically favoured side-on approach to the metal-oxo bond. The oxidizing power of Pacman metalloporphyrins can be enhanced by introducing electron-withdrawing groups on to the porphyrin periphery. A general scheme for the catalytic cycle for oxidation of aliphatic C–H bonds by Pacman system is presented in Figure 12.6.

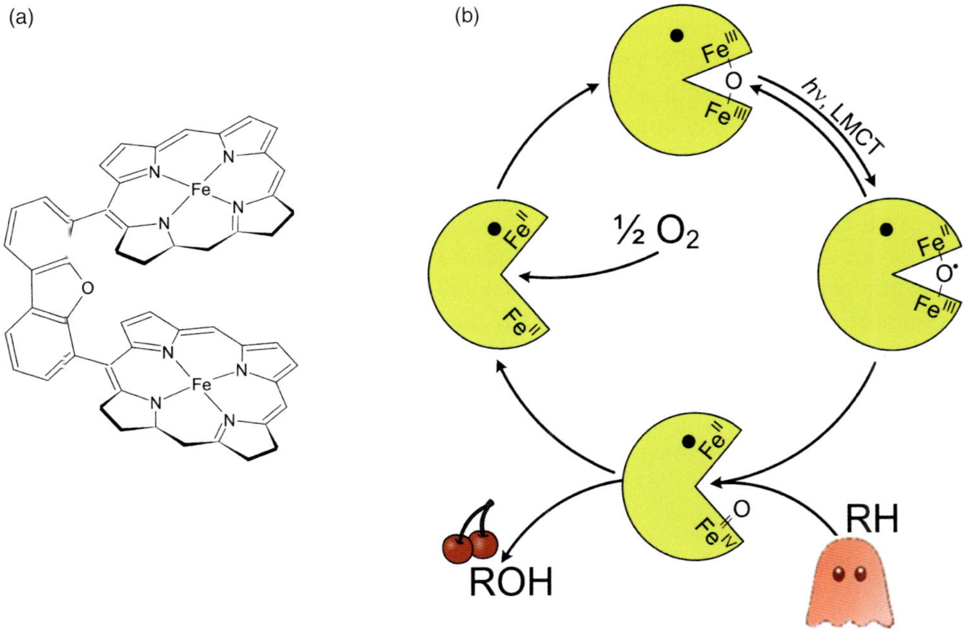

Figure 12.6 *(a) The structure of Pacman-type cofacial bisporphyrin and (b) simplified illustration of the Pacman effect in a photocatalytic cycle. LMCT, ligand-to-metal charge transfer*

Light excitation of di-iron(III) μ-oxo bisporphyrins results in cleavage of thermally inert Fe^{III}–O–Fe^{III} bond to generate a $PFe^{II}/PFe^{IV} = O$ (P = porphyrin) intermediate. The ferryl cofacial intermediate can oxidize organic substrates including phosphines, sulphides, olefins, and even hydrocarbons, while it is concomitantly reduced to iron(II). The formed di-iron(II) bisporphyrin reacts with molecular oxygen to reproduce a catalyst. The great advantage of this system is catalyzing two-electron oxidation under mild conditions and without the need for an external co-reductant. Molecular oxygen is used as the terminal oxidant as well as the oxygen atom source [41].

There are several photocatalysts mimicking hydrogenase activity that are not based on metalloporphyrin systems. Among them there are mixed-valence complexes of rhodium or iridium, [41] as well as complex systems encompassing photosensitizers (eg ruthenium complexes) attached to a catalytic bimetallic centre [43]. The design of more sophisticated systems approaches that of photosynthetic processes [44].

12.4 Towards Mimicking the Photosynthetic Processes

An increasing demand for cheap and clean energy stimulates development of novel chemical systems capable of efficient solar energy conversion. Until now, natural photosynthesis has been the cleanest and most efficient process known. It involves a large array of chromophores (antennae) to harvest visible light, charge separation systems and redox centres. Photosynthetic systems of green plants are very complex assemblies of light-harvesting antennae, energy and electron transfer systems (photosystems I and II), redox centres capable of water oxidation, and protomotoric-force-powered ATP synthase [45–47]. Photosynthesis is powered by visible photons and the light-harvesting antennae are tuned to the maximum light intensity available in a habitat. Electronic excitation in the antenna migrates from chromophore to chromophore and ultimately to the reaction centre, where is converted into chemical energy in the form of charge separation across the membrane.

Even a relatively simple bacterial photosynthetic system is very complex and its synthetic imitation is a challenging task. Mimicking of the natural photosynthetic process requires synthetic models of all the crucial components and linking them together into a working molecular assembly. All the elements (antenna, charge separation, and reaction centres) may involve transition metals. Application of metal complexes facilitates mimicking of this complex chemical system due to rich and versatile photochemical processes typical for transition metal complexes (see section 6.4 in Chapter 6) [48].

12.4.1 Light Harvesting Antennae

An antenna for light harvesting is an organized multicomponent system, in which an array of chromophoric molecules absorbs the incident light and channels the excitation energy to the common acceptor components [49–57]. The antenna effect can be obtained only in supramolecular arrays suitably organized in space, time, and energy domains. Each molecular component must absorb the incident light, and the excited state obtained in this way must transfer electronic energy to the nearby component before undergoing radiative or non-radiative deactivation [58].

One of the obvious candidates for light-harvesting antennae chromophores is porphyrin and phthalocyanine arrays [59–67], synthetic analogues of natural chlorophyll arrays present in photosystems of green plants. Porphyrins are characterized by high molar absorption coefficients and fast energy/electron transfer to other components of the system, and they can be easily used as building blocks in large supramolecular systems [65, 68, 69]. Widely studied antenna systems consist of an array of various metalloporphyrin units linked with free-base porphyrin or phthalocyanine by bridges that ensure good energy transfer and weak electronic coupling (eg diarylethynes) [65, 70]. Arrays comprised of weakly coupled pigments are ideally suited for energy transfer studies, as the energy transfer process does not depend on the state of other chromophores [71]. In the systems containing Mg or Zn porphyrin energy donors and free base porphyrin acceptor the energy transfer occurs with a lifetime of 25–115 ps and high quantum efficiency (>95%) [71]. The energy transfer quantum efficiency depends on geometrical arrangement of chromophores and varies from 76% for a cyclic array with peripheral acceptor, 84% for a cyclic array with integral acceptor, 91% for a tri-branched array with central acceptor to 98% for starburst arrays with the acceptor in the centre (Figure 12.7) [71].

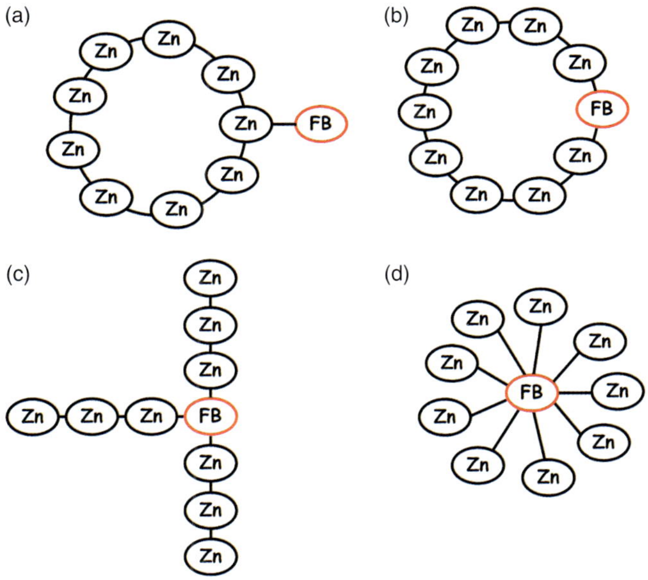

Figure 12.7 Various architectures of the porphyrin arrays for light harvesting. (a) Cyclic array with peripheral acceptor; (b) cyclic array with integral acceptor; (c) tri-branched array with central acceptor; (d) starburst array with central acceptor. FB denotes the free base porphyrin and Zn the zinc–porphyrin complex. (Adapted from Van Patten et al. [71])

Another class of extensively studied artificial antennae encompasses metallodendrimers and multimetal arrays based on polypyridine ligands [55, 57, 72–75]. Especially promising, due to high molar absorptivity, photostability, and excited state properties, are polypyridine complexes of ruthenium(II) and osmium(II). Irradiation of the above-mentioned complexes within their MLCT absorption bands leads to formation of long-lived excited states, which can efficiently transfer electronic energy to other metal centres of the same or lower excited state energy. This process can by used in a dendritic structure to focus energy within the core of the dendrimer or transport it from the inner part of the dendrimer to the edge. The electronic energy transfer within metallodendrimers may follow two independent pathways: superexchange along the dendritic framework or direct coulombic interaction via the dendritic folding. The direction of energy transfer can be tuned with the structure of the bridging and terminal ligands, and by proper selection of metal ions present in the branching points and on the edge of the dendrimer. The efficiency of the energy transfer is close to unity. The excitation of the mononuclear complexes to the ^{1}MLCT state is followed by immediate intersystem crossing to the lowest ^{3}MLCT state, which activates luminescence with high quantum yield. In the case of multinuclear supramolecular assemblies the energy transfer is so efficient that the luminescence of the unit with higher ^{3}MLCT energy cannot be observed and the quantum yield of energy transfer is unity [76, 77].

12.4.2 Charge-Separation Systems

Once the solar energy has been harvested and focused in the reaction centre, it must be converted into a more useful form, ie chemical energy. This goal can be achieved in the systems that allow formation of the charge-separated state with a sufficiently long lifetime. The simplest system to achieve this goal is a dyad molecule (Figure 12.8a), where donor (D) and acceptor (A) counterparts are covalently linked to form one molecule [78]. Photoinduced charge separation occurs here in a single step. It is obvious that in such a structure charge recombination will occur readily. To slow down the recombination it is necessary to introduce an additional reaction step, in which the charges are moved far apart. The simplest way to do so is to introduce the secondary donor (D′) or secondary acceptor (A′) (Figure 12.8b). One can envision more complex systems containing an array of donors and acceptors, separating charges as fast and as far as possible (Figure 12.8c).

In natural systems it is achieved by the carotene-containing system as a terminal electron donor and quinone as a terminal electron acceptor, with a series of porphyrin moieties in between. Artificial systems based on the same structural units are characterized with relatively short lifetimes of charge-separated states. The design of artificial photosynthetic systems is, however, not restricted to naturally occurring building blocks. Redox active sites may, in principle, give new interesting properties to the synthetic assemblies and expand their usefulness as potential solar energy conversion devices. The C_{60} bucky ball seems to be especially well suited for application as a component of an artificial photosynthetic device. The synthetic assemblies constitute good electron acceptors (up to six electrons) with a reduction

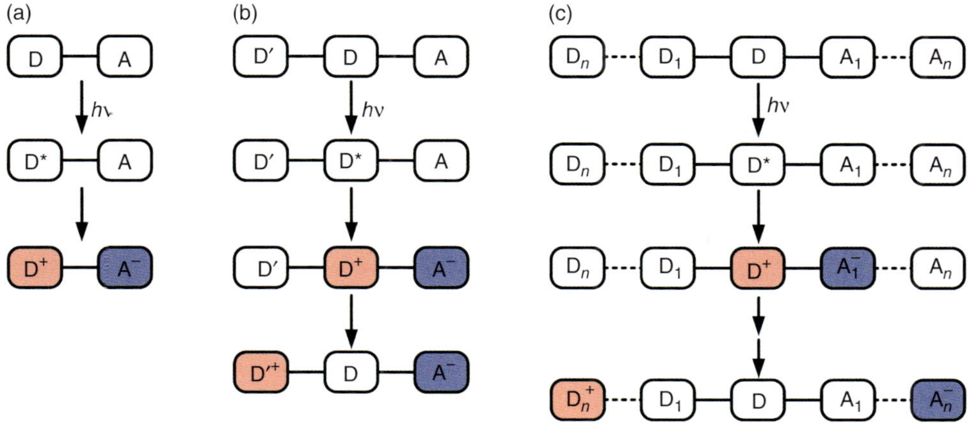

Figure 12.8 *Mimicking of biological photoinduced electron transfer in donor–acceptor arrays. (Adapted from Chambron et al. [78])*

potential very similar to that of quinones. Solubility of fullerenes in lipid membranes facilitates construction of biomimetic structures for charge separation systems [53, 79]. Metalloporphyrin–fullerene arrays are also used in solid state photovoltaic devices [60]. The most successful artificial systems contain porphyrin chromophore (or chromophores) linked with fullerene as an electron acceptor [53, 59–62, 79]. Among different systems containing series of electron donors (porphyrins, ferrocene) and acceptors (pyromellitimide, quinones, fullerene), the longest lifetime (380 ms) of a charge-separated state was observed for the tetrad Fc–ZnP–H_2P–C_{60} [62]. The lifetime of this state is more than one order of magnitude longer than in any other synthetic system and is comparable to that observed for the bacterial photosynthetic centres. Recently, even longer lifetime has been recorded for the zinc chlorin–fullerene dyad at −150 °C [64]. The quantum yield of the charge-separated state amounts to only 0.12.

12.4.3 Biomimetic Reaction Centres

In natural photosynthetic systems the excitation energy migrates from the antenna and initiates a cascade of electron transfer reactions between the reaction centre cofactors. Photosystem II contains the primary donor, P_{680} chlorophyll, which in the excited state transfers one electron to the nearby pheophytine moiety. The resulting P_{680}^+ cation is a strong oxidant and abstracts one electron from the tyrosine residue. It in turn oxidizes the manganese cluster, which comprises four manganese atoms. Four sequential oxidation steps then induce oxidation of two water molecules and generation of molecular oxygen.

This part of the photosynthetic chain can be mimicked by ruthenium–tyrosinate and manganese–tyrosinate complexes [80–82]. In the excited state the ruthenium–polypyridine complex abstracts one electron from the tyrosinate moiety, yielding the tyrosyl radical. More advanced models contain a ruthenium–polypyridine antenna linked with manganese phenolate or manganese–polyamine complexes containing mono-, di-, and trimeric Mn centres. Subsequent excitation of the Ru^{II} antenna results in oxidation of the Mn^{II} centre to Mn^{III} and Mn^{IV}. Di- and trinuclear complexes can provide more electrons to the system.

The other approach towards artificial photosynthesis does not model the natural photosystems directly, but instead links natural and synthetic components to achieve a system capable of photochemical water splitting [83].

References

1. Bellelli A, Brunori M, Brzezinski P, Wilson MT. Photochemically induced electron transfer. *Methods* 2001;**24**:139–52.
2. Millett F, Durham B. Design of photoactive ruthenium complexes to study interprotein electron transfer. *Biochemistry* 2002;**41**:11315–24.
3. Pieroni O, Fissi A, Angelini N, Lenci F. Photoresponsive photopeptides. *Acc Chem Res* 2001;**34**:9–17.
4. Lion-Dagan M, Willner I. Nitrospiran-modified alfa-chymotrypsin, a photostimulated biocatalyst in an organic solvent: effects of bioimprinting. *J Photochem Photobiol A: Chem* 1997;**108**:247–252.
5. Shipway AN, Willner I. Electronically transduced molecular mechanical and information functions on surfaces. *Acc Chem Res* 2001;**34**:421–32.
6. Willner I. Photoswitchable biomaterials: en route to optoelectronic systems. *Acc Chem Res* 1997;**30**:347–356.
7. Harvey AJ, Abell AD. Azobenzene-containing, peptidyl alfa-ketoesters as photobiological switches of alfa-chymotrypsin. *Tetrahedron* 2000;**59**:9763–71.
8. Fiamengo R, Crego-Calama M, Reinhoudt DN. Synthetic self-assembled models with biomimetic functions. *Curr Opin Chem Biol* 2001;**5**:660–73.
9. Willner I, Rubin S. Control of the structure and functions of biomaterials by light. *Angew Chem Int Ed Eng* 1996;**35**:367–85.
10. Doron A, Portnoy M, Lion-Dagan M, Katz E, Willner I. Amperometric transduction and amplification of optical signals recorded by a phenoxynaphthacenequinone monolayer electrode: photochemical and pH-gated electron transfer. *J Am Chem Soc* 1996;**118**:8937–44.
11. Willner I, Willner B. Electrical communication of redox proteins by means of electron relay-tethered polymers in photochemical, electrochemical and photoelectrochemical systems. *React Polym* 1994;**22**:267–79.
12. Zahavy E, Willner I. Photoinduced electron transfer in eosin-modified Co(II)-protoporphyrin IX reconstituted myoglobin and α- or β-hemoglobin subunits: photocatalytic transformations by the reconstituted photoenzymes. *J Am Chem Soc* 1996;**118**: 12499–514.
13. Willner I, Willner B. Vectorial photoinduced electro-transfer in tailored redox-active proteins and supramolecular nanoparticle arrays. *Coord Chem Rev* 2003;**245**:139–51.
14. Weber S. Light-driven enzymatic catalysis of DNA repair: a review of recent biophysical studies on photolyase. *Biochim Biophys Acta* 2005;**1707**:1–23.
15. Essen LO, Klar T. Light-driven DNA repair by photolyases. *Cell Mol Life Sci* 2006;**63**:1266–77.
16. Heyes DJ, Heathcote P, Rigby SEJ, Palacios MA, van Grondelle R, Hunter CN. The first catalytic step of the light-driven enzyme protochlorophyllide oxidoreductase proceeds via a charge transfer complex. *J Biol Chem* 2006;**281**:26847–53.
17. Heyes DJ, Hunter CN. Making light work of enzyme catalysis: protochlorophyllide oxidoreductase. *Trends Biochem Sci* 2005;**30**:642–9.

18. Odaka M, Fujii K, Hoshino M, et al. Activity regulation of photoreactive nitrile hydratase by nitric oxide. *J Am Chem Soc* 1997;**119**:3785–91.
19. Endo I, Odaka M. What evidences were elucidated about photoreactive nitrile hydratase? *J Mol Catal B: Enzym* 2000;**10**:81–6.
20. Greene SN, Chang CH, Richards NGJ. The role of post-translational modification in the photoregulation of Fe-type nitrile hydratase. *Chem Commun* 2002:2386–7.
21. Tai LA, Hwang KC. Photoreactivation of alloxanthine-inhibited xanthine oxidase. *Photochem Photobiol* 2001;**73**:439–46.
22. Tai LA, Hwang KC. Control of xanthine oxidase activity by light. *Angew Chem Int Ed* 2000;**39**:3886–8.
23. Kleiner O, Butenandt J, Carell T, Batschauer A. Class II DNA photolyase from *Arabidopsis thaliana* contains FAD as cofactor. *Eur J Biochem* 1999;**264**:161–7.
24. Sanders DB, Wiest O. A model for the enzyme-substrate complex of DNA photolyase and photodamaged DNA. *J Am Chem Soc* 1999;**121**:5127–34.
25. Kavakli HI, Sancar A. Analysis of the role of intraprotein electron transfer in photoreactivation by DNA photolyase in vivo. *Biochemistry* 2004;**43**:15104–10.
26. Nagashima S, Nakasako M, Dohmae N, et al. Novel nonheme iron center of nitrile hydratase with a claw setting of oxygen atoms. *Nature Struct Biol* 1998;**5**:347–52.
27. Yamada H, Kobayashi M. Nitrile hydratase and its application to industrial production of acrylamide. *Biosci Biotechnol Biochem* 1996;**60**:1391–400.
28. Wyatt JM, Knowles CJ. The development of a novel strategy for the microbial treatment of acrylonitrile effluents. *Biodegradation* 1995;**6**:93–7.
29. Endo I, Odaka M, Yohda M. An enzyme controlled by light: the molecular mechanism of photoreactivity in nitrile hydratase. *Trends Biotechnol* 1999;**17**:244–8.
30. Hamachi I, Tanaka S, Tsukiji S, Shinkai S, Oishi S. Design and semisynthesis of photoactive myoglobin bearing ruthenium tris(2,2′-bipyridine) using cofactor-reconstitution. *Inorg Chem* 1998;**37**:4380–8.
31. Ipe BI, Niemeyer CM. Nanohybrids composed of quantum dots and cytochrome P450 as photocatalysts. *Angew Chem Int Ed* 2006;**45**:504–7.
32. Willner I, Katz E, eds. *Bioelectronics. From theory to applications*. Weinheim: Wiley-VCH Verlag, 2005.
33. Willner I, Lion-Dagan M, Marx-Tibbon S, Katz E. Bioelectrocatalyzed amperometric transduction of recorded optical signals using monolayer-modified Au-electrodes. *J Am Chem Soc* 1995;**117**:6581–92.
34. Willner I, Blonder R, Katz E. Reconstitution of apo-glucose oxidase with a nitrospiropyran-modified FAD cofactor yields a photoswitchable biocatalysts for amperometric transduction of recorded optical signals. *J Am Chem Soc* 1996;**118**:5310–11.
35. Kirby AJ. Enzyme mechanisms, models and mimics. *Angew Chem Int Ed Eng* 1996;**35**:707–74.
36. Shiragami T, Matsumoto J, Inoue H, Yasuda M. Antimony porphyrin complexes as visible-light driven photocatalyst. *J Photochem Photobiol C: Photochem Rev* 2005;**6**:227–48.
37. Maldotti A, Andreotti L, Molinari A, Carassiti V. Photochemically driven models of oxygenases based on the use of iron. *J Biol Inorg Chem* 1999;**4**:154–61.
38. Knör G. Photocatalytic reactions of porphyrin-based multielectron transfer sensitizers. *Coord Chem Rev* 1998;**171**:61–70.
39. Takagi S, Suzuki M, Shiragami T, Inoue H. Photochemical P-450 oxygenation of cyclohexene with water sensitized by dihydroxy-coordinated (tetraphenylporphyrinato) antimony(V) hexafluorophosphate. *J Am Chem Soc* 1997;**119**:8712–13.
40. Knör G. Bionic catalyst design: a photochemical approach to artificial enzyme function. *ChemBioChem* 2001;**2**:593–6.
41. Rosenthal J, Bachman J, Dempsey JL, et al Oxygen and hydrogen photocatalysis by two-electron mixed-valence coordination compounds. *Coord Chem Rev* 2005;**249**:1316–26.
42. Rosenthal J, Luckett TD, Hodgkiss JM, Nocera DG. Photocatalytic oxidation of hydrocarbons by a bis-iron(III)-μ-oxo Pacman porphyrin using O_2 and visible light. *J Am Chem Soc* 2006;**128**:6546–6547.

43. Sun L, Akermark B, Ott S. Iron hydrogenase active site mimics in supramolecular systems aiming for light-driven hydrogen production. *Coord Chem Rev* 2005;**249**:1653–63.
44. Herrero C, Lassalle-Kaiser B, Leibl W, Rutherford AW, Aukauloo A. Artificial systems related to light driven electron transfer processes in PSII. *Coord Chem Rev* 2008;**252**:456–68.
45. Möbius K. Primary processes in photosynthesis: what do we learn from high-field EPR spectroscopy. *Chem Soc Rev* 2000;**29**:129–39.
46. Pullerits T, Sundström V. Photosynthetic light-harvesting pigment-protein complexes: toward understanding how and why. *Acc Chem Res* 1996;**29**:381–9.
47. Abrahamsson MAL. Electron transfer in ruthenium-manganese complexes for artificial photosynthesis. Thesis, Uppsala: Acta Universitatis Upsaliensis, 2001.
48. Hennig H. Homogeneous photo catalysis by transition metal complexes. *Coord Chem Rev* 1999;**182**:101–23.
49. Calzaferri G, Pauchard M, Maas H, Huber S, Khatyr A, Schaafsma T. Photonic antenna system for light harvesting, transport and trapping. *J Mater Chem* 2002;**12**:1–13.
50. Calzaferri G. Organic-inorganic composites as photonic antenna. *Chimia* 2001;**55**: 1009–13.
51. Maas H, Calzaferri G. Constructing dye zeolite photonic nanodevices. *Spectrum* 2003;**16**:18–24.
52. Sykora M, Maxwell KA, DeSimone JM, Meyer TJ. Mimicking the antenna-electron transfer properties of photosynthesis. *Proc Natl Acad Sci USA* 2000;**97**:7687–91.
53. Imahori H, Mori Y, Matano Y. Nanostructured artificial photosynthesis. *J Photochem Photobiol C: Photochem Rev* 2003;**4**:51–83.
54. Kawa M. Antenna effects of aromatic dendrons and their luminescence applications. *Top Curr Chem* 2003;**228**:193–204.
55. Dirksen A, De Cola L. Photoinduced processes in dendrimers. *C R Chimie* 2003; **6**:873–82.
56. Hofkens J, Schroeyers W, Loos D, et al. Triplet states as non-radiative traps in multichromophoric entities: single molecule spectroscopy of an artificial and natural antenna system. *Spectrochim Acta* 2001;A**57**:2093–107.
57. Balzani V, Ceroni P, Juris A, et al. Dendrimers based on photoactive metal complexes. Recent advances. *Coord Chem Rev* 2001;**219**–221:545–72.
58. Balzani V, Credi A, Venturi M. Photoprocesses. *Curr Opin Chem Biol* 1997;**1**:506–13.
59. Imahori H, Tamaki K, Araki Y, et al. Stepwise charge separation and charge recombination in ferrocene-meso, meso-linked porphyrin dimer-fullerene triad. *J Am Chem Soc* 2002;**124**:5165–74.
60. Imahori H, Kashiwagi Y, Hasobe T, et al. Porphyrin and fullerene-based artificial photosynthetic materials for photovoltaics. *Thin Solid Films* 2004;**451**–452:580–8.
61. Imahori H, Tamaki K, Yamada H, et al. Photosynthetic electron transfer using fullerenes as novel acceptors. *Carbon* 2000;**38**:1599–605.
62. Imahori H, Guldi DM, Tamaki K, et al. Charge separation in a novel artificial photosynthetic reaction center lives 380 ms. *J Am Chem Soc* 2001;**123**:6617–28.
63. Lammi RK, Ambroise A, Balasubramanian T, et al. Structural control of photoinduced energy transfer between adjacent and distant sites in multiporphyrin arrays. *J Am Chem Soc* 2000;**122**:7579–91.
64. Ohkubo K, Kotani H, Shao J, et al. Production of an ultra-long-lived charge-separated state in a zinc chlorin-C_{60} dyad by one step photoinduced electron transfer. *Angew Chem Int Ed Eng* 2004;**43**:853–6.
65. Choi M-S, Yamazaki T, Yamazaki I, Aida T. Bioinspired molecular design of light-harvesting multiporphyrin arrays. *Angew Chem Int Ed Eng* 2004;**43**:150–8.
66. Gust D, Moore TA, Moore AL. Mimicking photosynthetic solar energy transduction. *Acc Chem Res* 2001;**34**:40–8.
67. Hwang I-W, Kamada T, Ahn TK, et al. Porphyrin boxes constructed by homochiral self-sorting assembly: Optical separation, exciton coupling, and efficient excitation energy migration. *J Am Chem Soc* 2004;**126**:16187–98.

68. Wojaczynski J, Latos-Grazynski L. Poly- and oligometalloporphyrins associated through coordination. *Coord Chem Rev* 2000;**204**:113–71.
69. Shinmori H, Kajiwara T, Osuka A. Supramolecular assembly of light harvesting porphyrin hexamer. *Tetrahedron Lett* 2001;**42**:3617–20.
70. Li J, Lindsey JS. Efficient synthesis of light-harvesting arrays composed of eighth porphyrins and one phthalocyanine. *J Org Chem* 1999;**64**:9101–8.
71. Van Patten PG, Shreve AP, Lindsey JS, Donohoe RJ. Energy transfer modelling for the rational design of multiporphyrin light-harvesting arrays. *J Phys Chem B* 1998;**102**: 4209–16.
72. Balzani V, Campagna S, Puntoniero F, et al. Dendrimers made of Ru(II) and Os(II) polypyridine subunits as artificial light-harvesting antennae. *CR Chimie* 2003;**6**:883–93.
73. Balzani V, Ceroni P, Maestri M, Vicinelli V. Light-harvesting dendrimers. *Curr Opin Chem Biol* 2003;**7**:657–65.
74. Gittins PJ, Twyman LJ. Dendrimers and supramolecular chemistry. *Supramol Chem* 2003;**15**:5–23.
75. Serroni S, Campagna S, Puntoniero F, et al. Dendrimers made of Ru(II) and Os(II) polypyridine subunits as artificial light harvesting antennae. *CR Chimie* 2003;**6**:883–93.
76. Balzani V, Credi A, Venturi M. Photochemistry and photophysics of coordination compounds. *Coord Chem Rev* 1998;**171**:3–16.
77. Balzani V, Campagna S, Denti G, Juris A, Serroni S, Venturi M. Designing dendrimers based on transition-metal complexes. Light-harvesting properties and predetermined redox patterns. *Acc Chem Res* 1998;**31**:26–34.
78. Chambron J-C, Collin J-P, Dalbavie JO, et al. Rotaxanes and other transition metal-assembled porphyrin arrays for long-range photoinduced charge separation. *Coord Chem Rev* 1998;**178**–180:1299–312.
79. Gust D, Moore TA, Moore AL. Photochemistry of supramolecular systems containing C_{60}. *J Photochem Photobiol B: Biol* 2000;**58**:63–71.
80. Tommos C, Hoganson CW, Di Valentin M, et al. Manganese and tyrosyl radical function in photosynthesis oxygen evolution. *Curr Opin Chem Biol* 1998;**2**:244–52.
81. Rüttinger W, Dismukes GC. Synthetic water-oxidation catalysts for artificial photosynthetic water oxidation. *Chem Rev* 1997;**97**:1–23.
82. Vrettos JS, Limburg J, Brudvig GW. Mechanism of water oxidation: combining biophysical studies of photosystem II with inorganic model chemistry. *Biochim Biophys Acta* 2001;**1503**:229–45.
83. Sakamoto M, Kamachi T, Okura I, Ueno A, Mihara H. Photoinduced hydrogen evolution with peptide dendrimer-multi-Zn(II)-porphyrin, viologen and hydrogenase. *Biopolymers* 2001;**59**:103–9.

13

Photoinduced Electron Transfer in Proteins

Leben ist die Daseinsweise der Eiweißkörper, und diese Daseinsweise besteht wesentlich in der beständigen Selbsterneuerung der chemischen Bestandteile dieser Körper. [Life is the existence of protein structures, and this existence consists essentially in the constant self-renewal of the chemical components of these structures.]

Friedrich Engels

Biochemical reactions involving electron transfer (ET) between chemical species are considered to be the most important and fundamental processes in life. The redox active proteins and enzymes are employed in a number of biological events such as photosynthesis, respiration, metabolism and synthesis of bioactive compounds, metabolism of toxic materials, and many others processes. Moreover, the unique properties of enzymes, such as ability to cleave bonds in a region-specific and stereospecific manner, open the door for application of enzymes in detoxification of environmental pollutants and biotechnological drug production [1–3]. A common feature of a large fraction of redox proteins is the presence of metal ions or clusters, usually providing an electron sink. The rates of ET to and from metal centres depend on the protein matrix [4, 5]. Attempts to elucidate the physical and chemical aspects of these complex phenomena require both theoretical and experimental approaches. It is thus essential to develop convenient tools that allow triggering and follow redox processes in proteins in a controlled way. Photocontrol of the activity of naturally occurring enzymes by modulation of the redox state of their active sites via photoinduced electron injection or abstraction is one of the most attractive targets in recent protein engineering [6, 7].

In this chapter we review the application of metal complexes and light in the study of biological electron transfer events.

13.1 Photochemical Methodology

Photochemical methods offer a convenient tool to study intra- and interprotein ET because of their time resolution and selectivity. Various mechanistic and design approaches based on photochemistry of metal complexes have been undertaken. Most of the studies on protein electron transfer processes have been done for haemoproteins using among others ruthenium complex as a photosensitizer, modified haemoproteins in which haem iron is substituted by another metal (mainly Zn), and CO-bonded haem proteins [6, 7].

13.1.1 Photoactive Ruthenium Complexes

General Idea
Ruthenium complexes can be employed as photosensitizers in several ways:

- as unbound, free complexes
- attached to protein or its cofactor
- tethered to enzyme substrate.

Ruthenium complexes with polypyridyl ligands are substitutionally inert and possess favourable photochemical properties that make them attractive photosensitizers. They easily undergo photoexcitation with high quantum yields and have a long lifetime emission state. The change between Ru^{II} and Ru^{III} is associated with small structural changes, which correspond to very small reorganizational barriers. Moreover, it is possible to vary redox potential value in a relatively large range by selection of ligands to achieve the appropriate driving force for ET processes. Importantly, the photoactive ruthenium complexes can be applied for both photochemical reduction and oxidation of haem proteins.

The photoexcited $[Ru^{II}L_3]^{2+}$ complexes (where L is a polypyridyl-type ligand) can directly reduce ferric haem group in enzymes, and back ET from ferrous haem to $[Ru^{III}L_3]^{3+}$ can be avoided by addition of a sacrificial electron donor, eg EDTA or aniline, which reduces Ru^{III} to Ru^{II} (Figure 13.1a) [8]. When a stronger reducing agent is needed the photoexcited $[Ru^{II}L_3]^{2+}$ can be reductively quenched by an electron donor with lower potential, eg *p*-methoxy-*N,N*-dimethylaniline or *N,N*-dimethylaminobenzoate, yielding $[Ru^{II}L_2L^-]^+$ [9]. Although some authors suggest formation of Ru^+ species [10–12], in the presence of highly electrophilic ligands (eg polypyridyls) the redox process seems to be mainly localized on the aromatic ligand [13]. Subsequently, the metal centre in the protein can be reduced by the $[Ru^{II}L_2L^-]^+$ complex (Figure 13.1a). Furthermore, the $[Ru^{II}L_3]^{2+}$ complex can be used for photoinduced oxidation of haem proteins [14, 15]. Photoexcited $[Ru^{II}L_3]^{2+}$ can be oxidatively quenched by a suitable sacrificial acceptor ($[CoBr(NH_3)_5]^{2+}$ or $[Ru(NH_3)_6]^{3+}$) yielding $[Ru^{III}L_3]^{3+}$, and subsequently the photogenerated oxidant [16] reacts with Fe^{II} of the haem protein. Alternatively, for some ruthenium complexes (eg with bpz ligands, Figure 13.2) the excited state of ruthenium can be quenched by direct electron transfer from $^*[Ru^{II}L_3]^{2+}$ to the metal centre of a protein, thus forming

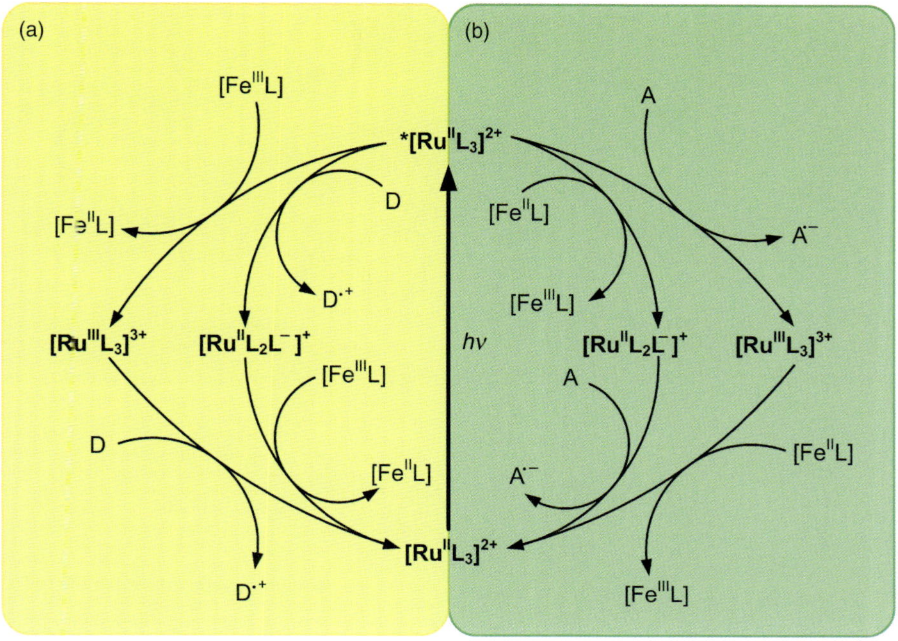

Figure 13.1 *Scheme for (a) the reduction or (b) the oxidation of the iron centre of haemoproteins with photoexcited ruthenium complex, using flash quench procedure. (Adapted from Szaciłowski et al. [124])*

$[Ru^{II}L_2L^-]^+$ and the oxidized metal centre. The back ET from $[Ru^{II}L_2L^-]^+$ to Fe^{III} can be inhibited by addition of an electron acceptor (eg $[CoCl(NH_3)_5]^{2+}$), yielding $[Ru^{II}L_3]^{2+}$ and ferric state protein. The oxidative pathway is shown in Figure 13.1b. The presented procedure is called a flash quenched method [17] and can be applied for any ruthenium complex and redox active protein. The flash quenched method allows the observation of ET rate over an extremely wide range by choosing the different ruthenium complexes and adding a proper quencher.

Free Ruthenium Complexes

The $[Ru(bpy)_3]^{2+}$ complex has been widely used to study ET in haem proteins for years and still has application in the photochemical reduction of haem proteins [8, 9, 14, 15, 18]. To improve the binding properties of ruthenium complexes to protein and increase the yield of photoreduction or photo-oxidation processes, novel binuclear ruthenium complexes – namely $[\{Ru(bpy)_2\}_2(bphb)]^{4+}$, $[\{Ru(bpy)_2\}_2(qpy)]^{4+}$ and $[\{Ru(bpz)_2\}_2(qpy)]^{4+}$ for ligands see Figure 13.2) – have been designed [20–24]. They have a net charge of +4, which makes them appropriate for binding strongly to the negatively charged domains of proteins and donate/abstract electrons to/from proteins, specifically on irradiation with good quantum yields. These high efficiencies are the consequence of longer lifetimes of the metal-to-ligand charge transfer (MLCT) states (about 1.0 μs), compared with the $[Ru(bpy)_3]^{2+}$ complex (about 0.62 μs) [25]. Binding properties and selectivity of

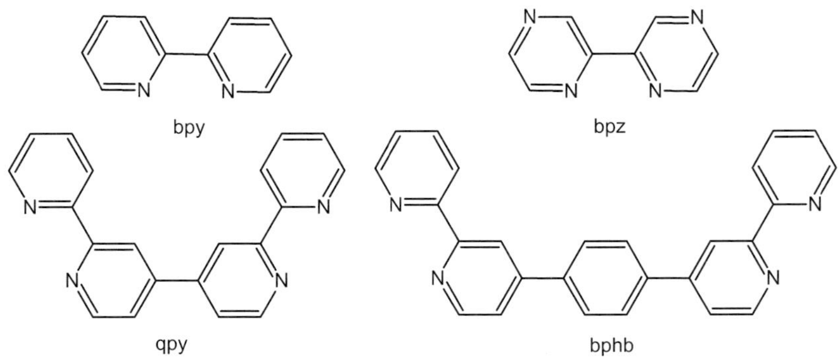

Figure 13.2 Structure of selected ligands appearing in this chapter. (See Abbreviations for the systematic names)

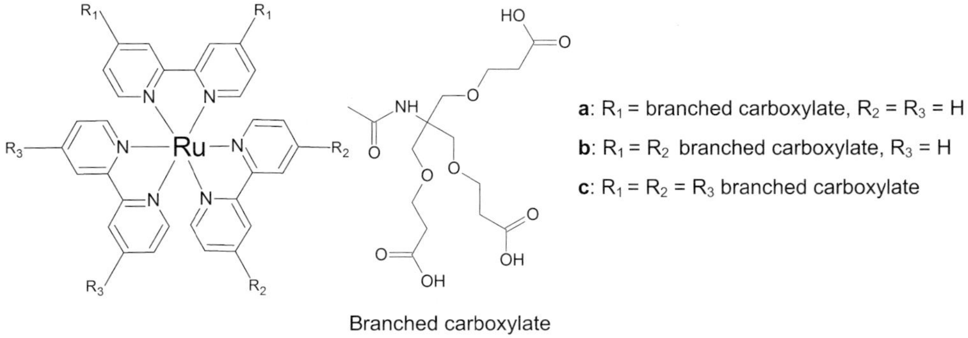

Figure 13.3 Structure of Ru(bpy)$_3$-based artificial receptors [26]

ruthenium complexes can be further improved by careful design of ligands bearing dendritic carboxylate moieties (Figure 13.3) [26]. The affinity to cytochrome *c* protein increases with increasing number of introduced carboxylate groups. In addition they exhibit much higher selectivity in binding to cytochrome *c* than to other proteins such as myoglobin, horseradish peroxidase, or cytochrome b_{562}. It has been found that the last compound *b* (Figure 13.3b), which has an asymmetrical charge distribution, possesses the best photoreduction efficiency.

Ruthenium Complexes Covalently Bound to Proteins or Their Cofactors
A more elaborate method requires the covalent attachment of ruthenium complex to protein or its cofactor. The general procedure is based on photoexcitation of the Ru-protein derivative leading to formation of the MLCT excited state, $*[Ru^{II}L_3]^{2+}$, which can be quenched in the same way as free ruthenium complex (see Figure 13.1) [6, 7, 17, 27]. The great advantage of ruthenium complex covalently appended to a protein is that no external reagents are required to induce the ET reactions. Two important criteria must be attained in designing ruthenium-modified proteins in order to examine an actual rate of intermolecular ET in biological systems. First,

the modified protein should interact with a redox partner in the same way as a native one, ie it should not interfere in the binding interaction. Second, the rate of photoinduced ET from $*[Ru^{II}L_3]^{2+}$ to the protein's metal centre must be fast compared with the subsequent ET from photoactivated protein to another protein or other redox partner [7].

Attachment to Specific Amino Acid

One approach to modify proteins encompasses covalent attachment of the photosensitizer to a specific amino acid (eg cysteine, histidine, or lysine) of the protein. The labelling of proteins with ruthenium complexes has been known for a long time [28–30]; however, the design of more effective ruthenium complexes for photoinduced ET and development of modern experimental techniques allows better understanding of the principles and mechanisms of ET occurring in complex biological systems. Different procedures were employed to modify proteins; in one a lysine amino group is labelled with a ruthenium reagent containing a reactive *N*-hydroxysuccinamide [31] or bromomethyl group [32]. Similarly, the sulfhydryl group of cysteine is modified by reaction with ruthenium complex containing bromomethyl group [29, 33]. Labelling of the histidine group involves reaction of the protein with the $[RuL_2(solv)_2]^{2+}$ complex (where L is an organic ligand and solv a solvent molecule) followed by reaction with imidazole or other ligand [27, 34]. A ruthenium complex is usually placed out of the binding domain at the surface amino acids of the protein molecule. A different type of protein modified with various ruthenium complexes has been prepared and the ET event in those systems extensively studied [21, 35–44].

Attachment to Haem

Another strategy includes covalent attachment of the photosensitizer to the protein using cofactor reconstitution [45]. A protohaemin molecule, the prosthetic group of most haemoproteins, can be extracted under certain conditions from the active pocket, giving an apoprotein or apoenzyme [46]. The resultant molecule can subsequently react with chemically modified cofactors, leading to the formation of the diversity of new proteins with distinguished properties. Using this methodology Hamachi et al. [45] prepared the $[Ru(bpy)_3]$–protohaemin unit by binding of $[Ru(bpy)_3]^{2+}$ complex through various spacers to protoporphyrin IX, and then incorporating it into the haem crevice of the apoenzyme, as shown in Figure 13.4. Several proteins have been prepared in this way (eg myoglobin, cytochrome b_{562}) and their photoactivation properties and ET reaction mechanisms employing the flash quenched method were examined [47–50]. The semisynthetic myoglobins reconstituted with such cofactors have redox potential values of the same order as native myoglobins, and are therefore suitable models for study the ET processes [45]. Moreover, Hamachi et al. have shown that the binding of the oxygen molecule to myoglobin can be regulated by switching the light on and off in the presence of air [47].

Attachment of Ruthenium Complex to Enzyme Substrate or Ligand

The strategy based on the attachment of the photosensitizer via an aliphatic spacer to compounds that bind to the enzyme active site (Figure 13.5) has been recently

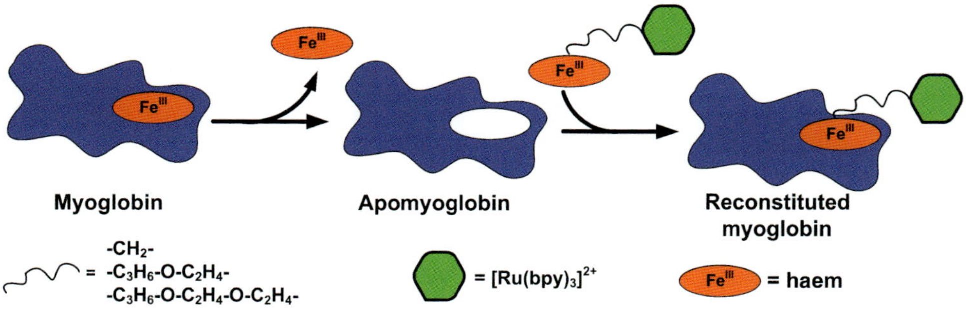

Figure 13.4 *Reconstitution of ruthenium-tris(bipyridine)-appended haem to apomyoglobin [47–49]. (Adapted from Szaciłowski et al. [124])*

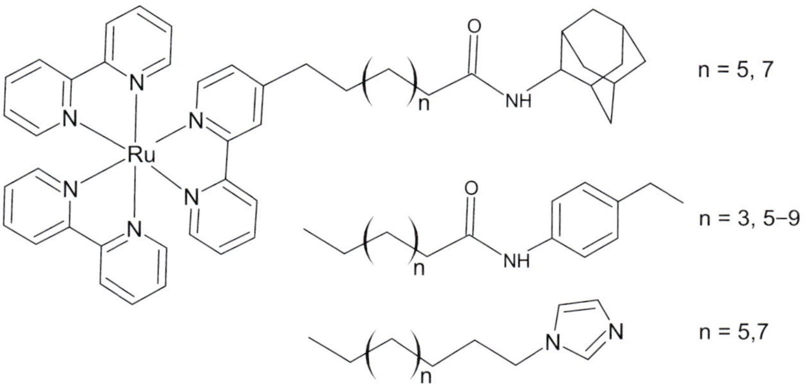

Figure 13.5 *Ruthenium complex linked to enzyme substrates (adamantane, ethylbenzene) or ligand (imidazole) [51–53]*

applied to a group of cytochromes P450, particularly P450*cam* [51–53]. The ruthenium complex $[Ru(bpy)_3]^{2+}$ has been attached via a hydrocarbon chain to substrates (eg adamantane, ethylbenzene) or ligands (eg imidazole), which have high affinity for the cytochrome P450 haem pocket. The association of Ru–substrate compounds with cytochrome P450 is sufficiently strong and specific to recognize submicromolar concentrations of cytochrome P450 in the presence of other haem proteins. Binding of the Ru^{II}–substrate compounds to cytochrome P450 protein can be detected by the decrease of the $*[Ru^{II}L_3]^{2+}$–substrate/ligand's lifetime and the monophasic luminescence decay profile usually becomes biphasic. The substrate/ligand with attached photosensitizer can reduce or oxidize the protein's haem group on photoexcitation in the presence of reductive or oxidative quencher (see Figure 13.1) [51]. This strategy is particularly useful when the reactive redox centre is deeply buried in proteins and the ET between the active site and the external redox agent, even when covalently attached to the protein, is inefficient.

13.1.2 Metal-Substituted Haemoproteins

Another approach involves modification of haemoproteins, in which haem iron is substituted by other metals, mainly by Zn. The photoexcited states of haem decay rapidly by thermal dissipation because the iron quenches not only porphyrin fluorescence but also fluorescence of tryptophan and tyrosine residues [6, 54]. However, replacement of the iron with other transition metals such as Zn^{II}, or Mg^{II} enhances the fluorescence of porphyrin and populates a long-lived triplet state via intersystem crossing from the singlet excited state (see Chapter 4) [55, 56]. Metal-substituted haemoproteins in triplet states can act as good reducing agents. The overall scheme of ET reactions in Zn-substituted protein (ZnP) and unmodified haem protein (FeP) is presented in Figure 13.6. Photoexcitation of metal-substituted porphyrin yields the singlet excited state of the chromophore, which can be converted to the triplet state, acting as an electron donor to another unmodified haem. This results in reduction of haem (FeP) and formation of π-radical cation of the metalloporphyrin ($ZnP^{\bullet+}$, see Figure 13.6). The π-radical cation is a strong oxidant that rapidly recaptures the lost electron, thereby returning to the ground state. In the absence of any other cofactor acting as an electron acceptor, decay of the triplet exited states occurs via phosphorescence, thermal dissipation, delayed fluorescence, and quenching by solvent components, mainly molecular oxygen.

The replacement of native haem with metal-substituted porphyrin can be performed in two ways. In the first, iron ions are removed from the protein by treating native protein with anhydrous HF, followed by insertion of the appropriate metal to metal-free protein [57–59]. In the second, haem is removed either chemically or by recombination (preparing a proper recombinant protein), and then protein is reconstituted with metal-substituted porphyrins [60–64]. The Zn-substituted metalloproteins such as cytochrome *c* [65–67], myoglobin [59, 61, 62, 64, 68, 69], and haemoglobin [64, 68, 70] have been extensively used to study photoinduced ET (PET) between modified proteins and their physiological redox partners. Interestingly, in haemoglobin with α and β subunits it was possible to determine ET parameters for

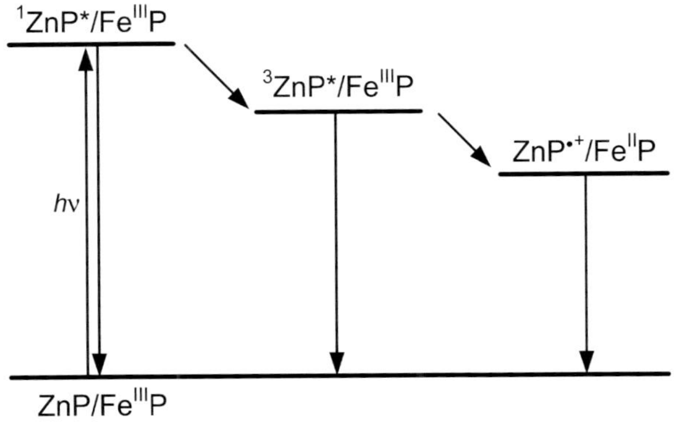

Figure 13.6 Electron transfer between Zn-substituted porphyrin (ZnP) and unmodified porphyrin in an oxidized state in haemoproteins (FeP) [6]

both subunits by preparing hybrids [Zn, Fe] (only haem in one subunit is substituted by Zn) [70].

A more sophisticated method involves the combination of metal-substituted proteins with electron acceptors covalently attached to the amino acid residue located at the protein surface. The $[Ru(NH_3)_5]^{3+}$ complex attached to the histidine residue had been used in Zn-substituted myoglobin [71, 72] or cytochrome b_{562} [73] as an electron acceptor. This design allows the distance between donor and acceptor to be fixed and is very useful for experimental analysis of intramolecular electron transfer in proteins.

13.1.3 Photoinduced Ligand Dissociation

Previously discussed methods of PET studies in proteins assume preservation of the protein redox centre properties. An alternative protocol is based on tuning the redox potential of the metal centre in protein. This method involves binding of certain ligands to the redox-active metal site in proteins and photolabilization of the ligand [74]. Carbon monoxide has a high affinity towards ferrous haem but does not bind to ferric haem and CO binding to ferrous centre increases the apparent redox potential of the protein. The haem–CO complexes are light sensitive and can photodissociate with high quantum yield ($\Phi \approx 1$) [75]. The light-induced CO dissociation results in lowering of the redox potential and rapidly (within nanoseconds) converts a good electron acceptor (the CO complex) into an electron donor (five-coordinate ferrous species) leading to ET towards the oxidized acceptor protein (Figure 13.7). The recombination between CO and the five-coordinate ferrous centre can compete with the ET process. On one hand, the recombination reaction can be slowed down by reducing CO concentration. On the other, the concentration of CO

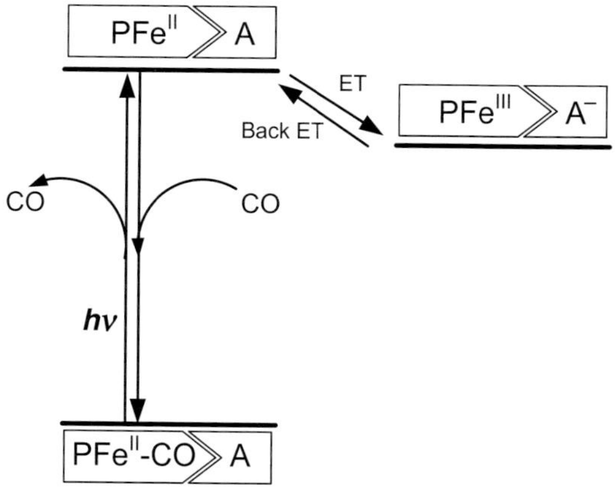

Figure 13.7 *Light-induced electron transfer (ET) between CO-bound porphyrin (PFe^{II}–CO) in haemoprotein and the oxidized acceptor protein (A). Acceptor protein can be the same protein possessing an additional active site, which can accept electrons or an externally bound protein in an oxidized state [82]*

cannot be too low because at very low CO concentrations the contribution of non-carbonylated ferrous haem changes the redox properties of the system. The intramolecular electron transfer triggered by photodissociation of the CO ligand has been observed for cytochrome *c* oxidase [76, 77], cytochrome cd_1 [78–80], and inducible nitric oxide synthase oxyFMN (oxygenase domain, flavin mononucleotide) [81]. Alternatively, the intermolecular ET has been investigated between carboxymethyl cytochrome *c* and cytochrome *c* oxidase [82, 83] or plastocyanin [84].

13.2 Biochemical Applications

13.2.1 Mechanisms of Electron Transfer

Understanding the ET phenomenon in proteins has been the subject of extensive research over the last decades and is discussed in detail elsewhere [85–87]. Here, we focus on information gained mainly from photochemical studies. The intramolecular ET processes in proteins can be described by a few determinants such as driving force, reorganization energy, distance, and pathway between donor and acceptor. The inverted effect (decrease of ET rate with increasing driving force) in a wide range of driving forces has been investigated using ruthenium-modified protein, and agreement with theory has also been proved for ET in proteins [27, 88]. Extensive work on the systems where donors and acceptors are held at fixed distances and orientations has shown that the ET rate decreases exponentially with the distance between redox partners [89–92], as expected for tunnelling reactions [90, 91, 93]. Thus, it is likely that the redox centres in these proteins are coupled electronically through the chemical bond framework of the intervening medium [89, 90, 93–95]. A great deal of insight into the pathway of ET and the potential role of amino acid residues in mediating ET between redox centres (tunnelling effect) has been gained by a combination of the mutation procedure with PET [38, 96, 97]. The same method can be used for investigating the role of the amino acid residues in the active site [41].

The interprotein electron transfer reactions are much more complicated because they involve at least three steps:

(1) formation of a reactant complex of the donor and acceptor protein
(2) ET within the reactant complex
(3) dissociation of the product complex.

It has often been assumed that the second step, intracomplex ET, takes place within a static reactant complex. However, a number of cases have been observed in which the reactant complex is not static, but undergoes dynamic configurational fluctuations (defined as protein–protein orientations in the bound complex) [65] or conformational changes (defined as a tertiary structure of each protein) [40]. If the reaction is rate limited by dynamic fluctuation in configuration rather than by the actual ET event, it is said to be gated [21, 22, 37, 98, 99]. Kostic and co-workers [100] found that the rate constants for intracomplex ET between zinc–cytochrome *c* and either plastocyanin or cytochrome b_5 decrease significantly with increasing viscosity, indicating gating by dynamic configurational fluctuations. Durham et al. [101] reported the evidence supporting configurational gating in the complex of ruthenium-labelled cytochrome *c* and plastocyanin. Gated ET in model systems consisting of various ruthenium-labelled peptides and ferricytochrome *c* has been recently studied by Ogawa et al. [102] as a probe of the configurational dynamics of the peptide–protein complex. Studies on kinetics and structure under the same conditions are needed to probe the interactions promoting electron tunnelling between proteins. A protein crystal containing photoactive donors and acceptors at specific lattice sites is an ideal medium for investigating the dependence of tunnelling rates on the structure. Gray and co-workers [97] have measured and analyzed the kinetics of ET between native and Zn-substituted cytochrome *c* molecules in crystals of known structures. The results indicate that van der Waals' interactions and water-mediated hydrogen bonds are effective coupling elements for tunnelling across protein–protein interfaces. Integration of photosensitizers into protein crystals provides a powerful tool for studying the dynamics of biochemical reactions.

The important step in an ET reaction is formation of the complex between the two proteins (protein docking). Researchers have defined a new dynamic docking paradigm for ET, in which interprotein interactions can involve numerous weakly bound configurations; of these only a small subset are ET active [67, 68, 103, 104]. Interestingly, the most reactive conformations are not the most favourable for binding [103, 104]. The investigation of the effect of ionic strength on ET reactions of metalloproteins can provide information on whether the complex is stabilized by the electrostatic and/or non-polar/hydrophobic interactions, and unravel the influence of these interactions on regulating protein–protein recognition and reactivity [38, 39, 42, 63, 67, 105]. These studies have allowed determination of the location of binding domains engaged in redox reactions. It has been found and/or confirmed that several ET complexes share a common binding motif, consisting of a central non-polar contact domain surrounded by a peripheral electrostatic domains containing oppositely charged residues on the two proteins. These groups of complexes include cyt *c*–cyt *c* peroxidase [35, 96, 106], cyt *c*–cyt bc_1 [39, 42], and cyt *c*–cyt *c* oxidase complex [38].

The final step of the ET event in proteins involves the dissociation of a product complex. The application of photochemical methods has allowed for evaluation the dissociation constants for several systems [35, 38, 96].

13.2.2 Cross-Linking of Proteins

Large multiprotein complexes often participate in ET reactions but are also responsible for many other biological processes. Currently, the design of photoactive reagent, which would cross-link the associated proteins, seems to be a promising method for probing the dynamics of protein–protein interactions [107–111]. The photogenerated excited state of $[Ru(bpy)_3]^{2+}$ in the presence of the electron acceptor (ammonium peroxodisulfate) yields a Ru^{III} complex, the sulphate radical, and sulphate anion. The proposed mechanism of photoinitiated protein cross-linking involves the formation of a tyrosyl radical in reaction with Ru^{III} complex. This is a crucial step of the cross-linking reaction. Subsequently the tyrosyl radical attacks the nearby nucleophile (lysine or cysteine residue) and the hydrogen atom is abstracted by the sulphate radical. Alternatively, if another tyrosine (or tryptophan) residue is in the vicinity, coupling with the tyrosyl radical occurs and sulphate radical is engaged in hydrogen abstraction. The formation of tyrosyl radical in $[Ru(bpy)_3]^{2+}$/ $S_2O_8^{2-}$-mediated protein cross-linking is fast and efficient compared with traditional cross-linking techniques [112–114] and can be used in crude extracts as well as in purified proteins. The great advantage of this method is the possibility of switching the cross-linking at a desired time, which allows studying dynamic processes.

13.2.3 Analyzing Intermediates and Testing New Inhibitors

Highly oxidized iron–porphyrin intermediates are implicated in the catalytic cycle of many haem enzymes. The application of photochemical methods allows for trapping of short-lived active intermediates in a redox reaction proceeding in a protein matrix. This tool has been used for generation and resolving of structures of intermediates involved in catalytic cycles of metalloenzymes such as peroxidase [15], cytochrome P450 [14], and cytochrome *c* oxidase [20] or proteins, eg myoglobin [49, 50], which have high-valent iron porphyrins as an active intermediate.

On the other hand, photochemical methods are a convenient tool for testing enzyme inhibitors by probing their influence on ET reaction. Millett et al. have studied the effect of famoxadone, a new cytochrome bc_1 inhibitor, on PET in cytochrome bc_1 complex [23, 24]. They found a significant decrease in the ET rate in the presence of this inhibitor, and detailed studies allow for a better insight into conformational changes induced by famoxadone and the dynamics of the overall process. Gray and co-workers have discovered that the Ru diimine wire-type complexes prevent ET between two domains of nitric oxide synthase (NOS) which suggests a novel design for NOS inhibitors [115].

13.2.4 Folding of Proteins

Spontaneous refolding of proteins occurs in a wide range of timescales: from microseconds to hours. The rapid photochemical methods described above can be utilized to investigate the fast protein-folding processes. The refolding process can be triggered by PET from photosensitizer to unfolded protein [116–119]. Alternatively, the photochemical method, based on PET, can be used to probe the progress and mode of folding or denaturation processes under solvent conditions favouring the folded or unfolded protein [120, 121]. Studies of folding processes can throw some light on the mystery of polypeptide self-assembly into native structures. This knowledge can be helpful in the design of artificial enzymes [122].

Methods that exploit metal complexes and their photochemistry have facilitated answers to fundamental questions about biological ET events and their overall mechanisms in biocatalysis. The knowledge can be employed in industry for biotechnological synthesis of different compounds, as well as in pharmacy for design of new inhibitors of enzymes, the activity of which gives undesirable effects, eg leads to development of various diseases. All the methods described above are complementary to each other and, to exclude specific interactions associated with a feature of one method, the use of at least two of them is suggested. The approach of a covalently appended photoactive inorganic redox complex to the study of interprotein ET between redox centres in non-covalent protein–protein complexes has the potential advantage of being used to study any protein pair, including those in which neither partner contains a haem. However, the attachment of a large and often highly charged ruthenium complex can strongly perturb the delicate balance of interactions that govern complex formation, thereby having a significant effect on the strength and even the mode of protein–protein binding [123]. Haem substitution procedures are successful in characterizing direct haem–haem reactivity; however, the ET rate constants measured by this method are non-physiological because of very different redox potentials of the species involved in the processes [55].

References

1. Lewis DFV. *Cytochromes P450 – Structure, Function and Mechanism*. London: Taylor & Francis, 1996.
2. Kellner DG, Maves SA, Sligar SG. Engineering cytochrome P450s for bioremediation. *Curr Opin Biotechnol* 1997;**8**:274–8.
3. Schenkmann JB, Greim H. *Handbook of Experimental Pharmacology*. Berlin: Springer-Verlag, 1993.
4. Stephens PJ, Jollie DR, Warshel A. Protein control of redox potentials of iron-sulfur clusters. *Chem Rev* 1996;**96**:2491–513.
5. Holm RH, Kennepohl P, Solomon EI. Structural and functional aspects of metal sites in biology. *Chem Rev* 1996;**96**:2239–14.
6. Bellelli A, Brunori M, Brzezinski P, Wilson MT. Photochemically induced electron transfer. *Methods* 2001;**24**:139–52.
7. Millett F, Durham B. Design of photoactive ruthenium complexes to study interprotein electron transfer. *Biochemistry* 2002;**41**:11315–24.
8. Contzen J, Jung C. Changes in secondary structure and salt links of cytochrome P-450_{cam} induced by photoreduction: A Fourier transform infrared spectroscopy study. *Biochemistry* 1999;**38**:16253–60.
9. Jensen TJ, Gray HB, Winkler JR, Kuznetsov AM, Ulstrup J. Dynamic ionic strength effects in fast bimolecular electron transfer between a redox metalloprotein of high electrostatic charge and an inorganic reaction partner. *J Phys Chem B* 2000;**104**: 11556–62.
10. Mines GA, Bjerrum MJ, Hill MG, et al. Rates of heme oxidation and reaction in Ru(His33)cytochrome *c* at very high driving forces. *J Am Chem Soc* 1996;**118**:1961–5.
11. Bellelli A, Brunori M, Brzezinski P, Wilson MT. Photochemically induced electron transfer. *Methods* 2001;**24**:139–52.
12. Millett F, Durham B. Design of photoactive ruthenium complexes to study interprotein electron transfer. *Biochemistry* 2002;**41**:11315–24.

13. Marcaccio M, Paolucci F, Paradisi C, et al. Electrochemistry of multicomponent systems, redox series comprising up to 26 reversible reduction processes in polynuclear ruthenium(II) bipyridine-type complexes. *J Am Chem Soc* 1999;**121**:10081–91.
14. Low DW, Winkler JR, Gray HB. Photoinduced oxidation of microperoxidase-8: Generation of ferryl and cation-radical porphyrins. *J Am Chem Soc* 1996;**118**:117–20.
15. Berglund J, Pascher T, Winkler JR, Gray HB. Photoinduced oxidation of horseradish. *J Am Chem Soc* 1997;**119**:2464–9.
16. Hoffman MZ, Bolleta F, Moggi L, Hug GL. Rate constants for the quenching of excited states of metal complexes in fluid solution. *J Phys Chem Ref Data* 1989;**18**:219–543.
17. Liang N, Mauk AG, Pielak GJ, Johnson JA, Smith M, Hoffman BM. Regulation of interprotein electron transfer by residue 82 of yeast cytochrome c. *Science* 1988;**240**: 311–13.
18. Nakano R, Sato H, Shimizu T. Tris(2,2′-bipyridyl)ruthenium(II)-mediated photoinduced electron transfer of engineered cytochrome P450 1A2. *J Photochem Photobiol B: Biol* 1996;**32**:171–6.
19. Bergamini G, Saudan C, Ceroni P, et al. Proton-driven self-assembled systems based on cyclam-cored dendrimers and $[Ru(bpy)_2(CN)_4]^{2-}$. *J Am Chem Soc* 2004;**126**:16466–71.
20. Zaslavsky D, Sadoski RC, Wang K, Durham B, Gennis RB, Millett F. Single Electron Reduction of cytochrome c oxidase compound F: Resolution of partial steps by transient spectroscopy. *Biochemistry* 1998;**37**:14910–16.
21. Sadowski RC, Engstrom G, Tian H, et al. Use of photoactivated ruthenium dimer complex to measure electron transfer between the Rieske iron-sulfur and cytochrome c_1 in the cytochrome bc_1 complex. *Biochemistry* 2000;**39**:4231–6.
22. Engstrom G, Xiao K, Yu C-A, Yu L, Durham B, Millett F. Photoinduced electron transfer between the Rieske iron-sulfur protein and cytochrome bc_1 complex. *J Biol Chem* 2002;**277**:31072–8.
23. Xiao K, Engstrom G, Rajagukguk R, et al. Effect of famoxadone on photoinduced electron transfer between the iron-sulfur center and cytochrome c_1 in the cytochrome bc_1 complex. *J Biol Chem* 2003;**278**:11419–26.
24. Rajagukguk S, Yang S, Yu C-A, Yu L, Durham B, Millett F. Effect of mutations in the cytochrome *b* of loop on the electron-transfer reactions of the Rieske iron-sulfur protein in the cytochrome bc1 complex. *Biochemistry* 2007;**46**:1791–8.
25. Kalyanasundaram K, Grätzel M. Applications of functionalized transition metal complexes in photonic and optoelectronic devices. *Coord Chem Rev* 1998;**77**:347–414.
26. Takashima H, Shinkai S, Hamachi I. $Ru(bpy)_3$-based artificial receptors towards a protein surface: selective binding and efficient photoreduction of cytochrome *c*. *J Chem Soc Chem Commun* 1999:2345–6.
27. Mines GA, Bjerrum MJ, Hill MG, et al. Rates of heme oxidation and reaction in Ru(His[33])cytochrome *c* at very high driving forces. *J Am Chem Soc* 1996;**118**: 1961–5.
28. Willie A, Stayton PS, Sligar SG, Durham B, Millett F. Genetic engineering of redox donor sites: measurement of intracomplex electron transfer between ruthenium-65-cytochrome b_5 and cytochrome *c*. *Biochemistry* 1992;**31**:7237–42.
29. Geren L, Hahm S, Durham B, Millett F. Photoinduced electron transfer between cytochrome *c* peroxidase and yeast cytochrome *c* labeled at Cys 102 with (4-bromomethyl-4′-methylbipyridine)[bis(bipyridine)]ruthenium^{2+}. *Biochemistry* 1991;**30**:9450–7.
30. Scott JR, Willie A, MacLean M, et al. Intramolecular electron transfer in cytochrome b_5 labeled with ruthenium(II) polypyridine complexes: rate measurements in the Marcus inverted region. *J Am Chem Soc* 1993;**115**:6820–4.
31. Pan LP, Durham B, Wolinska J, Millett F. Preparation and characterization of singly labeled ruthenium polypyridine cytochrome *c* derivatives. *Biochemistry* 1988;**27**: 7180–4.
32. Liu R-Q, Geren L, Anderson P, et al. Design of ruthenium–cytochrome *c* derivatives to measure electron transfer to cytochrome *c* peroxidase. *Biochimie* 1995;**77**:549–61.

33. Geren LM, Beasley JR, Fine BR, et al. Design of a ruthenium-cytochrome *c* derivative to measure electron transfer to the initial acceptor in cytochrome *c* oxidase. *J Biol Chem* 1995;**270**:2466–72.
34. Durham BD, Pan LP, Hahm S, Long J, Millett F. Electron-transfer kinetics of singly labelled ruthenium(II) polypyridine cytochrome *c* derivatives. In: Johnson MK, King RB, Kurtz DM, Kutal C, Norton ML, Scott RA (eds), *Electron Transfer in Biology and the Solid States: Inorganic compounds with unusual properties*. ACS Advances in Chemistry Series. Washington DC: American Chemical Society, 1990: 180–93.
35. Wang K, Mei H, Geren L, et al. Design of a ruthenium-cytochrome *c* derivative to measure electron transfer to the radical cation and oxyferryl heme in cytochrome *c* peroxidase. *Biochemistry* 1996;**35**:15107–19.
36. Scott JR, Fairris JL, McLean M, et al. Intramolecular electron-transfer reactions of cytochrome b_5 covalently bonded to ruthenium(II) polypyridine complexes: reorganizational energy and pressure effects. *Inorg Chim Acta* 1996;**243**:193–200.
37. Mei H, Wang K, Peffer N, et al. Role of configurational gating in intracomplex electron transfer from cytochrome *c* to the radical cation in cytochrome *c* peroxidase. *Biochemistry* 1999;**38**:6846–54.
38. Wang K, Zhen Y, Sadoski R, et al. Definition of the interaction domain for cytochrome *c* on cytochrome *c* oxidase. II. Rapid kinetic analysis of electron transfer from cytochrome *c* to *Rhodobacter sphaeroides* cytochrome oxidase surface mutants. *J Biol Chem* 1999;**274**:38042–50.
39. Tian H, Sadoski R, Zhang L, et al. Definition of the interaction domain for cytochrome *c* on the cytochrome bc_1 complex. *J Biol Chem* 2000;**275**:9587–95.
40. Contzen J, Kostka S, Kraft R, Jung C. Intermolecular electron transfer in cytochrome P450cam covalently bound with tris(2,2′-bipyridyl)ruthenium(II): structural changes detected by FTIR spectroscopy. *J Inorg Biochem* 2002;**91**:607–17.
41. Wang K, Geren L, Zhen Y, et al. Mutants of the Cu_A site in cytochrome *c* oxidase of *Rhodobacter sphaeroides*: II. Rapid kinetic analysis of electron transfer. *Biochemistry* 2002;**41**:2298–304.
42. Engstrom G, Rajagukguk R, Saunders A, et al. Design of a ruthenium-labeled cytochrome *c* derivative to study electron transfer with the cytochrome bc_1 Complex. *Biochemistry* 2003;**42**:2816–24.
43. Gradinaru C, Crane BR. Comparison of intra- vs intermolecular long-range electron transfer in crystals of ruthenium-modified azurin. *J Phys Chem B* 2006;**110**:20073–6.
44. Zheng X, Stuchebrukhov AA. Electron tunneling in the His^{126} Ru-modified azurin: tunneling jumps between protein strands via hydrogen bonds. *J Phys Chem B* 2003;**107**:9579–84.
45. Hamachi I, Shinkai S. Chemical modification of the structures and functions of proteins by the cofactor reconstitution method. *Eur J Org Chem* 1999:539–49.
46. Asakura T. Hemoglobin porphyrin modification. *Methods Enzymol* 1978;**52**:447–55.
47. Hamachi I, Tanaka S, Tsukiji S, Shinkai S, Oishi S. Design and semisynthesis of photoactive myoglobin bearing ruthenium tris(2,2′-bipyridine) using cofactor-reconstitution. *Inorg Chem* 1998;**37**:4380–8.
48. Hamachi I, Tanaka S, Tsukiji S, Shinkai S, Shimizu M, Nagamune T. Photocontrol of the redox state of ruthenium-tris(bipyridine)-appended cytochrome b_{562}. *J Chem Soc Chem Commun* 1997:1735–6.
49. Hamachi I, Tsukiji S, Shinkai S, Oishi S. Direct observation of the ferric-porphyrin cation radical as an intermediate in the phototriggered oxidation of ferric- to ferryl-heme tethered to $Ru(bpy)_3$ in reconstituted myoglobin. *J Am Chem Soc* 1999;**121**:5500–6.
50. Immoos CE, Di Bilio AJ, Cohen MS, van Der Veer W, Gray HB, Farmer PJ. Electron-transfer chemistry of Ru-linker(heme)-modified myoglobin: rapid intraprotein reduction of photogenerated porphyrin cation radical. *Inorg Chem* 2004;**43**:3593–6.
51. Wilker JJ, Dmochowski IJ, Dawson JH, Winkler JR, Gray HB. Substrates for rapid delivery of electrons and holes to buried active sites in proteins. *Angew Chem Int Ed Eng* 1999;**38**:90–2.

52. Dmochowski IJ, Crane BR, Wilker JJ, Winkler JR, Gray HB. Optical detection of cytochrome P450 by sensitizer-link substrates. *Proc Natl Acad Sci USA* 1999;**96**: 12987–90.
53. Dunn AR, Dmochowski IJ, Bilwes AM, Gray HB, Crane BR. Probing the open state of cytochrome P450cam with ruthenium-linker substrates. *Proc Natl Acad Sci USA* 2001;**98**:12420–5.
54. Henry ER, Eaton WA, Hochstrasser RM. Molecular dynamics simulations of cooling in laser-excited heme proteins. *Proc Natl Acad Sci USA* 1986;**83**:8982–6.
55. Zemel H, Hoffman BM. Long-range triplet-triplet energy transfer within metal-substituted hemoglobins. *J Am Chem Soc* 1981;**103**:2103–7.
56. Tsukahara K, Okada M, Asami S, Nishikawa Y, Sawai N, Sakurai T. Photoinduced electron-transfer reactions of zinc and magnesium myoglobins. *Coord Chem Rev* 1994;**132**:223–8.
57. Vanderkooi JM, Erecinska M. Cytochrome *c* interaction with membranes: absorption and emission spectra and binding characteristics of iron-free cytochrome *c*. *Eur J Biochem* 1975;**60**:199–207.
58. Alleyne TA, Wilson MT. Zinc cytochrome *c* fluorescence as a probe for conformational changes in cytochrome *c* oxidase. *Biochem J* 1987;**247**:475–84.
59. Furukawa Y, Matsuda F, Ishimori K, Morishima I. Investigation of the electron-transfer mechanism by cross-linking between Zn-substituted myoglobin and cytochrome b_5. *J Am Chem Soc* 2002;**124**:4008–4019.
60. Bellelli A, Brzezinski P, Arese M, Cutruzzola F, Silvestrini MC, Brunori M. Electron transfer in zinc-reconstituted nitrite reductase from *Pseudomonas aeruginosa*. *Biochem J* 1996;**319**:407–10.
61. Nocek JM, Sishta BP, Cameron JC, Mauk AG, Hoffman BM. Cyclic electron transfer within the [Zn-myoglobin, cytochrome b_5] complex. *J Am Chem Soc* 1997;**119**: 2146–55.
62. Hayashi T, Hitomi Y, Ogoshi H. Artificial protein-protein complexation between a reconstituted myoglobin and cytochrome *c*. *J Am Chem Soc* 1998;**120**:4910–15.
63. Furukawa Y, Ishimori K, Morishima I. Electron transfer in Zn-substituted cytochrome P450cam. *Biochemistry* 2000;**39**:10996–1004.
64. Liang Z-X, Jiang M, Ning Q, Hoffman BM. Dynamic docking and electron transfer between myoglobin and cytochrome b_5. *J Biol Inorg Chem* 2002;**7**:580–8.
65. Zhou JS, Tran ST, McLendon G, Hoffman BM. Photoinduced electron transfer between cytochrome *c* peroxidase (D37K) and Zn-substituted cytochrome *c*: probing the two-domain binding and reactivity of the peroxidase. *J Am Chem Soc* 1997;**119**:269–77.
66. Tremain SM, Kostic NM. Molten-globule and other conformational forms of zinc cytochrome *c*. Effect of partial and complete unfolding of the protein on its electron-transfer reactivity. *Inorg Chem* 2002;**41**:3291–301.
67. Furukawa Y, Ishimori K, Morishima I. Oxidation-state-dependent protein docking between cytochrome *c* and cytochrome b_5: high-pressure laser flash photolysis study. *Biochemistry* 2002;**41**:9824–32.
68. Wheeler KE, Nocek JM, Cull DA, Yatsunyk LA, Rosenzweig AC, Hoffman BM. Dynamic docking of cytochrome b_5 with myoglobin and a-hemoglobin: heme-neutralization 'squares' and the binding of electron-transfer-reactive configurations. *J Am Chem Soc* 2007;**129**:3906–17.
69. Takashima H, Tara C, Namikawa S, Kato T, Araki Y, Ito O, Tsukahara K. Photoinduced intramolecular electron-transfer reactions of reconstituted met- and zinc-myoglobins appending acridine and methylacridinium ion as DNA-binders. *J Phys Chem B* 2006;**110**:26413–23.
70. Naito NR, Huang H, Sturgess AW, Nocek JM, Hoffman BM. Binding and electron transfer between cytochrome b_5 and the haemoglobin α- and β-subunits through the use of [Zn, Fe] hybrids. *J Am Chem Soc* 1998;**120**:11256–62.
71. Sugiyama Y, Takahashi S, Ishimori K, Morishima I. Pressure effects on electron transfer rates in zinc/ruthenium modified myoglobin. *J Am Chem Soc* 1997;**119**:9582–3.

72. Furukawa Y, Ishimori K, Morishima I. Pressure dependence of the intramolecular electron transfer reaction in myoglobin reinvestigated. *J Phys Chem B* 2000;**104**: 1817–25.
73. Hamachi I, Takashima H, Tsukiji S, Shinakai S, Nagamune T, Ogoshi H. Electron transfer from Zn-protoporphyrin IX to ruthenium amine attached at His63 of reconstituted cytochrome b_{562}. *Chem Lett* 1999:551–2.
74. Willner I. Photoswitchable biomaterials: en route to optoelectronic systems. *Acc Chem Res* 1997;**30**:347–56.
75. Brunori M, Giacometti GM, Antonini E, Wyman J. Protein states and protein quakes. *Proc Natl Acad Sci USA* 1973;**70**:3141–4.
76. Boelens R, Wever R, van Gelder BF. Electron transfer between hemes in mammalian cytochrome *c* oxidase. *Biochim Biophys Acta* 1982;**682**:264–72.
77. Ädelroth P, Sigurdson H, Hallén S, Brzezinski P. Kinetic coupling between electron and proton transfer in cytochrome *c* oxidase: simultaneous measurements of conductance and absorbance changes. *Proc Natl Acad Sci USA* 1996;**93**:12292–7.
78. Wilson EK, Bellelli A, Liberti S, et al. Internal electron transfer and structural dynamics of cd1 nitrite reductase revealed by laser CO photodissociation. *Biochemistry* 1999;**38**:7556–64.
79. Sjögren T, Svensson-Ek M, Hajdu J, Brzezinski P. Proton-couple structural changes upon binding of carbon monoxide to cytochrome cd1: A combined flash photolysis and X-ray crystallography study. *Biochemistry* 2000;**39**:10967–74.
80. Wilson EK, Bellelli A, Cutruzzola F, Zumft WG, Gutierrez A, Scrutton NS. Kinetics of CO binding and CO photodissociation in *Pseudomonas stutzeri* cd_1 nitrite reductase: probing the role of extended N-termini in fast structural relaxation upon CO photodissociation. *Biochem J* 2001;**355**:39–43.
81. Feng C, Thomas C, Holliday MA, et al. Direct measurement by laser flash photolysis of intramolecular electron transfer in a two-domain construct of murine inducible nitric oxide synthase. *J Am Chem Soc* 2006;**128**:3808–11.
82. Brzezinski P, Wilson MT. Photochemical electron injection into redox-active proteins. roc *Natl Acad Sci USA* 1997;**94**: 6176–9.
83. Jasaitis A, Rappaport F, Pilet E, Liebl U, Vos MH. Activationless electron transfer through the hydrophobic core of cytochrome oxidase. *Proc Natl Acad Sci USA* 2005;**102**:10882–6.
84. Karpefors M, Wilson MT, Brzezinski P. Photoinduced electron transfer from carboxymethylated cytochrome *c* to plastocyanin. *Biochim Biophys Acta* 1998;**1364**: 385–9.
85. Page CC, Moser CC, Dutton PL. Mechanism for electron transfer within and between proteins. *Curr Opin Chem Biol* 2003;**7**:551–6.
86. Gray HB, Winkler JR. Long-range electron transfer. *Proc Natl Acad Sci USA* 2005;**102**:3534–9.
87. Stubbe J, Nocera DG, Yee CS, Chang MCY. Radical initiation in the class I ribonucleotide reductase: long-range proton-coupled electron transfer? *Chem Rev* 2003;**103**: 2167–201.
88. Marcus RA, Sutin N. Electron transfer in chemistry and biology. *Biochim Biophys Acta* 1985;**811**:265–322.
89. Langen R, Chang I-J, Germanas JP, Richards JH, Winkler JR, Gray HB. Electron tunneling in proteins: coupling through a beta strand. *Science* 1955;**268**:1733–5.
90. Gray HB, Winkler JR. Electron transfer in proteins. *Annu Rev Biochem* 1996;**65**:537–61.
91. Williams RJP. The medium in electron transfer proteins. *J Biol Inorg Chem* 1997;**2**:373–7.
92. Moser CC, Keske JM, Warneke K, Farid RS, Dutton PL. Nature of biological electron transfer. *Nature* 1992;**355**:796–802.
93. Beratan DN, Betts JN, Onuchic JN. Protein electron transfer rates set by the bridging secondary and tertiary structure. *Science* 1991;**252**:1285–8.

94. Daizadeh I, Gehlen JN, Stuchebrukhof AA. Calculation of electronic tunneling matrix element in proteins: comparison of exact and approximate one-electron methods for Ru-modified azurin. *J Chem Phys* 1997;**106**:5658–66.
95. Winkler JR. Electron tunneling pathways in proteins. *Curr Opin Chem Biol* 2000;**4**:192–8.
96. Mei H, Wang D, McKee S, et al. Control of formation and dissociation of the high-affinity complex between cytochrome *c* and cytochrome *c* peroxidase by ionic strength and the low-affinity binding site. *Biochemistry* 1996;**35**:15800–6.
97. Tezcan FA, Crane BR, Winkler JR, Gray HB. Electron tunneling in protein crystals. *Proc Natl Acad Sci USA* 2001;**98**:5002–6.
98. Hoffman BM, Ratner MA. Gated electron transfer: when are observed rates controlled by conformational interconversion? *J Am Chem Soc* 1987;**109**:6237–43.
99. Davidson VL. Unraveling the kinetic complexity of interprotein electron transfer reactions. *Biochemistry* 1996;**36**:14035–9.
100. Ivkovic-Jensen MM, Kostic NM. Effects of viscosity and temperature on the kinetics of the electron-transfer reaction between the triplet state of zinc cytochrome *c* and cupriplastocyanin. *Biochemistry* 1997;**36**:8135–44.
101. Harris MR, Davis DJ, Durham B, Millett F. Temperature and viscosity dependence of the electron-transfer reaction between plastocyanin and cytochrome *c* labeled with a ruthenium(II) bipyridine complex. *Biochim Biophys Acta* 1997;**1319**:147–54.
102. Liu L, Hong J, Ogawa MY. Gated electron transfer as a probe of the configurational dynamics of peptide-protein complexes. *J Am Chem Soc* 2004;**126**:50–1.
103. Liang Z-X, Nocek JM, Kurnikov IV, Beratan DN, Hoffman BM. Electrostatic control of electron transfer between myoglobin and cytochrome b_5: effect of methylating the heme propionates of Zn-myoglobin. *J Am Chem Soc* 2000;**122**:3552–3.
104. Liang Z-X, Nocek JM, Huang K, et al. Dynamic docking and electron transfer between Zn-myoglobin and cytochrome b_5. *J Am Chem Soc* 2002;**124**:6849–59.
105. Grove TZ, Kostic NM. Metalloprotein association, self-association, and dynamics governed by hydrophobic interactions: simultaneous occurrence of gated and true electron-transfer reactions between cytochrome and cytochrome *c*6 from *Chlamydomonas reinhardii*. *J Am Chem Soc* 2003;**125**:10598–607.
106. Mei HK, Geren L, Miller MA, Durham B, Millett F. Role of the low-affinity binding site in electron transfer from cytochrome *c* to cytochrome *c* peroxidase. *Biochemistry* 2002;**41**:3968–76.
107. Fancy DA, Denison C, Kim K, et al. Scope, limitations and mechanistic aspects of the photo-induced cross-linking of proteins by water-soluble metal complexes. *Chem Biol* 2000;**7**:697–708.
108. Fancy DA, Kodadek T. Chemistry for the analysis of protein-protein interactions: Rapid and efficient cross-linking triggered by long wavelength light. *Proc Natl Acad Sci USA* 1999;**96**:6020–4.
109. Xie Y, Denison C, Yang S-H, Fancy DA, Kodadek T. Biochemical characterization of the TATA-binding protein-Gal4 activation domain complex. *J Biol Chem* 2000;**275**: 31914–20.
110. Andreev OA, Reshetnyak Y, Goldfarb RH. Evidence of inter- and intra-molecular crosslinking of tyrosine residues of carmodulin induced by photo-activation of ruthenium(II). *Photochem Photobiol Sci* 2002;**1**:834–6.
111. Meunier S, Strable E, Finn MG. Crosslinking of and coupling to viral capsid proteins by tyrosine oxidation. *Chem Biol* 2004;**11**:319–29.
112. Schere PE, Krieg UC. Cross-linking reagents as tools for identifying components of the yeast mitochondrial protein import machinery. *Methods Cell Biol* 1991;**34**:419–26.
113. Mattson G, Conklin E, Desai S, Nielander G, Savage MD, Morgensen S. A practical approach to cross-linking. *Mol Biol Rep* 1993;**17**:167–83.
114. Norcum MT, Warrington JA. Structural analysis of the multienzyme aminoacyl-tRNA synthetase complex: A three domain model based on reversible chemical cross-linking. *Protein Sci* 1998;**7**:79–87.

115. Dunn AR, Belliston-Bittner W, Winkler JR, Getzoff ED, Stuehr DJ, Gray HB. Luminescent ruthenium(II)- and rhenium(I)-diimine wires bind nitric oxide synthase. *J Am Chem Soc* 2005;**127**:5169–73.
116. Pascher T, Chesick JP, Winkler JR, Gray HB. Protein folding triggered by electron transfer. *Science* 1996;**271**:1558–60.
117. Wittung-Stafshede P, Gray HB, Winkler JR. Rapid formation of a four-helix bundle. Cytochrome b_{562} folding triggered by electron transfer. *J Am Chem Soc* 1997;**119**: 9562–3.
118. Wittung-Stafshede P, Lee JC, Winkler JR, Gray HB. Cytochrome b_{562} folding triggered by electron transfer: Approaching the speed limit for formation of a four-helix-bundle protein. *Proc Natl Acad Sci USA* 1999;**96**:6587–90.
119. Lee JC, Gray HB, Winkler JR. Cytochrome *c′* folding triggered by electron transfer: Fast and slow formation of four-helix bundles. *Proc Natl Acad Sci USA* 2001;**98**: 7760–4.
120. Lee JC, Chang I-J, Gray HB, Winkler JR. The cytochrome *c* folding landscape revealed by electron-transfer kinetics. *J Mol Biol* 2002;**320**:159–64.
121. Chang I-J, Lee JC, Winkler JR, Gray HB. The protein-folding speed limit: Intrachain diffusion times set by electron-transfer rates in denatured $Ru(NH_3)_5$(His-33)-Zn-cytochrome *c*. *Proc Natl Acad Sci USA* 2003;**100**:3838–40.
122. Telford JR, Wittung-Stafshede P, Gray HB, Winkler JR. Protein folding triggered by electron transfer. *Acc Chem Res* 1998;**31**:755–63.
123. Miller MA, Geren L, Han GW, et al. Identifying the physiological electron transfer site of cytochrome *c* peroxidase by structure-based engineering. *Biochemistry* 1996;**35**: 667–73.
124. Szaciłowski K, Macyk W, Drzewiecka-Matuszek A, Brindell M, Stochel G. Bioinorganic photochemistry. Frontiers and mechanisms. *Chem Rev* 2005;**105**:2647–94.

14

Nucleic Acid Photocleavage and Charge Transport

The information encoded in your DNA determines your unique biological characteristics, such as sex, eye color, age and Social Security number
Dave Barry

DNA plays a key role in cell life and in pathological processes, so the elucidation of the mechanisms of its interactions with small molecules and proteins, understanding the molecular basis of cancer and others diseases, as well as mechanisms leading to DNA damage are of great importance. The design of nucleic acid cleavage agents is an important area of investigation for the development of new therapeutic agents and tools in biochemistry. The natural enzymes that cleave DNA are very useful in many applications; however, their large size and limiting range of sequence-recognition capabilities encourage the search for synthetic cleavers. Much attention is being paid to the design of the cleaving agent activated by light [1, 2]. The use of light-inducible reactions offers a unique possibility of initiating the desired activity in selected targets at a chosen time. In this respect a proper design of metal complexes by tuning their structural and electronic properties results in selective metallophotocleavers for the study of DNA and RNA structures, and DNA–protein interactions. Photocleavers can be applied as a spectroscopic tool in monitoring charge transport phenomenon in DNA or as therapeutic agents (see Chapter 17).

In this chapter we review the application of metal complexes and light in the studies of nucleic acid cleavage and charge transport.

14.1 Mechanisms and Strategies for Advanced Metallophotocleavers

Photocleavers can cause an immediate strand break or initiate a cleavage of nucleic acid that needs further chemical manipulation such as incubation with hot

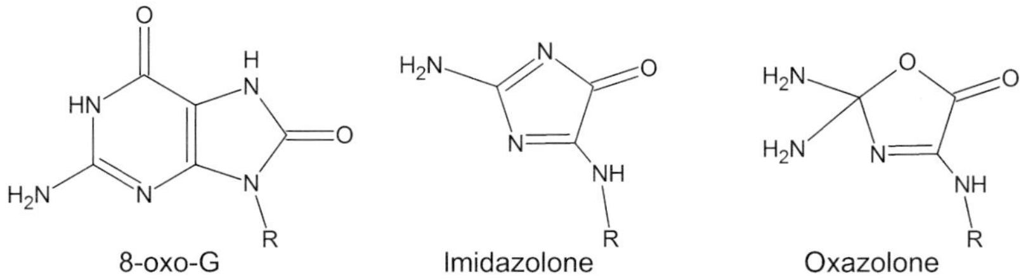

Figure 14.1 *Main products of oxidation damage of guanine residue in DNA caused by photocleavage agents targeting nucleobases [4]*

piperidine or aniline to recognize the sites and the extent of damage [1]. Nucleic acids have two possible targets for photocleavers: sugar moieties or nucleobases. The DNA photocleavage arising from oxidation of deoxyribose proceeds via hydrogen atom abstraction from the sugar ring, resulting in the formation of sugar radicals. The hydrogen atoms on carbons 1′, 3′, 4′, and 5′ are thermodynamically favoured for abstraction due to the presence of heteroatoms at α-positions. The sugar radical can decompose on various pathways yielding low-molecular-mass products and DNA fragments [3]. Photocleavage agents, targeting nucleobases, oxidize mainly guanine with formation of a series of different products (oxidized guanine, G^{ox}) such as 7,8-dihydro-8-oxoguanine (8-oxo-G), imidazolone, or oxazolone (Figure 14.1) [4].

A special class of photocleavers is metal complexes, so-called metallophotocleavers. The photoreactivity of metal complexes in the presence of nucleic acids and their mode of binding can be tuned by selection of either metal centre or ligands. Changing the metal may result in distinctive photophysical properties and altering the geometry of complex (square planar, tetrahedral, octahedral, etc), which may affect its interaction with nucleic acids. Varying ligands bound to the same metal centre may have a great influence on the redox potential of complexes in the ground and excited states and nucleic acid affinity. A wide range of metallophotocleavers has been designed and the set comprises metal complexes of various centres, such as Ru, Rh, Co, Cu, W, or V with different ligands, among which the polypyridyls are the most commonly used.

14.1.1 Ruthenium Complexes

The dramatic growth in design and development of synthetic nucleic acid photocleavage agents based on ruthenium complexes has been observed over the last decade. Various strategies have been used to improve the sequence selectivity of cleavage, which can arise from the preferential binding or preferential reactivity at a certain site. Several mechanisms can be engaged in photoactivated DNA cleavage by ruthenium complexes:

- oxidation of the base by singlet oxygen formed *via* energy transfer from the triplet excited state of the photocleaver

- direct electron transfer from the base to the excited state of the photocleaver
- oxidation of the base by Ru^{III} complex formed *in situ* by oxidative quenching of the triplet excited state of the corresponding Ru^{II} complex.

One of the mechanisms leading to DNA photocleavage proceeds via generation of singlet oxygen ($^1O_2$) by energy transfer from the triplet exited state of the metal complex to triplet oxygen ($^3O_2$). Photoexcited ruthenium(II) complexes with simple polypyridyl ligands such as bpy or phen (see Figure 14.2), can sensitize formation of $^1O_2$, which preferentially oxidizes nearby guanine residue [5,6].To improve the efficiency of the DNA photocleavage by those sensitizers, a series of derivatives containing ligands with good intercalating properties, eg $[Ru(bpy/phen)_2(ddz/bpip/ppip)]^{2+}$ (see Figure 14.2), has been tested as selective DNA photocleavers [7–9]. For complexes with bpip ligand, not only singlet oxygen but also hydroxyl radicals were found to play an important role in the cleavage mechanism. It is interesting that the exchange of bpip ligand into ppip one that differs only by the absence of the methylene group (Figure 14.2), results in change of the cleavage mechanism – in

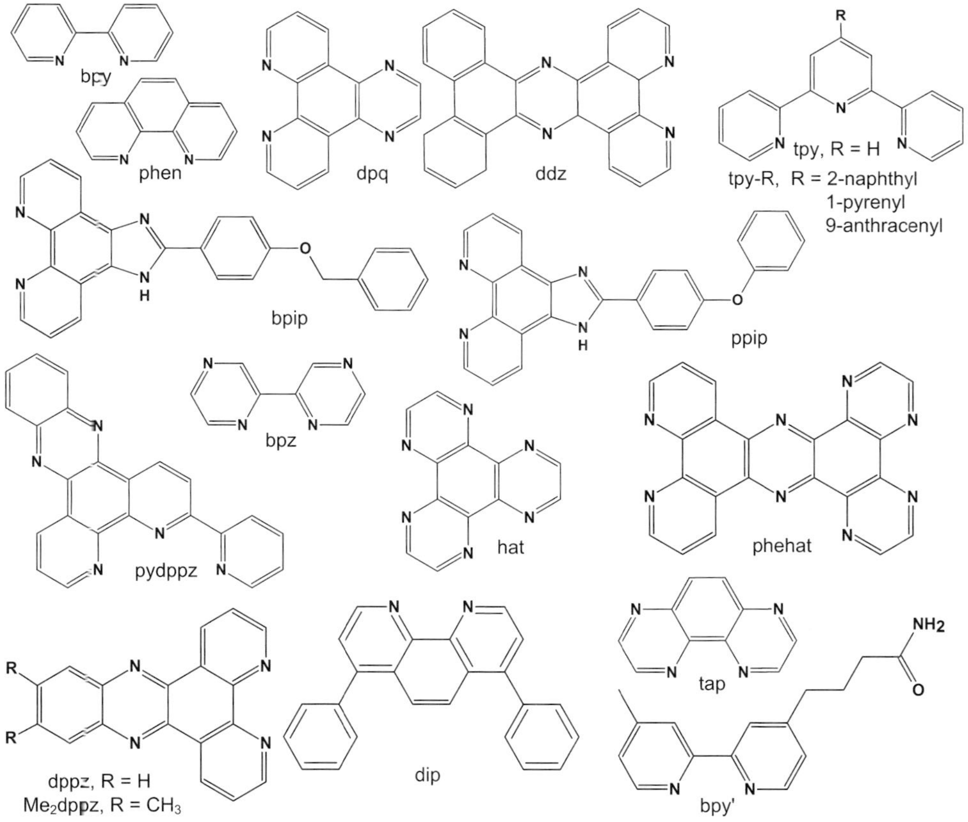

Figure 14.2 *Structures of selected ligands for ruthenium complexes. (See Abbreviations on page xii for the systematic names)*

the ppip case only hydroxyl radical is responsible for photocleavage [8, 9]. Recently, a series of ruthenium(II) complexes possessing tridentate terpyridine-type ligands ($[Ru(tpy)(tpy\text{-}R/pydppz)]^{2+}$, see Figure 14.2) have been tested as metallophotocleavers and it was recognized that some of them are characterized by very high quantum yield of $^1O_2$ generation [10, 11]. For some complexes, it is difficult to distinguish whether the cleavage mechanism is mediated by generation of singlet oxygen or by oxidation of DNA through direct electron transfer. In some cases both pathways are observed as for the $[Ru(bpz)_3]^{2+}$ complex (Figure 14.2) [12]; however, one of the processes can dominate.

Direct oxidation of guanine residues in DNA may be obtained with ruthenium complexes possessing high oxidation power in their ^{3}MLCT states. To obtain complexes that are sufficiently strong oxidants in their excited states it is necessary to use strong π-accepting ligands, such as bpz, hat, phehat, or tap (Figure 14.2) [13]. Complexes of the type $[Ru(hat/tap/bpy/phen/dppz/dip)(hat/tap)_2]^{2+}$ are able to abstract an electron from a guanine unit (G) of DNA under visible irradiation [13–15]. The electron transfer generates an oxidized $G^{\bullet+}$ and a reduced complex $[Ru^{II}L_2(L^{\bullet-})]^+$ as postulated by Kirsch-De Mesmaeker et al. (the percentage of protonated reduced species depends on pH) [15]. The oxidized radical of guanine can return to its non-radical form by reaction with reduced complex or can undergo further reaction with O_2 or H_2O leading to formation of stable products (oxidized guanine, G^{ox}, see above) resulting in a strand break as a final step. For some complexes such as $[Ru(bpy/phen)_n(hat/tap)_{3-n}]^{2+}$ (where $n = 0$ or 1) [16, 17] and $[Ru(bpz)_3]^{2+}$ [12] the formation of irreversible photoadducts is also observed. The nature of the photoadducts is quite different from the platinum dark adducts previously studied in the literature. The tap-containing complexes form a covalent linkage with the 2-NH_2 group of the guanine residues and the hat-containing one forms a covalent bond with the exocyclic O-6 atom of G on visible light irradiation [15, 17–19]. The overall mechanism is schematically presented in Figure 14.3. The formation of photoadducts is independent of the mode of interaction with DNA, both surface binders (complexes with tap ligands) and intercalators (complexes with hat ligands) being able to generate photoadducts with a covalent bond between the metallic

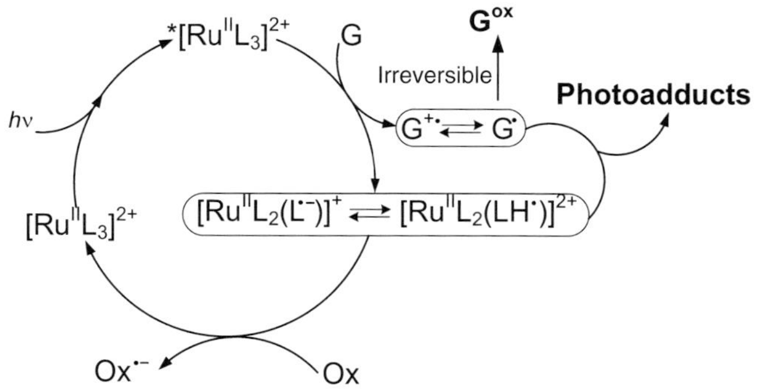

Figure 14.3 *Molecular mechanism for the photoinduced electron transfer from guanine residues to excited* Ru^{II} *complex. Ox denotes an oxidant (eg* O_2*)*

compound and a DNA base [19]. To increase the DNA–complex affinity metal complexes with extended planar ligands have been designed. Such ligands result in an increased surface area of intercalative interactions. A series of ruthenium(II) complexes with phehat and dppz ligands (see Figure 14.2) such as $[Ru(phehat)(phen)_2]^{2+}$ or $[Ru(tap)_2(dppz)]^{2+}$ was synthesized and applied as metallophotocleavers [18, 20, 21]. These complexes possess high affinity towards DNA with preservation of their photoreactivity. They undergo photoinduced electron transfer from the G residue to the excited state of ruthenium complexes; however, formation of photoadducts is not observed.

The described complexes photoreact with DNA in a non-specific way and, to direct the photoactive complex to a specific sequence, the complexes have been anchored to oligonucleotides. Synthetic oligonucleotides that have the sequence complementary to a part of messenger RNA (mRNA) may recognize and react with the target mRNA sequence. In this way they inhibit the expression of the targeted sequence. This strategy, based on modified synthetic oligonucleotides, the so-called the anti-sense or anti-gene strategy, has a few drawbacks such as the poor stability after hybridization of synthetic oligonucleotides with target DNA and destruction of the system by some enzyme. In this context, the use of a photoactive ruthenium(II) complex such as $[Ru(dip)(tap)_2]^{2+}$ (Figure 14.2) attached to an oligonucleotide allows for irreversible cross-linking of the two strands on irradiation. Electron transfer to the photoexcited complex from DNA initiates the formation of adducts between the metal complex and a specific G residue in the target DNA strand [22–25].

Barton and co-workers have developed another method for oxidation of G residues by ruthenium(II) complexes possessing intercalating ligand(s) (eg $[Ru(dppz)(phen)_2]^{2+}$) that have insufficient oxidation potential in their ^{3}MLCT state but are strong ground-state oxidants when oxidized to the corresponding Ru^{III} complexes [26]. Photoexcited ruthenium(II) complexes can be quenched by the electron transfer to a non-intercalating electron acceptor such as methyl viologen, $[Ru(NH_3)_6]^{3+}$ or $[RuCl(NH_3)_5]^{2+}$ resulting in formation of a Ru^{III} intercalator. This species, which is a potent ground-state oxidant, can be reduced back to Ru^{II} by reaction with the reduced quencher or by electron transfer from a nearby G base, which is the most easily oxidizable one [26, 27]. The formation of the neutral radical $G^{\bullet}$ proceeds via deprotonation of G^{ox}. This intermediate can return to its non-radical form by reaction with reduced quencher or undergo further reaction with O_2 or H_2O, leading to formation of stable products (G^{ox}, see above). The presented procedure, shown schematically in Figure 14.4, is called the 'flash quench method' and was originally invented to study protein-mediated electron transfer [28]. Importantly the yield of oxidative damage can be tuned by the choice of quencher. Similar method has been applied for oxidation of G residue by a $[Ru(phen)_3]^{2+}$ complex using *E,Z*1,4-bis[2-(1-methylpyridinium-4-yl)vinyl]benzene as a quencher [29].

To get inside the oxidative damage of DNA caused by photoactive ruthenium(II) intercalators, complexes can be attached to one end of double-stranded DNA. The great advantage of this method is that intercalation of tethered metal complexes takes place at a defined distance from the oxidation site. The ruthenium(II) complex of the type $[Ru(bpy')(Me_2dppz)(phen)]^{2+}$ (structure of ligands presented in Figure

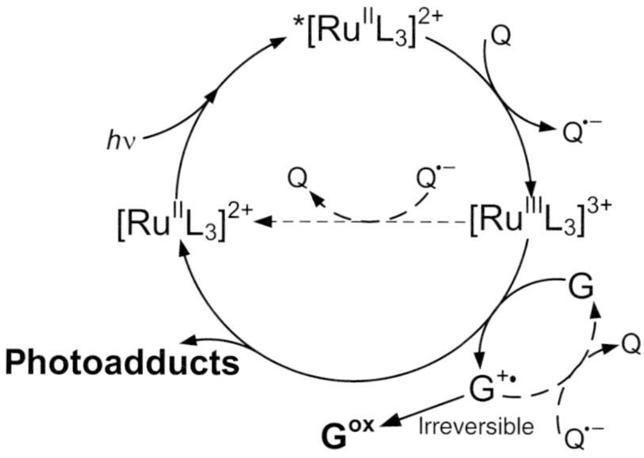

Figure 14.4 *Molecular mechanism of the flash quench method applied for photo-oxidation of guanine by generated* in situ *Ru^{III} complex. Q denotes a quencher*

14.2), tethered to one strand of the nucleic acid, has been prepared and used to study long-range oxidation of guanine in DNA by applying the flash quench method [27, 30].

Turro and co-workers recently have designed the supramolecular version of the flash quench method using a $[Ru(bpy)_3]^{2+}$-like moiety covalently linked to an electron acceptor chain composed of two viologen units [31]. On excitation of the Ru^{II} unit the formed ^{3}MLCT is quenched by the proximal viologen subunit and subsequently the electron is transferred to the terminal viologen subunit. A long-lived charge-separated state with a Ru^{III} centre is formed, and the Ru^{III} species is able to abstract an electron from the G residue.

14.1.2 Rhodium Complexes

Another class of metallophotocleavers comprises rhodium(III) complexes, which can specifically bind to the DNA, targeting a specific sequence or recognizing special structures (eg triply bonded sites, base-pair mismatches) [1, 32]. The strategy is based on shape-selective recognition by proper design of ligands [32]. A series of photoactive Rh^{III} complexes containing intercalating phi ligand(s) (see Figure 14.5) have been designed and tested as cleaving agents. The irradiation of intercalated Rh^{III}-phi complexes in the ultraviolet region leads to generation of the cation radical on the phi ligand via ligand-to-metal charge transfer. The cation radical ligand (either still coordinated or dissociated) can afterwards abstract a hydrogen from an adjacent sugar moiety (most probably the C2′-H followed by H migration to form the C3′ radical). The sugar radical undergoes degradation leading to direct DNA strand cleavage [33]. Alternatively, Rh^{III}-phi complexes can be used for photo-oxidation of nucleobases in the DNA, mainly G and adenine (A) residues. The near-UV and visible excitation of Rh^{III}-phi complexes leads to population of their ligand-centred (LC) excited states, which are powerful oxidants. The $*[Rh^{III}(phi)(bpy/phen)]^{3+}$ oxidizes purine DNA bases (G and A) with concomitant formation of Rh^{II} complex

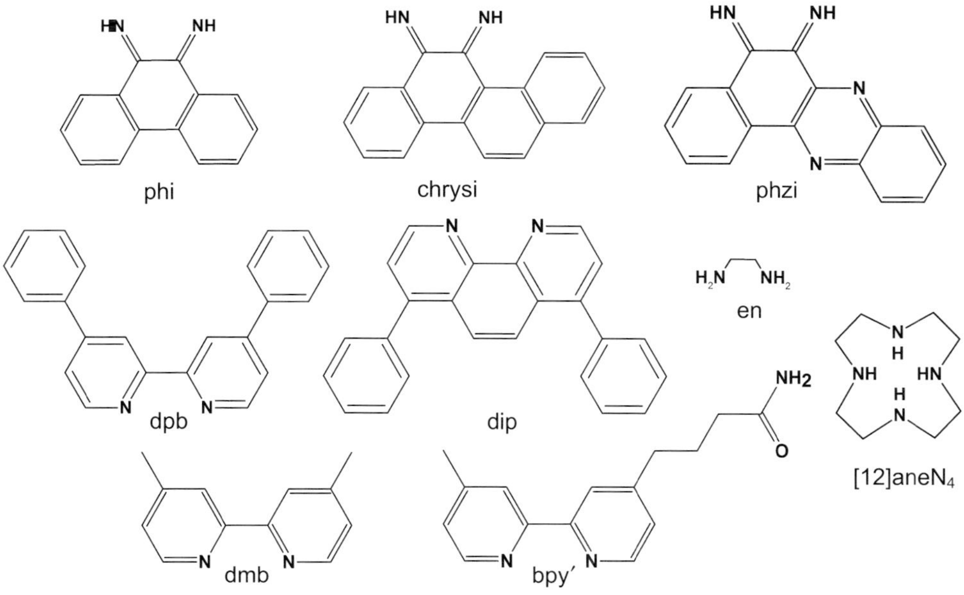

Figure 14.5 Structure of selected ligands for rhodium complexes. (See Abbreviations on page xiii for the systematic names)

(see Figure 14.6). The extent of direct oxidative strand cleavage via hydrogen abstraction from sugar moiety versus oxidation of DNA bases depends not only on the wavelength of excitation but also on the pH of solution, and in some cases these processes proceed in parallel [30, 34, 35].

The cleavage by hydrogen abstraction is non-selective because identical deoxyribose moieties are present along the DNA molecule. The sequence selectivity can, however, be reached by designing photocleavers that can bind to DNA in a sequence-selective manner. The phi ligand is inserted between base-pairs whereas the ancillary, non-intercalating ligands are oriented in the major groove. Modification of the ancillary ligands allows a high sequence specificity and reactivity of Rh^{III}-phi complexes with DNA to be obtained, resulting in new photoactive, sequence-specific metallointercalators [33, 36–40]. $[Rh(phen)_2(phi)]^{3+}$ has been found to target triple base sites occurring in natural and synthetic RNAs and can therefore be used as a chemical to probe triply bonded sites in folded RNA molecules [32, 38, 39]. Site-specific recognition of a palindromic octamer 5′-CT**C**TAGAG-3′ was observed for Δ-$[Rh(dpb)_2(phi)]^{3+}$ (for ligand structure, see Figure 14.5) which induced photocleavage at the cytosine residue indicated by emboldening. This synthetic metallointercalator binds to nucleic acid with a level of specificity mimicking DNA-binding proteins and thus can inhibit the reaction of DNA with proteins site specifically [32]. A family of amine rhodium complexes of the type $[Rh(phi)L]^{3+}$ (where L = $3NH_3$, 2en,[12]aneN_4; Figure 14.5) was found to cleave strongly at 5′-GC-3′ sites in DNA. The site specificity was related to the possibility of hydrogen bond formation between the axial amines of the metal complex and the O-6 of the G residue [32, 40].

Another example of shape-selective recognition has been observed for rhodium complexes with chrysi [41, 42], phzi [43], or dip [44, 45] ligands (see Figure 14.5), which bind preferentially in the destabilized regions at or close to base mismatches. The $[Rh(bpy)_2(chrysi)]^{3+}$ complex was found to cleave DNA specifically at mismatch sites around single mispaired bases in DNA sequence. The size of the chrysi ligand makes it sterically too bulky to insert into standard B-form DNA; however, it has access to destabilized mismatches [42, 46, 47]. A similar facility possesses complexes with a phzi ligand sterically similar to the chrysi one; again, however, the presence of an aromatic heterocycle offers the possibility of greater stabilization through stacking within the mismatched site, which results in improved binding affinity of the complex [43]. Base-pair mismatches can occur in DNA as a result of either errors in replication or externally induced mutations. The design of new agents for base-pair mismatch detection based, on combination of the shape-selective recognition metal complexes with their photocleavage activity, has great therapeutic potential and can be useful in genetic diagnosis. Another complex, namely $[Rh(dip)_3]^{3+}$ (see Figure 14.5), was found to cleave RNA backbone on irradiation specifically at G·U (U is uracil) mismatch, as well as at an exposed G residue in the larger loop region. A unique stacking arrangement appears to be well suited to interactions with a planar ligand (dip) [44, 45]. The specific photocleavage ability of this complex can be employed to gain greater knowledge of the secondary and tertiary structures of RNA (detection of base mismatches, assigning regions of double- or single-stranded RNA or exposed residues).

The ability of rhodium complexes to photo-oxidize DNA bases was applied to study charge transport in DNA (see section 14.2 for detailed discussion). Rhodium(III) complexes such as $[Rh(bpy')(phi)_2]^{3+}$, $[Rh(dmb)(phi)_2]^{3+}$ (structures of ligands presented in Figure 14.5) have been attached to one end of double-stranded DNA and light-induced oxidation of G residues investigated [48–51]. The irradiation of rhodium complex with near UV–visible light ($\lambda = 365\,nm$) results in generation of a potent photo-oxidant ($*[Rh^{III}(phi)_2(L)]^{3+}$) that is able to oxidize the G bases in DNA. It was found that these complexes can promote long-range oxidative damage of the DNA by preferential oxidation of the 5′-G residue in 5′-GG-3′ guanine doublets located remotely from the intercalated complex. In addition, if the thymine dimers are present in DNA, rhodium complexes are capable of repairing this lesion by photo-oxidation [48]. The occurrence of both 5′-GG-3′ site and thymine dimers on the same duplex leads to competitive reactions: repair and oxidative damage, schematically presented in Figure 14.6.

14.1.3 Other Metal Complexes

In the search for new advanced metallophotocleavers not only has ligand modification in ruthenium and rhodium complexes been studied but also the possibility of the application of other transition metal ions such as Co, Cu, W, or V has been considered. Among them is a group of Co^{III} complexes with mixed-polypyridine ligands, such as bpy or phen, as the ancillary ligands and one ligand with good intercalating properties, such as dppz or its derivatives (qdppz, dicnq), ligands based on the

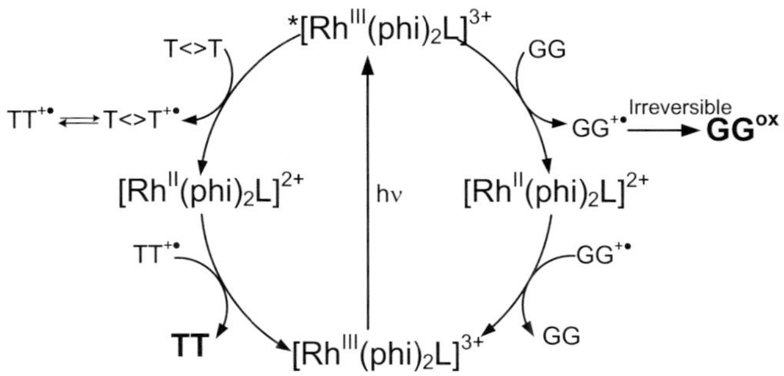

Figure 14.6 The proposed mechanism for thymine dimer (T<>T) repair and guanine oxidation by excited state of Rh^{III} complex [48]

imidazophenanthroline moiety (ip, pip, cnoip, hnoip, hipip, hnaip), or asymmetrical ligands containing the triazine moiety (pdtb, pdta, pdtp) [52–58]. The structure of these ligands is shown in Figure 14.7. The exact mechanism of the photocleavage reaction for these complexes is not fully understood. It has been suggested that cobalt(III) complexes are capable of photogenerating various radicals, including reactive oxygen species as hydroxyl and superoxide anion radicals, which can cleave DNA [55–58]. Also the cobalt(II) complex with tridentate ligand 2,6-bis(benzimidazol-2-yl)pyridine has been reported as a photocleaver of DNA; however, the detailed mechanism of this process is not clear [59]. Some cobalt(III) complexes with non-pyridyl ligands, eg bleomycin and its synthetic analogues, have been studied as potential DNA photocleavers [1, 60]. The tentative mechanism assumes generation of hydroxyl radicals that abstract hydrogen atoms from deoxyribose to give sugar radicals and water. All presented metallophotocleavers based on cobalt ions involve irradiation in the UV region, which can be a disadvantage for any medical application; however, they can be used as chemical footprinting agents.

Recently, a series of Cu^{II} complexes with one polypyridyl ligand (phen, dpq, dppz or dmp, see Figure 14.2) and one non-polypyridyl ligand such as amino acid (*S*-methyl-L-cysteine, L-methionine, L-arginine) [61–63], aromatic sulphur containing Schiff base [64, 65], substituted thiosemicarbazone [66], or scorpionate [67] were reported to be efficient DNA photocleavers. The reactive species involved in DNA cleavage was found to be singlet oxygen formed on visible or UV light irradiation of Cu^{II} complexes. Some of these complexes are characterized by very efficient singlet oxygen formation on irradiation with red light, which makes them very good candidates for PDT application. A new class of DNA photocleavers based on copper(II) complexes containing 9-diazo-4,5-diazafluorene ligands has been suggested as a potential model for the action of kinamycin anticancer antibiotics [68]. It has been proposed that visible light irradiation of this complex leads to generation of radical intermediates on metal-bound ligands, which are responsible for the DNA damage. Another group of Cu^{II} and Cu^{I} complexes with ligands containing enediyne units has been found to photocleave DNA [69]. The proposed mechanism involves formation of diradical intermediates as a consequence of photo-Bergman

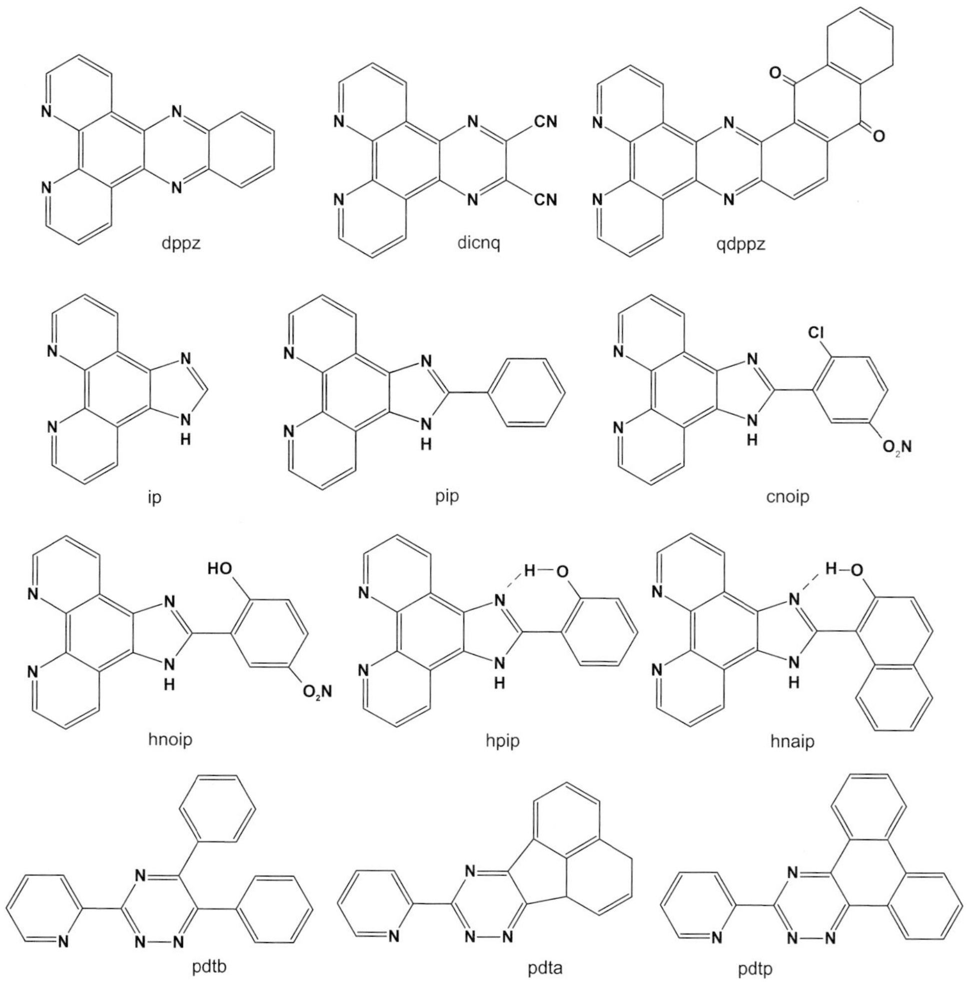

Figure 14.7 *Structures of selected ligands for cobalt complexes. (See Abbreviations on page xiii for the systematic names)*

cyclization of copper metalloenediynes. Diradicals are able to cleave DNA via hydrogen atom abstraction from the sugar moiety [69].

Complexes of several other metals have been also tested as photocleavers of DNA. Some cyclopentadienyl (Cp) tungsten complexes of the type [W(CO)$_3$(Cp)(CH$_3$/Ph)] with covalently attached DNA recognition elements (netropsin analogue) have been tested as metallophotocleavers. In this group of complexes, cleavage of the DNA is performed by photogenerated methyl or phenyl radicals on UV-A light irradiation. The attachment of the netropsin analogue to the [W(CO)$_3$(Cp)(CH$_3$/Ph)] complex resulted in an increase in efficiency of the strand scission whereas the increased selectivity is observed only for phenyl derivatives due to production of a non-diffusible phenyl radical that cleaves preferentially at T residues within the binding site [70]. A series of vanadium(V) bisperoxo complexes of the type $[VO(O_2)_2L]$ (where L is a polypyridyl ligand such as bpy, phen, terpy, or its derivatives, or oxalate, citrate, nitrilotriacetate, pyridine-2-carboxylate, pyridine-2,6-dicarboxylate iminodiacetate, etc) was reported to cleave DNA on photoactivation in UV-A light (365 nm) [71–74]. There has been much speculation about the mechanism of photocleavage; Hirot et al. [71] and subsequently Kwong et al. [72] suggested the formation of singlet oxygen on UV-A irradiation, and it was proposed that the vanadium complex provides its own freely dissociated singlet oxygen. In contrast, the report of Sam et al. [73] has strongly supported the hydroxyl radical as the active agent. Another class, oxovanadium(V) complexes bearing *N*-salicylidene derivatives as a tridentate ligand, was found to cleave DNA when exposed to UV-A light [75]. It has been proposed that an *o*-quinone-methide vanadium-bound hydroxyl species formed on irradiation is responsible for DNA cleavage. Recently, another group of complexes based on oxovanadium(IV) ions of the type [VO(B)(L)] (where B is dpq or dppz, see Figure 14.2, and L is *N*-salicidene-L-methionate or tryptophanate) was reported as metallophotocleavers able to cleave DNA on irradiation with red light (750 nm) [76]. The mechanism of cleavage is also related to generation of singlet oxygen. A large group of photometallocleavers includes metalloporphyrins, discussed in Chapter 19. The photocleavage proceeds via generation of singlet oxygen by energy transfer from the triplet excited state of metal complexes to triplet molecular oxygen.

A different group of photocleavers represents the uranyl(VI) ion, UO_2^{2+}, or its complexes with different ligands. The uranyl ion binds to DNA via electrostatic interaction, preferentially to A/T-rich regions in the minor groove [77, 78]. Mechanistic studies indicate that the cleavage occurs via hydrogen abstraction from the sugar moiety on visible light irradiation; however, the detailed mechanism has not been completely elucidated [79]. Some studies have shown an important role of singlet oxygen in the cleavage reaction [80]. Uranyl ions can be employed as a probe for a variety of nucleic acid structures as well as an indicator for metal ion binding to RNA [81–83]. A new approach, based on immobilization of uranyl ion to a polymeric matrix, has been reported recently. The system obtained, insoluble in aqueous media, enables cleavage of DNA in the presence of visible light and sunlight and can be reused after washing. This is an example of heterogeneously active, reusable photonucleases [80].

14.1.4 Di- and Trinuclear Complexes

Recently di- and trinuclear complexes targeting special DNA geometries or topologies have been designed. The dinuclear ruthenium(II) complex $[\{Ru(phen)_2\}_2(hat)]^{4+}$ (ligand structures are shown in Figure 14.2) has been regarded as an excellent photoprobe for denatured DNA because its interaction with denatured DNA is different to that with the double-stranded form [84]. The dinuclear complex possesses the same photoreactivity with nucleotides (particularly guanosine monophosphate, GMP) as that of the corresponding monometallic complex [16, 17], ie the same correlation of a photoinduced electron transfer from GMP to the excited complex with formation of photoadducts. This correlation is also found in the photoreaction with denatured DNA but is not observed in double-stranded DNA. This could originate from the lack of free access to the G residues buried inside double-stranded DNA as opposed to free GMP or more accessible G residues of denatured DNA.

Recently, a series of dirhodium(II/II) complexes of the type $[Rh_2(\mu\text{-}OOCCH_3)_4(dppz)L]^{n+}$ (where L = CH_3OH, bpy, or dppz; for ligand structures see Figure 14.2) has been designed to achieve oxygen-independent DNA photocleavage with visible light [85–87]. The irradiation of those complexes with visible light leads to photocleavage of DNA but the nature of the reactive excited state is still unclear. It was tentatively suggested that dppz ligand is involved in the reactive excited state. The ability of the dppz ligand's intercalation between DNA bases gives complexes an affinity for DNA; however, the introduction of the second dppz ligand prevents the intercalation and leads to intermolecular π-stacking of the complex, enhanced by polyanionic DNA. The photocleavage potency was found to be greater for complexes possessing one dppz ligand and was explained in terms of the ability to intercalate.

Photocleavage activity was also reported for tetra-anionic diplatinum(II) complexes with pyrophosphito ligands. The photocleavage proceeds via hydrogen atom abstraction, which occurs on collision between an excited state of complex and DNA, and is strongly enhanced by magnesium ion, probably due to screening of the electrostatic repulsion between the two reactants [88].

A strong emphasis has been placed on developing photosensitizers that absorb low energy light to eliminate the interfering absorbance by other chromophores present in living systems. Mixed-metal supramolecular complexes that couple ruthenium(II) or osmium(II) light absorbers to a central rhodium(III) core ($[\{(bpy)_2M(dpp)\}_2RhCl_2]^{5+}$ (M = Ru or Os) or $[\{(tpy)_2RuCl(dpp)\}_2RhCl_2]^{3+}$ (for ligand structure, see Figure 14.2) can promote light-activated DNA cleavage when irradiated with low-energy visible light [89, 90]. It has been proposed that the lowest metal-to-metal charge transfer (MMCT) excited state of the triad promotes DNA photocleavage; however, a detailed mechanism is still unclear [89, 90].

14.2 Photoinduced DNA-Mediated Charge Transport

Charge migration through DNA has been extensively examined [51, 91–96]. Nowadays it has been established that DNA-mediated charge transport is feasible but it is still unclear how this process occurs. The application of metal complexes and light had a great impact on investigation of this phenomenon. The rate and efficiency of electron transfer through DNA depend on the employed oxidant and unique features of DNA itself.

The construction of well-defined DNA assemblies containing donors and acceptors at discrete sites on the helix allows the probing of DNA as a model of molecular wire. Barton et al. have designed a DNA duplex with a $[Ru(bpy/phen)_2(dppz)]^{2+}$ derivative as an electron donor and $[Rh(phi)_2(bpy/phen)]^{3+}$ derivative as an electron acceptor tethered by flexible linker to the opposite ends of the assembles [91, 96, 97]. Both complexes were free to intercalate but separated by a fixed distance. The intercalation sites for both complexes were characterized by employing duplex DNA bearing a tethered ruthenium or rhodium complex. For ruthenium complexes

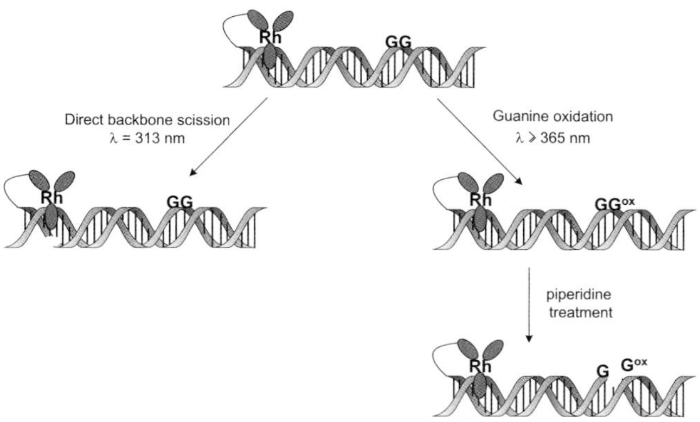

Figure 14.8 *Irradiation of rhodium(III) complex tethered to DNA duplex at high energy ($\lambda = 313$ nm) permits determination of the site of intercalation whereas irradiation at lower energy ($\lambda \geq 365$ nm) permits determination of the site of long-range oxidation*

the intercalation site was marked using their sensitivity in the local environment whereas for rhodium complexes the photocleavage experiment was applied (Figure 14.8; explanation of the mechanism is given in section 14.1 and 14.2). The photoexcited donor (ruthenium(II) complex) may transfer an electron to the ground state acceptor (rhodium(III) complex), which results in quenching of the donor emission and therefore allows monitoring of the DNA-mediated electron transfer between metallointercalators. The products of this forward electron transfer reaction (Ru^{III} and Rh^{II} complexes) then undergo thermal recombination to regenerate the ground state pair (Ru^{II} and Rh^{III} complexes). A similar method can be applied to the donors and acceptors that are bound non-covalently to the DNA duplex; however, the experiments can be carried out only at low reactant loading and the donor–acceptor distance is not so well fixed as in the case of tethered complexes [97, 98].

The DNA-mediated charge transport phenomenon can be exploited in terms of utilization of the DNA not only as a conduit for a long-range electron transfer but also as a reactant in the electron transfer process. Radicals can migrate through the DNA duplex to react at a site remote from the oxidant. The long-range oxidation of G bases within nucleic acid molecules, the major target for oxidative damage, can be investigated by application of strategically placed tethered Ru^{II} or Rh^{III} complexes. Rh^{III} complexes such as $[Rh(bpy')(phi)_2]^{3+}$or $[Rh(dmb)(phi)_2]^{3+}$ (structures of ligands are presented in Figure 14.5) can be attached to one end of double-stranded DNA and light-induced oxidation of G residues studied [48–51]. The direct strand cleavage of the DNA promoted by irradiation at 313 nm of rhodium-modified duplex marks the site of intercalation whereas irradiation at 365 nm triggers charge transport through DNA, leading to photo-oxidation of G residues (see Figure 14.8; explanation of the mechanism is given in section 14.1 and 14.2) [27, 99, 100]. The position and yield of oxidative damage are revealed by treatment of the modified DNA with piperidine, which induces strand cleavage at the damage site. Another way to study long-range oxidation of G in DNA is using tethered Ru^{II} complexes such as $[Ru(bpy')(Me_2dppz)(phen)]^{2+}$ (structures of ligands presented in Figure 14.2) as a photo-oxidant, and applying the flash quench method (described in detail in section 14.1) [27, 30].

An entirely new approach to the study the DNA charge transport involves attachment of oligonucleotides to TiO_2 nanoparticles through a dopamine anchor [101, 102]. The photoexcitation of these TiO_2/dopamine/oligonucleotide triads results in the charge separation across the nanoparticle interface followed by the oxidation of double-stranded DNA via holes if the redox properties of the attached DNA are more negative than the oxidation potential of TiO_2/dopamine. This sequential charge separation does not occur when single-stranded oligonucleotides are attached.

The long-range charge transport through the DNA double helix has been found to be sensitive to the sequence [99, 103] and stacking of the intervening bases, eg mismatches [27, 101, 104, 105] or bulges [106]. It has been shown that variation in stacking can lead to significant change in efficiency and yield of long-range charge transfer through DNA. In particular the pyrimidine–pyrimidine mismatches, which cause pronounced destabilization of the helix due to lack of proper hydrogen bonding and small stacking surface area, result in significant reduction of yields of oxidative damage. The presence of bulges also diminishes the yield of oxidative damage. This leads to the conclusion that an effectively coupled aromatic π-array is essential for an efficient long-range charge transfer through DNA. Many investigations have been conducted to investigate the influence of the base composition on rate and efficiency of the charge transfer through DNA. These studies have shown that DNA-mediated charge transport can occur over a significant distance (eg 20 nm), and for the same applied photo-oxidant the rate of this process can be either strongly or weakly distance dependent; this effect is modulated by intervening sequence and sequence-dependent conformational dynamics [51, 99]. In addition to sequence and stacking of the intervening bases, charge transport through DNA is effected by its structure. The yield of oxidative damage caused by long-range charge transport was studied not only for DNA duplex (B-form) but also for different DNA structures such as DNA/RNA hybrids (A-like structure), DNA four-way junctions [107], double-crossovers [49], triple strands [108], or quadruplexes [109]. It has been demonstrated that protein binding to DNA can both activate and inhibit charge transfer through the helix, depending on the nature of the interactions and its influence on DNA base stacking [110–114]. In general, the interactions that disturb the DNA π-stacking array reduce the charge transport through DNA whereas the interactions that make the π-stack rigid promote this process. This modulation by protein binding may be an issue of physiological importance because in vivo DNA does not float free in solution but is packaged in nucleosome core particles (NCPs). Recent investigations have demonstrated that the long-rage oxidative damage in NCPs as well as in nuclei is feasible, and this could help us to understand the biochemical mechanism for DNA damage and repair [100, 115].

The mechanism of long-range charge transport through DNA is still unclear in spite of many experiments conducted in this field. There are two general mechanistic possibilities for charge transport from the donor to the acceptor through the DNA strand: superexchange (tunnelling) and hole hopping [92–94, 109]. In a tunnelling mechanism, the charge tunnels from the donor to the acceptor through the bridge formed by the DNA orbitals. These orbitals are energetically higher than the donor and acceptor, so the charge tunnels through the bridge without formally occupying it. In the hopping model, the donor and acceptor orbitals are close in energy to the bridge, so charge occupies the bridge hopping between discrete molecular orbitals. Based on these two theories several other mechanisms have been proposed. Barton et al. have proposed the 'domain hopping' model, in which charge is transiently delocalized over sequence-dependent domains defined by local structures and this domain hops along the bridge [103]. Schuster and co-workers have suggested the phonon-assisted polaron hopping mechanism. The hole injection results in formation of a transient polaron, which causes base-pairs in and near the structural distortion to leave or join the polaron [92].

The use of DNA molecules as wires in electronic systems may open a new opportunity in nanoelectronics. DNA has the appropriate molecular recognition features and well-characterized self-assembly. There is evidence to suggest that DNA is only a marginally better electron conductor than proteins [116–118]. As a result, many studies have focused on various methods of DNA modification leading to improvement in its conductive properties. It is possible to enhance the conductivity of DNA by coating it with a thin film of metal atoms, but the molecular recognition properties of the DNA are then destroyed. An effective approach to this problem is the incorporation of metal ions into the DNA double helix [118–121]. Preliminary results suggest that a metal ion–DNA complex may be a much better conductor than B-DNA, because the former shows a metallic conduction whereas the latter behaves like a wide-band gap semiconductor [118].

References

1. Armitage B. Photocleavage of nucleic acid. *Chem Rev* 1998;**98**:1171–200.
2. Szaciłowski K, Macyk W, Drzewiecka-Matuszek A, Brindell M, Stochel G. Bioinorganic Photochemistry. Frontiers and Mechanisms. *Chem Rev* 2005;**105**:2647–94.
3. Pogozelski WK, Thomas DT. Oxidative strand scission of nucleic acids: routes initiated by hydrogen abstraction from sugar moiety. *Chem Rev* 1998;**98**:1089–107.
4. Burrows CJ, Muller JG. Oxidative nucleobase modifications leading to strand scission. *Chem Rev* 1998;**98**:1109–51.
5. Mei H-Y, Barton JK. Tris(tetramethylphenanthroline)ruthenium(II): a chiral probe that cleaves A-DNA conformations. *Proc Natl Acad Sci USA* 1988;**85**:1339–43.
6. Fleisher MB, Waterman KC, Turro NJ, Barton JK. Light-induced cleavage of DNA by metal complexes. *Inorg Chem* 1988;**25**:3549–51.
7. Hergueta-Bravo A, Jimenez-Hernandez ME, Montero F, Oliveros E, Porellana G. Singlet oxygen-mediated DNA photocleavage with Ru(II) polypyridyl complexes. *J Phys Chem B* 2002;**106**:4010–17.
8. Tan L-F, Wang F, Chao H, Zhou Y-F, Weng C. Ruthenium(II) mixed-ligand complexes containing 2-(4′-benzyloxy-phenyl)imidazo[4,5-*f*][1,10]phenentholine: Synthesis, DNA-binding and photocleavage studies. *J Inorg Biochem* 2007;**101**:700–8.
9. Tan L-F, Chao H, Li H, et al. Synthesis, characterization, DNA-binding and photocleavage studies of $[Ru(bpy)_2(PPIP)]^{2+}$ and $[Ru(phen)_2(PPIP)]^{2+}$. *J Inorg Biochem* 2005;**99**: 513–520.
10. Liu Y, Hammitt R, Lutterman DA, Thummel RP, Turro C. Marked differences in light-switch behavior of Ru(II) complexes possessing a tridentate DNA intercalating ligand. *Inorg Chem* 2007;**46**:6011–21.
11. Ding H-Y, Wang X-S, Song L-Q, et al. Aryl-modified ruthenium bis(terpyridine) complexes: Quantum yield of $^1O_2$ generation and photocleavage on DNA. *J Photochem Photobiol A: Chem* 2006;**177**:286–94.

12. E. G, Paillous N, Vicendo P. Mechanism of DNA damage photosensitized by trisbipyrazyl ruthenium complex. Unusual role of Cu/Zn superoxide dismutase. *Photochem Photobiol* 2000;**72**:583–9.
13. Moucheron C, Kirsch-De Mesmaeker A, Kelly JM. Photoreactions of ruthenium and osmium complexes with deoxyribonucleic acid. *J Photochem Photobiol B: Biol* 1997;**40**:91–106.
14. Lecomte JP, Kirsch-De Mesmaeker A, Kelly JM. Photochemistry Ru(II)-tris-1,4,5,8-tetraazaphenantrene with mononucleotides. *Bull Soc Chim Belg* 1994;**103**:193–200.
15. Elias B, Kirsch-De Mesmaeker A. Photo-reduction of polyazaaromatic Ru(II) complexes by biomolecules and possible applications. *Coord Chem Rev* 2006;**250**:1627–1641.
16. Kelley EE, Domann FE, Buettner GR, Oberley LW, Burns CP. Increased efficacy in vitro Photofrin photosensitization of human oral squamous cell carcinoma by iron and ascorbate. *J Photochem Photobiol B: Biol* 1997;**40**:273–77.
17. Jacquet L, Davies RJH, Kirsch-De Mesmaeker A, Kelly JM. Photoaddition of $Ru(tap)_2(bpy)^{2+}$ to DNA: a new mode of covalent attachent of metal complexes to duplex DNA. *J Am Chem Soc* 1997;**119**:11763–8.
18. Moucheron C, Kirsch-De Mesmaeker A, Choua S. Photophysics of $Ru(phen)_2(PHEHAT)^{2+}$: a novel 'light switch' for DNA and photooxidant for mononucleotides. *Inorg Chem* 1997;**36**:584–92.
19. Blasius R, Moucheron C, Kirsch-De Mesmaeker A. Photoadducts of metallic compounds with nucleic acids – role played by the photoelectron transfer process and by the TAP and HAT ligands in the Ru^{II} complexes. *Eur J Inorg Chem* 2004:3971–9.
20. Sentagne C, Chambron JC, Sauvage JP, Paillous N. Tuning the mechanism of DNA cleavage photosensitized by ruthenium dipyridophenazine complexes by varying the structure of the two nonintercalating ligands. *J Photochem Photobiol B: Biol* 1994;**26**:1994.
21. Haq I, Lincoln P, Suh D, Norden B, Chowdhry BZ, Chaires JB. Interaction of 'DELT'- and 'LAMBDA'-$[Ru(phen)_2DPPZ]^{2+}$ with DNA: A calorimetric and equilibrium binding study. *Am Chem Soc* 1995;**117**:4788–96.
22. Ortmans I, Content S, Boutonnet N, et al. Ru-labeled oligonucleotide for photoinduced reactions on targeted DNA guanines. *Chem Eur J* 1999;**5**:2712–21.
23. Garcia-Fresnadillo D, Boutonnet N, Schumm S, et al. Luminescence quenching of Ru-labeled oligonucleotide by target complementary strands. *Biophys J* 2002;**82**:978–87.
24. Lentzen O, Constant J-F, Defrancq E, Moucheron C, Dumy P, Kirsch-De Mesmaeker A. Photoadduct leading to crosslinking in Ru^{II}-derivatized oligonucleotides. *Nucleosides, Nucleotides Nucleic Acid* 2003;**22**:1487–9.
25. Lentzen O, Defrancq E, Constant J-F, et al. Determination of DNA guanine sites forming photo-adducts with Ru(II)-labeled oligonucleotides; DNA polymerase inhibition by resulting photo-crosslinking. *J Biol Inorg Chem* 2004;**9**:100–8.
26. Stemp EDA, Arkin MR, Barton JK. Oxidation of guanine in DNA by $Ru(phen)_2dppz^{3+}$ using the flash-quench technique. *J Am Chem Soc* 1997;**119**:2921–5.
27. Arkin MR, Stemp EDA, Pulver SC, Barton JK. Long-range oxidation of guanine by Ru(III) in duplex DNA. *Chem Biol* 1997;**4**:389–400.
28. Chang IJ, Gray HB, Winkler JR. High-driving-force electron-transfer in metalloproteins – intramolecular oxidation of ferrocytochrome-*c* by $Ru(2,2'\text{-bpy})_2(lm)(His\text{-}33)^{3+}$. *J Am Chem Soc* 1991;**113**:7056–7.
29. Juskowiak B, Dominiak A, Takenaka S, Takagi M. Isomerization of DNA-bound distilbazolium ligand induced by electron transfer from photoexcited tris(1,10-phenanthroline)Ru(II). *Photochem Photobiol* 2001;**74**:391–400.
30. Hall DB, Holmlin RE, Barton JK. Oxidative DNA-damage through long-range electron-transfer. *Nature* 1996;**382**:731–5.
31. Fu PK-L, Bradley PM, van Loyen D, Durr H, Bossmann SH, Turro C. DNA photocleavage by a supramolecular Ru(II)-viologen complex. *Inorg Chem* 2002;**41**:3808–10.
32. Erkkila KE, Odom DT, Barton JK. Recognition and reaction of metallointercalators. *Chem Rev* 1999;**99**:2777–95.

33. Sitlani A, Long EC, Pyle AM, Barton JK. DNA Photocleavage by phenanthrenequinone diimine complexes of rhodium(III): shape-selective recognition and reaction. *J Am Chem Soc* 1992;**114**:2303–12.
34. Turro C, Evenzahav A, Bossmann SH, Barton JK, Turro NJ. Excited stats properties of $Rh(pai)_2(phen)^{3+}$ and related complexes: a strong photooxidant. *Inorg Chim Acta* 1996;**243**:101–8.
35. Turro C, Hall DB, Chen W, Zuilhof H, Barton JK, Turro NJ. Solution photoreactivity of phenanthrenequinone diimine complexes of rhodium and correlations with DNA photocleavage and photooxidation. *J Phys Chem A* 1998;**102**:5708–15.
36. Terbrueggen RH, Johann TW, Barton JK. Functionalized rhodium intercalators for DNA recognition. *Inorg Chem* 1998;**37**:6874–83.
37. Kielkopf CL, Erkkila KE, Hudson BP, Barton JK, Rees DC. Structure of photoactive rhodium complex intercalated into DNA. *Nat Struct Biol* 2000;**7**:117–21.
38. Lim AC, Barton JK. $Rh(phen)phi^{3+}$ as a shape-selective probe of triple helices. *Biochemistry* 1998;**37**:9138–46.
39. Lim AC, Barton JK. Targeting the Tat-binding site of bovine immunodeficiency virus TAR RNA with a shape-selective rhodium complex. *Bioorg Med Chem* 1997;**5**:1131–6.
40. Shields TP, Barton JK. Sequence-selective DNA recognition and photocleavage: a comparison of enantiomers of $Rh(en)_2phi^{3+}$. *Biochemistry* 1995;**34**:15037–48.
41. Jackson BA, Barton JK. Recognition of DNA base mismatches by a rhodium intercalator. *J Am Chem Soc* 1997;**119**:12986–7.
42. Kisko JL, Barton JK. Recognition of DNA base pair mismatches by a cyclomethylated Rh(III) intercalator. *Inorg Chem* 2000;**39**:4942–9.
43. Junicke H, Hart JR, Kisko J, Glebov O, Kirsch IR, Barton JK. A rhodium(III) complex for high-affinity DNA base-pair mismatch recognition. *Proc Natl Acad Sci USA* 2003;**100**:3737–42.
44. Chow CS, Barton JK. Recognition of G-U mismatches by tris(4,7-diphenyl-1,10-phenanthroline)rhodium(III). *Biochemistry* 1992;**31**:5423–9.
45. Chow CS, Cunningham PR, Lee KS, Meroueh M, SantaLucia J Jr, Varma S. Photoinduced cleavage by rhodium complex at G*U mismatches and exposed guanines in large and small RNAs. *Biochimie* 2002;**84**:859–68.
46. Jackson BA, Barton JK. Recognition of base mismatches in DNA by 5,6-chrysenequinone diimine complexes of rhodium(III): A proposed mechanism for preferential binding in destabilized region of the double helix. *Biochemistry* 2000;**39**: 6176–82.
47. Jackson BA, Alekseyev VY, Barton JK. A versatile mismatch recognition agent: specific cleavage of a plasmid DNA at a single base mispair. *Biochemistry* 1999;**38**:4655–62.
48. Dandliker PJ, Núnez ME, Barton JK. Oxidative charge transfer to repair thymine dimers and damage guanine bases in DNA assemblies containing tethered metallointercalators. *Biochemistry* 1998;**37**:6491–502.
49. Odom DT, Dill EA, Barton JK. Robust charge transport in DNA double crossover assemblies. *Chem Biol* 2000;**7**:475–81.
50. Hall DB, Barton JK. Sensitivity of DNA-mediated electron transfer to the intervening π-stack: a probe for the integrity of the DNA base stack. *J Am Chem Soc* 1997;**119**: 5045–6.
51. Delaney S, Barton JK. Long-range DNA charge transport. *J Org Chem* 2003;**68**:6475–83.
52. Arounaguiri S, Maiya BG. Inorg Chem 1996;**35**:4267–4270.
53. Sastri CV, Eswaramoorthy D, Giribabu L, Maiya BG. DNA interactions of new mixed-ligand complexes of cobalt(III) and nickel(II) that incorporate modified phenanthroline ligands. J Inorg Biochem 2003;**94**:138–45.
54. Barve A, Ghosh S, Kumbhar A, Kumbhar A, Puranik V. Synthesis, DNA-binding and photocleavage studies of cobalt(III) polypyridyl complexes. *J Inorg Biochem* 2003;**96**:100.
55. Zhang QL, Liu JG, Liu JZ, et al. Effect of intramolecular hydrogen-bond on the DNA-binding and photocleavage properties of polypyridyl cobalt(III) complexes. *Inorg Chim Acta* 2002;**339**:34–40.

56. Zhang QL, Liu JG, Liu J, et al. DNA-binding and photocleavage studies of cobalt(III) mixed-polypyridyl complexes containing 2-(2-chloro-5-nitrophenyl)imidazo [4,5-f][1,10] phenanthroline. *J Inorg Biochem* 2001;**85**:291–6.
57. Zhang QL, Liu JG, Chao H, Xue GQ, Ji LN. DNA-binding and photocleavage studies of cobalt(III) polypyridyl complexes: $[Co(phen)_2IP]^{3+}$ and $[Co(phen)_2PIP]^{3+}$. *J Inorg Biochem* 2001;**83**:49–55.
58. Wang X-L, Chao H, Li H, et al. DNA interactions of cobalt(III) mixed-polypyridyl complexes containing asymmetric ligands. *J Inorg Biochem* 2004;**98**:1143–50.
59. Vaidyanathan VG, Nair BU. Photooxidation of DNA by a cobalt(II) tridentate complex. *J Inorg Biochem* 2003;**94**:121–6.
60. Farinas ET, Tan JD, Mascharak PK. Photoinduced DNA cleavage reactions by designed analogues of Co(III)-bleomycin: the metalated core is the primary determinant of sequence specificity. *Inorg Chem* 1996;**35**:2637–43.
61. Patra AK, Dhar S, Nethaji M, Chakravarty AR. Visible light-induced nuclease activity of a ternary mono-phenanthroline copper(II) complex containing L-methionine as a photosensitizer. *Chem Commun* 2003:1562–3.
62. Patra AK, Bhowmick T, Ramakumar S, Chakravarty AR. Metal-based netropsin mimics showing AT-selective DNA binding and DNA cleavage activity at red light. *Inorg Chem* 2007;**46**:9030–2.
63. Patra AK, Nethaji M, Chakravarty AR. Synthesis, crystal structure, DNA binding and photo-induced DNA cleavage activity of (*S*-methyl-L-cysteine)copper(II) complexes of heterocyclic bases. *J Inorg Biochem* 2007;**101**:233–44.
64. Dhar S, Senapati D, Das PK, Chattopadhyay P, Nethaji M, Chakravarty AR. Ternary copper complexes for photocleavage of DNA by red light: direct evidence for sulfur-to-copper charge transfer and d-d band involvement. *J Am Chem Soc* 2003;**125**:12118–24.
65. Reddy PAN, Santra BK, Nethaji M, Chakravarty AR. Metal-assisted light induced DNA cleavage activity of 2-metrhylthiophenylsalicyamide Schiff base copper(II) complexes having planar heterocylic bases. *J Inorg Biochem* 2004;**98**:377–86.
66. Thomas AM, Naik AD, Nethaji M, Chakravarty AR. Synthesis, crystal structure and photoinduced DNA cleavage activity of ternary copper(II)-thiosemicarbazone complexes having heterocyclic bases. *Inorg Chim Acta* 2004;**357**:2315–23.
67. Dhar S, Chakravarty AR. Photosensitizer in a molecular bowl: steric protection enhancing the photonuclease activity of copper(II) scorpionates. *Inorg Chem* 2005;**44**:2582–4.
68. Eppley HJ, Lato SM, Ellington A, Zaleski JM. Transition metal kinamycin model as a DNA photocleaver for hypoxic environments: bis(9-diazo-4,5-diazafluorene)copper(II) nitrate. *J Chem Soc Chem Commun* 1999;**1999**:2405–6.
69. Benites PJ, Holmberg RC, Rawat DS, et al. Metal-ligand charge-transfer-promoted photoelectronic Bergman cyclization of copper metalloenediynes: photochemical DNA cleavage via C-4′ H-atom absorption. *J Am Chem Soc* 2003;**125**:6434–46.
70. Hurley AL, Maddox MP III, Scott TL, Flood MR, Mohler DL. Photoinduced DNA cleavage by cyclopentadienyl metal complexes conjugated to DNA recognition elements. *Org Lett* 2001;**3**:2761–4.
71. Hirot C, Goodisman J, Dabrowiak JC. Cleavage of DNA by the insulin-mimetic compound, $NH_4[VO(O_2)_2(phen)]$. *Biochemistry* 1996;**35**:12354–62.
72. Kwong DWJ, Chan OY, Wong RNS, Musser SM, Vaca L. DNA-photocleavage activities of vanadium(V)-peroxo complexes. *Inorg Chem* 1997;**36**:1276–77.
73. Sam M, Hwang JH, Chanfreau G, Abu-Omar MM. Hydroxyl radical is the active species in photochemical DNA strand scission by bis(peroxo)vanadium(V) phenanthroline. *Inorg Chem* 2004;**43**:8447–55.
74. Kwong DWJ, Chan C-K, Shek LK, Wong RNS. Effects of supercoiling on the sequence-specific photo-modification of DNA by a vanadium(V)-peroxo complex. *J Inorg Biochem* 2005;**99**:2062–73.
75. Chen C-T, Lin J-S, Kuo J-H, et al. Site-selective DNA photocleavage involving unusual photoinitiated tautomerization of chiral tridentate vanadyl(V) complexes derived from *N*-salicylidene α-amino acids. *Org Lett* 2004;**6**:4471–4.

76. Sasmal PK, Patra AK, Nethaji M, Chakravarty AR. DNA cleavage by new oxovanadium(IV) complexes of *N*-salicylidene α-amino acids and phenanthroline bases in the photodynamic therapy window. *Inorg Chem* 2007;**46**:11112–21.
77. Sonnichsen SH, Nielsen PE. Enhanced uranyl photocleavage across the minor grove of all (A/T)(4) sequences indicates a similar narrow minor groove conformation. *J Mol Recognit* 1996;**9**:219–27.
78. Mollegaard NE, Bailly C, Waring MJ, Nielsen PE. Effects of diaminopurine and inosine substitutions on A-tract induced DNA curvature. Importance of the 3′-A-tract junction. *Nucleic Acids Res* 1997;**25**:3497–502.
79. Nielsen PE, Hiort C, Sonnichsen SH, Buchardt O, Dahl O, Norden B. DNA binding and photocleavage by uranyl(VI) (UO_2^{2+}) salts. *J Am Chem Soc* 1992;**114**:4967–75.
80. Madhavaiah C, Verma S. Reusable photonucleases: plasmid scission by a uranyl ion impregnated adenine homopolymer in the presence of visible light and sunlight. *J Chem Soc Chem Commun* 2003:800–1.
81. Wittberger D, Berens C, Hammann C, Westhof E, Schroeder R. Evaluation of uranyl photocleavage as a probe to monitor ion binding and flexibility in RNAs. *J Mol Biol* 2000;**300**:339–52.
82. Bassi GS, Mollegaard NE, Murchie AIH, Lilley DMJ. RNA folding and misfolding of the hammerhead ribozyme. *Biochemistry* 1999;**38**:3345–54.
83. Mollegaard NE, Nielsen PE. Increased temperature and 2-methyl-2,4-pentanediol change the DNA structure of both curved and uncurved adenine/thymine-rich sequences. *Biochemistry* 2003;**42**:8587–93.
84. van Gijte O, Kirsch-De Mesmaeker A. The dinuclear ruthenium(II) complex $[\{Ru(Phen)_2\}_2(HAT)]^{4+}$ (HAT = 1,4,5,8,9,12-hexaazatriphenylene), a new photoreagent for nucleobase and photoprobe for denatured DNA. *J Chem Soc Dalton Trans* 1999:951–6.
85. Bradley PM, Angeles-Boza AM, Dunbar KR, Turro C. Direct DNA photocleavage by a new intercalating dirhodium(II/II) complex: comparison to $Rh_2(\mu\text{-}O_2CCH_3)_4$. *Inorg Chem* 2004;**43**:2450–2.
86. Angeles-Boza AM, Bradley PM, Fu PK-L, Wicke SE, Bacsa RR, Dunbar KR, Turro C. DNA binding and photocleavage in vitro by new dirhodium(II) dppz complexes: correlation to cytotoxicity and photocytotoxicity. *Inorg Chem* 2004;**43**:8510–19.
87. Angeles-Boza AM, Bradley PM, Fu PK-L, et al. Photocytotoxicity of a new Rh_2(II,II) complex: increase in cytotoxicity upon irradiation similar to that of PDT agent hematoporphyrin. *Inorg Chem* 2005;**44**:7262–4.
88. Breiner KM, Daugherty MA, Oas TG, Thorp HH. An anionic diplatinum DNA photocleavage agent: chemical mechanism and footprinting of 'lambda' repressor. *J Am Chem Soc* 1995;**117**:11673.
89. Swavey S, Brewer KJ. Visible light induced photocleavage of DNA by a mixed-metal supramolecular complex: $[\{(bpy)_2Ru(dpp)\}_2RhCl_2]^{5+}$. *Inorg Chem* 2002;**41**:6196–8.
90. Holder AA, Swavey S, Brewer KJ. Design aspects for the development of mixed-metal supramolecular complexes capable of visible light induced photocleavage of DNA. *Inorg Chem* 2004;**43**:303–8.
91. Murphy CJ, Arkin MR, Jenkins Y, et al. Long-range photoinduced electron transfer through a DNA helix. *Science* 1993;**262**:1025–9.
92. Schuster GB. Long-range charge transfer in DNA: transient structural distortions control the distance dependence. *Acc Chem Res* 2000;**33**:253–60.
93. Giese B. Long-distance charge transport in DNA: the hopping mechanism. *Acc Chem Res* 2000;**33**:631–6.
94. Lewis FD, Letsinger RL, Wasielewski MR. Dynamics of photoinduced charge transfer and hole transport in synthetic DNA hairpins. *Acc Chem Res* 2001;**34**:159–70.
95. Treadway CR. Charge transport through a molecular π-stack: double helical DNA. *Chem Phys* 2002;**281**:409–28.
96. Núnez ME, Barton JK. Probing DNA charge transport with metallointercalators. *Curr Opin Chem Biol* 2000;**4**:199–206.

97. Holmlin RE, Dandliker PJ, Barton JK. Charge transfer through the DNA base stack. *Angew Chem Int Ed Eng* 1997;**36**:2714–30.
98. Arkin MR, Stemp EDA, Holmlin RE, et al. Rates of DNA-mediated electron transfer between metallointercalators. *Science* 1996;**273**:475–80.
99. Núnez ME, Hall DB, Barton JK. Long-range oxidative damage to DNA: effect of distance and sequence. *Chem Biol* 1999;**6**:85–97.
100. Núnez ME, Noyes KT, Barton JK. Oxidative charge transport through DNA in nucleosome core particles. *Chem Biol* 2002;**9**:403–15.
101. Rajh T, Saponjic Z, Liu J, et al. Charge transfer across DNA. *Nano Lett* 2004;**4**:1017–23.
102. Liu J, de la Garza L, Zhang L, et al. Photocatalytic probing of DNA sequence by using TiO_2/dopamine-DNA triads. *Chem Phys* 2007;**339**:154–63.
103. Williams TT, Odom DT, Barton JK. Variation in DNA charge transport with nucleotide composition and sequence. *J Am Chem Soc* 2000;**122**:9048–9.
104. Bhattacharya PK, Barton JK. Influence of intervening mismatches on long-range guanine oxidation in DNA duplex. *J Am Chem Soc* 2001;**123**:8649–56.
105. Yoo J, Delaney S, Stemp EDA, Barton JK. Rapid radical formation by DNA charge transport through sequences lacking intervening guanines. *J Am Chem Soc* 2003;**125**:6640–1.
106. Hall DB, Barton JK. Sensitivity of DNA-mediated electron transfer to intervening π-stack: a probe for the integrity of the DNA. *J Am Chem Soc* 1997;**119**:5045–6.
107. Odom DT, Dill EA, Barton JK. Charge transport through DNA four-way junctions. *Nucleic Acids Res* 2001;**29**:2026–33.
108. Kan Y, Schuster GB. Radical cation transport and reaction in triplex DNA: long-range guanine damage. *J Am Chem Soc* 1999;**121**:11607–14.
109. Delaney S, Barton JK. Charge transport in DNA duplex/quadruplex conjugates. *Biochemistry* 2003;**42**:14159–65.
110. Rajski SR, Kumar S, Robert RJ, Barton JK. Protein-modulated DNA electron transfer. *J Am Chem Soc* 1999;**121**:5615–16.
111. Nakatani K, Dohno C, Ogawa A, Saito I. Suppression of DNA-mediated charge transport by BamHI binding. *Chem Biol* 2002;**9**:361–6.
112. Rajski SR, Barton JK. How different DNA-binding proteins affect long-range oxidative damage to DNA? *Biochemistry* 2001;**40**:5556–64.
113. Wagenknecht HA, Rajski SR, Pascaly M, Step EDA, Barton JK. Direct observation of radical intermediates in protein-dependent DNA charge transport. *J Am Chem Soc* 2001;**123**:4400–7.
114. Boon EM, Pope MA, Williams SD, David SS, Barton JK. DNA-mediated charge transport as a probe of MutY/DNA interaction. *Biochemistry* 2002;**41**:8464–70.
115. Núnez ME, Holmquist GP, Barton JK. Evidence for DNA charge transport in the nucleus. *Biochemistry* 2001;**40**:12465–71.
116. Porath D, Bezryadin A, de Vries S, Dekker C. Direct measurement of electrical transport through DNA molecules. *Nature* 2000;**403**:635–8.
117. Gomez-Navarro C, Moreno-Herrero F, de Pablo PJ, Gomez-Herrero J, Baro AM. Contactless experiments on individual DNA molecules show no evidence for molecular wire behavior. *Proc Natl Acad Sci USA* 2002;**99**:8484–7.
118. Rakitin A, Aich P, Papadopoulos C, et al. Metallic conduction through engineered DNA: DNA nanoelectronic building blocks. *Phys Rev Lett* 2001;**86**:3670–3.
119. Zhang LZ, Cheng P. Study of Ni(II) ion-DNA interactions with methylene blue as fluorescent probe. *J Inorg Biochem* 2004;**98**:569–74.
120. Bregadze V, Kutsishvili I, Chkhaberidze J, Sologashvili K. DNA as a mediator for proton, electron and energy transfer induced by metal ions. *Inorg Chim Acta* 2002;**339**:145–59.
121. Aich P, Skinner RJS, Wettig SD, Steer RP, Lee JS. Long range molecular wire behaviour in a metal complex of DNA. *J Biomol Struct Dynam* 2002;**20**:93–8.

Part V

Towards Applications

15
Light and Biomatter

Following the light of the sun, we left the Old World.
Christopher Columbus

The electromagnetic spectrum provides a diverse set of photonic tools for probing, manipulating, and interacting with biological systems. Many of these tools are used in biomedicine to detect and treat disease and to advance scientific knowledge. Biomedical photonics is a relatively new field, having a potential to revolutionize the biological sciences and medicine [1–4].

The term 'light', discussed in this book, includes the following spectral ranges of electromagnetic spectrum: near infrared (NIR), visible, and ultraviolet (UV) A, B, and C bands (Figure 15.1).

Light interacts with living matter at various biological organization levels: (1) the biomolecule, (2) subcellular structures, (3) cells, (4) biotissues, (5) organs, and (6) the whole organism. All these structures fall within different sections of the spatial spectrum (Figure 15.2).

All types of light–biomatter interaction can be divided into: (1) non-resonant processes involving light scattering by molecular vibrations or by refractive index irregularities in the exposed biotissues or its reflection from boundaries between media differing in refractive indices (interfaces); and (2) resonant processes, involving the absorption of light and the resultant excitation of biomolecules.

Light interaction with multilayered and multicomponent tissues is a very complex process (Figure 15.3) [5–8]. Biological tissues are optically inhomogeneous – absorbing media are characterized by the average refractive index higher than that of air causing a partial reflection of radiation at the tissue–air interface. Only a certain part of incident light can penetrate the tissue. Multiple scattering and absorption are responsible for light beam broadening and its possible decay as it travels through tissue, whereas bulk scattering is a major effect causing dispersion of a large fraction of radiation in backward directions. Cellular organelles, such as

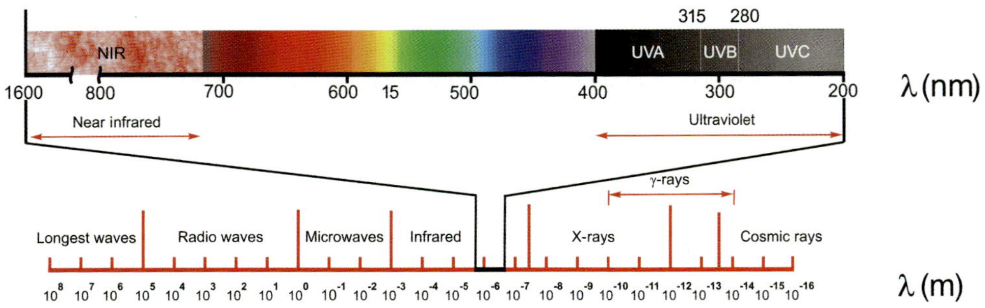

Figure 15.1 *The electromagnetic spectrum with the discussed regions including UV-C (100–280 nm), UV-B (280–315 nm), and UV-A (315–400 nm), the visible range (400–760 nm), and near infrared (760–1400 nm)*

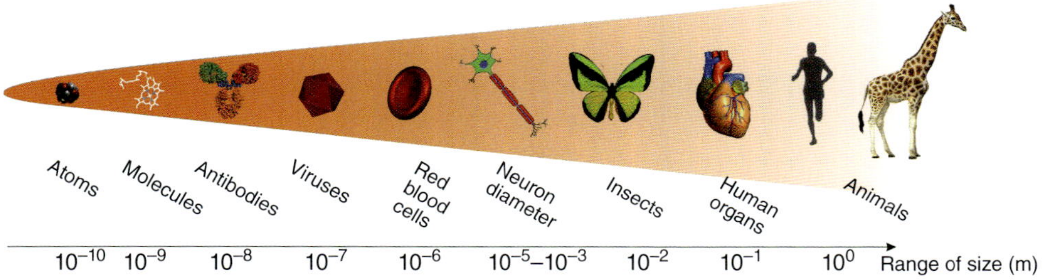

Figure 15.2 *Size spectrum of biostructures*

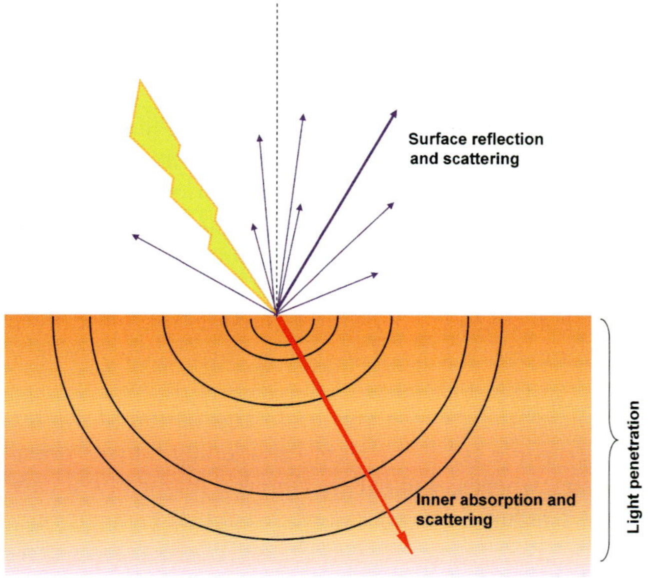

Figure 15.3 *Light–tissue interactions*

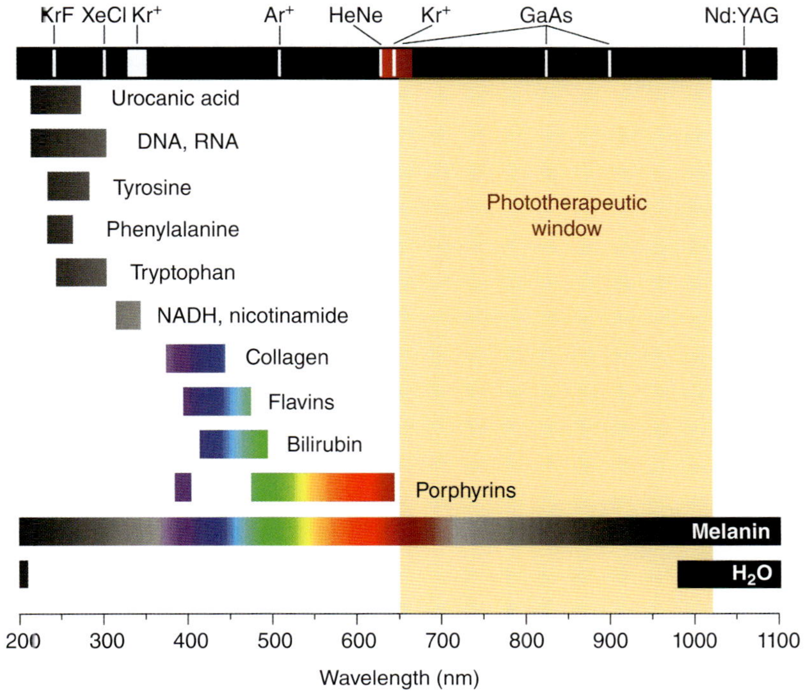

Figure 15.4 *Spectral characteristics of endogenous chromophores, the phototherapeutic window and common lasers*

mitochondria, are the main scatterers in various tissues [9–15]. Light absorption by tissues is a function of their molecular composition. The absorption spectrum depends on the type of predominant absorption centres and water content of tissue (Figure 15.4). Some of the important absorbers of UV include DNA, the aromatic amino acids (tryptophan and tyrosine), proteins, melanins, and porphyrins. Absolute values of absorption coefficients for typical tissues lie in the range 10^{-2}–$10^{4}\,cm^{-1}$ [5, 9–15]. UV and IR parts of light are readily absorbed which accounts for a small contribution of scattering and the inability of radiation to penetrate deep inside tissue (only across one or a few cell layers). At the wavelengths 600–1600 nm scattering prevails over absorption. Depth of light penetration depends on the tissue type; however, in the case of most tissues light of the spectral range 600–700 nm penetrates 50–200% deeper than light of the range 400–500 nm [16, 17].

The principal chromophores (absorbing molecules) for optical radiation in skin are water, melanin, and haemoglobin. Absorption spectra of the major chromophores in skin throughout the UV, visible, and IR ranges are shown in Figure 15.5. Absorption depends on the concentration of present chromophores. About 4–7% of light is reflected upon striking the skin surface because of the sudden change in refractive index between the air and the stratum corneum [5]. The remaining light penetrates the skin and can be either absorbed or scattered by molecules, particles,

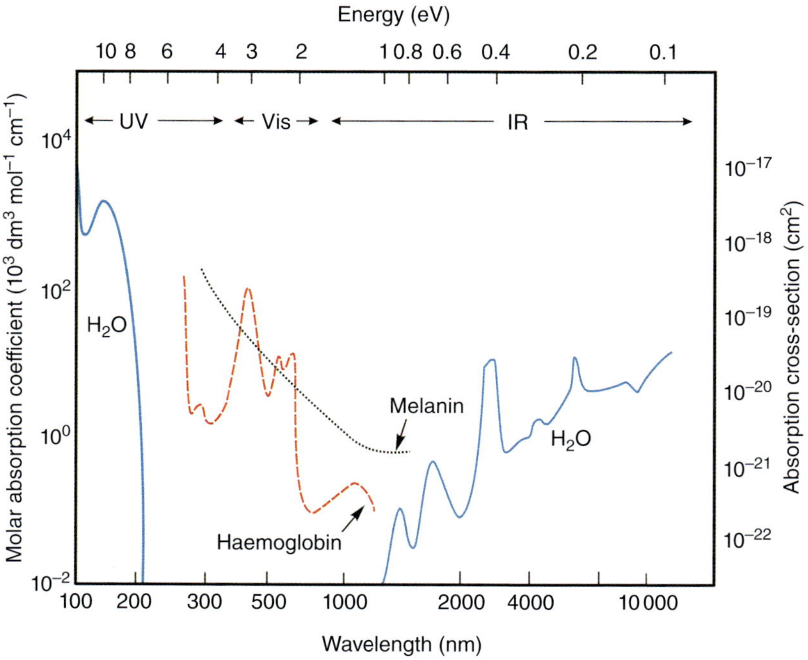

Figure 15.5 *Absorption spectra of water and main chromophores of the skin*

and structures in the tissue. Absorption is the dominating process limiting penetration of UV and visible light through the epidermis. Proteins, melanin, urocanic acid, and DNA absorb UV wavelengths shorter than about 320 nm. For wavelengths from 320 nm to 1200 nm, especially at shorter wavelengths, melanin is the main light absorber in the epidermis. Water is the dominant skin chromophore for mid- and far-IR wavelengths. In the dermis, scattering by collagen fibres is very important for determining optical penetration of light. In general, scattering is an inverse function of wavelength, which accounts in part for a large increase in optical penetration into the dermis, with increasing wavelength from UV through visible and NIR. The absorption coefficient of the dermis is low throughout the visible and NIR spectrum, except in blood vessels. Blood absorbs efficiently blue, green, and yellow light and weakly in the range 800–1000 nm. An optical 'window' between 600 and 1300 nm enables the deepest light penetration. The least penetrating wavelengths fall in the far-UV and far-IR due to strong protein and water absorption, respectively. For visible and NIR wavelengths, the exposure spot size also affects the effective penetration depth of a laser beam into skin due to scattering [5–7].

Various light sources are available for biophotonic applications [2, 16, 18–20]. Their choice is determined by the target type and location, the spectral range and the light power density, as well as the dose to be delivered. The most common coherent light sources include: argon and argon-pumped lasers, metal vapour-pumped dye lasers, solid state lasers, and optical parametric oscillator lasers. There are several types of the non-coherent light sources such as halogen lamps, xenon arc lamps,

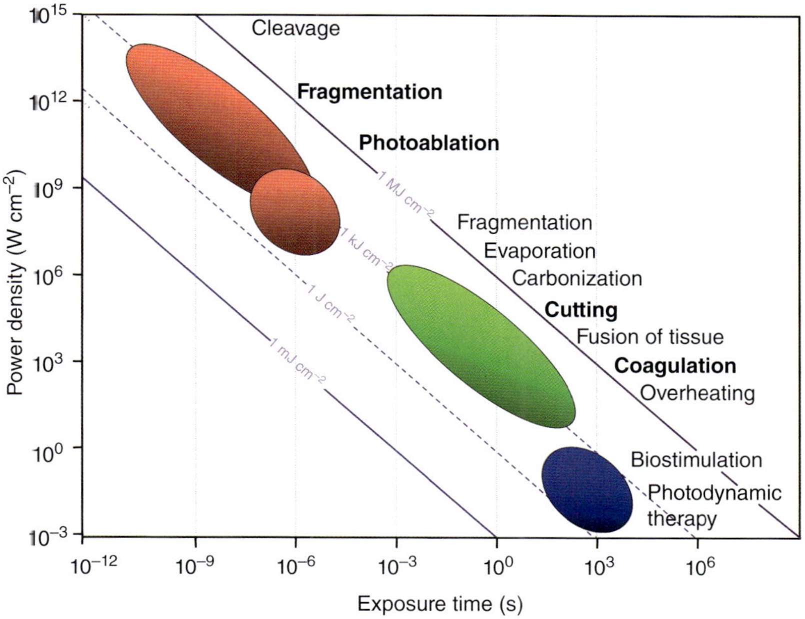

Figure 15.6 *Effects of laser radiation interaction with tissue. Adapted from Boulnois [21]*

metal halide lamps, phosphorus-coated sodium lamps, fluorescent lamps and light-emitting diodes (LEDs). Introduction of two novel light sources – LEDs and femtosecond solid state lasers – strongly influenced strategies of modern phototherapy development. Although lasers are the most commonly used devices for treatment, non-coherent light sources become more and more popular because of their price and simplicity of handling. Instead of a high-intensity monochromatic light emitted by lasers, lamps support high-intensity light distributed over a broader spectral range. As the lamp cannot be easily coupled with small optic fibres without significant loss of its output power, the applications of lamps have been limited to irradiation of surface targets, whereas lasers are also used for endoscopic purposes [16, 18]. Laser radiation parameters (wavelength, spectral bandwidth, intensity, duration) can be varied over very wide limits, thus making it possible to implement various types of light–biomatter interaction (linear and nonlinear, single and multiple photon, coherent and non-coherent, thermal and non-thermal, etc). Thus various effects in biotissues can be induced (photochemical modification, thermal destruction, explosive ablation, optical breakdown, shock pressure waves, photodisruption, etc) [21].

Generally, there are three main types of interaction between laser radiation and tissue (Figure 15.6) depending on irradiation power density and irradiation time: photochemical (biostimulation, phototherapy), thermal (coagulation, vaporization and photoablation), and electromechanical (photodestruction) [6, 21]. All these effects are used in medicine. The most frequently used mechanism of photon energy conversion in laser medicine is heating. Local heating of irradiated samples occurs with all the other effects (cutting, vaporization, coagulation, and ablation)

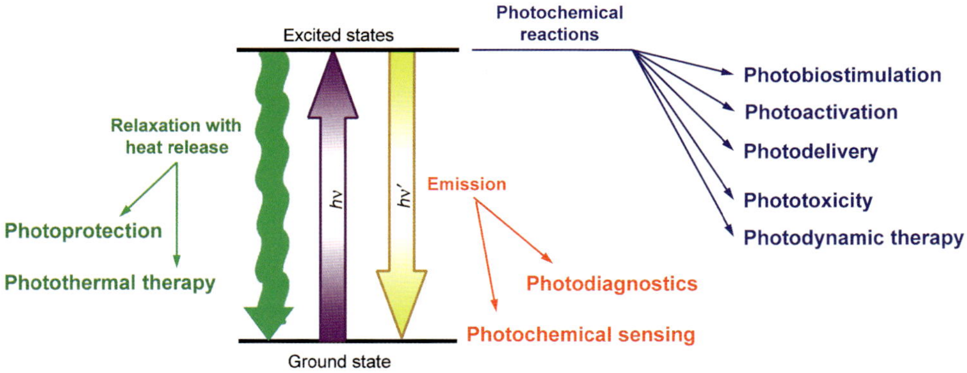

Figure 15.7 Application of photophysical and photochemical processes in biology, biomedicine, and medicine

[22]. At low light intensities the photochemical conversion of energy absorbed by a photoacceptor prevails [19, 22, 23]. This type of reaction is well known for specialized photoreceptor pigment species such as rhodopsin. In medicine light absorption by non-specific photoacceptor molecules is used rather extensively. The photoexcitation of endogenous or exogenous chromophores can result either in the re-emission of light (this may be used in photodiagnosis) or in partial conversion of the absorbed radiation into thermal or chemical energy (Figure 15.7). The last two processes generate photothermal and photochemical sensitization, respectively, leading to the irreversible damage of cells and tissues; hence they can be used for phototherapeutic purposes. The efficacy and scope of photodiagnostic and phototherapeutic techniques improve as knowledge about the influence of biological microenvironments on mechanisms of photoprocesses is developed. New approaches have been developed for controlling the biodistribution of photosensitizing agents in vivo and the depth of incident light penetration into tissues. In particular, even the localized and sublethal damage of cells can often be modulated to generate intracellular signalling in order to either induce a programmed cell death (apoptosis) or accelerate cell metabolism (photobiostimulation) [19, 23].

References

1. Vo-Dinh T. *Biomedical Photonics Handbook*. New York: CRC Press, 2003.
2. Palumbo G, Pratesi R. *Laser and Current Optical Techniques in Biology*. Cambridge: Royal Society of Chemistry, 2004.
3. Grossweiner L, Grossweiner J, Rogers G, Jones L. *The Science of Phototherapy: An introduction*. New York: Springer, 2005.
4. Lim HW, Soter NA. *Clinical Photomedicine*. New York: CRC Press, 1993.
5. Tuchin V. Light–tissue interaction. In: Vo-Dinh T, ed. *Biomedical Photonics Handbook*. New York: CRC Press, 2003: 3.1–3.26.
6. Letokhov VS. Laser light in biomedicine and the life sciences: From the presence to the future. In: Vo-Dinh T, ed. *Biomedical Photonics Handbook*. New York: CRC Press, 2003.

7. Mobley J, Vo-Dinh T. Optical properties of tissue. In: Vo-Dinh T, ed. *Biomedical Photonics Handbook*. New York: CRC Press, 2003: 2.1–2.77.
8. Batschauer A. *Photoreceptors and Light Signalling*. Cambridge: RSC, 2003.
9. Chance B. Optical method. *Annu Rev Biophys Chem* 1991;**20**(1):1–28.
10. Mueller G, Chance B, Alfano R, eds. *Medical Optical Tomography: Functional imaging and monitoring*. Bellingham, WA: SPIE Press, 1993.
11. Tuchin VV, ed. *Selected Papers on Tissue Optics Applications in Medical Diagnostics and Therapy*. Bellingham, WA: SPIE Press, 1994.
12. Minet O, Mueller G, Beuthan J, eds. *Selected Papers on Optical Tomography, Fundamentals and Applications in Medicine*. Bellingham, WA: SPIE Press, 1998.
13. Tuchin VV. *Tissue Optics: Light scattering methods and instruments for medical diagnosis. SPIE Tutorial Texts in Optical Engineering*. Bellingham, WA: SPIE Press, 2000.
14. Tuchin VV, ed. *Handbook of Optical Biomedical Diagnostics*. Bellingham, WA: SPIE Press, 2002.
15. Benaron D, Bigio I, Sevick-Muraca E, Yodh AG, eds. Special issue honoring Professor Britton Chance. *J Biomed Opt* 2000;**5**:115.
16. Brancaleon L, Moseley H. Laser and non-laser light sources for photodynamic therapy. *Lasers Med Sci* 2002;**17**:173–86.
17. Graczykowa A. *Fotodynamiczna meoda rozpoznawania i leczenia nowotworów.* [*Photodynamic Method of Diagnosis and Therapy of Tumors* (in Polish).] Warsaw: Bellona, 1999.
18. Kalka K, Merk H, Mukhtar H. Photodynamic therapy in dermatology. *J Am Acad Dermatol* 2000;**42**:389–413.
19. Karu T. Low-power laser therapy. In: Vo-Dinh T, ed. *Biomedical Photonics Handbook*. New York: CRC Press, 2003: 48.1–48.25.
20. Vo-Dinh T. Basic Instrumentation in Photonics. In: Vo-Dinh T, ed. *Biomedical Photonics Handbook*. New York: CRC Press, 2003.
21. Boulnois J-L. Photophysical processes in recent medical laser developments. A review. *Lasers Med Sci* 1986;**1**:47–51.
22. Gibson K, Kernohan W. Lasers in medicine – a review. *J Med Eng Technol* 1993;**17**(2):51–7.
23. Karu T. Primary and secondary mechanisms of action of visible to near-IR radiation on cells. *J Photochem Photobiol B: Biol* 1999;**49**:1–17.

16

Fluorescent and Chromogenic Sensing and Labelling

Equipped with his five senses, man explores the universe around him and calls the adventure Science.
Edwin Powell Hubble

The crucial recognition events of chemistry, biology, and materials science occur at a molecular level. Information about these events can be conveniently transmitted to us via light signals emitted by purpose-built molecular devices. There are three different families of fluorescent (and chromogenic) sensors (Figure 16.1) [1]:

- sensors that do not associate with the substrate, but the fluorescence quenching occurs on collision
- complexing fluorophores
- fluorophores linked with receptors.

The last family of devices encompasses usually complex assemblies of several molecular moieties: (1) receptor that binds the target species; (2) reporter that changes its optical properties upon target binding; and (3) linker that provides communication between receptor and reporter moieties. Besides this sensory role, such molecular devices are in fact molecular information-processing units. The response of such sensors can be easily described using Boolean logic; furthermore, these molecular logic devices operate at all the three information levels: syntactic, semantic, and pragmatic [2, 3].

Information at the syntactic level is concerned with the formal interrelationship of the elements of information, the rules of corresponding language, the capacity of communication channels, and the design of coding systems for information transmission, processing, and storage. The meaning of information and its practical meaning are neglected at this level. The semantic level relates information to its meaning and semantic units (words and groups of words) are assigned more or less precisely to

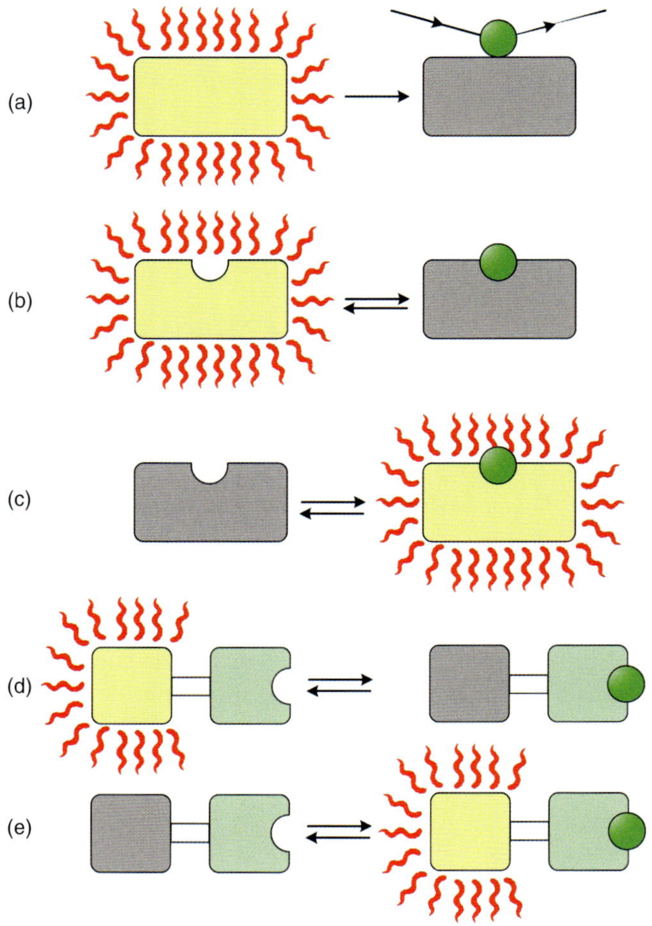

Figure 16.1 *Main classes of fluorescent molecular sensors of ions and molecules based on: (a) quenching of luminescence by collisions with analyte; (b) quenching or (c) increase of fluorescence of the complexing fluorophore; (d, e) the same processes in fluorophore-receptor assemblies [1]*

their meaning. For correct information processing at the syntactic level, semantics is not necessary. At the pragmatic level the information is related to its practical value. It strongly depends on the context and may be of economic, political, or psychological importance. Let us consider application of a fluorescent chemosensor in medical analysis. It performs logic operations (eg fluorescence switching in the presence of analyte) at the syntactic level. The result of this operation at a semantic level reports concentration of electrolytes, which in turn, at a pragmatic level, informs on the patient's condition.

The advantages of molecular luminescence for sensing and switching can be summarized as follows: high sensitivity of detection down to a single molecule, 'on–off' switch ability, feasibility of communication between human and molecule, subnanometre spatial resolution with submicrometre visualization, and submillisecond temporal resolution. Furthermore, many of the structural features that reduce fluorescence efficiency such as double-bond torsion, low energy $n\pi^*$ levels, 'heavy' atoms, weak bonds, and opportunities for photoinduced electron transfer (PET) or electronic energy transfer, have been delineated [4]. Apart from electronic level-related issues, the other important factors are substrate binding ability of the receptor and selectivity of this interaction [5].

Most chemosensors are based on molecular recognition, which can be defined as a process involving both selection and bonding of a substrate within the receptor. Usually the notions of binding and recognition are misunderstood. It should be emphasized that mere binding is not recognition; recognition can be formulated as binding with a purpose and should include a well-defined pattern of intermolecular interaction [6].

The highest stability of the sensor–analyte complex is achieved when the substrate fits perfectly in the 'hole' within the receptor. The analyte does not have to fit the receptor cavity perfectly; efficient binding can be achieved by careful design of the receptor. In the case of metal ion sensors the receptor must contain a proper type and number of donor atoms; angular orientation and directionality of lone electron pairs are also of crucial importance [5].

Metal ions play a quadruple role in fluorescent and chromogenic chemosensors: they are the most common substrates; some metal complexes are involved as receptors for anions and neutral molecules and some metal ions (of Ru, Re, Os, Eu, Tb) constitute fluorescent reporters. Furthermore, in some applications metal ions are structural elements of fluorescent chemosensors [7].

16.1 Cations as Targets in Biochemical Sensing

Colorimetric pH sensing has a long tradition and numerous chromogenic sensors such as phenolphthalein, bromothymol blue, methyl red, and many others, have been developed. These indicators are involved in protonation–deprotonation equilibria between two (or more) forms of different spectral properties. Much higher sensitivity can be obtained with fluorescent proton sensors. Furthermore, this technique can be widely used in bioanalytical chemistry, cellular biology, and medicine. Application of fluorescence imaging techniques provides spatial information on pH.

Information on pH can be obtained from fluorimetric determination of protonated and deprotonated forms of proton chemosensor (equation 16.1),

$$\mathrm{pH} = \mathrm{p}K + \log\frac{[\mathrm{B}]}{[\mathrm{A}]} + \log\frac{f_{\mathrm{B}}}{f_{\mathrm{A}}} \tag{16.1}$$

where pK is the equilibrium constant of protonation of the sensor, [A] and [B] represent concentrations of the protonated and neutral forms of an indicator, and f_{A} and f_{B} are the corresponding activity coefficients. Due to the last factor, the equilibrium constant depends not only on temperature, but also on ionic strength (at least one form of the sensor is ionic). Highly charged indicators such as pyranine are very sensitive to ionic strength. For less charged indicators, the effect is less significant, eg the p*K* value of fluorescein is almost independent of ionic strength. Furthermore, the p*K* equilibrium constant is sensitive to specific interactions depending on the chemical nature of the indicator and the surrounding medium (eg buffer constituents, composition of the physiological medium, vicinity of interfaces in microheterogeneous media). Structural changes of the medium also influence the properties of the indicator and should be taken into account. In particular, when probing microheterogeneous media (eg the vicinity of interfaces of micelles or lipid bilayers), the p*K* value differs significantly from the true p*K* value measured in dilute aqueous solutions [1].

Depending on the photophysical properties the pH fluorescent sensors can be divided into three classes: (1) chemosensors that undergo photoinduced proton transfer, (2) chemosensors that undergo photoinduced electron transfer, and (3) those sensors that undergo neither proton nor electron transfer.

Most of the fluorophores belonging to the first group are much more acidic in the excited state than in the ground state (ie excitation results in weakening of the O–H bond; see Figure 16.2a); in other words the p*K* value in the excited state is much lower than that in the ground state, so in a pH range around the ground state p*K*, the emitting form is always the basic form because excitation of the acidic form is followed by excited-state deprotonation. The fluorescence spectrum is thus unchanged, in contrast to the excitation spectrum [1].

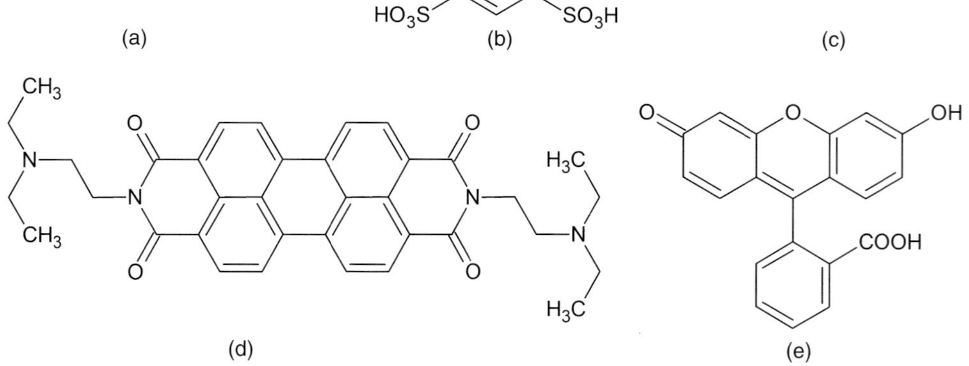

Figure 16.2 *Typical structures of fluorescent proton sensors: (a) hydroxycoumarin, (b) pyranine, (c) anthracene derivative, (d) perylenebisimide, and (e) fluorescein*

Operation of other sensors is based on photoinduced electron transfer. The fluorescence quantum yield of these fluorophores is very low when they are in the non-protonated form because of internal quenching by electron transfer. Protonation (which suppresses electron transfer) induces a very large enhancement of fluorescence. The bandshapes of the excitation and fluorescence spectra are independent of pH [1]

The last group of fluorescent sensors is based on neither photoinduced proton transfer nor photoinduced electron transfer. The best-known example of this kind of molecular device is fluorescein (Figure 16.2e). The evolution of the fluorescence spectrum versus pH should be similar to that of the absorption spectrum. In other words, when increasing the pH, the absorption and emission bands of the acidic form should decrease with a concomitant increase in the absorption and emission bands of the basic form [1].

Protons are relatively simple targets for sensor molecules and do not require engineered receptors, however, achievement of selective interactions with other chemical species requires much more elaborate receptors. In the most cases cations are bound via electrostatic or coordinative interactions within the receptors: alkali metal cations, which are rather poor central ions and form only very weak coordination bonds, are usually bound within crown ethers, azacrown macrocycles, cryptands, podands, and related types of receptor moieties with oxygen and nitrogen donor atoms [8]. Most of the common cation sensors are based on the photoinduced electron transfer (PET) mechanism, so the receptor moiety must have its redox potential (HOMO energy) adjusted to quench luminescence of the fluorophore (Figure 16.3).

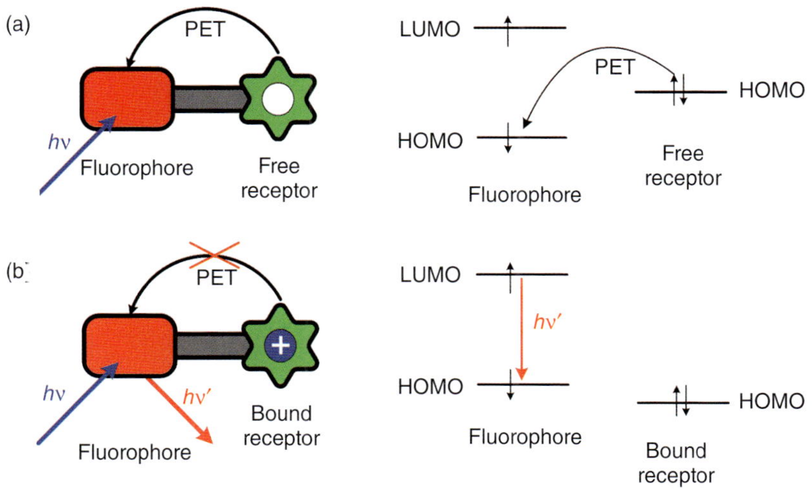

Figure 16.3 *Principle of the PET (photoinduced electron transfer) chemosensor. (a) The HOMO level of the unbound receptor acts as an electron donor and effectively quenches the fluorescence of the sensor. (b) Upon coordination of the substrate, the energy of the HOMO level of the receptor is decreased due to electrostatic or coordinative interaction with cationic species and PET quenching is no longer possible*

The principle of the PET-based sensor is shown in Figure 16.3. In most cases the sensor consists of a receptor, which selectively and reversibly binds a substrate, a fluorophore that provides optical communication between the sensor and the outer world, and a spacer (linker) that binds both components together and provides electronic communication between the receptor and the fluorophore, if the process is thermodynamically and kinetically feasible [9]. In most instances the kinetic restrictions are minor although there are some exceptions. Importantly, the rate of electron transfer is much faster than the luminescence when PET is thermodynamically allowed (Figure 16.3a). Luminescence and PET are the two main competitors that deactivate the excited state of the fluorophore. Binding of the substrate to the receptor drastically alters the thermodynamics to the endoergonic situation (Figure 16.3b). In the simplest case the change of the energetics of the system is caused by electrostatic interactions between the receptor (electron-rich moiety) and the analyte (cationic).

Another large category of optical chemosensors is based on photoinduced charge transfer (PCT) process. In contrast to PET-based sensors the receptor and fluorophore (chromophore) moieties are connected in a way that provides extensive orbital delocalization between these two parts. One end of such a molecule is electron rich, whereas the other is electron poor; upon interaction with the environment the electron distribution may significantly change, thus changing the optical properties of the switch. In such 'push–pull' systems the excitation leads to redistribution of electron density and generation of a dipole moment. If the receptor binds a charged trigger species, the additional charge interacts with a photogenerated dipole, thus modifying the fluorescence spectrum. Repulsive interaction results in a hypsochromic shift (Figure 16.4a) of absorption and emission bands, whereas attraction results in a bathochromic shift (Figure 16.4b).

The main advantage of PCT switches over PET switches consists in the possibility of usage of several wavelengths to analyze the state of the sensor, which is essential especially in intracellular cation sensing [10].

Furthermore, a large family of fluorogenic chelating ligands is also used as chemosensors (see Figure 16.1c). These include various derivatives of quinoline, dansyl sulphonamides, and other compounds [11]. In this case, fluorescence quenching results from intra- and intermolecular (eg with solvent) hydrogen bonds. Binding of the analyte (eg Zn^{2+}) destroys hydrogen bonds and the luminescence of the ligand is restored.

16.1.1 Cations Common in Biological Systems

One of the first alkali metal cation chemosensors was based on the molecular architecture of the anthracene derivative (Figure 16.2c) [9]. Substitution of the simple tertiary amino group with an azacrown macrocycle resulted in fluorescence of the chemosensor responding to protons and potassium cations (Figure 16.5a). Further increase in selectivity of the sensor can be achieved via reduction of Brønsted's basicity of the macrocycle. On the other hand, this modification must retain the electron donor character of the receptor in order to preserve the PET mode of

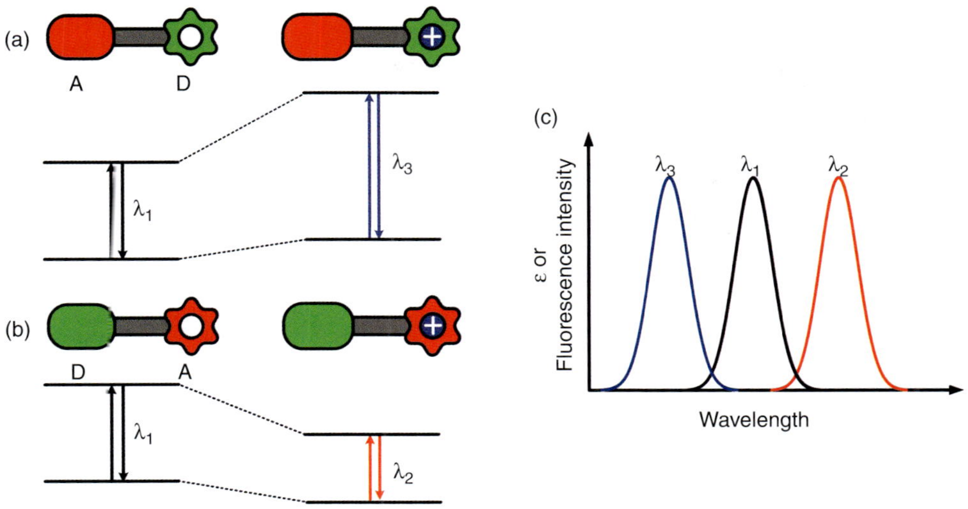

Figure 16.4 *Principle of the PCT (photoinduced charge transfer), chemically driven, luminescent, molecular sensor based on the donor–spacer–acceptor architecture. (a) Binding of analyte trigger to the donor (green) moiety results in hypsochromic shift of absorption (emission) band; (b) binding of the same analyte to the acceptor moiety (red) results in bathochromic shift of corresponding transition*

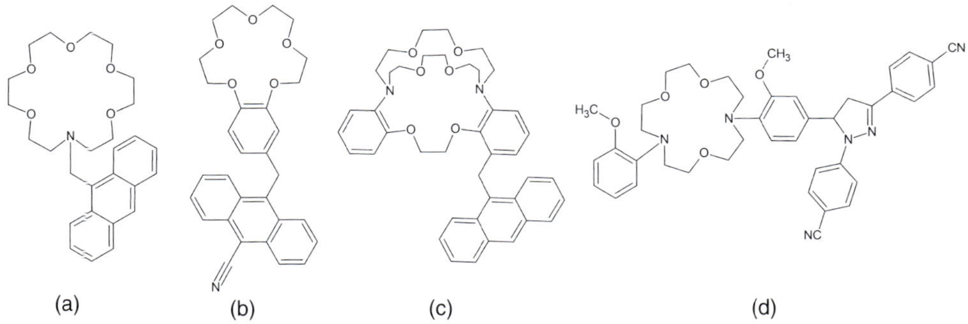

Figure 16.5 *Development of the crown receptor in alkali metal fluorescent sensors: (a) azacrown, (b) benzocrown, (c) cryptand, and (d) pseudocryptand receptors*

operation. This can be achieved by application of benzocrown ethers, which are good ligands for alkali metal cations, whereas the presence of dialkoxybenzene moiety provides proper HOMO energy for the receptor. An example of such a sensor, selective for Na^+, is shown in Figure 16.5b. Thus, fluorescent signalling systems obtained show a specific interaction with alkali metal cations with virtually no interference with protons. The crown ether receptor can be further extended to yield the three-dimensional cryptand receptor (Figure 16.5c). This modification improves the coordination ability of the ligand and reduces the possibility of axial coordination of solvent molecules to the target ion, although introduction of the nitrogen atoms may result in increased affinity towards protons. An efficient way to diminish the basicity of the tertiary nitrogen atoms is to couple them with aromatic rings, as was done in the case of the sensor shown in Figure 16.5c. This particular assembly brings another beneficial effect: on binding of the target ion the conformational changes electronically decouple the nitrogen atoms and the aromatic rings, resulting in more efficient fluorescence switching compared with a purely electrostatic effect [9]. This effect is especially important when sensing large cations such as K^+ and Rb^+. Although efficient, the cryptand receptors are relatively difficult to synthesize. A good alternative is the pseudocryptand receptors containing diazacrown macrocyclic receptor supplemented with additional donor atoms on side arms (Figure 16.5d). This molecular architecture combines the beneficial steric and electronic properties of cryptands and the synthetic accessibility of azacrown ethers [9].

Although macrocyclic receptors containing N and O donor atoms are suitable for detection of alkali metal cations, the detection of alkali earth cations required application of chelating receptors based on amino acids with EDTA-like spatial arrangement, eg bis(*o*-amino-phenoxy)-ethane-*N,N,N′,N′*-tetraacetic acid (BAPTA) [9, 12]. The most common calcium PET sensors are shown in Figure 16.6.

In biological systems, where autofluorescence of the sample may be a severe obstacle for application of PET sensors, the PCT-based sensors are very useful. INDO-1 and FURA-2 are the best examples of such molecular devices (Figure 16.7).

In biological studies, apart from alkali and alkaline earth cations, zinc sensing is very important, especially in neuroscience. Binding of zinc and other (often toxic) transition metal ions requires receptors of different structure and coordination properties. Polypyridines, dendritic pyridines, and thiacrown ethers are the receptors of choice.

Fluorescent sensing of zinc cations often involved quinoline and dansyl sulphonamide fluorogenic chelating ligands [11]. The most successful sensors are based on sulphonated aminoquinoline moiety and application of the TSQ sensor is usually recognized as a milestone in the development of fluorescent probes for Zn^{2+} ions in biological systems (Figure 16.8). The fluorescence of the ligand is too weak to be observed, whereas its zinc(II) complex, on ultraviolet (UV) excitation at pH 7.4, emits strong fluorescence with a maximum at 495 nm. As a pH-independent fluorescent probe, TSQ has proved to be a selective, non-toxic, fluorescent sensor for the detection of biological zinc(II), and will remain the most widely used compound in biological zinc detection. Further improvement of this breakthrough sensor was achieved via introduction of carboxylic or ester groups to extend the 6-methoxyl functionality (Figure 16.8). This modification rendered the probe membrane permeable. A significant limitation to the wide application of quinoline-based zinc sensors is the necessity for excitation with UV light, which is strongly absorbed by some biomolecules and may have a deleterious effect on some biostructures. Furthermore, sensors with higher selectivity were required for biological studies. The former problem can be solved by application of visible light-active fluorophores and the PET principle of fluorescence modulation, whereas the latter can be solved by application of another zinc chelator, eg bis(2-pyridylmethyl) amine (DPA) or *N,N,N′,N′*-tetrakis(2-pyridylmethyl)-ethylenediamine (TPEN), neutral chelators highly specific for Zn^{2+} ions.

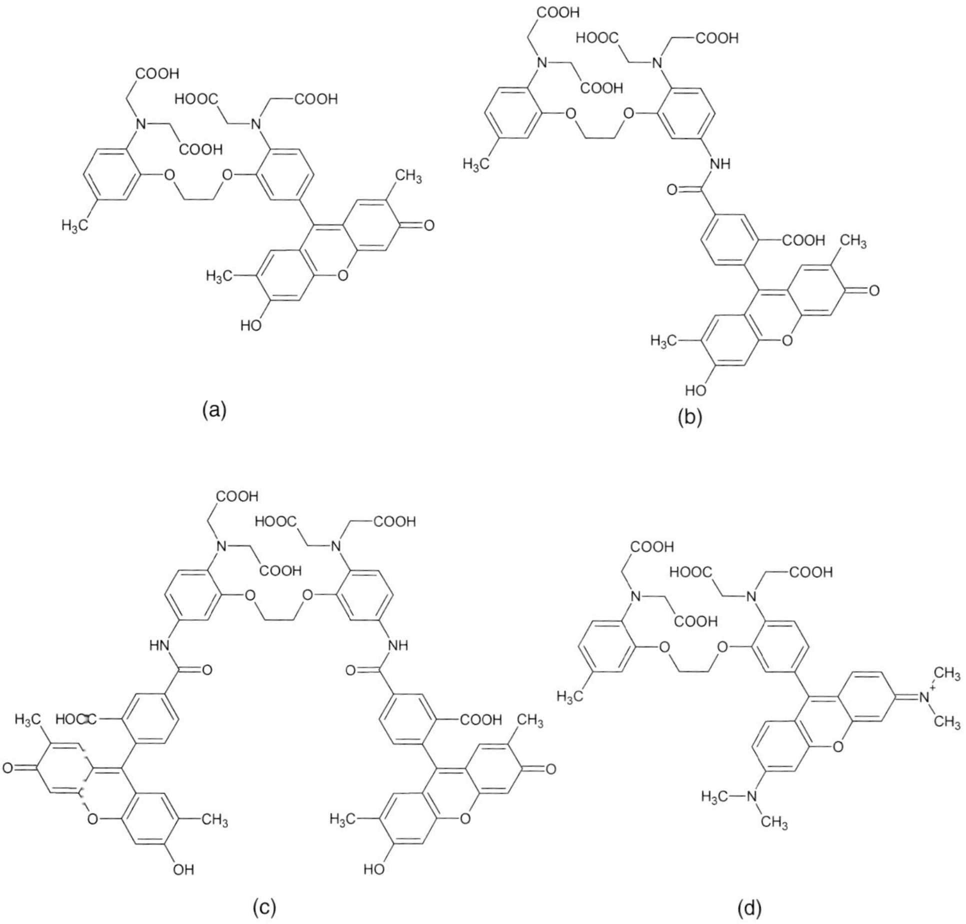

Figure 16.6 *Molecular structures of typical calcium PET (photoinitiated electron transfer) sensors: (a) FLUO-3, (b) Calcium Green 1, and (c) Calcium Green 2 are commonly used in medical analytics*

Application of the PET principle in fluorescent sensing of zinc ions resulted in a whole family of different sensors (Figure 16.9). As a proof of principle the anthracene derivative (Figure 16.9a) was tested as a zinc fluorescent sensor. It binds zinc in a 1:1 ratio and this results in a 77-fold increase in its fluorescence quantum yield, but the fluorophore required UV excitation. Visible light can be used for the first PET zinc sensor, Newport Green (Figure 16.9b). Coordination of Zn^{2+} ions results in a fourfold increase of fluorescence intensity. Due to only one chelating moiety the sensitivity of Newport Green is relatively low and requires micromolar zinc concentrations. Higher sensitivity can be achieved when two zinc-binding groups are present in the sensor molecule, as is the case for the Zinpyr family of chemosensors (Figure 16.9c). Unfortunately, they can form both 1:1 and 1:2 complexes, which makes the quantitative assay much less reliable, although the sensors are suitable

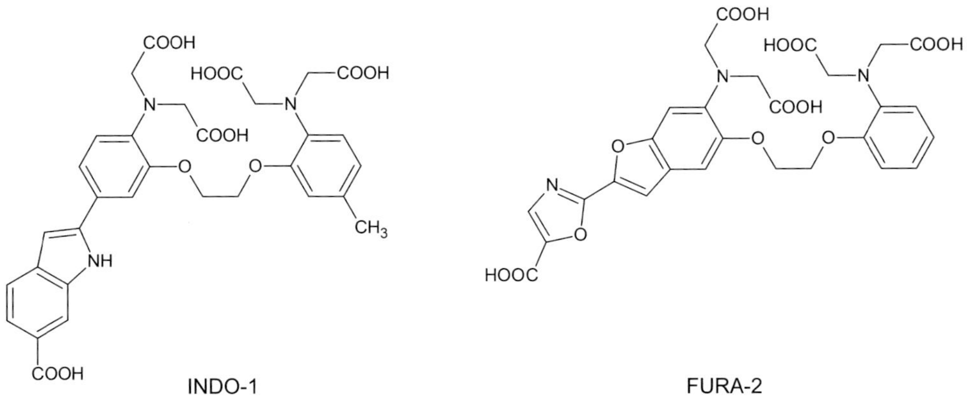

INDO-1 FURA-2

Figure 16.7 *Molecular structures of the most common calcium PET (photoinitiated electron transfer) sensors*

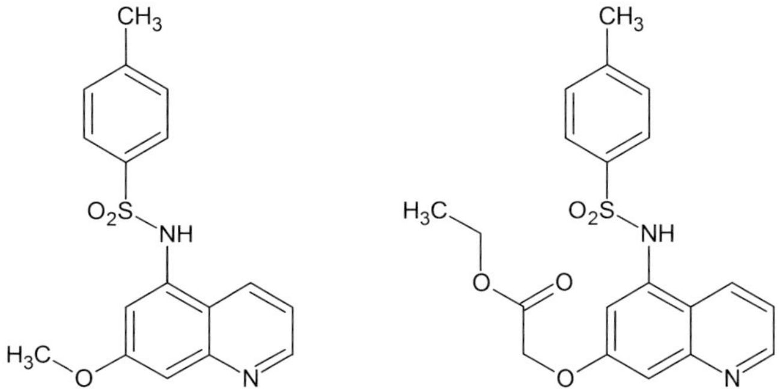

TSQ Zinquin

Figure 16.8 *Molecular structure of TSQ and Zinquin fluorescent sensors*

for intracellular zinc imaging. The speciation problem can in turn be solved by introduction of a secondary fluorophore (eg coumarin). The resulting sensor Coumazine-1 (Figure 16.9d) can be used for ratiometric determination of zinc ions in biological samples [11].

Linear and cyclic polyamines [13], as well as linear oligoesters [14], are also efficient chelators for zinc ions, used especially for zinc imaging in cultured cells. Depending on the target cell and/or intercellular structure, various sensor architectures are applied.

Another metal present in biological media in relatively high concentrations is iron. Most iron ions are strongly coordinated within proteins, although a small fraction of iron ions is loosely bound with carboxylates and phosphates, phospholipids, glycans, and sulphonates. This fraction is usually called labile iron and constitutes a pool of iron ions accessible for incorporation into functional metalloproteins. In contrast to closed shell cations, rather than increasing the fluorescence quantum

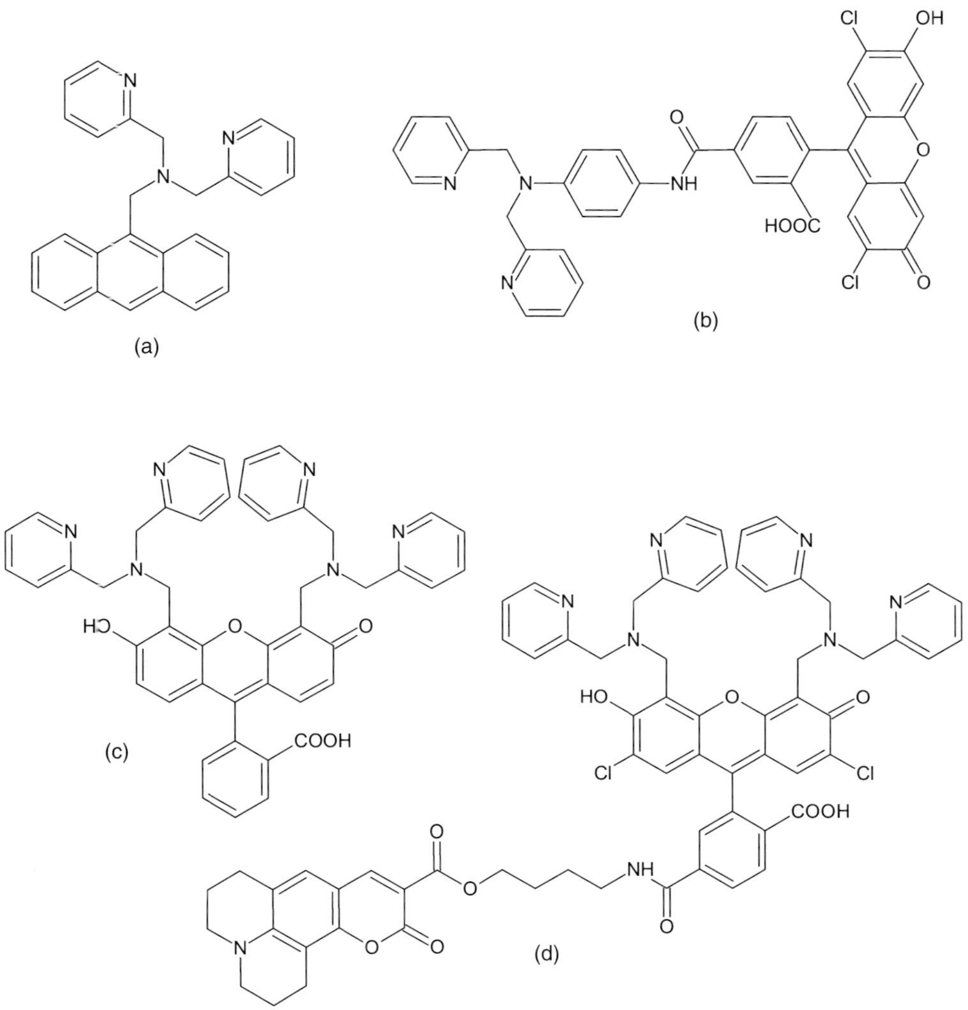

Figure 16.9 *Molecular structures of* Zn^{2+} *fluorescent sensors based on the bis(2-pyridylmethyl)amine (DPA) chelator*

yield iron ions usually completely quench the luminescence of the fluorescent label used in the chemosensor. The quenching processes involve both energy and electron transfer, but first of all the paramagnetic character of iron ions bound within the sensor is responsible for the non-radiative deactivation of excited states [15–17]. Despite these differences, the general architecture of fluorescent iron sensors follows the scheme presented above. The sensors contain fluorescent reporter (usually fluorescein or nitrobenzodiazole moieties) and a chelating group: phenanthroline (Figure 16.10a), branched amino acids (Figure 16.10b), or desferrioxamine (Figure 16.10c,d) [16]. Unbound sensor molecules are characterized with high fluorescence quantum yields (0.7–0.9, comparable with unmodified fluorophores); on iron binding the quenching amounts to 70–90%.

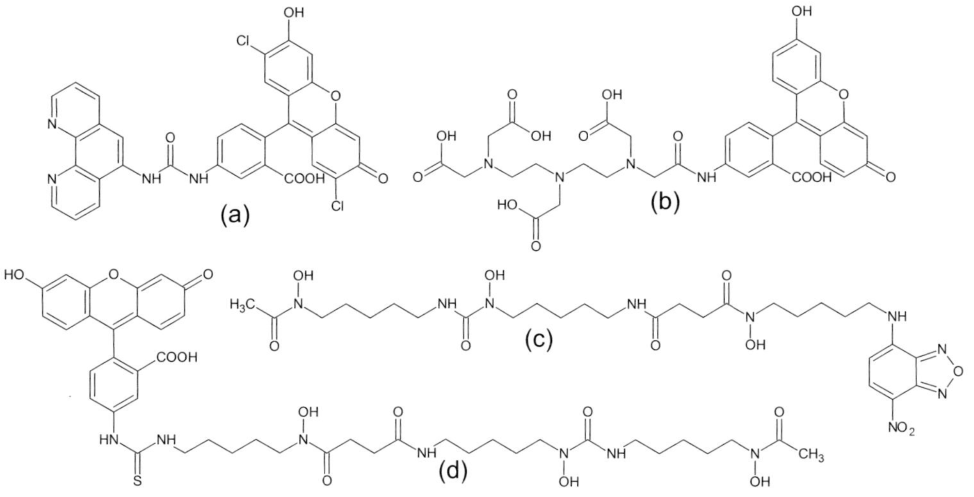

Figure 16.10 *Molecular structures of Fe^{2+} fluorescent sensors*

16.1.2 Fluorescent Detection of Toxic Cations

Apart from biologically important ions, such as Na^{+}, Ca^{2+}, Zn^{2+}, and Fe^{2+}, fluorescent sensing is essential for detection and quantification of toxic elements, such as mercury and cadmium. Some cadmium sensors, especially those with long polyamine chains are also sensitive to zinc, their spectral response is virtually identical in the case of both ions and therefore they cannot be used for determination of cadmium in the presence of zinc ions. Much higher selectivity is achieved with softer receptors containing sulphur donor atoms, although some other architectures or uncommon coordination modes are also very efficient. Unexpectedly the anthracene derivative containing amino acid moieties is much more sensitive for Cd^{2+} than for Zn^{2+} cations (Figure 16.11a). Introduction of additional phenyl rings results in more complex behaviour (Figure 16.11b). Both Zn^{2+} and Cd^{2+} interact with the sensor. Coordination of zinc ions switches the monomer emission on, whereas coordination of cadmium results in strong excimer emission [18].

A thiacrown receptor coupled with anthracene fluorophore (Figure 16.11c) yields a selective copper sensor. Although it binds other heavy metal ions (eg silver) only copper induces strong fluorescence quenching due to good matching of redox potentials. Cu^{2+} within the tetrathiamacrocycle is the primary electron acceptor that adopts one electron from the anthracene moiety on excitation, thus leading to complete fluorescence quenching [19].

Lead(II) cations can be selectively detected with the azacrown derivative shown in Figure 16.11d. This ligand can coordinate many heavy metal cations, but only lead induces strong changes in its absorption and emission parameters. Introduction of sulphur atoms to the macrocycle receptor makes this molecule sensitive to Hg^{2+} exclusively (Figure 16.11e) [20, 21]. A similar receptor coupled with diazobenzene (Figure 16.11f) and deposited on silica (or glass) surface makes a very good colorimetric indicator for Hg^{2+} ions, which in the presence of NO_3^- changes colour from

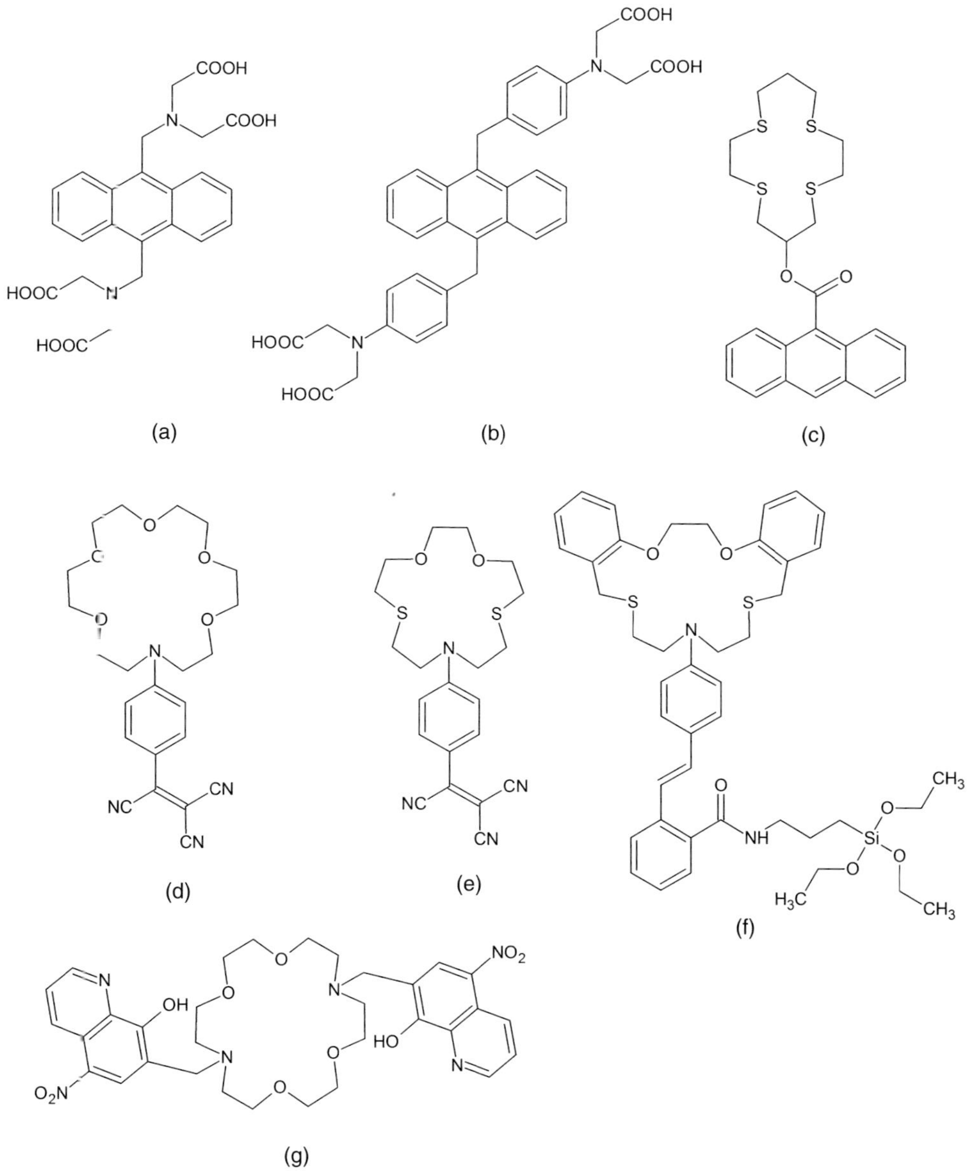

Figure 16.11 *Molecular structures of Cd^{2+}, Pb^{2+}, Cu^{2+}, and Hg^{2+} fluorescent sensors*

yellow–orange to purple–red. Another fluorescent mercury sensor is based on the quinoline derivative shown in Figure 16.11g. This ligand coordinates many transition metal ions, but only the Hg^{2+} complex is luminescent; moreover, the equilibrium constant for Hg^{2+} complex is two orders of magnitude higher than for any other cation [22]. Surprisingly the central azacrown moiety does not act as a receptor, but provides only a scaffold and serves as proton acceptor.

16.2 Fluorescent and Chromogenic Sensing of Anions

Although the operation of most cation sensors is based on metal–ligand interaction and the resulting changes in the electronic structure of the reporter moiety, anion sensing is more complex. First of all, there are numerous anions that are only poor ligands, and even good coordinating ligands (eg carboxylates and phosphates) exhibit very similar coordinating properties irrespective of the structure of the anion's side chain. Therefore, some new strategies for specific and selective binding of anions have been developed. The receptors for anions are usually based on combination of the following interactions: hydrogen bonding, electrostatic interaction, and Lewis acid–base complexes [23, 24]. The last involve both coordination of the target anion to the coordinatively unsaturated Lewis acid (metal centre, boron, or silicon atoms) or displacement of weakly bound ligand by the analyte. Therefore anion sensing and recognition in the most cases involve coordinately unsaturated complexes of a transition metal or lanthanide [23–29], protonated azapodands or azamacrocycles, and substituted ureas, thioureas, and guanidines [30], with the conformation preferably binding only one anion [24, 26, 31]. The sensors from the last group usually contain a metal-based fluorophore (lanthanide or ruthenium(II) polypyridine complexes). The output of the sensor depends on the perturbation of the excited state properties on the metal-based fluorophore by the analyte. The nature of excited state bonding between the receptor and the analyte may change the order of energy levels and switch between luminescent and non-luminescent forms [24]. Metal-centred (MC) and metal-to-ligand charge transfer (MLCT) excited states are susceptible to stabilization or destabilization through interactions that perturb the energetics of the metal- or ligand-based orbitals. Changes in the coordination environment of a metal centre affect the energetics of the metal-based orbitals, which in turn can alter the energy of MC, MLCT, or ligand-to-metal charge transfer (LMCT) transition involving these orbitals. Depending on the complexity of the excited state manifold of the complex, the result of such processes can range from subtle changes in emission energies to switching between emissive and non-emissive forms [24]. The latter is especially interesting from the analytical chemistry point of view because it can yield the highest signal-to-noise ratio.

16.2.1 Common Anions

Among all the inorganic anions, phosphate, and phosphoric acid esters earned the highest attention due to their biological importance [32] or extreme toxicity (sarin, soman, tabun, and other nerve warfare agents) [33]. Also fluoride anion is a common target for fluorescent anion sensing due to health issues, and its involvement in the production of chemical and nuclear weapons [34]. In most cases the selectivity of the fluorescent and chromogenic anion sensors is relatively low and several species are bound by the receptors with comparable equilibrium constants. Fluoride anion is an exception among halide and chalcogenide anions, and there are several fluorescent sensors of high selectivity.

Probably the most ubiquitous anion receptors are urea and thiourea moieties. Depending on the substituents, they strongly interact with fluoride (F^-), hydrogen phosphate (HPO_4^{2-}), and carboxylates ($RCOO^-$). Fluorescent anion sensors contain a fluorescent reporter (usually anthracene, pyrene, or naphthalene moieties, ruthenium polypyridyl complexes), whereas chromogenic sensors contain strong chromophoric groups: nitrophenol, azobenzene, or indoaniline [30, 31, 35].

PET-based anion sensors are relatively rare; there are several reports of such molecular devices. The most common architecture encompasses a fluorophore (anthracene or naphthalene moieties) coupled with one or more (thio)urea groups (Figure 16.12a,b), and shows significant fluorescence quenching upon binding of fluoride, phosphate, and acetate [30]. Another possibility involves two (or more) fluorophores coupled with one molecular receptor (Figure 16.12c,d). Interestingly, subtle changes in the geometry of the receptor can increase the phosphate-binding constant by an order of magnitude as was observed in the case of the sensor shown in Figure 16.12d [30]. High selectivity for phosphate can also be achieved using a bisimidazolium receptor (Figure 16.12e) [36]. Slight modification of this receptor yields a chemosensor suitable for pyrophosphate detection (Figure 16.12f) [37]. Application of ruthenium bipyridine fluorophore brings significant improvement to the operation of the anion sensors. Instead of luminescence quenching (which may be induced by numerous processes) an increase in emission quantum yields is observed [31]. Simple sensor molecules (Figure 16.13a,b) cannot distinguish between hydrogen phosphate and chloride, but large supramolecular assemblies (Figure 16.13c,d) show much higher selectivity. Operation of these sensors is based on several parallel processes. Electrostatic interaction of the anion with the ruthenium complex induces an increase of excited state-ground state separation (observed as a hypsochromic shift of emission band), thus leading to a decrease of both the Franck–Condon factors and the non-radiative decay rate. Furthermore, the increased rigidity of the molecular assembly inhibits the non-radiative decay of the ^{3}MLCT excited state because it hampers amide oscillation modes [31].

Simple amide (Figure 16.13a) or macrocyclic amide (Figure 16.13b) receptors do not provide satisfactory selectivity of the sensor and the responses to chloride and hydrogen phosphate are comparable. Larger macrocyclic amide receptors based on calixarene architecture are much preferred to hydrogen phosphate presumably due to multiple hydrogen bonds between the anion and macrocyclic receptor [31]. An interesting modification of this type of sensor was presented by Beer (Figure 16.13d) [38]. Incorporation of two benzo(15-crown-5) moieties enables external control over selectivity of the sensor. In the absence of potassium cations the receptor's cavity is large and can easily accommodate a large phosphate anion. In the presence of potassium, a sandwich-type complex is formed that brings about dramatic modification of the receptor. It looses its sensitivity towards phosphate, but at the same time becomes selective to chloride anions. High selectivity towards phosphate can be also achieved via incorporation of metallocene moieties (Figure 16.13e).

Further changes in operation of these sensors can be easily achieved by substitution of side chains: incorporation of two imidazole moieties (Figure 16.13f) results in a sensor that specifically binds phosphodiesters [31], whereas incorporation of a

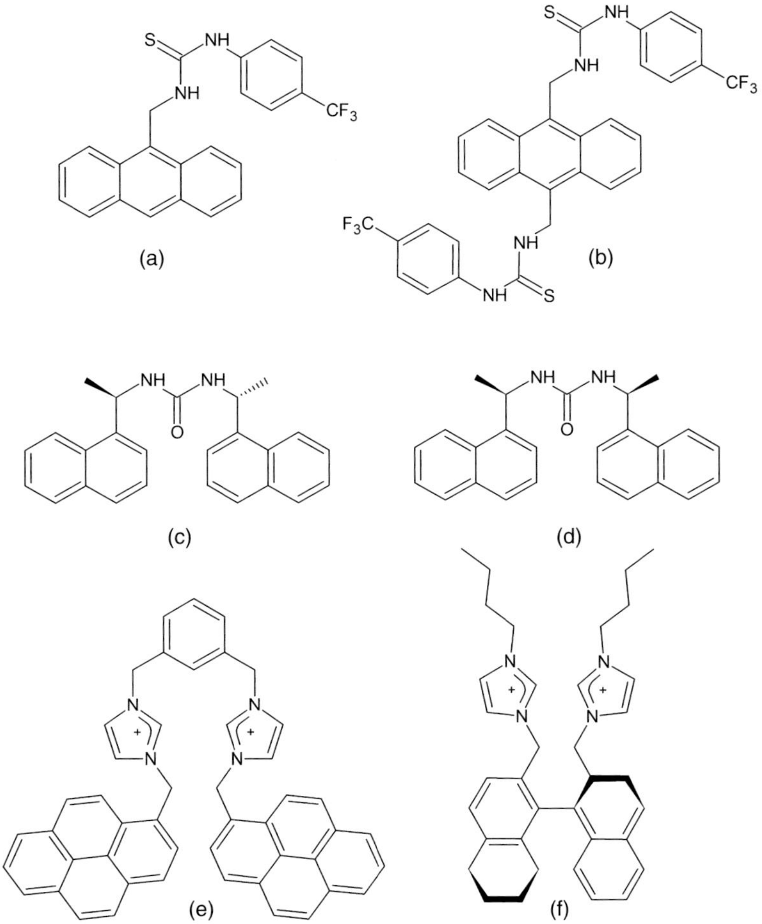

Figure 16.12 *Molecular structures of selected PET (photoinitiated electron transfer)-based fluorescent phosphate sensors*

1,4,8,11-tetraazacyclotetradecane moiety yields receptors sensitive to ATP (Figure 16.13g) [39].

More elaborate receptors for ATP contain two zinc ions complexed by dipicolylamine (DPA) ligands appended to an anthracene fluorophore (Figure 16.14a) [40]. This sensor is also reported to respond to phosphorylated amino acids and peptides [41]. On binding of the target molecule the fluorescence of the anthracene fluorophore is switched on. A zinc–DPA complex was also used in pyrophosphate sensors (Figure 16.14b) [42]. The interaction with pyrophosphate switched the excimer emission on due to stabilization of the dimeric structure. Various phosphates can also be determined using imidazolium-based sensors. Specific interaction with guanosine triphosphate quenches the luminescence of the anthracene derivative (Figure 16.14c) [43], whereas the luminescence of the 2-phenyl-3*H*-indole is enhanced (Figure 16.14d) [44].

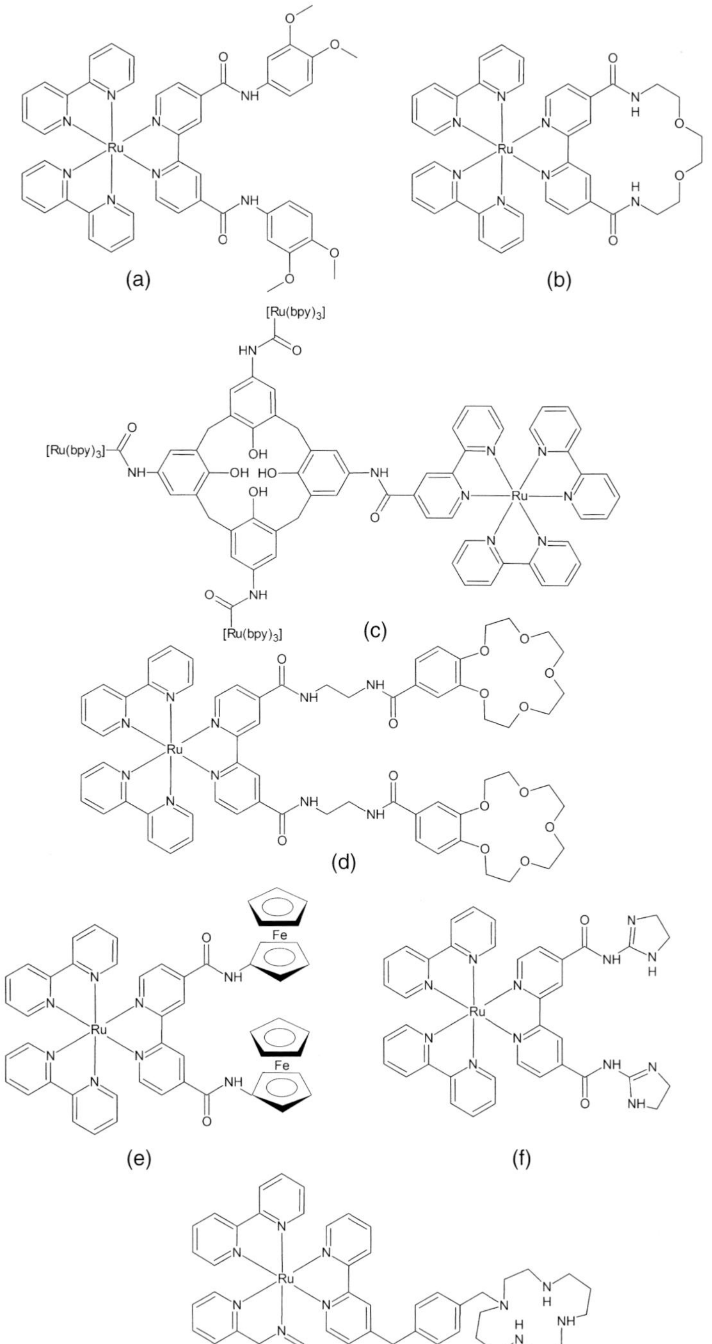

Figure 16.13 *Structures of Ru(bpy)$_3$-based fluorescent phosphate and phosphoric acid ester sensors*

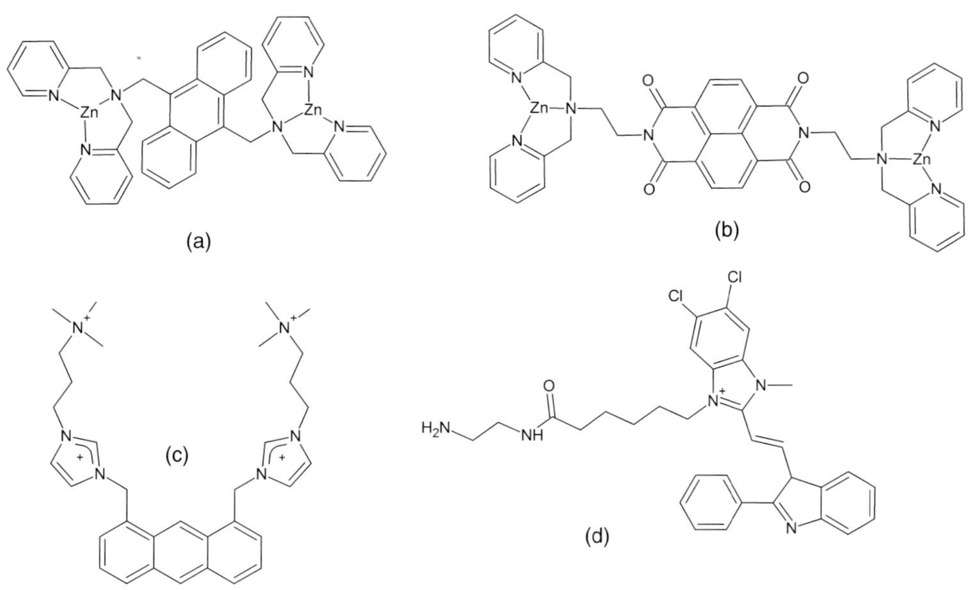

Figure 16.14 *Structures of the fluorescent (a, b) ATP and (c, d) GTP sensors*

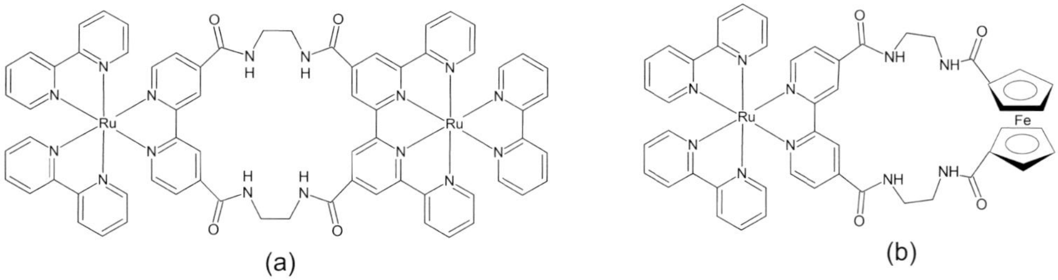

Figure 16.15 *Structures of Ru(bpy)$_3$-based fluorescent chloride sensors*

Introduction of another ruthenium bipyridine moiety or bridging metallocene (ferrocene, cobaltocenium) results in sensors that specifically bind chloride anions (Figure 16.15). The structural modification of the amide receptor results in a decease in cavity dimensions and significant rigidity of the macrocycle. Therefore it cannot accommodate hydrogen phosphate anions, but only much smaller Cl^- [45, 46].

16.2.2 Toxic Anions

Another important target for fluorescent sensors are highly toxic phosphates, found in various pesticides and chemical warfare agents, commonly known as *nerve gases* (Figure 16.16) [47]. There are two main families of these chemical warfare agents: G and V family. The former contains electrophilic substituent at phosphorus atom (fluoride or cyanide), whereas in the latter short aliphatic chains are connected directly to the phosphorus atom. Due to their extremely high toxicity most studies are performed on less toxic models, which mimic the most important molecular features of the analytical targets (Figure 16.16c). There are two main analytical strategies towards the nerve gases: make use of their high affinity for lanthanide cations and chemical reactivity of some phosphonate esters.

Phosphorylation and subsequent cyclization of aliphatic alcohols were utilized by Pilato et al. for construction of the first fluorescent detector of G-class nerve gases [48]. Virtually non-fluorescent ($\Phi = 0.00001$) platinum dithiolene complex reacts with fluoro-, chloro-, and cyanophosphonates, yielding an unstable ester, which undergoes rapid cyclization to produce a complex with coordinated planar quinolizinium cation (Figure 16.17a). The resultant complex is fluorescent ($\Phi = 0.01$ in dimethylsulphoxide, $\Phi = 0.2$ upon immobilization). Analogous reactions were used in the case of all-organic fluorogenic sensors of nerve gases (Figure 16.17b,c,d) [49]. The reactions of these compounds with nerve gas mimics the results in formation of fluorescent quinolizinium derivatives. The naphthalene derivative (Figure 16.17d) is fluorescent in the open form as well, which makes it suitable for

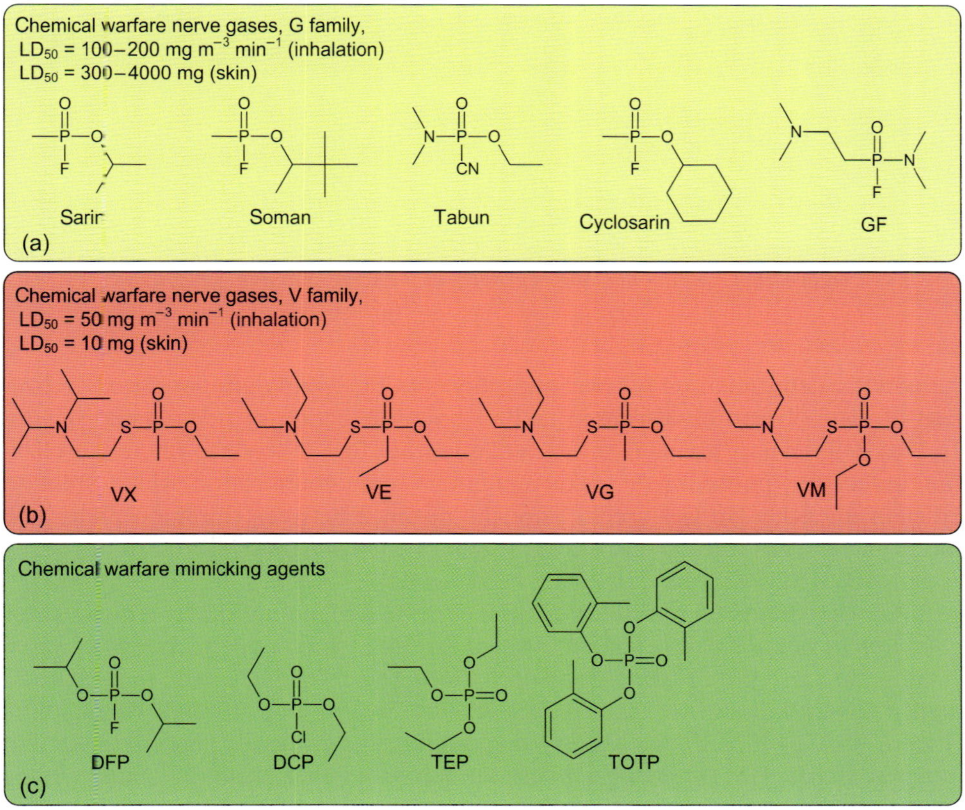

Figure 16.16 Structure and toxicity data of most dangerous chemical warfare agents and their low-toxicity mimicking agents; LD_{50}-lethal dose

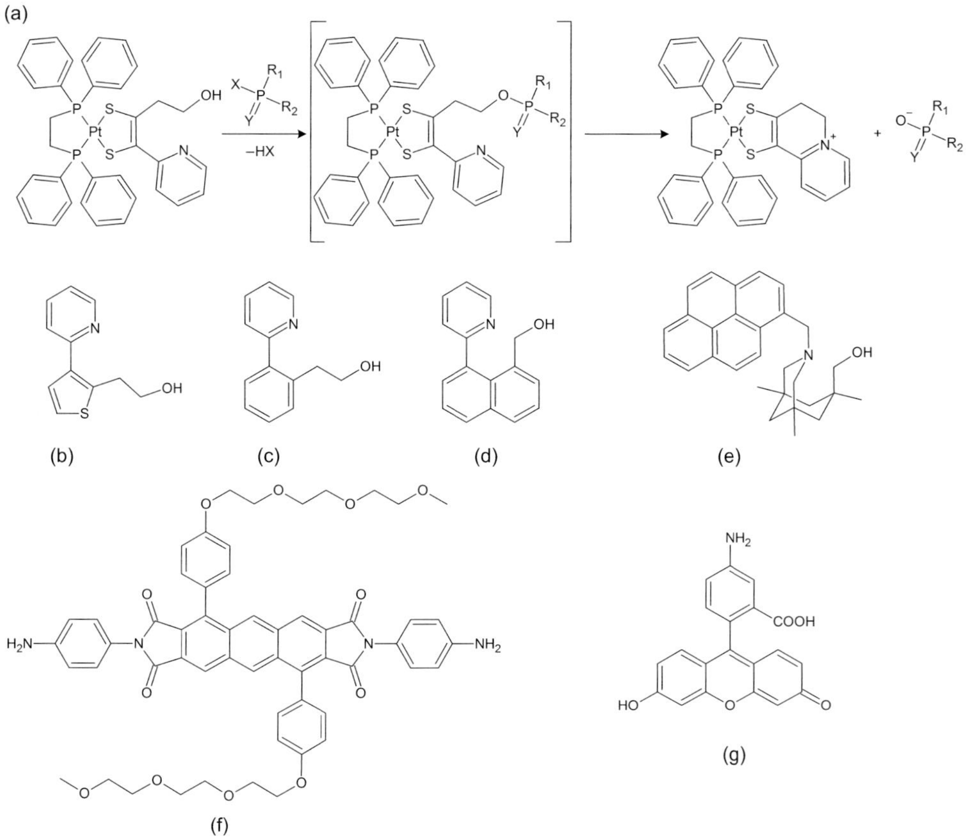

Figure 16.17 *Structures and reactivity of fluorogenic nerve gas sensors*

ratiometric determination of toxic phosphonates. Phosphorylation-induced cyclization of alcohols can be also used in PET-based detection of nerve gases (Figure 16.17e) [50]. Tertiary amine is an efficient quencher for anthracene, whereas cyclization induced by phosphonate yields quaternary ammonium cation, which is not an efficient quencher. In this architecture the distance between the fluorophore and the quencher is a critical issue: incorporation of a longer linker decreases the utility of the sensor, because the quenching becomes less effective.

Another approach for the development of PET-based molecular indicators for detection of nerve agents utilizes phosphorylation of an amine functionality, which is integrated into the structure of a fluorophore. This principle is used in the case of highly substituted anthracene bisimide (Figure 16.17f) [51] and fluoresceinamine (Figure 16.17g) [52]. These sensory compounds can be easily incorporated into polymer matrices [51] or deposited onto the surface of silica beads [52], thus yielding solid state detectors.

The strong affinity of aliphatic phosphates for lanthanide cations was explored in the context of fluorescent warfare sensing. Complexes of fluorescent ligands

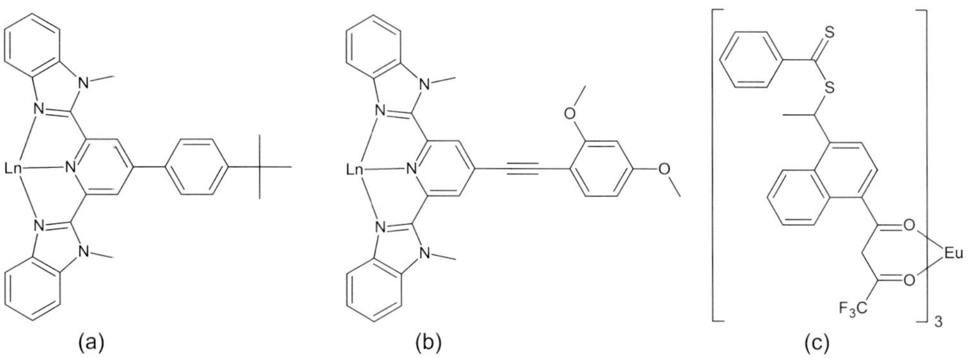

Figure 16.18 *Lanthanide-based fluorescent nerve gas sensors*

(Figure 16.18a, b) [53] with lanthanum(III) and europium(III) show ligand-centred (green) and metal-centred (red) emission, respectively. On organophosphate binding the ligands are liberated and only blue luminescence of the free ligand can be observed. These complexes can also be deposited onto hydrophobic silica particles, yielding solid state sensors. The drawback of the sensor is its sensitivity to aliphatic amine vapour, which results in similar luminescence changes. The performance of this device can be improved by comparison of its response with the luminescence of the zinc complex (yellow), which responds to amines, whereas phosphates do not interact with this material. A strong increase in luminescence intensity on binding of the phosphate was observed in the case of some europium complexes (Figure 16.18c) [54]. Selectivity of these sensors can be achieved via molecular imprinting [55] of the ethylene glycol dimethylacrylate/methyl methacrylate and divinyl benzene/styrene copolymers [54].

Fluoride anion strongly interacts with various inorganic and organic boron and silicon compounds. These reactions are the basis for several fluoride sensors. Interaction of fluoride with boron compounds results in electron density redistribution and may also induce structural changes. Formation of fluoride complex by ferrocene derivative (Figure 16.19a) results in a decrease of oxidation potential by 200 mV [56]. Trianthrylboron (Figure 16.19b) is a strongly coloured compound: visible light absorption results from strong interaction between empty π-orbitals of aromatic moieties via empty p-orbitals of the central boron atom [57]. Coordination of fluoride inhibits this interaction and the resultant fluoroborate becomes colourless [58]. Further extension of the system results in stepwise colour change on interaction with increasing concentrations of fluoride. Other anions do not induce any significant spectral change of these sensors. The fluorescent boron compound (Figure 16.19d) dramatically changes its fluorescence upon interaction with fluoride [59]. Interaction of fluoride with the silicon analogue of trianthrylboron, trianthrylfluorosilane (Figure 16.19e), results in a dramatic increase of fluorescence quantum yield [60]. Strong, through-space interaction between anthryl groups results in a bathochromic shift of the anthracene absorption spectrum and low fluorescence quantum yields. On coordination of fluoride, a difluorosilicate anion is formed. Change in

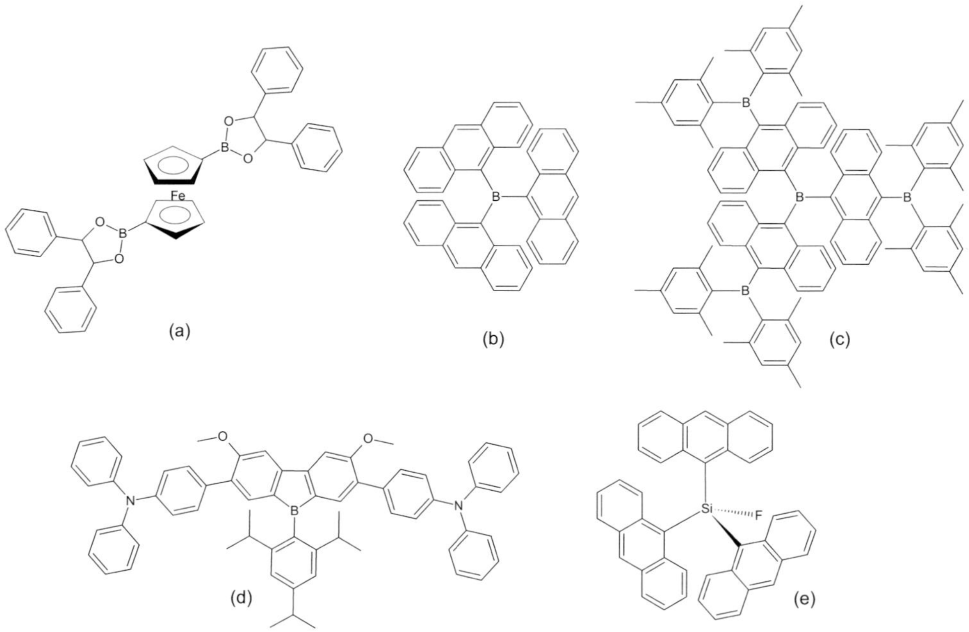

Figure 16.19 (a) Electrochemical, (b, c) colorimetric, and (d, e) fluorescent organoboron sensors for fluoride anions

geometry renders the through-space interaction less efficient and strong fluorescence is observed [60].

16.3 Optical Detection of Neutral Molecules

Analogous strategies are applied to sensing of neutral molecules, which can serve as ligands. There are numerous amino acid and amine sensors based on Zn^{II} and Cu^{II} complexes [26, 61] with fluorescent ligands [24, 31, 62]. It is possible, however, to detect non-coordinating molecules (eg hydrocarbons) based on the same principle of operation. Hydrocarbons do not coordinate to lanthanide cations, but a clever design allows efficient sensing. The sensor consists of two binding sites linked together: a macrocycle for lanthanide and a cyclodextrin for hydrophobic substrates. The fluorescence enhancement is based on the AETE (absorption/energy-transfer/emission) effect and the analyte acts as a photonic antenna for the lanthanide, europium, and terbium complexes with azacrown-like ligands covalently attached to cyclodextrin-based receptors [31]. Lanthanide luminescence quenching by molecular oxygen constituted a basis for several chemosensing systems [27, 63].

More practical oxygen and volatile, organic, luminescent chemosensors are based on a transition metal complex embedded in suitable polymers or anchored to solid surfaces [64–78]. The sensors are realized by immobilizing a luminescent indicator in a gas-permeable polymeric matrix. The most common luminophores are aromatic hydrocarbons, metalloporphyrins, and transition metal complexes. Ruthenium(II) polypyridyls [69], ruthenium(II) [74], rhodium(II) [70], platinum(II) [70–72, 74, 75], and palladium(II) [70, 74, 75] porphyrins show excellent properties: long fluorescence lifetimes, absorption and emission in the visible range, high luminescence quantum yields, and remarkable sensitivity towards molecular oxygen. Further improvement of sensitivity and selectivity of these chemosensors can be achieved via modification of a solid matrix. Many different polymers, both inorganic and organic, have been tested. The best performance and photostability of the sensor were observed in the case of fluorinated polymers. Due to high electron affinity of fluorine atoms, they do not undergo porphyrin-catalyzed photo-oxidation and in addition are characterized by a high oxygen permeability [70–72].

Various volatile organic substances were found to change the optical properties of free base porphyrins [79], metalloporphyrins [80], solvatochromic dyes [81], and pH indicators [4]. Interaction of these species immobilized on solid supports results in changes in optical properties of these materials. Metalloporphyrins are a natural choice for the detection of metal ligating vapours because of their open coordination sites for axial ligation, their large spectral shifts upon ligand binding, and their intense coloration [82, 83]. As a result of the strong tendency to large colour changes induced in metalloporphyrins on ligand binding, it was possible to develop an easy colorimetric technique that minimizes the need for extensive signal transduction hardware. This represents the first example of a colorimetric array detector for vapour-phase ligands. Simply by taking the difference before and after exposure of scanned images of the array, one obtains unique colour-change signatures for analytes; these signatures can be used for both qualitative recognition and quantitative analysis [67]. The system constitutes a very efficient artificial olfactory device [77, 84]. This method allows closely related toxic organics (thiols, phosphines, amines) [67], aliphatic alcohols [64], and other solvents [77] to be distinguished, and for checking the quality of food and spices [65]. Further improvement of this photonic olfactory system was achieved by introduction of size-specific coordination centres, namely zinc porphyrins containing variable numbers of bulky siloxyl substituents [85–87]. This modification allows more precise control over ligation of the target molecules to metal centres within porphyrin macrocycles owing to steric and supramolecular interactions [67, 86, 88]. Small modification of the device allows detection of a whole variety of organic compounds in aqueous solutions [89]. In this case hydrophobic porphyrins, shape-selective porphyrins, polarity indicators, and pH indicators are deposited on hydrophobic polymeric substrate (Figure 16.20a) [89–91].

In all cases colour digital photographs of the sensor matrices are taken before and after exposure to the analyzed mixture. Resulting photographs are analyzed using commercially available graphical software, and difference images are produced (Figure 16.20b) [67, 86]. This simple device enables precise analysis of complex mixtures of volatile compounds in gas and liquid phases. Unique patterns are generated for every chemical species interacting with the sensors, and analysis of complex mixtures becomes straightforward. The device is usually referred to as an optoelectronic nose [88, 92].

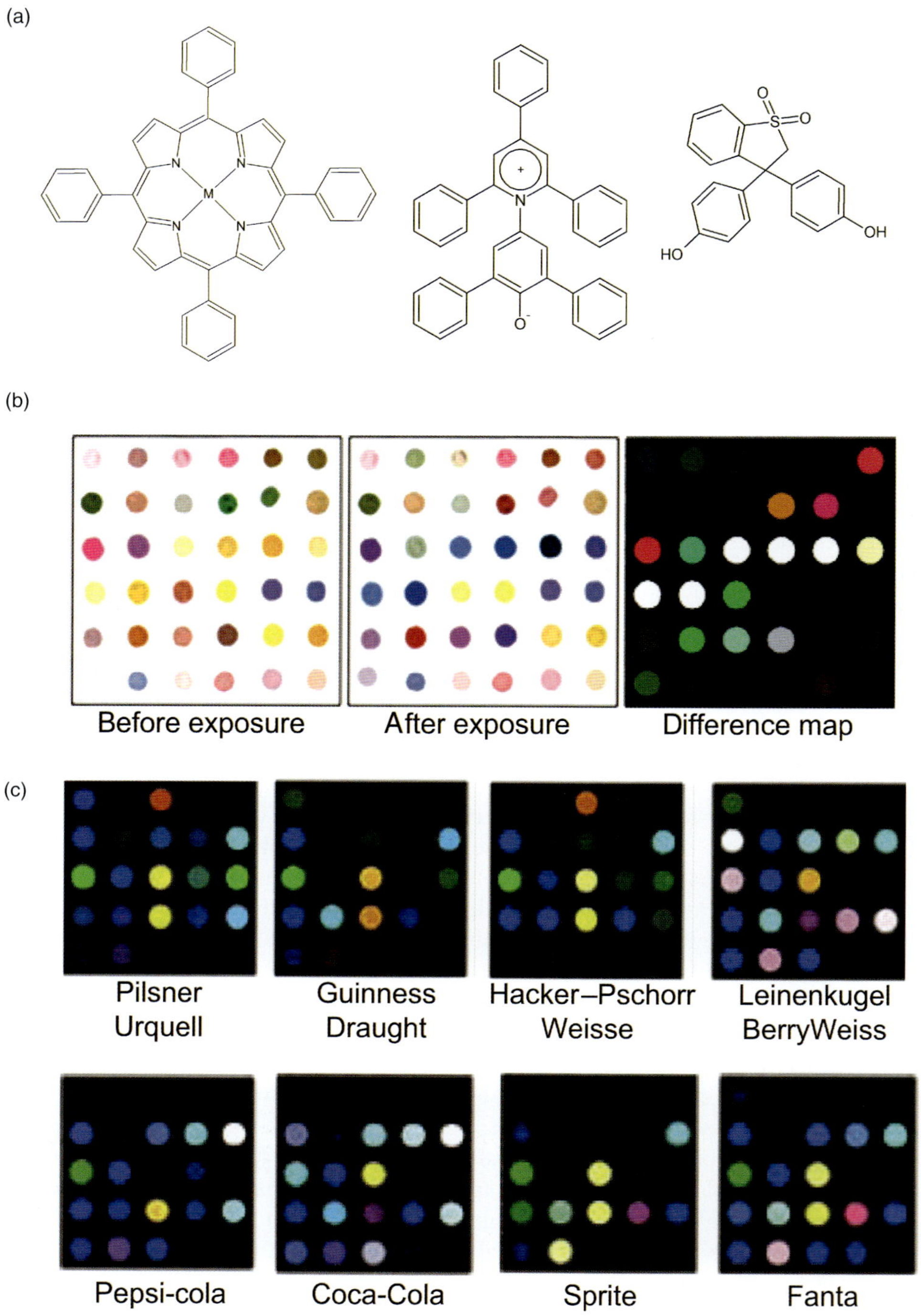

Figure 16.20 *Examples of sensor molecules: (a) tetraphenyl porphyrin, solvatochromic Reichardt's dye and pH indicator (phenol red); (b) image of the 36-dye matrix before exposure, after exposure, and the resulting difference map; and (c) colour change profiles for common beverages. (Reprinted with permission from Zhang and Suslick [90], Suslick et al. [91], and Janzen et al. [147]. © American Chemical Society 2006–2007)*

The same principle was observed in the organic vapour chemosensors based on cyano-bridged rhenium–cobalt cluster complexes, $[Co_2(H_2O)_4][Re_6S_8(CN)_8]\cdot 12H_2O$ and $[Co_2(H_2O)_4][Co(H_2O)_3]_4[Re_6Se_8(CN)_6]_3\cdot 44H_2O$ [93]. The colour of the solid sample of these cluster complexes changes from orange (without organic analyte) to red, violet, green, and blue on interaction with various organic solvents (Figure 16.21). It is caused by a change in coordination geometry of cobalt ions within the crystal framework. This interpretation was confirmed by magnetic studies of these materials [93].

One of the most important biomolecules for which fluorescent sensing [94–96] is of great importance is nitric oxide [97–101]. Nitric oxide can react with several organic dyes, switching on their fluorescence as a result of a triazole ring closure reaction [94]. There are also useful and selective NO optical sensors based on transition metal complexes (Figures 16.22 and 16.23) [94–96].

One of the first metal-containing NO fluorescent sensors contains an Fe(II) macrocyclic complex connected with quinoline ligand on a pendant arm (Figure 16.22). Coordination of NO to the iron results in labilization of the quinoline moiety and fluorescence decrease [94]. Other NO sensors are based on Fe [102], Mn [95], and Co [103–105] tropocoronand complexes. One example is shown in Figure 16.23. The tropocoronand ligands incorporate substituted dimethylaminonaphthalene sulphonate fluorophores. Free ligands exhibit strong fluorescence, which is largely

Figure 16.21 *Powder samples of compounds $[Co_2(H_2O)_4][Re_6S_8(CN)_8]\cdot 12H_2O$ (upper) and $[Co_2(H_2O)_4][Co(H_2O)_3]_4[Re_6Se_8(CN)_6]_3\cdot 44H_2O$ (lower) deposited on filter paper and doused with selected solvents: water, tetrahydrofuran, and diethyl ether, from left to right, respectively) (Reprinted with permission from Bauvais et al. [93]. © 2000 American Chemical Society)*

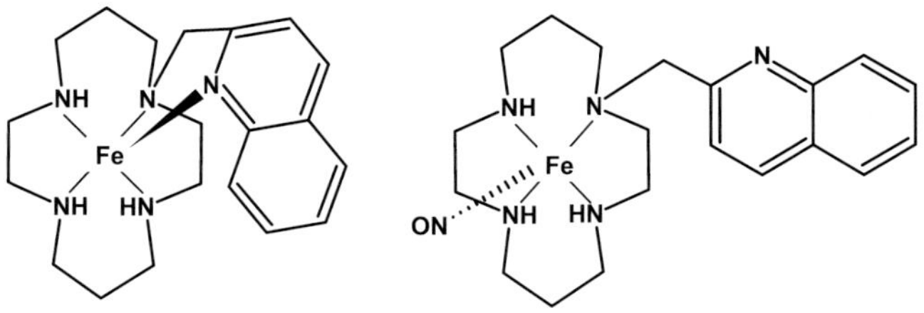

Figure 16.22 *Fluorescent nitric oxide (NO) sensor: NO coordination to iron quenches the fluorescence [94]*

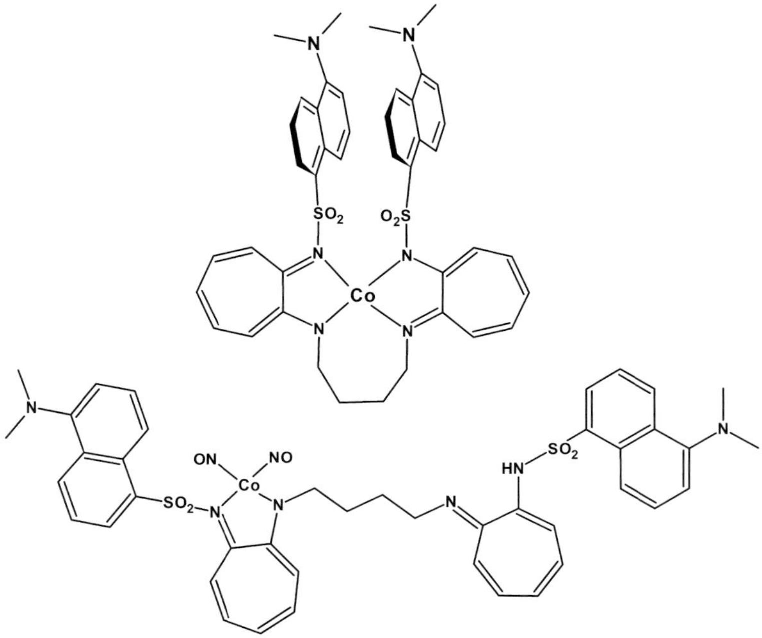

Figure 16.23 *Fluorescent nitric oxide sensor. The Co tropocoronand complex does not emit due to efficient excimer formation (top). Coordination of two nitric oxide molecules changes the geometry and excimer formation is geometrically inhibited (bottom) [94]*

diminished on coordination to transition metal cations due to excimer formation [94]. The reaction with NO liberates one arm of the ligand. The new conformation reduces interaction between dansyl fluorophores and hence disturbs the excimer formation, which results in large increase of the fluorescence quantum yield. NO can be also determined with an absorptiometric chemosensor based on a Mn(II) porphyrin complex. The tetrakis(*N*-ethylpyridinium-2-yl)porphyrin complex of manganese reacts with NO in the presence of ascorbate, yielding the Mn-nitrosylated species. The difference in absorption spectra of the free and nitrosylated complex allows fast and accurate spectrophotometric determination of NO in aqueous solutions at a submicromolar level [106].

16.4 Nanoparticles in Biochemical Sensing and Labelling

Other, entirely different methods of optical sensing and imaging utilize not metal complexes, but metal and semiconductor nanoparticles [107–111]. Strong light absorption of metal nanoparticles suspended in dielectric materials results from plasmon resonance, whereas that of semiconductor nanoparticles results from valence band-to-conduction band transitions, usually called interband transitions (see section 7.3 in Chapter 7). Excitation of semiconducting nanoparticles may result in efficient luminescence. Optical properties of these particles depend not only on their chemical composition, but also on the size, shape, and surface 'decoration' of these objects [108, 112–121]. By controlling their structure at a nanometre scale, one can precisely control and tailor properties of nanoparticles [122] to make them more suitable for biological systems, eg modify their surface layer for enhanced aqueous solubility, biocompatibility, and biorecognition. There are known surface modified nanoparticles attached to such biomolecules as sugars [123–125], peptides [126], proteins [127–129], and DNA [114–116, 130–132], which can act as recognition sites or 'position tuners' [133]. These nanoparticle conjugates are used for assembly of new materials, development of homo- and heterogeneous assays, multicolour fluorescent labelling, and high-throughput detection and imaging [122]. The replacement of conventional chromo- and fluorophores with nanoparticles results in higher quantum efficiencies, greater scattering or absorbance cross-section, optical activity at biocompatible wavelengths, and significantly increased chemical and photochemical stability [122].

The wavelength and half-width of the plasmon resonance absorption depend strongly on the nanoparticle size and shape [113]. This phenomenon was used for specific sensing of metal ions [134] and DNA sequencing [114–116, 130, 131, 135–137]. Molar absorption coefficients of metallic nanoparticles within their plasmon bands are at least three orders of magnitude higher than those of any inorganic or organic chromophores (approximately 2.4×10^8 $mol^{-1} dm^3 cm^{-1}$), allowing use of nanomolar concentrations of nanoparticles. DNA surface-modified gold nanoparticles show a characteristic plasmon absorbance similar to that of unmodified nanoparticles. Addition of complementary DNA strands causes aggregation of nanoparticles and hence the colour change. The reaction is extremely sensitive to the DNA sequence and one pair mismatch, deletion, or insertion changes the aggregation equilibrium dramatically. Due to a very high selectivity and sensitivity the nanoparticle-based method of DNA sequence detection was successfully used for anthrax detection [114].

The same protocol based on aggregation of metallic particles may be used for detection and determination of various biomolecules and even simple inorganic ions. Antibodies, enzymes, biotin, streptavidin, or lectins can be used for nanoparticle modification [138] (Figure 16.24).

Interaction of biomolecules with the surface of metallic nanoparticles may also result in local changes in refraction index. This in turn may result in delicate modification of plasmon resonance frequency and yield detectable analytical signal [138].

Semiconductor nanoparticles are powerful fluorescent probes, which can be used for labelling biomolecules. Quantum dots have several advantages over

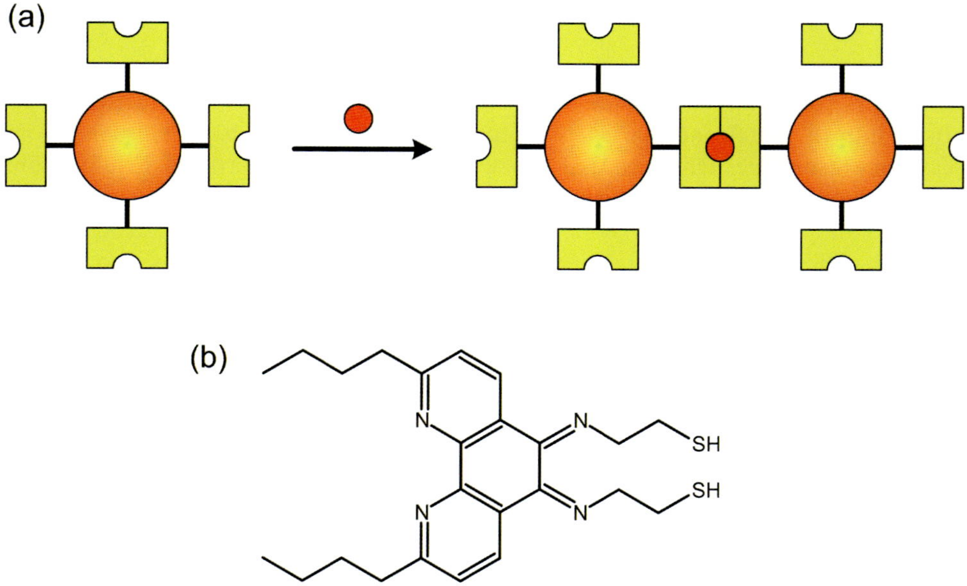

Figure 16.24 *(a) A generalized scheme for chemical sensing based on analyte-assisted aggregation of metallic nanoparticles and (b) molecular structure of a ligand used in lithium binding. (Adapted from Murphy et al. [138])*

conventional fluorescent dyes. Their emission and absorption spectra are easily tuned by changes in their size and material composition. Their absorption spectra are broad, although they have very narrow emission lines. Fluorescence quantum yields are relatively high (0.3–0.5) and the photostability against photobleaching is two orders of magnitude better than that of any other fluorescent labels. The fluorescence lifetime is long (a few hundred nanoseconds), which allows time-delayed fluorescence measurements, and can be used to suppress the autofluorescence of biological matrices [108]. Among the disadvantages of semiconductor quantum dots, large dimensions and intermittent luminescence (blinking) should be mentioned [139].

All the above-mentioned advantages make the semiconductor quantum dots very efficient markers and labels applied to biological systems. The rich surface chemistry of oxide and chalcogenide semiconductors allows immobilization of almost any biomolecule on to their surfaces [107, 108]. These interactions may alter the electronic properties of semiconductors (eg band edge potential, work function, surface states) [140, 141]. These effects are, however, not frequently utilized in bioinorganic studies. Semiconductor nanoparticles for biological applications are usually of core-shell architecture, which prevents interaction of the inner, luminescent part with the environment. This makes the semiconductor quantum dots efficient and robust molecular scale light sources, which can be covalently attached to any biological structure and hence serve as labels or markers [142, 143].

More sophisticated designs of nanoparticle-based sensors include various photoprocesses such as electronic energy transfer and PET.

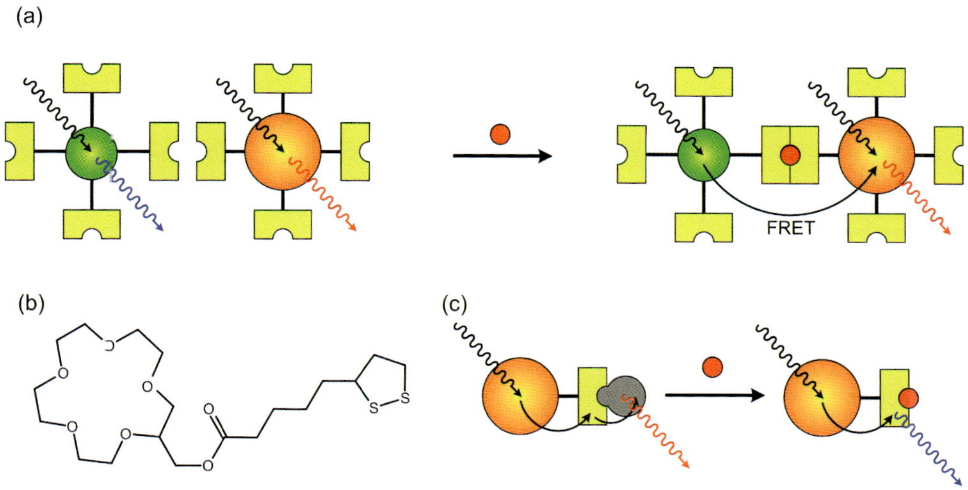

Figure 16.25 Semiconductor nanoparticle-based fluorescent sensors: (a) Förster resonant energy transfer (FRET) between two nanoparticles induced by analyte, (b) crown ether receptor for potassium ions, and (c) operation principle of maltose fluorescent sensor. (Adapted from Chen et al. [144] and Medintz et al. [146])

Although aggregation of metallic nanoparticles perturbs their electronic structure and especially the plasmon resonance frequency, the semiconductor quantum dots are much more resistant to such effects.

The crown ether ligand can be immobilized onto CdSe/ZnS core/shell quantum dots via thiolate anchors (Figure 16.25a,b). This assembly allows potassium-induced aggregation of nanoparticles, resulting in Förster energy transfer between quantum dots of two different sizes (Figure 16.25a). In the absence of potassium ions, solutions containing mixture of two kinds of quantum dots show dual fluorescence. On addition of K^+ ions the formation of sandwich-type complexes between K^+ and 15-crown-5 moieties results in formation of various aggregates. The average distance between individual quantum dots decreases, allowing efficient energy transfer [144]. Similar aggregation of gold nanoparticles via π–π stacking of aromatic surface groups results in strong electronic coupling between individual nanocrystals [145].

More sophisticated designs involved semiconductor quantum dots with fluorescent protein receptors immobilized on the surface [146]. The binding site of the protein receptor is occupied with an efficient fluorophore. On excitation a series of FRET (Förster resonant energy transfer) processes takes place; excitation energy is transferred from the core of the quantum dot to the fluorescent protein and subsequently to the fluorophore. On substrate binding only one FRET step takes place and luminescence of the receptor is observed [146]. In the simplest sensor architecture the protein contains bound quencher. Upon interaction with analyte the quencher is liberated and luminescence of the quantum dot is observed (Figure 16.25c).

In conclusion, semiconductor quantum dots play the role of mechanical supports, photonic antennae, and fluorophores. Such versatility results from unique properties of semiconductor nanoparticles.

References

1. Valeur B. *Molecular Fluorescence. Principles and applications*. Weinheim: Wiley-VCH, 2002.
2. Waser R. Properties of information. In: Waser R, ed. *Nanoelectronics and information technology*. Weinheim: Wiley-VCH Verlag GmbH, 2003:11–23.
3. Szaciłowski K. Digital information processing in molecular systems. *Chem Rev* 2008;**108**:3481–548.
4. de Silva AP, Gunaratne HQN, Gunnlaugsson T, et al. Signaling recognition events with fluorescent sensors and switches. *Chem Rev* 1997;**97**:1515–66.
5. Comba P. Metal ion selectivity and molecular modelling. *Coord Chem Rev* 1999;**185–186**:81–98.
6. Lehn J-M. Supramolecular chemistry – scope and perspectives. Molecules – supermolecules – molecular devices. In: Frängsmyr T, Malmström BG, eds. *Nobel Lectures, Chemistry 1981–1990*. Singapore: World Scientific Publishing Co., 1992.
7. Szaciłowski K, Macyk W, Drzewiecka-Matuszek A, Brindell M, Stochel G. Bioinorganic photochemistry: frontiers and mechanisms. *Chem Rev* 2005;**105**:2647–94.
8. Gokel GW, Leevy WM, Weber ME. Crown ethers: sensors for ions and molecular scaffolds for materials and biological models. *Chem Rev* 2004;**104**:2723–50.
9. de Silva AP, Fox DA, Huxley AJM, Moody TS. Combining fluorescence, coordination chemistry and electron transfer for signalling purposes. *Coord Chem Rev* 2000;**205**: 41–57.
10. Callan JF, de Silva AP, Magria DC. Luminescent sensors and switches in the early 21st century. *Tetrahedron* 2005;**61**:8551–88.
11. Jiang P, Guo Z. Fluorescent detection of zinc in biological systems: recent development on the design of chemosensors and bioreceptors. *Coord Chem Rev* 2004;**248**:205–29.
12. Valeur B, Leray I. Design principles of fluorescent molecular sensors for cation recognition. *Coord Chem Rev* 2000;**205**:3–40.
13. Pina F, Bernardo MA, García-Espana E. Fluorescent sensors containing polyamine receptors. *Eur J Inorg Chem* 2000:2143–57.
14. Lim NC, Freake HC, Brücker C. Illuminating zinc in biological systems. *Chem Eur J* 2005;**11**:38–49.
15. Zeng Z, Jewsbury RA. Fluorimetric determination of iron using 5-(4-methoxyphenyazo) -8-(4-toluenesulfonamido)-quinoline. *Analyst* 2000;**125**:1661–5.
16. Espósito BP, Epsztejn S, Breuer W, Cabantchik ZI. A review of fluorescence methods for assessing labile iron in cells and biological fluids. *Anal Biochem* 2002;**304**:1–18.
17. Chang JH, Jeong Y, Shin Y-K. Selective photoinduced electron transfer-based cyanoanthryl fluorescent chemosensors for paramagnetic metal cations. *Bull Korean Chem Soc* 2003;**24**:119–22.
18. Gunnlaugsson T, Leeb TC, Parkesh R. Highly selective fluorescent chemosensors for cadmium in water. *Tetrahedron* 2004;**60**:11239–49.
19. De Santis G, Fabrizzi L, Licchelli M, Mangano C, Sacchi D, Sardone N. A fluorescent chemosensor for the copper(II) ion. *Inorg Chim Acta* 1997;**257**:69–76.
20. Jiménez D, Martínez-Máñez R, Sancenón F, et al. Multi-channel receptors and their relation to guest chemosensing and reconfigurable molecular logic gates. *Eur J Inorg Chem* 2005:2393–403.
21. Jiménez D, Martínez-Máñez R, Sancenón F, Soto J. Electro-optical triple-channel sensing of metal cations via multiple signalling patterns. *Tetrahedron Lett* 2004;**45**:1257–9.
22. Prodi L, Bargossi C, Montalti M, et al. An effective fluorescent chemosensor for mercury ions. *J Am Chem Soc* 2000;**122**:6769–70.
23. Gale PA. Anion-pair receptor chemistry: highlights from 2000 and 2001. *Coord Chem Rev* 2003;**240**:191–221.
24. Rogers CW, Wolf MO. Luminescent molecular sensors based on analyte coordination to transition metal complexes. *Coord Chem Rev* 2002;**233**–234:341–50.

25. Beer PD, Hayes EJ. Transition metal and organometallic anion complexation agents. *Coord Chem Rev* 2003;**240**:167–89.
26. Fabrizzi L, Lichelli M, Rabaioli G, Taglietti A. The design of luminescent sensors for anions and ionizable analytes. *Coord Chem Rev* 2000;**205**:85–108.
27. Parker D. Luminescent lanthanide sensors for pH, O_2 and selected anions. *Coord Chem Rev* 2000;**205**:109–30.
28. Beer PD, Cadman J. Electrochemical and optical sensing of anions by transition metal receptors. *Coord Chem Rev* 2000;**205**:131–55.
29. de Silva AP, Fox DA, Huxley AJM, McClenaghan ND, Roiron J. Metal complexes as components of luminescent signalling systems. *Coord Chem Rev* 1999;**185–186**: 297–306.
30. Gunnlaugsson T, Glynn M, Tocci GM, Kruger PE, Pfeffer FM. Anion recognition and sensing in organic and aqueous media using luminescent and colorimetric sensors. *Coord Chem Rev* 2006;**250**:3094–117.
31. Keefe MH, Benkstein KD, Hupp JT. Luminescent sensor molecules based on coordinated metals: a review of recent developments. *Coord Chem Rev* 2000;**205**:201–28.
32. Aoki S, Kimura E. Recent progress in artificial receptors for phosphate anions in aqueous solution. *Rev Mol Biotechnol* 2002;**90**:129–55.
33. Royo S, Martínez-Máñez R, Sancenón F, Costero AM, Parra M, Gil S. Chromogenic and fluorogenic reagents for chemical warfare nerve agents' detection. *Chem Commun* 2007:4839–47.
34. Thomas III SW, Joly GD, Swager TM. Chemical sensors based on amplifying fluorescent conjugated polymers. *Chem Rev* 2007;**107**:1339–86.
35. Suksai C, Tuntulani T. Chromogenic anion sensors. *Top Curr Chem* 2005;**255**:163–98.
36. Lee HN, Singh NJ, Kim SK, et al. New imidazolium systems bearing two pyrene groups as fluorescent chemosensors for anions and anion induced logic gates. *Tetrahedron Lett* 2007;**48**:69–172.
37. Kim SK, Singh NJ, Kwon J, et al. Fluorescent imidazolium receptors for the recognition of pyrophosphate. *Tetrahedron* 2006;**62**:6065–72.
38. Beer PD. Transition-metal receptor systems for the selective recognition and sensing of anionic guest species. *Acc Chem Res* 1998;**31**:71–80.
39. Padilla-Tosta ME, Lloris JM, Martinez-Manez R, et al. A fluorescent chemosensor based on ruthenium(II) terpyridine core containing peripheral amino groups that selectively sense ATP in an aqueous environment. *Inorg Chem Commun* 2000;**3**:45–8.
40. Ojida A, Park S, Mito-oka Y, Hamachi I. Efficient fluorescent ATP-sensing based on coordination chemistry under aqueous neutral conditions. *Tetrahedron Lett* 2002;**43**: 6193–5.
41. Ojida A, Mito-oka Y, Inoue M, Hamachi I. First artificial receptors and chemosensors toward phosphorylated peptide in aqueous solution. *J Am Chem Soc* 2002;**124**: 6256–8.
42. Lee HN, Xu Z, Kim SK, et al. Pyrophosphate-selective fluorescent chemosensor at physiological formation of a unique excimer upon addition of pyrophosphate. *J Am Chem Soc* 2007;**129**:3828–9.
43. Kwon JY, Singh NJ, Kim HN, Kim SK, Kim KS, Yoon J. Fluorescent GTP-sensing in aqueous solution of physiological pH. *J Am Chem Soc* 2004;**126**:8892–3.
44. Wang S, Chang Y-T. Combinatorial synthesis of benzimidazolium dyes and its diversity directed application toward GTP-selective fluorescent chemosensors. *J Am Chem Soc* 2006;**128**:10380–1.
45. Beer PD, Cadman J. Electrochemical and optical sensing of anions by transition metal based receptors. *Coord Chem Rev* 2000;**205**:131.
46. Beer PD, Hayes EJ. Transition metal and organometallic anion complexation agents. *Coord Chem Rev* 2003;**240**:167–89.
47. Sadik OA, Land WH Jr, Wang J. Targeting chemical and biological warfare agents at the molecular level. *Electroanalysis* 2003;**15**:1149–59.

48. Houten KAV, Heath DC, Pilato RS. Rapid Luminescent detection of phosphate esters in solution and the gas phase using (dppe)Pt{S_2C_2(2-pyridyl)(CH_2CH_2OH)}. *J Am Chem Soc* 1998;**120**:12359–60.
49. Zhang S-W, Swager TM. Fluorescent detection of chemical warfare agents: functional group specific ratiometric chemosensors. *J Am Chem Soc* 2003;**125**:3420–1.
50. Dale TJ, Rebek J Jr. Fluorescent sensors for organophosphorus nerve agent mimics. *J Am Chem Soc* 2006;**128**:4500–1.
51. Ilhan F, Tyson DS, Meador MA. Synthesis and chemosensory behavior of anthracene bisimide derivatives. *Chem Mater* 2004;**16**:2978–80.
52. Bencic-Nagale S, Sternfeld T, Walt DR. Microbead chemical switches: an approach to detection of reactive organophosphate chemical warfare agent vapors. *J Am Chem Soc* 2006;**128**:5041–8.
53. Knapton D, Burnworth M, Rowan SJ, Weder C. Fluorescent organometallic sensors for the detection of chemical warfare agent mimics. *Angew Chem Int Ed* 2006;**45**: 5825–9.
54. Southard GE, Van Houten KA, Ott EW Jr, Murray GM. Luminescent sensing of organophosphates using europium(III) containing imprinted polymers prepared by RAFT polymerization. *Anal Chim Acta* 2007;**581**:202–7.
55. Piletsky SA, Turner NW, Laitenberger P. Molecularly imprinted polymers in clinical diagnostics – Future potential and existing problems. *Med Eng Phys* 2006;**28**:971–7.
56. Aldridge S, Bresner C, Fallis IA, Coles SJ, Hursthouse MB. Multidentate Lewis acids: synthesis, structure and mode of action of a redox-based fluoride ion sensor. *Chem Commun* 2002:740–1.
57. Yamaguchi S, Akiyama S, Tamao K. Tri-9-anthrylborane and its derivatives: new boron-containing π-electron systems with divergently extended π-conjugation through boron. *J Am Chem Soc* 2000;**122**:6335–6.
58. Yamaguchi S, Akiyama S, Tamao K. Colorimetric fluoride ion sensing by boron-containing π-electron systems. *J Am Chem Soc* 2001;**123**:11372–5.
59. Yamaguchi S, Shirasaka T, Akiyama S, Tamao K. Dibenzoborole-containing π-electron systems: remarkable fluorescence change based on the 'on/off' control of the p_π-π^* conjugation. *J Am Chem Soc* 2002;**124**:8816–17.
60. Yamaguchi S, Akiyama S, Tamao K. Photophysical properties changes caused by hypercoordination of organosilicon compounds: from trianthrylfluorosilane to trianthryldifluorosilicate. *J Am Chem Soc* 2000;**122**:6793–4.
61. Fabrizzi L, Francese G, Lichelli M, Perotti A, Taglietti A. Fluorescent sensor for imidazole and histidine. *J Chem Soc Chem Commun* 1997:581–2.
62. Kimura E, Aoki S, Koike T, Shiro M. A tris(Zn(II)-1,4,7,10-tetraazacyclododecane) complex as a new receptor for phosphate dianions in aqueous solution. *J Am Chem Soc* 1997;**119**:3068–76.
63. Parker D, Senanayake PK, Williams JAG. Luminescent sensors for pH, O_2, halide and hydroxide ions using phenantroline as a photosensitizer in macrocyclic europium and terbium complexes. *J Chem Soc Perkin Trans 2* 1998:2129–39.
64. Akrayas, Salleh MM, Yahaya M. Enriching the selectivity of metalloporphyrins chemical sensors by means of optical technique. *Sens Actuators* 2002;**B85**:191–6.
65. Salleh MM, Akrayas, Yahaya M. Optical sensing of capsicum aroma using four porphyrin derivatives this films. *Thin Solid Films* 2002;**417**:162–5.
66. Potera C. Smelling in color. *Environ Health Persp* 2001;**109**:A129–31.
67. Rakow NA, Suslick KS. A colorimetric sensor array for odour visualization. *Nature* 2000;**406**:710–12.
68. Delmare D, Bien-Charreton C. Grafting of cobalt porphyrins in sol-gel matrices: application to the detection of amines. *Sens Actuators* 2000;**B62**:136–42.
69. Ertekin K, Kocak S, Ozer MS, Aycan S, Cetinkaya B. Enhanced optical oxygen sensing using a newly synthesized ruthenium complex together with oxygen carriers. *Talanta* 2003;**61**:573–9.

70. Vasil'ev V, Borisov SM. Optical oxygen sensor based on phosphorescent metal porphyrins immobilized in perfluorinated ion-exchange membrane. *Sens Actuators* 2002;**B82**: 272–6.
71. Amao Y, Miyashita T, Okura I. Platinum tetrakis(pentafluorophenyl)porphyrin immobilized in polytrifluoroethylmetacrylate film as a photostable optical oxygen detection material. *J Fluorine Chem* 2001;**107**:101–6.
72. Amao Y, Miyashita T, Okura I. Novel optical oxygen sensing material: platinum octaethylporphyrin immobilized in a copolymer film of isobutyl methacrylate and tetrafluoropropyl methacrylate. *React Funct Polym* 2001;**47**:49–4.
73. Brunel M, Ozkul C, Sanchez F. Nonlinear optical devices for oxygen concentration measurements. *Sens Actuators* 2000;**B62**:238–43.
74. DiMarco G, Lanza M. Optical solid state oxygen sensors using metalloporphyrin complexes immobilized in suitable polymeric matrices. *Sens Actuators* 2000;**B63**:42–8.
75. Amao Y, Miyashita T, Okura I. Optical oxygen detection based on luminescence change of metalloporphyrins immobilized in poly(isobutylmetacrylate-co-trifluoroethylmethacrylate) film. *Anal Chim Acta* 2000;**421**:167–74.
76. Bariain C, Matias IR, Fernandez-Valdivielso C, Arregui FJ, Rodriguez-Mendez ML, de Saja JA. Optical fiber sensor based on luthetium bisphthalocyanine for the detection of gases using standard telecommunication wavelengths. *Sens Actuators* 2003;**B93**: 153–8.
77. Di Natale C, Salimbeni D, Paolesse R, Macagnano A, D'Amico A. Porphyrins-based optoelectronic nose for volatile compounds detection. *Sens Actuators* 2000;**B65**:220–6.
78. D'Amico A, Di Natale C, Paolesse R, Macagnano A, Mantini A. Metalloporphyrins as basic material for volatile sensitive sensors. *Sens Actuators* 2000;**B65**:209–15.
79. Burrell AK, Officer DL, Plieger PG, Reid DCW. Synthetic routes to multiporphyrin arrays. *Chem Rev* 2001;**101**:2751–96.
80. Satake A, Kobuke Y. Dynamic supramolecular porphyrin systems. *Tetrahedron* 2005;**61**:13–41.
81. Reichardt C. Solvatochromic dyes as solvent polarity indicators. *Chem Rev* 1994;**94**:2319–58.
82. Purrello R, Gurrieri S, Lauceri R. Porphyrin assemblies as chemical sensors. *Coord Chem Rev* 1999;**190–192**:683–706.
83. Arnold DP, Blok J. The coordination chemistry of tin porphyrin complexes. *Coord Chem Rev* 2004;**248**:299–319.
84. Röck F, Barsan N, Weimar U. Electronic nose: current status and future trends. *Chem Rev* 2008;**108**:705–25.
85. Sen A, Suslick KS. Shape-sensitive discrimination of small organic molecules. *J Am Chem Soc* 2000;**122**:11565–6.
86. Suslick KS, Rakow NA, Sen A. Colorimetric sensor arrays for molecular recognition. *Tetrahedron* 2004;**60**:11133–8.
87. Bhyrappa P, Vaijayanthimala G, Suslick KS. Shape-selective ligation to dendrimer-metalloporphyrins. *J Am Chem Soc* 1999;**121**:262–3.
88. Suslick KS, Bailey DP, Ingison CK, et al. Seeing smells: Development of an optoelectronic nose. *Quim Nova* 2007;**30**:677–81.
89. Zhang C, Suslick KS. A colorimetric sensor arrays for organics in water. *J Am Chem Soc* 2005;**127**:11548–9.
90. Zhang C, Suslick KS. Colorimetric sensor array for soft drink analysis. *J Agric Food Chem* 2007;**55**:237–42.
91. Zhang C, Bailey DP, Suslick KS. Colorimetric sensor array for the analysis of beers: A feasibility study. *J Agric Food Chem* 2006;**54**:4925–31.
92. Suslick KS. An optoelectronic nose: seeing smells by means of colorimetric arrays. *Mat Res Soc Bull* 2004:720–5.
93. Bauvais LG, Shores MP, Long JR. Cyano-bridged Re_8Q_8 (Q=S, Se) cluster-cobalt(II) framework materials: versatile solid chemical sensors. *J Am Chem Soc* 2000;**122**: 2763–72.

94. Burdette SC, Lippard SJ. ICCC34 – golden edition of coordination chemistry reviews. Coordination chemistry for the neurosciences. *Coord Chem Rev* 2001;**216–217**:333–61.
95. Franz KJ, Lippard SJ. Disproportionation of nitric oxide promoted by a Mn tropocoronand. *J Am Chem Soc* 1998;**120**:9034–40.
96. Franz KJ, Singh N, Spingler B, Lippard SJ. Aminotroponimines as ligands for potential metal-based nitric oxide sensors. *Inorg Chem* 2000;**39**:4081–92.
97. Ignarro LJ. Nitric oxide as a unique signalling molecule in the vascular system: a historical overview. *J Physiol Pharm* 2002;**53**:503–14.
98. Butler AR, Flitney FW, Rhodes P. The role of nitric oxide in animal physiology. *Persp Bioinorg Chem* 1996;**3**:251–77.
99. Furchgott RF. Endothelium-derived relaxing factor: discovery, early studies and identification as nitric oxide. *Angew Chem Int Ed* 1999;**38**:1870–80.
100. Ignarro LJ. Nitric oxide: a unique endogenous signalling molecule in vascular biology. *Angew Chem Int Ed* 1999;**38**:1882–92.
101. Murad F. Discovery of the biological effects of nitric oxide and its role in cell signalling. *Angew Chem Int Ed* 1999;**38**:1856–68.
102. Franz KJ, Lippard SJ. NO disproportionation reactivity of Fe tropocoronand complexes. *J Am Chem Soc* 1999;**121**:10504–12.
103. Jaynes SB, Ren T, Masschelein A, Lippard SJ. Stereochemical control of reactivity in cobalt(III) alkyl complexes of the tropocoronand ligand system. *J Am Chem Soc* 1993;**115**:5598–99.
104. Jaynes SB, Doerrer LH, Liu SC, Lippard SJ. Synthesis, tuning of the stereochemistry, and physical properties of cobalt(II) tropocoronand complexes. *Inorg Chem* 1995;**34**:5735–44.
105. Doerrer LH, Bautista MT, Lippard SJ. Tuning of electronic properties and reactivity in four-coordinate cobalt(III) complexes by the tetraazamacrocyclic tropocoronand ligand. *Inorg Chem* 1997;**36**:3578–9.
106. Spasojević I, Banitić-Haberle I, Fridovich I. Nitrosation of Mn(II) tetrakis(N-ethylpyridinium-2-yl-)porphyrin: a simple and sensitive spectrophotometric assay for nitric oxide. *Nitric Oxide* 2000;**4**:526–33.
107. Rosi NL, Mirkin CA. Nanostructures in biodiagnostics. *Chem Rev* 2005;**105**:1547–62.
108. Niemeyer CM. Nanoparticles, proteins and nucleic acids: biotechnology meets material science. *Angew Chem Int Ed Eng* 2001;**40**:4128–58.
109. Daniel M-C, Astruc D. Gold nanoparticles: assembly, supramolecular chemistry, quantum size-related properties, and applications toward biology, catalysis, and nanotechnology. *Chem Rev* 2004;**104**:293–346.
110. Stewart ME, Anderton CR, Thompson LB, et al. Nanostructured plasmonic sensors. *Chem Rev* 2008;**108**:494–521.
111. Hormola J. Surface Plasmon resonance sensors for detection of chemical and biological species. *Chem Rev* 2008;**108**:462–93.
112. Liz-Marzán LM. Nanometals: formation and color. *Materials Today* 2004;**7**:26–31.
113. Soenissen C. Plasmons in metal nanostructures. PhD dissertation, Ludwig-Maximilians-Universitaet, Munich, 2001.
114. Mirkin CA. Programming the assembly of two and three-dimensional architectures with DNA and nanoscale inorganic building blocks. *Inorg Chem* 2000;**39**:2258–72.
115. Mirkin CA, Taton TA. Semiconductors meet biology. *Nature* 2000;**405**:626–7.
116. Storhoff JJ, Elghanian R, Mucic RC, Mirkin CA. One-pot differentiation of polynucleotides with single base imperfections using gold nanoparticle probe. *J Am Chem Soc* 1998;**120**:1959–64.
117. Moriarty P. Nanostructured materials. *Reports Progr Phys* 2001;**64**:297–381.
118. Henglein A. Small particle research: physicochemical properties of extremely small colloidal metal and semiconductor particles. *Chem Rev* 1989;**89**:1861–73.
119. Nirmal M, Brus LE. Luminescence photophysics in semiconductor nanocrystals. *Acc Chem Res* 1999;**32**:404–14.

120. Alvosatos AP. Semiconductor clusters, nanocrystals and quantum dots. *Science* 1996;**271**:933–7.
121. Schmidt G. Large clusters and colloids: metals in the embryonic state. *Chem Rev* 1992;**92**:1709–27.
122. Otsuka H, Nagasaki Y, Kataoka K. PEGylated nanoparticles for biological and pharmaceutical applications. *Adv Drug Deliv Rev* 2003;**55**:403–19.
123. Rojas TC, de la Fuente JM, Barrientos AG, Penades S, Ponsonnet L, Fernandez A. Gold glyconanoparticles as building blocks. *Adv Mater* 2002;**14**:585–8.
124. de la Fuente JM, Barrientos AG, Rojas et al. Gold glyconanoparticles as water soluble building blocks. *Angew Chem Int Ed Eng* 2001;**40**:2258–61.
125. Otsuka H, Akiyama Y, Nagasaki Y, Kataoka K. Quantitative and reversible lecitin-induced association of gold nanoparticles modified with α-lactosyl-ω-mercapto-poly(ethylene glycol). *J Am Chem Soc* 2001;**123**:8226–30.
126. Whaley SR, English DS, Hu EL, Barbara PF, Belcher AM. Selection of peptides with semiconductor binding specificity for directed nanocrystal assembly. *Nature* 2000;**405**: 665–8.
127. Chan WCW, Nie S. Quantum dot bioconjugates for ultrasensitive nonisotopic detection. *Science* 1998;**281**:2016–18.
128. Mattoussi H, Mauro JM, Goldman ER, et al. Self-assembly of CdSe-ZnS quantum dots bioconjugates using as engineered recombinant protein. *J Am Chem Soc* 2000;**122**: 12142–50.
129. Bruchez Jr M, Moronne M, Gin P, Weiss S, Alvisatos AP. Semiconductor nanocrystals as fluorescent biological labels. *Science* 1998;**281**:2013–15.
130. Mitchell GP, Mirkin CA, Letsinger RL. Programmed assembly of DNA functionalized quantum dots. *J Am Chem Soc* 1999;**121**:8122–3.
131. Park S-J, Lazarides AA, Mirkin CA, Letsinger RL. Directed assemble of directed materials from protein and oligonucleotide-modified nanoparticle building blocks. *Angew Chem Int Ed Eng* 2001;**40**:2909–12.
132. Cao YW, Jin R, Mirkin CA. DNA-modified core-shell Ag/Au nanoparticles. *J Am Chem Soc* 2001;**123**:7961–2.
133. Uchiyama S, Iwai K, de Silva AP. Multiplexing Sensory molecules map protons near micellar membranes. *Angew Chem Int Ed Eng* 2008;**47**:4667–9.
134. Lu Y, Liu J, Li J, Bruesehoff PJ, Pavot CM-B, Brown AK. New highly sensitive and selective catalytic DNA biosensor for metal ions. *Biosens Bioelectron* 2003;**18**: 529–40.
135. Mirkin CA, Letsinger RL, Mucic RC, Storhoff JJ. A DNA based method for rationally assembling nanoparticles into macroscopic materials. *Nature* 1996;**382**:607–9.
136. Elghanian R, Storhoff JJ, Mucic RC, Letsinger RL, Mirkin CA. Selective colorimetric detection of polynucleotides based on the distance-dependent optical properties of gold nanoparticles. *Science* 1997;**277**:1078–81.
137. Taton TA, Mirkin CA, Letsinger RL. Scanometric DNA array detection with nanoparticle probes. *Science* 2000;**289**:1757–60.
138. Murphy CJ, Gole AM, Hunyadi SE, et al. Chemical sensing and imaging with metallic nanorods. *Chem Commun* 2008:544–57.
139. Michalet X, Pinaud F, Lacoste TD, et al. Properties of fluorescent semiconductor nanocrystals and their application to biological labeling. *Single Mol* 2001;**2**:261–76.
140. Janata J, Josowicz M. Chemical modulation of work function as a transduction mechanism for chemical sensors. *Acc Chem Res* 1998;**31**:241–8.
141. Ashkenasy C, Cahen D, Cohen R, Shanzer A, Vilan A. Molecular engineering of semiconductor surfaces and devices. *Acc Chem Res* 2002;**35**:121–8.
142. Alivisatos P. The use of nanocrystals in biodetection. *Nature Biotech* 2004;**22**:47–52.
143. Lee J, Kim J, Park E, Jo S, Song R. PEG-ylated cationic CdSe/ZnS QDs as an efficient intracellular labeling agent. *Phys Chem Chem Phys* 2008;**10**:1739–42.
144. Chen C-Y, Cheng C-T, Lai C-W, et al. Potassium ion recognition by 15-crown-5 functionalized CdSe/ZnS quantum dots in H_2O. *Chem Commun* 2006:263–5.

145. Pradhan S, Ghosh D, Xu L-P, Chen S. Interparticle charge transfer mediated by π-π stacking of aromatic moieties. *J Am Chem Soc* 2007;**129**:10622–3.
146. Medintz IL, Clapp AR, Mattoissi H, Goldman ER, Fisher B, Mauro JM. Self-assembled nanoscale biosensors based on quantum dot FRET donors. *Nature Mat* 2003;**2**:630–8.
147. Janzen MC, Ponder JB, Bailey DP, Ingison CK, Suslick KS. Colorimetric sensor arrays for volatile organic compounds. *Anal Chem* 2006;**78**:3591–600.

17

Therapeutic Strategies

Es gibt nur eine Gesundheit end eine Menge von Krankheiten.
[There is only one health but the plethora of diseases.]
Wilhelm von Humboldt

Light is essential to our health and is useful as a tool in therapy, diagnosis, and preventive medicine [1–5] (Figure 17.1). There are many beneficial effects of sunlight on the body, from warmth to recovery from depression. One of the most important functions of light is transmission of information about the environment, enabling spatial orientation and motion. Photoreceptor systems affect growth, hormonal stimulation, clocking mechanism, etc. A well-known example of the positive light effect is the photosynthesis of vitamin D upon exposure to UV components of sunlight.

The history of light therapy dates back to the ancient Egyptians, Hindus, Romans, and Greeks; all of them created temples to worship the therapeutic powers of sunlight, for healing the body, as well as the mind and the soul. In Greek mythology, Apollo, the god of healing, who taught medicine to humans, is called the god of light. In prehistoric times the healing power of light was often attributed to mythological, religious, and supernatural powers [6].

Modern phototherapy is generally considered to have been started by Finsen, who at the end of the nineteenth century treated tubercular conditions of skin with heat-filtered light from a carbon arc lamp [7]; for this innovation he was awarded a Nobel Prize in 1903. The direct use of light as a therapeutic agent is currently important in the treatment of vitamin D deficiency, neonatal jaundice, autoimmune system diseases, manic depression, etc. The sequence of events leading to the therapeutic effect of the indirect phototherapy starts with the absorption of light by an administered drug. Subsequently, the photoexcited drug molecule can undergo a number of primary processes, such as photochemical reactions of the drug itself, photoreactions with endogenous molecules, or energy or electron transfer. The use of light

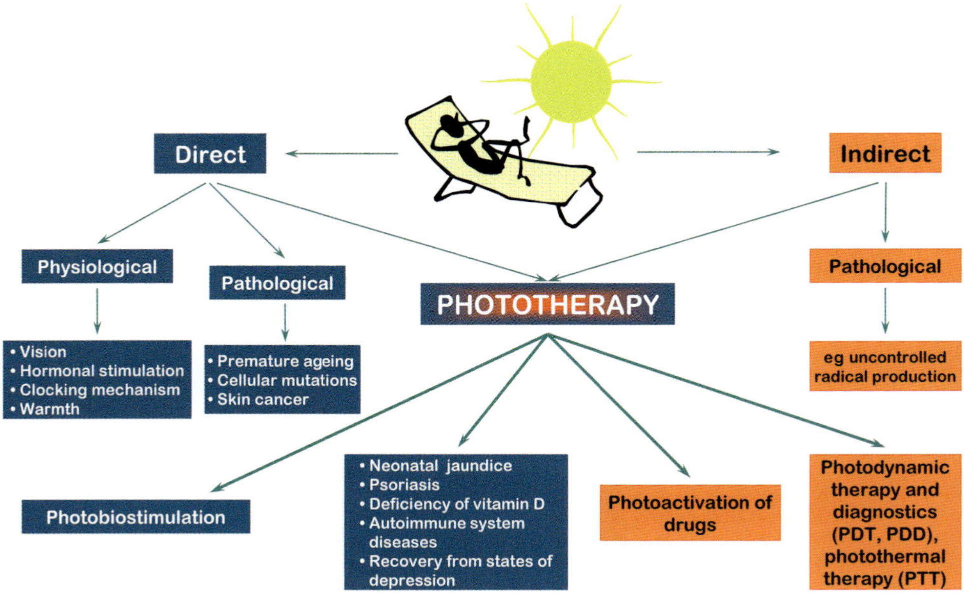

Figure 17.1 *Influence of light on human health*

and exogenous dyes to treat medical conditions can be traced back to ancient Egypt and Greece, where vitiligo was treated with the combination of orally ingested psoralen-containing plants and sunlight. At the very beginning of the twentieth century the term 'photodynamic therapy' was introduced into specialized literature [8]. Nowadays one of the most active research fields of indirect phototherapy is the photodynamic therapy of tumours. At the start photomedicine was focused mainly on treatment for cancer and dermatological diseases. Now it is obvious that several new areas of application have become accessible to this young field of medicine. Photomedical methodologies have the intrinsic advantage of dual selectivity because the photoprocess is the result of the combined effect of two factors: light and the photoactive chromophore. Photomedicine extends nowadays to such different fields as phototherapy, photodiagnosis, phototargeting, photodelivery, photoprotection, etc.

The progress in various fields of science and technology has always led to new useful applications in medicine. The human body is a uniquely organized functional system including biomolecular, cellular, and tissue structures, measuring from a fraction of a nanometre to a few metres (see Figure 15.2, Chapter 15). As medicine penetrates to even deeper organizational levels (cellular and molecular) in the human organism (levels that are closely similar in all living creatures), it evolves into biomedicine and becomes more closely connected with all the life sciences.

Biomedical photonics has the potential to revolutionize medicine as we know it, especially by application of nanotechnology and nanomaterials, leading to diagnostics and therapy at the molecular level.

Inorganic photochemistry is used in different branches of medicine. Metal ions or metal compounds important for our health are of both endogenous and exogenous origin [9–17]. Endogenous metal ions and complexes are required for many critical processes such as respiration, much of metabolism, development, neural transmission, muscle contraction, signal transduction, gene expression, protection against toxic, and mutagenic agents [9, 18]. Exogenous compounds can be introduced into the organism through diet and via interaction with the environment in order to induce a predetermined alteration of the system, which is the case in medical applications. It is notable that some organic pharmaceuticals or pollutants may be directed towards metal targets in the body or require metal binding in order to function. The biological activity of metal complexes depends on the metal, its ligands, or both. On the other hand, the size and shape of metal or metal oxide structures play an important role in nanomedical applications.

Light can interact with biologically and medically important metals and metal compounds both in vivo or in vitro. These interactions can be controlled or not, bringing positive (physiological, therapeutic, diagnostic, etc) or negative (pathological, toxic, etc) effects [5, 16, 17, 19–22]. Interaction of light and inorganic species is already used or considered for various therapeutic strategies. Some of these strategies (photobiostimulation, photoactivation of drugs, and photodynamic and photothermal therapies) are discussed in the following chapters.

17.1 Photobiostimulation

The first publication on low-power laser therapy (then called laser biostimulation or photobiostimulation) appeared more than 30 years ago [23]. Nowadays photobiostimulation is considered as a part of light therapy as well as part of physiotherapy. In fact light therapy is one of the oldest therapeutic methods used by humans (first as sun therapy, then colour light therapy, and UV therapy). The use of lasers and light-emitting diodes (LEDs) as light sources was the next step in the technological development of light therapy. Clinical application of low-power laser therapy is diverse [23–26] in physiotherapy to treat a variety of acute and chronic musculoskeletal aches and pains, in dentistry to treat inflamed oral tissues and heal diverse ulcerations, in dermatology to treat oedema, indolent ulcers, burns, and dermatitis, in rheumatology to relieve pain and treat chronic inflammation and autoimmune diseases, in sports medicine to reduce swelling and haematoma, relieve pain, improve mobility, and treat acute soft tissue injuries, and in other treatment. Low-power laser therapy is also used in veterinary medicine. The field is characterized by a variety of methodologies and uses various light sources (lasers, LEDs, lamps) supporting light of a broad range of wavelengths and maximal output power [27, 28].

The question is how radiation from therapeutic lasers and LEDs works at the cellular and organism level and what the optimal light parameters are for different uses of these light sources. It has been suggested [24] that coherent properties of laser light are not manifested at the molecular level by light interaction with biotissue. Some additional (therapeutic) effects from coherent and polarized radiation can occur only in deeper layers of bulk tissue [24]. At the cellular level, biological responses are determined by absorption of light by photoacceptor molecule. After promotion of electronically excited states, primary molecular processes from these states can lead to a measurable biological effect at the cellular level. For the identification of photoacceptor compounds action spectra are used. A graph representing photoresponse as a function of wavelength λ, wave number λ^{-1}, frequency v, or photon energy is called an action spectrum. The action spectrum resembles the absorption spectrum of the photoacceptor molecule [24, 29, 30]. Irradiation of cells in the range 380–1000 nm in the absence of an exogenous dye can alter the whole metabolism by activation of some cellular native components [31, 32]. Irradiation can activate or inactivate enzymes [33–40], may induce electron transfer and, as a consequence folding of proteins [41, 42], cause topical heating (eg haem) [43], generate superoxide anion and singlet oxygen [32], or have a biostimulating effect [31, 44] by affecting the redox chain in mitochondria [32]. Low-energy and low-intensity optical irradiation has stimulating effects on various biological systems. Irradiation of blood vessels induces local vasodilatation which results in improved microcirculation and therefore improved blood supply in organs. Moreover, several effects are observed on cellular and subcellular levels: (1) increased activity of lysosomes and phagocytes, (2) enhanced cell division and growth, and (3) activation of protein and cytokine synthesis.

Some of those beneficial effects can be associated with photochemistry of metal-containing biomolecules, such as cytochrome *c* oxidase, superoxide dismutase, nitric oxide synthase, oxyhaemoglobin, and nitrosohaemoglobin [24, 45]. Photothermolysis requires several visible-light-absorbing endogenous chromophores (eg haem, melanin). The radiationless decay of their excited states leads to heat evolution. The highly localized hyperthermal effect can induce important chemical (eg bond cleavage) and mechanical (eg ultrasonic shock) damage of biological systems [46]. Selective photothermolysis with oxyhaemoglobin as the molecular target allows the treatment of some epidermal and dermal lesions such as port-wine stains [47, 48] and chronic plaque psoriasis [49]. This therapy is limited to treatment of pigmented or highly vascularized tissues. Recently it has been proven that nitrosylhaemoglobin (Hb-NO) may be one of the possible targets for visible light in the systemic blood [50]. Laser light irradiation at the Soret band leads to NO release both in Hb-NO solution [51] and red blood suspensions [50]. NO released from the nitrosyl complex may cause vasodilatation in vivo.

The primary mechanisms of light action after absorption of light quanta and the promotion of electronically excited states have not been established. Various mechanisms have been suggested and, as it has been specified by T. Karu [24], the suggestions made to date are summarized in Figure 17.2. The first historically proposed mechanism was the 'singlet-oxygen hypothesis' [52]. Certain photoabsorbing molecules, such as porphyrins and flavoproteins (some respiratory-chain components belong to these classes of compounds), can be reversibly converted to photosensitizers [53]. Based on visible-laser-light action on RNA synthesis rates in HeLa cells and spectroscopic data for porphyrins and flavins, the hypothesis was put forward – the absorption of light quanta by these molecules was responsible for singlet-oxygen generation and, therefore, for stimulation of the RNA synthesis rate

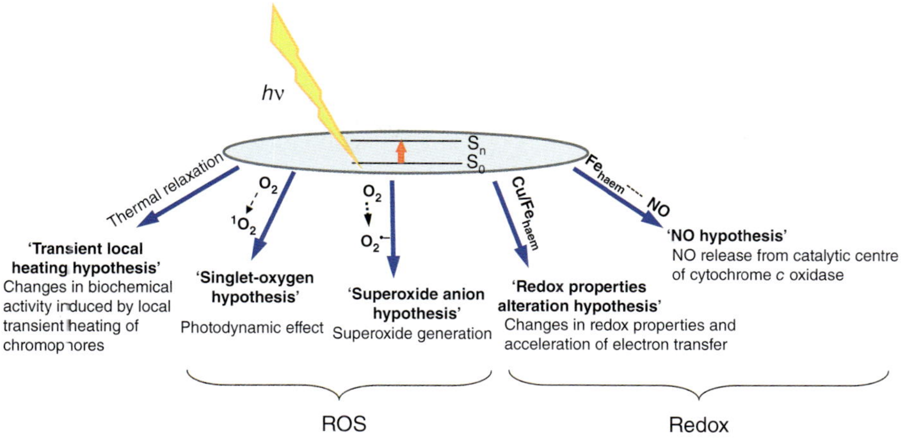

Figure 17.2 *Biostimulation: hypothesis on primary biochemical processes after light absorption. (Adapted from Karu [24])*

[52] and the DNA synthesis rate [54]. This possibility has been considered for some time as a predominantly suppressive reaction when cells are irradiated at higher doses and intensities [55, 56]. The 'redox properties alteration hypothesis' was the next proposed mechanism [57]. Photoexcitation of certain chromophores in the cytochrome *c* oxidase molecule (Cu^A and Cu^B or haems) [58] influences the redox state of these centres and, consequently, the rate of electron flow in the molecule [57]. Proving the regulation of the activity of cytochrome *c* oxidase under physiological conditions by nitric oxide (NO) [59] led to the next hypothesis of photobiological primary mechanism, called the 'NO hypothesis'. In 1993 it was suggested [60] that activation of the respiratory chain by irradiation would also increase production of superoxide anions ('superoxide anion hypothesis'). It has been shown that the production of $O_2^{-\bullet}$ depends primarily on the metabolic state of the mitochondria [61]. The mechanism suggested most recently is called 'transient local heating hypothesis' [62]. When electronic states are excited with light, a noticeable fraction of the excitation energy is inevitably converted to heat, which causes a local transient increase of temperature. The local transient rise of temperature of the absorbing biosystem may cause structural (eg conformational) changes and trigger biochemical activity (cellular signalling or secondary dark reactions) [45, 62, 63].

17.2 Photoactivation of Drugs

The strategy based on light application as a trigger of pharmacological activity of the prodrug enables precise control of the drug action in both a temporal and spatial domain, which reduces the side effects and restricts the drug activity to the desired tissue and/or organ. Careful design of the exogenous photoactive molecule together with extensive photophysical and photochemical studies enable excited state engineering and channelling of the excitation energy in vivo to the desired mode of action (see Figure 15.6 in Chapter 15) [64, 65]. Thus, in photodiagnosis (PD) (also called 'photodynamic diagnosis', PDD), widely used in clinical detection of cancer lesions [66–68], the photoactive molecule re-emits the excitation energy as fluorescence. During photothermal therapy (PTT) the excited photoactive molecule undergoes internal conversion to the ground state, leading to local heating of the tissue [43]. Excitation of the photoactive molecule can also induce different types of photoreactions including the direct photoactivation of a pharmaceutical or its indirect photodynamic action. The latter, the so-called photodynamic effect, is used in photodynamic therapy (PDT) (compare section 17.3, Chapter 17), which is one of the most promising anticancer therapies still under investigation [7, 19, 68–70]. The photodynamic effect has been also applied in antimicrobial therapies and blood sterilization [71, 72] (see Chapter 18).

Photoreactions leading to biologically active products and proceeding directly from the metallopharmaceutical excited state constitute one possibility of photoactivation. Depending on the character of the excited state (CT, LF, or IL) various types of photoreactions (redox, dissociation, substitution, and rearrangement) occur and can be used for medical applications [5, 21]. Photoactivation of metallopharmaceuticals may thus consist of photodissociation of the coordinated ligand (which may be also a consequence of the photoredox process), redox transformation of the metal centre (eventually accompanied with changes in the first coordination sphere), isomerization etc. All these processes can finally lead to formation of pharmacologically active species.

Active Ligand

Inorganic NO donors are good examples of prospective pharmaceuticals activated via their photodissociation, photosubstitution, and photoredox processes.

NO plays an important role in mammalian biology, including functions in the cardiovascular system, neurotransmission, and the immune response [73–79]. The reactive NO species (RNOS), which are generated, transferred, and consumed within biological and medical processes, may be viewed as $NO^{\bullet}$, NO^{-}, and NO^{+} species. The release of the active forms of NO from the NO donor drug may proceed on spontaneous, enzyme-catalyzed, or light-induced pathways.

Among various NO donors only one metal nitrosyl complex is used in clinical practice (sodium nitroprusside, $Na_2[Fe(CN)_5NO]\cdot 2H_2O$) under the commercial name Nipride [80]. It can be administered under very restricted conditions (very low concentration, protection of infusion fluid from light) as an emergency drug in the case of hypertension crises that are no longer influenced by other drugs. From various $[M(CN)_xNO_y]^{n-}$ complexes (where $M = Cr^{I}$, Mn^{I}, Mn^{II}, Fe^{I}, Fe^{II}, Fe^{III}) only iron complexes appeared to be pharmacologically active [5, 81]. These complexes have the highest predisposition to undergo a nucleophilic attack followed by NO^{+} release which is consistent with one of the two mechanisms suggested for in vivo metabolism of nitroprusside. The second mechanism involves its reduction, which is then followed by a nucleophilic attack [5, 81, 82]. Photochemical release of NO from nitrosylpentacyanoferrate(II) has been shown to lead to the abatement of pharmacological activity of nitroprusside and prolonged irradiation with visible light can even cause phototoxicity due to release of cyanide ligand [81, 83]. Thermally stable $[Fe(CN)_5N(O)SR]^{3-}$ complexes formed in the reactions of nitrosylpentacyanoferrate(II) with biologically important S-nucleophiles undergo photodecomposition via unstable $[Fe^{III}(CN)_5\{N(O)SR^{\bullet-}\}]^{3-}$ charge-separated states, which in turn decompose to the nitrosothiyl anion radical. Photogenerated nitrosothiyl anion radical may decompose to the thiolate anion and nitric oxide or undergo oxidation, yielding nitrosothiol [84].

There are attempts to use other iron nitrosyls, both monomeric complexes and clusters [85–90]. Photochemical liberation of NO follows three main reaction pathways: (1) photooxidation-substitution, (2) photoreduction, and (3) ligand rearrangement or decomposition as a result of photoreaction.

The photooxidation mode is typical for nitrosyl complexes containing linear nitrosyl ligands (formally NO^+). This was the case for nitrosylpentacyanoferrate and its thiol derivatives described above. Various iron complexes with polypyridine–carboxamide ligands were found to be good NO donors on visible light irradiation. These complexes are also reported to S-nitrosate thiols with good yields [91]. Nitrosyl complexes of ruthenium have been found to be good NO donors [92–94]. Two NO complexes, trichloronitrosylruthenium $[RuCl_3(NO)]$ and dipotassium pentachloronitrosylruthenate $K_2[RuCl_5(NO)]$, have been shown to be thermally stable but photolabile, releasing NO on exposure to near-UV light [95, 96]. They are water soluble and thereby unlikely to cross biological membranes. Among various nitrosyl ruthenium complexes, an important class is represented by the salen complexes [97]. They undergo photochemically activated labilization of NO. Other transition metal macrocyclic complexes, such as $[Ru^{II}X(NO)(cyclam)]^{2+}$ (where X = halogen) and $[Cr^{III}(ONO)_2(cyclam)]^+$, were also found to photoactivate NO [98, 99]. Recently a series of new ruthenium nitrosyls with very high quantum yields of NO release was reported [100–102]. These complexes contain various N-ligands (tetraazacyclopentadecane, polypyridyls or pyridine ligands, amino acids, and alkyl phosphites). Irradiation within the ligand-to-metal charge transfer (LMCT) absorption band results in photolabilization of coordinated nitrosyl with quantum yields up to 0.6 [100–102]. A family of ruthenium sulphoxide–nitrosyl complexes represents dual biological activity. They serve not only as NO donors, but the sulphoxide–ruthenium fragment shows a strong anti-tumour activity in animal tumour models [93, 103].

The photoreduction mode is important in complexes containing bent (formally NO^-) nitrosyl ligand. A good example of this class of NO donors is iron-sulphur nitrosyl monomers and clusters. The most studied complexes belonging to the $[Fe_xS_y(NO)_z]^{n-}$ family are Roussin's red salt, $[Fe_2S_2(NO)_4]^{2-}$, and Roussin's black salt, $[Fe_4S_3(NO)_7]^-$. These compounds contain a large number of nitrosyl ligands within a single anion and are known to be highly effective nitrovasodilators with unusual pharmacological profiles [85, 86, 104]. Easy interconversion of $[Fe_4S_3(NO)_7]^-$, $[Fe_2S_2(NO)_4]^{2-}$, $[Fe_2(SR)_2(NO)_4]$, and $[Fe(SR)_2(NO)_2]^-$ makes the system susceptible to ligand-controlled NO photogeneration (Figure 17.3). The quantum yield of NO photogeneration depends strongly on oxygen concentration. Solubility of the Fe–S–NO cluster compounds is in turn controlled by the ligand structure. These properties of the prodrug may lead to tissue-selective NO photogeneration, which is important from a photomedical point of view [84, 105, 106].

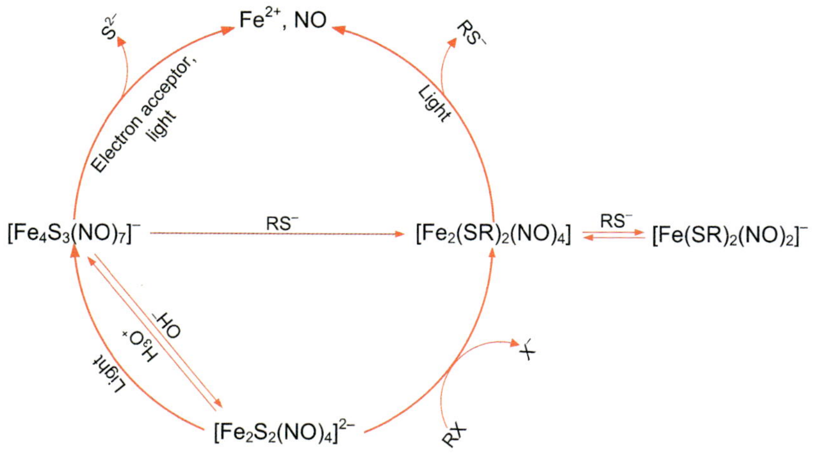

Figure 17.3 *Thermal and photochemical transformations of iron-sulfur-nitrosyl clusters. (Adapted from Szaciłowski et al. [5])*

Metalloporphyrins are an interesting class of light-sensitive nitrosyl complexes [107–111]. Depending on coordination mode (linear MNO vs bent MNO), the photoinduced denitrosylation process can be regarded as photo-oxidation or photoreduction. Metalloporphyrin and haem protein nitrosyl complexes were the subjects of photochemical studies for a long time [112–119]. It is known that these complexes release NO with quantum yields varying from virtually 0 to 1 when photons are absorbed by the porphyrin macrocyclic π-system. Electronic excited states in these systems are inherently short-lived because empty d-orbitals of transition metals can couple with the porphyrin π-orbitals, forming states of intermediate energy [119]. Only those excited states that relax via the CT state result in loss of NO [120, 121]. It was found that utility of the nitrosyl porphyrin complexes of the first-row transition metals in photochemical NO delivery to specific targets is severely hampered by their lability and oxygen sensitivity [119, 122]. In this respect the main attention was turned to the nitrosyl ruthenium–porphyrin complexes, which were anticipated to be more stable. This has directed investigations into the synthesis, structural characterization, and photochemical reactivities of several ruthenium–porphyrin complexes of the $[RuL(ONO)(NO)]^n$ type, where L = tpp, oep, tmtp, and fttp [87, 119, 122, 123].

Transition metal centres of various biomolecules are among the most important receptors of NO in all living organisms [50, 124–128] and these reactions are responsible for most physiological actions of NO, both direct and indirect, beneficial and deleterious. It creates the unique possibility of controlling natural physiological processes by light. Coordination of NO to metalloproteins results in activation (guanylyl cyclase, haemoxygenase) or deactivation (cytochromes, catalase, ferritin, nitrile hydratase) of their biological functions [34, 127]. Selective irradiation of chosen NO-metalloproteins induces photolabilization of NO [50]. This may have important medical applications and serve as a tool in elucidation of biochemical pathways involving NO and metalloproteins [109, 110, 129–131] Recently the same approach was used to control the activity of the industrially important enzyme, nitrile hydratase [34, 132–134].

Photodissociation of 4-aminopyridine from ruthenium polypyridyl complex was reported recently [135]. It was found that the complex does not interact with neurons, although the photoliberated ligand gives a strong physiological response. Photodissociation occurs at low-energy excitation. This research opens a new possibility for optical control of activity of selected groups of neurons for neurobiochemical studies.

A property of considerable importance for the use of photoactivatable agents in vivo is the ability to absorb light in the range of high tissue penetration and sufficient energy to labilize the metal–nitrosyl bond selectively. For mammalian tissue, near-infrared (NIR) light of λ between 700 and 1100 nm has the highest transmission [136, 137]. Therefore the development of compounds that can be subjected to a single (SPE) or two-photon (TPE) excitation by NIR light, resulting in the photochemical production of NO, is one of the current research topics [138]. TPE is especially interesting because it allows access to excited states which would be populated on near-UV excitation in the case of SPE [139, 140]. Another key advantage of TPE for pro-drug activation is spatial selectivity achieved from the intensity-squared dependence of the two photon absorption processes [140]. TPE has drawn attention to new strategies for PDT (see section 17.3, Chapter 17) and for the photochemical release (uncaging) of bioactive substances [141]. Recently a dye-derived iron sulphur–nitrosyl cluster $[Fe_2(\mu\text{-}RS)_2(NO)_4]$ with pendant chromophore designed to have a high two-photon absorption cross-section has been reported [142]. Under continuous photolysis at 436 nm it decomposes with a modest quantum yield, releasing NO.

Active Metal Centre or Complex

Biological activation of Pt^{IV} complexes via their photoreduction to Pt^{II} species has been suggested recently [143–147]. The photoactive Pt^{IV} analogues of the anticancer drug cisplatin $[PtCl_2(NH_3)_2]$, such as $[Pt(OAc)_2I_2en]$ and $[Pt(OAc)_2(N_3)_2en]$ compounds, have recently been designed and investigated [143–147]. It has been shown that iodo complexes of Pt^{IV} can platinate a nucleotide (5′-GMP) and DNA under illumination. The cytotoxicity is slightly enhanced by photolysis of these compounds in comparison to the dark control [144, 145]. However, slow photoreactions and low stability in serum due to a high reactivity towards reducing agents limit the use of these complexes in anticancer therapy. Recently developed Pt^{IV} diazidediamine complexes, which are highly stable in human blood plasma and react very slowly with reducing agent (eg glutathione), seem to be more promising [147]. It has been reported that visible light photoactivation of these compounds results in formation of a very reactive Pt^{II} species that binds rapidly to nucleotides (5′-GMP, d(GpG)) and DNA forming cisplatin-nucleotide cross-links [146, 148]. Transcription mapping of plasmid DNA showed the platination sites to be similar to those obtained for cisplatin, mainly GG sequences [146]. Recent studies have revealed that photoactivation of Pt^{IV} diazidediamine complexes leads to inhibition of the growth of human bladder and ovarian cancer cells; they are not cross-resistant to cisplatin [148].

Photoinduced ligand loss and/or ligand substitution reactions can also find application in prospective drug activation. The cytotoxic effect of two ruthenium(II) complexes, *cis*- and *trans*-[$RuCl_2(dmso)_4$] against melanoma cells was found to be significantly increased by irradiation with UV-A light [149–151]. It was suggested that the observed enhancement of antiproliferative activity could arise from phototransformation of the complex into more active species by gradual substitution of dimethylsuphoxide (dmso) and Cl^- ligands by H_2O. The formed new species had considerable substitution lability relative to parent complexes, and therefore were proposed to be more reactive towards DNA. The enhancement of cytotoxic effect against several tumour cell lines (GN4, M109, KB) upon irradiation with UV light was also observed for the rhodium complex, *cis*-[$RhCl_2$(phen)(dppz)]$^+$ (see Figure 14.8) [152]. Moreover, this complex is able to photoinactivate the Sindbis virus. The nucleic acid is suggested as a target for the rhodium complex and a dual mode of the reaction with the nucleic acid, ie covalent binding and nicking, is proposed. The detailed mechanism is unknown but one can speculate that the loss of chloride ligands upon irradiation can be at least partially responsible for the formation of covalent bonds between aquated complex and DNA. Recently a new dinuclear ruthenium(II) arene complex [{[(η^6-indan)RuCl]}$_2$(μ-2,3dpp)](PF_6)$_2$ was synthesized and it was shown that its irradiation with UV-A light leads to loss of indan and chloride ligands and formation of highly reactive dinuclear ruthenium complexes [153]. These complexes can bind to DNA whereas indan can be used as a fluorescent marker. It was also demonstrated that the inhibition of RNA polymerase by the studied complex is enhanced by irradiation, suggesting formation of stable cross-links.

Another type of photochemical activation of metal complex is based on promoting the photocleavage of DNA often followed by photoadduct formation. These types of complexes were applied to photocleavage of DNA; complete mechanisms of their action are described in detail in Chapter 14. It has been shown that various ruthenium complexes capable of DNA photooxidation with concomitant formation of covalent adducts have significant effects on DNA functions and disturb gene expression [152, 154–158]. The photocleavage of DNA, as well as formation of photoadducts with DNA, can inhibit the binding of enzymes involved in transcription and replication processes or prevent the progression of these enzymes along DNA. As a consequence these actions inhibit the transcription and replication processes. The complexes that possess such activity could be good candidates for phototherapy. Several compounds acting as photocleavers have been found to inhibit growth of tumour cells. Ruthenium complexes of the [Ru(tap)$_2$(phen)]$^{2+}$ and [Ru(tap)$_2$(poq-Nmet)]$^{2+}$ type inhibit gene transcription by direct photoinduced electron transfer between the metal complex (containing π-deficient ligands) and DNA, with concomitant formation of covalent photoadducts with DNA [154]. The real advantage is that the activity is no longer based on the presence of oxygen. Another complex type, Λ-1-[Rh(mgp)$_2$(phi)]$^{3+}$, can inhibit site specifically the binding of transcription factor to the DNA [155]. The photoreaction of these complexes with DNA is non-specific and, in order to direct the photoactive complex to a specific sequence, the complexes have been anchored to oligonucleotide. Synthetic oligonucleotides that have sequences complementary to part of the messenger RNA (mRNA) may recognize and react with the target mRNA sequence. In that way they inhibit the expression of the targeted sequence. This strategy, the so-called anti-sense or anti-gene strategy, has a few drawbacks such as poor stability after hybridization of synthetic oligonucleotides with target DNA and destruction of the system by some enzymes. In this context the use of photoactive ruthenium(II) complexes, such as $[Ru(dip)(tap)_2]^{2+}$ attached to an oligonucleotide, allow irreversible cross-linking of the two strands on irradiation. Electron transfer to the photoexcited complex from DNA initiates the formation of adducts between the metal complex and a specific guanine residue of the target DNA strand [156–162]. Other types of photoadducts have been found for the $[Ru(CH_3CN)(dppz)(tpy)]^{2+}$ complex tethered to the oligonucleotide. Acetonitrile dissociates from the complex on irradiation, generating the aqua derivatives, which in dark reaction form a cross-link with a guanine base via a coordination bond [162, 163].

Recently, a photocytotoxic activity of a series of dirhodium(II/II) complexes, $[Rh_2(\mu\text{-}OOCCH_3)_4(dppz)L]^{n+}$ (where L = CH_3OH, bpy, or dppz), has been demonstrated in in vitro tests [164–166]. It was found that some of the derivatives show pronounced cytotoxicity on irradiation and can be good candidates for phototherapy. It is believed that they can photocleave DNA; however, the exact mechanism of photochemical reaction and the nature of the reactive excited states remain unclarified.

17.3 Photodynamic Therapy

PDT, also known as photochemotherapy, involves cell destruction caused by means of toxic singlet oxygen and/or other free radicals produced from a sequence of photochemical and photobiological processes [140, 167–170]. These processes are initiated by the reaction of a photosensitizer with tissue oxygen upon exposure to a specific wavelength of light in the visible or NIR region. Acridine was the earliest photosensitizer, which was reported in 1900 to kill paramecia [171], followed by eosin for skin cancer treatment in 1903 [172]. Although many chemicals have been later reported for photochemical therapy, porphyrin-based sensitizers [1, 7, 173–176] lead the role in clinical applications because of their preferential retention in cancer tissues and due to the high quantum yields of singlet oxygen generation. For deeply seated tumours, a second generation of photosensitizers, absorbing in the NIR region, have been introduced (among others there were core-modified porphyrins, chlorins, phthalocyanines, and naphthalocyanines).

During PTT the excited photoactive molecule undergoes internal conversion to the ground state, leading to local heating of the tissue [1, 2, 140, 177, 178]. Natural chromophores in the tissue [46, 179] or externally added dye molecules such as indocyanine green [180], naphthalocyanine [1, 7, 65], and porphyrins coordinated with transition metals [1, 7, 181] can play a role of photoabsorbing agents. The choice of the exogenous photothermal agent is made on the basis of the strong absorption cross-section and high efficiency of light-to-heat conversion.

17.3.1 Mechanisms of PDT and PTT

The molecular mechanisms involved in PDT and PTT are illustrated in Figure 17.4 (the Jabłoński diagram). The electronic ground state of the photosensitizer is a singlet state (S_0). (1) Upon light absorption the photosensitizer is excited to a short-lived first excited singlet state (S_1). The photosensitizer can return to the S_0 state by (2) emitting the absorbed energy as fluorescence or (3) internal conversion. (4) Alternatively, S_1 can undergo conversion (with a reasonably high quantum yield) to the first excited triplet state (T_1) by intersystem crossing. From the T_1 state the photosensitizer can also (5) release energy by emitting phosphorescence or (6) return to the ground state through non-radiative path. (7) The sufficiently long-lived T_1 state is able to take part in chemical reactions based on the electron (or hydrogen) transfer with (8) generation of radical intermediates involved in photodynamic reactions of type I. (9) Energy transfer from the triplet state of the photosensitizer to a suitable acceptor, most frequently oxygen, results in generation of highly reactive singlet oxygen involved in (10) photodynamic reactions of type II. Reactive oxygen species (ROS) and other radical intermediates are highly cytotoxic agents. In most cases the mechanism type II is dominant in PDT [182, 183]. However, type I reactions are most efficient at low oxygen and high substrate concentrations [184]. Photosensitizers should also act as catalysts, ie their ground state should be ultimately regenerated, otherwise fast 'photobleaching' (destruction of the absorbing chromophore) is generally observed on irradiation [185].

PDT in vivo acts through at least three principal modes: (1) direct cell killing by lethal oxidative damage of tumour cells (necrosis, apoptosis); (2) indirect cell killing due to photodynamic damage or shutdown of the (neo)vasculature with loss of oxygen and nutrients supply to the tumour; and (3) additional anti-tumour contributions from the inflammatory and immune responses [186]. There are many organelles such as cellular membranes, nuclei, mitochondria, which can be affected by photogenerated reactive species. In consequence apoptosis, necrosis, or their combination takes place [5, 140, 167, 182, 187, 188].

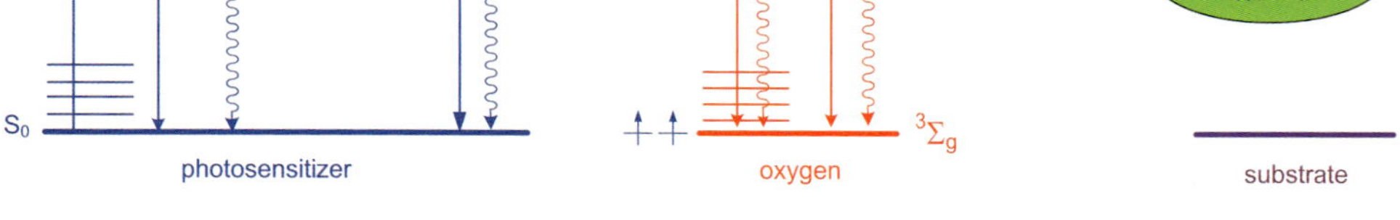

Figure 17.4 *Physical and chemical processes involved in (7–10) photodynamic therapy (PDT), (3, 6) photothermal therapy (PTT), and (2, 5) photodiagnosis (PD). (Adapted from Szaciłowski et al. [5])*

17.3.2 Photosensitizers

The phototherapeutic effects of photosensitizers depend on their physicochemical properties in the ground and excited states, pharmacokinetic and pharmacodynamic behaviour, and photoactivity in vivo [16, 19, 189]. A good photosensitizer should be a single substance with constant composition and a high degree of chemical purity, non-toxic in the dark, and sufficiently stable under physiological conditions, characterized by a high absorption coefficient within the phototherapeutic window (about 620–1000 nm) and as low as possible in the range 400–600 nm to avoid prolonged skin sensitivity towards solar irradiation after drug administration. As the efficiency of a PDT photosensitizer depends on photophysical properties of the first excited triplet state, this state should be generated with a high quantum yield, and have an appropriate energy and long lifetime to allow an efficient energy or electron transfer to an oxygen molecule. The photosensitizer must be photostable and it should not be easily oxidized by singlet oxygen or other ROSs generated *in situ*. It should be selectively accumulated into malignant and easily removed from healthy tissues. The amphiphilic character of the compound is advantageous, enabling its accumulation and acting in both membranes and physiological fluids [19, 65, 69, 190–194]. Where PDD applications are concerned, it should exhibit an intensive fluorescence, whereas for PTT purposes non-radiative conversion to the ground state should dominate [5].

The first-generation photosensitizers are based on chemically modified natural haematoporphyrins [189, 195]. Their weak light absorption in the phototherapeutic window, a relatively poor specificity of uptake, and retention with respect to malignant and healthy tissues, as well as prolonged skin photosensitivity (usually 2–3 months) [196], are their main drawbacks. The second-generation PDT sensitizers are mainly based on engineered, synthetic, and semisynthetic porphyrins, with various substituents, at both pyrrole rings and methylene bridges [16, 197]. They are structurally homogeneous compounds with long-wavelength absorption bands of high intensity. So far many compounds have been tested as potential photosensitizers for application in PDT [16, 19, 189]. They can be divided into five groups: organic dyes (eosin, rose Bengal, methylene blue), aromatic hydrocarbons (naphthalenes, anthracenes, biphenyls, quinines), polypyrrolic and metallopolypyrrolic compounds [16, 19], transition metal complexes, and semiconductors [190]. The third-generation photosensitizers consist of the photosensitizing moiety linked to biomolecules such as monoclonal antibodies, polypeptide chains, proteins, which allow their selective delivery [198]. This strategy overcomes difficulties in molecule recognition and specific binding to the tumour. Nowadays various porphyrins, chlorins, bacteriochlorins, phthalocyanines, naphthalocyanines, and texaphyrins are most common investigated photosensitizers (Figure 17.5a) [16, 189, 191]. Photosensitizers currently approved for clinical use belong to various groups: the first accepted photosensitizer is Photofrin, which is a mixture of haematoporphyrin monomers, dimers, and oligomers; then precursors of protoporphyrin IX – 5-aminolevulinic acid (Levulan) – and its methyl (Metvix), hexyl (Hexvix), and benzyl (Benzvix) ester derivatives; *m*-tetrahydroxyphenylchlorin (*m*THPC, Foscan); and luthetium texaphyrin complex (Lutex, Motexafin, Lutetium) [191].

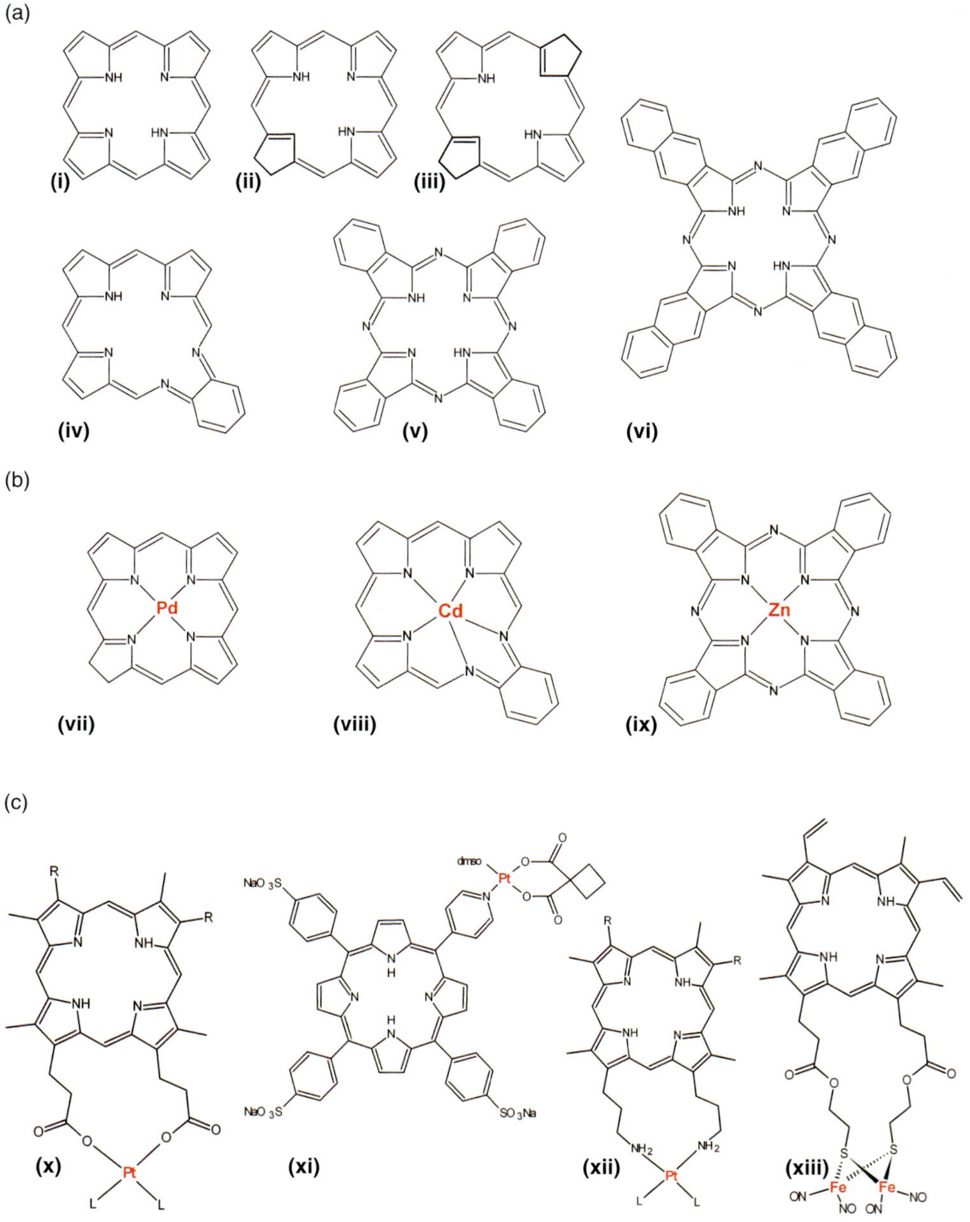

Figure 17.5 *Skeleton structures of (a) polypyrrolic photosensitizers: (i) porphyrin, (ii) chlorine, (iii) bacteriochlorin, (iv) phthalocyanine, (v) naphthalocyanine, (vi) texaphyrin; (b) examples of metalloderivatives; (c) photosensitizers incorporating the metallodrug moieties: (x–xii) cisplatin- and/or carboplatin-like structures, (xiii) iron–sulphur–nitrosyl cluster*

17.3.3 Inorganic Photosensitizers

Various metal compounds have been tested as possible photosensitizers for PDT and PTT. Some of them are described below.

17.3.3.1 Transition Metal Complexes

Metal complexes can act as prospective photosensitizers through both energy and electron transfer to oxygen molecules. Most of the studies on photogeneration of singlet oxygen involve polypyrrolic dyes (porphyrins, porphycenes, phthalocyanines, etc) and their metalloderivatives (see below), but there is a significant number of other metal complexes capable of photosensitized singlet oxygen generation. A low-energy triplet excited state of long lifetime is a prerequisite for this phenomenon. Mono- and dinuclear Ru^{II}, Cr^{III}, Os^{II}, Ir^{III}, and Pd^{II} complexes with polypyridines, 3,4-hydroxybenzoic acid or di- and tetrahydroxybenzaldehyde, and some other ligands, can be used as singlet oxygen photogenerators [190].

Other possible applications of metal complexes in photodynamic therapy do not involve singlet oxygen generation, but photoinduced redox processes. In this context ruthenium(II) complexes with heterocyclic ligands are promising agents for PDT [103, 199]. The excitation of Ru^{II} polypyridyls leads to the ^{1}MLCT state which undergoes an efficient (quantum yield close to unity) intersystem crossing to the ^{3}MLCT state, which in turn can be deactivated by (1) radiative emission, (2) radiationless decay with heat evolution, (3) radiationless thermal conversion to the higher-energy ^{3}MC state, and (4) chemical reaction (Figure 17.6). The molecule in

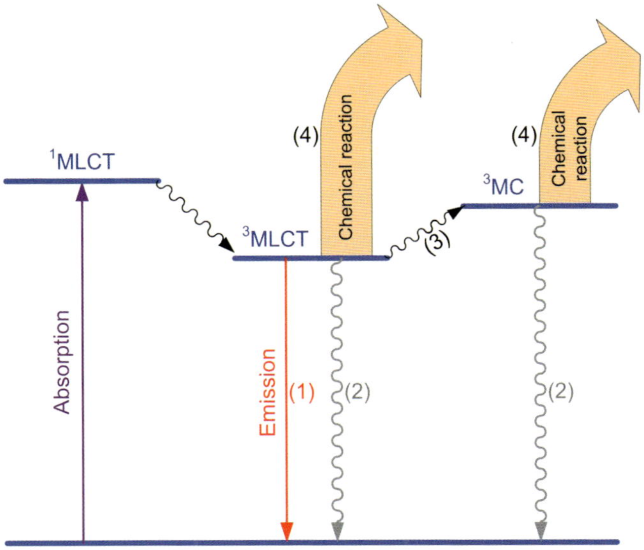

Figure 17.6 *Photophysical processes of the ruthenium(II) polypyridine complexes. MC, metal centred; MLCT, metal-to-ligand charge transition. (Adapted from Moucheron et al. [199])*

the ^{3}MLCT state is a better oxidant and a better reducer than in the ground state and may initiate some redox reactions. In contrast, the ^{3}MC state involves population of a higher-energy *d*-orbital of antibonding character, thereby weakening one or more Ru–N bonds, which may result in ligand substitution. The photoredox properties of Ru^{II} polypyridyls can be modulated by changing the nature or combination of the ligands. Photoreaction of $[RuL_3]^{2+}$ with DNA results in photocleavage of DNA strands with quantum yields correlated with redox potentials of the Ru^{II} complexes. In the case of analogous Os^{II} complexes the thermal activation to the ^{3}MC state is not possible, so selective photoredox reactivity of Os^{II} complexes is superior to that of Ru^{II} compounds. The oxidation power of ^{3}MLCT Os^{II} excited state is lower than that of ruthenium, so the osmium species can selectively oxidize guanine (the strongest reducer among nucleic bases), whereas other bases remain intact [5, 200].

17.3.3.2 Macrocyclic Photosensitizers Modified by Metal Ions

Insertion of a metal ion (through metallation, transmetallation, or template synthesis) to a polypyrrolic photosensitizer (see Figure 17.5b) can modify properties of its ground and excited state in such a way that not only photophysical and spectroscopic properties of the photosensitizer will be modulated, but also its hydrophobicity, degree of aggregation, stability, and consequently the route of the photosensitizer transport into the cell and its further distribution [19, 184]. Generally, the properties of compounds with a porphyrin or porphyrin-like skeleton are, in both the ground and excited state, affected by electronic configuration of the central metal ion. The properties of these compounds containing metal ions with d^0 or d^{10} configurations (closed shell) are determined essentially by the π electrons of the tetrapyrrolic ring with only minor perturbations from the electrons of the central ion. In contrast, in the case of metal ions with *d* orbitals not fully occupied (open shell), the *d*-electrons can significantly integrate with the π and π^* orbitals of the macrocycle [201, 202].

Ground State Properties

Stability. Generally metalloporphyrins and metallophthalocyanines are thermodynamically stable. Considering different polypyrrolic ligands, the stability of corresponding metallocomplexes decreases in a series: phthalocyanines > porphyrins > chlorins. Texaphyrins are capable to coordinate larger cations because of expanded cavity of a central hole. To asses thermodynamic stability of metalloporphyrins empirical 'stability index' $S_i = 100\chi(Z/r_i)$ (where χ is the Pauling electronegativity, Z the oxidation number, and r_i the effective ionic radius [pm] of a central metal ion) was suggested by Buchler [203]. It includes both covalent bonding capacity and electrostatic interactions. According to this the complexes of metal with high values of χ and Z and small r_i, such as Ru^{III}, Al^{III}, and Sn^{IV}, are more stable than these of low χ and Z and big r_i (eg Mg^{II}, Cd^{II}) [19]. The other effect with high impact on stability of porphyrins and porphyrin-like systems is the susceptibility to one-electron oxidation. The redox properties of metalloporphyrins and metallochlorins depend significantly on a nature of the coordinated metal ion and have been linearly correlated to χ [204]. Generally, complexes of metals of a higher positive charge exhibit higher half-wave potentials for ring oxidation. For metal ions with the same charge various effects are observed, eg Pd^{II}(tpp) ($E_{½} = 1.02$ V) is harder to oxidize than the free base ($E_{½} = 0.95$ V), in contrast to Mg^{II}(tpp) ($E_{½} = 0.54$ V) and Cd^{II}(tpp) ($E_{½} = 0.63$ V) [19, 205]. Thus insertion of a metal ion into the porphyrin ring influences its susceptibility to photobleaching during PDT [5].

Aggregation. In general, porphyrin and porphyrin-like compounds have a tendency to aggregate, causing a decrease of singlet oxygen generation and a reduction of a compound's photosensitizing efficiency [206]. The presence of additional ligands in an axial position obviously influences, both electronically and sterically, the environment of a metal complex. The presence of closed-shell metal ions, which can support octahedral d^2sp^3 hybrid orbitals (eg Zn^{II}), guarantees a reasonable yield of $^1O_2$ generation, because axial ligands generate steric hindrance to intermolecular aggregation. In addition, if the ligand is hydrophilic, the solubility of the complex in aqueous phases is enhanced, which is important because the amphiphilic character of the photosensitizer is advantageous for PDT.

Selectivity. The selectivity of photosensitizers can be also affected by central metal ions and substituents at the macrocycle, eg tetrasulphonated chloroaluminium phthalocyanine [Al(pcS_4)Cl] localizes well in the tumour, whereas tetrasulphonated zinc phthalocyanine [Zn(pcS_4)] (and also disulphonated zinc phthalocyanine [Zn(pcS_2)]) shows a relatively poor tumour localization [207].

UV-Vis Absorption Spectra. The main effect of the metal on the spectrum comes from the conjugation of its *d*-electrons with the π-electrons of the ring. Complexation of a metal ion by the porphyrin macrocycle causes strong charge transfer and electrostatic interactions [202]. Porphyrins with the central metal, which have the d^0 or d^{10} electronic configuration, show three-banded spectra (α, β, Soret). The α-band is localized within the range 570–630 nm whereas the Soret band is in the region 395–405 nm [208]. The lowest energy β-band shows a hypsochromic shift induced by metal ions [16]. The absorption spectra of chlorins and chlorophylls are characterized by a strong Soret band close to 400 nm, a relatively strong – compared with the fully unsaturated porphyrins – Q_y band around 660 nm, and two weak bands in a central part of visible part of electronic spectrum [204]. Metal coordination by chlorophylls results in a hypsochromic shift of the Q_y band. In the case of metal-substituted bacteriochlorophylls four bands are observed: B_y (~330–370 nm), B_x (~380–390 nm), Q_x (~520–610 nm), and Q_y (~750–780 nm). The most important for photodynamic therapy, Q_y bands are slightly affected by coordination of various metals [204]. The insertion of a metal (eg tin) to purpurin causes a red shift of the Q_y band by about 20–30 nm, in contrast to porphyrins, where a blue shift is observed upon metallization [5, 209].

Excited State Properties

Emission. The fluorescence properties of metalloporphyrin and porphyrin-like compounds depend on the character of central metal ion. Fluorescence of porphyrins is generally observed from the free-base or closed-shell metal complexes. The longest lifetime of fluorescence (τ_F) and its high quantum yield (Φ_F) are observed for the first- and second-row elements (eg Mg, Al) and decrease on going to the third-row elements (eg Zn). Both τ_F and Φ_F are significantly decreased in the fourth-row elements (eg Cd, In). It is caused by a spin-orbit coupling on the central atom, which increases the intersystem crossing. Complexes with open-shell central metals can be divided into three groups: one contains complexes of the second- and third-row transition metal ions such as Ru^{II}, Pt^{II}, Pd^{II}, which are diamagnetic and show an intense phosphorescence. The second group covers the first-row transition metal complexes (eg Cu) – luminescent and paramagnetic, whereas the third group contains non-luminescent complexes of Ni^{II}, Co^{II}, Fe, etc. The emission is not observed because of the enhancement of the internal conversion to the ground state [201]. For metal-substituted bacteriochlorophylls (Pd^{II}, Co^{II}, Ni^{II}, Cu^{II}, Zn^{II}, Mg^{II}, Cd^{II}, Mn^{II}), no emission was detected for Co^{II}, Ni^{II}, Cu^{II}, and Mn^{II} [210]. Fluorescent photosensitizers have an advantage of being at the same time therapeutic agents and diagnostic tools, which allow monitoring of the therapy progress. Water-soluble lutetium(III) texaphyrin complex has been proved to be an excellent fluorescent marker and it is also an efficient generator of singlet oxygen [211]. [Lu(tex)$(OAc)_2$]-mediated phototherapy was shown in an experimental fibrosarcoma tumour model to be highly selective for tumour microvasculature while leaving the normal vessels in the surrounding tissue unharmed [5, 212].

IC and ISC Quantum Yield. Introduction of a heavy metal ion strongly affects the parameters of intersystem crossing in the complex, via the enhancement of the spin-orbit coupling, which influences the formation and decay of triplet states. As a result a rate of the intersystem crossing is enhanced. In extreme cases, eg Pt^{II} porphyrins, it results in intense phosphorescence even at room temperature, which can be used in optical sensors for oxygen in vivo [213]. In a series of bacteriochlorins, short S_1 lifetimes (H_2bchl – 2.6 ns; [Zn(bchl)] – 2.1 ns; [Pd(bchl)] – 65 ps) result from an efficient intersystem crossing to the triplet state leading to the increase of triplet quantum yields: 76%, 85%, and >99%, respectively [214]. Recently, the palladium(II) derivative of bacteriochlorophyll was shown to be photodynamically active in vivo [215]. In case of open-shell metal ions the *d*-electrons can significantly integrate with the π- and π^*-orbitals of the macrocycle. In consequence, the triplet excited states of these compounds are usually short-lived and the quantum yields of triplet state formation are very low. This effect can be observed, eg in the case of Ni^{II} (d^8)-substituted bacteriochlorophyll, where the excited state lifetime is extremely short (<40 fs) compared with Mg bacteriochlorophyll (~10 μs), due to enhanced internal conversion rate [216]. Such dyes exhibit photophysical and photochemical features that appear to be optimal for photothermal sensitization.

A high molar extinction coefficient, a very low quantum yield of fluorescence, a short-lived (picosecond range), and low energy triplet state mean that these materials are predestined for PTT. Among them porphyrinoid dyes with paramagnetic transition metal ions constitute a large group of well-characterized compounds [217–219]. Vibrational de-excitation of the triplet excited states generates heat locally, which may induce local 'selective cooking' of the malignant tissue [220]. In this context the use of Cu^{II}-haematoporphyrin, a chromophore that is devoid of any appreciable photodynamic activity, for photothermally sensitized inactivation of amelanotic melanoma cells by irradiation at 532 nm was reported [221]. Natural dyes used in PTT include haemoglobin and melanin whereas only a few synthetic dyes have been studied. Semisynthetic Cu^{II}-haematoporphyrin and indocyanine green are believed to act through the PTT mechanism [221]. Ni^{II}-octabutoxynaphthalocyanine [$Ni(nc(OBu)_8)$], showed neither fluorescence nor $^1O_2$ and other ROSs generation upon photoexcitation. This material can be a potential photothermal sensitizer due to a high molar absorption coefficient in the far-red spectral region which corresponds to the phototherapeutic window [1, 222].

Singlet Oxygen Generation. Closed-shell metalloporphyrins show some advantages compared with free-base porphyrins, eg the increase of quantum yield of singlet oxygen generation. In this case the major non-radiative decay route is the intersystem crossing from the first excited singlet state to the first excited triplet state, whereas the internal conversion to the ground state is minor. The photophysical effects of metallization are notable in a series of phthalocyanines. Phthalocyanine derivatives with coordinated zinc and aluminium (diamagnetic, closed-shell) show high singlet oxygen formation yields, higher compared with metal-free phthalocyanine. Thus metallized phthalocyanines are more convenient for PDT use.

17.3.3.5 Semiconductors as Photosensitizers

Redox properties of excited semiconductor particles, especially TiO_2, are responsible for high efficiency of ROS generation (compare Chapter 7). TiO_2 itself shows a very weak or no toxicity in vitro and in vivo [223]. A significant cytotoxicity of TiO_2 particles irradiated with UV light has been reported in the context of PDT applications [190, 223–232]. Most in vitro studies concern unmodified TiO_2 excited with UV light. The phototoxic action was observed under these conditions on normal cells such as mouse lymphoma cell (L5178Y) [225], Chinese hamster cell CHL/IU [225], Chinese hamster ovary (CHO) [226, 229], and human skin fibroblasts (CRL1634) [228], as well as for human cancer cells: human monocytic leukaemia cells (U937) [227], HeLa cells [233], and human adenocarcinoma cells [230]. Photoinduced killing of T-24 human bladder cancer cells in the presence of TiO_2 was reported by Fujishima et al. [234–236].

Phototoxicity of TiO_2 depends on its morphology [225]. As an $OH^•$ radical is the strongest photogenerated oxidant, it probably remains the main agent responsible for cellular death mainly due to efficient lipid peroxidation [226, 230]. The unique role of other photogenerated ROS also cannot be contested [227, 228, 237]. ROSs may damage the cellular membrane [230, 238] including altered permeability to some ions (eg K^+ and Ca^{2+} [234]), and may cause further oxidative damage of cellular components, in particular confirmed targets such as nucleic acids [225, 227, 228, 239], peptides [240], and enzymes [241]. It was suggested that DNA lesions caused in the presence of photoexcited TiO_2 particles result mainly in chromosomal aberrations.

Until now the exact mechanism of cellular death induced by TiO_2 is unknown. Nevertheless it seems obvious that incorporation of TiO_2 particles into cells should

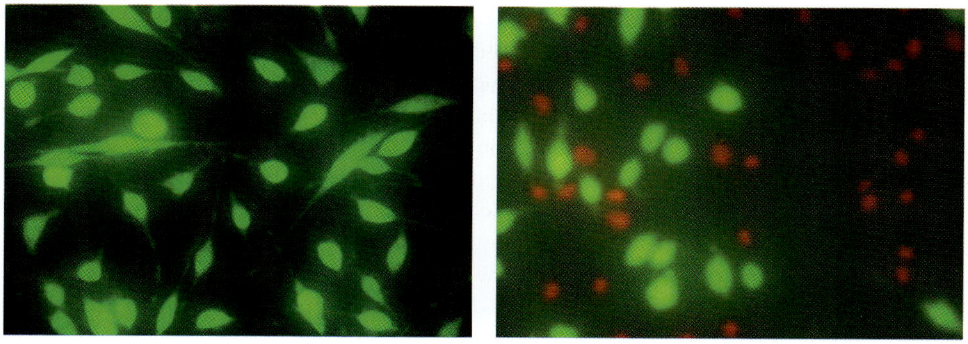

Figure 17.7 *S-91 cells stained with fluorescein diacetate and propidium iodide: (left) the control and (right) cells irradiated with visible light ($\lambda > 455$ nm, 15 min) in the presence of $\{[TiO_2]\text{-O-}PtCl_4(H_2O)\}^-$ (50 mg L^{-1}). Green-alive cells, red-dead cells*

strengthen the photodynamic effect [242]. The same effect may also be applied for photoinduced disinfection procedures because photoinactivation of microorganisms (PDI) in the presence of irradiated TiO_2 was also reported [243–246] (see Chapter 18). Besides unmodified TiO_2, TiO_2-based photocatalysts active under visible light may induce the PDI [247–249].

Recently the photodynamic effect of visible light active (photosensitized) TiO_2 has been reported [250, 251]. Tested material was simply prepared by chemisorption of Pt^{IV} chloride complex yielding $\{[TiO_2]\text{-O-}PtCl_4(H_2O)\}^-$ [252–255]. Effective decrease of survival fraction of mouse melanoma cells (S-91, Figure 17.7) and murine macrophages, RAW 264.7, was achieved on irradiation with visible light ($\lambda > 455$ nm). Lipid peroxidation and membrane perforation caused by ROSs resulted in necrotic death of the cells.

Inoue et al. assumed, that photogenerated singlet oxygen in the presence of irradiated TiO_2 may also be responsible for its photodynamic activity [256]. The energy transfer from recombination of photogenerated electron/hole pair to an adsorbed oxygen molecule in the case of TiO_2 is, however, not efficient [257]. This process requires a special surface modification of this semiconductor (see Chapter 21). Nanocrystalline silicon appears much more efficient in photogeneration of singlet oxygen, as described in Chapter 7. The role of nano-Si in PDT is described below.

17.3.4 Supporting Role of Metal Ions in Photodynamic Therapy

The effectiveness of photodynamic action of a photosensitizer is also affected by the presence of various metals ions in the cellular environment, in particular iron, which is essential for the proper functioning of all living cells. In the presence of molecular oxygen, labile iron complexes are able to cycle between the two most stable oxidation states thereby generating oxygen-derived free radicals such as the hydroxyl radical, which results in peroxidative tissue damage. The production of such highly reactive species is undesirable in normally functioning cell and thus a number of protective strategies have been adopted by cells to prevent their formation. As the protection mechanisms are not effective enough in the excess of iron ions, the synergistic action of the photosensitizer Photofrin and exogenous Fe^{III} in the presence of ascorbate was tested [258, 259]. The ascorbate reduces Fe^{III} to Fe^{II}, which readily donates an electron to lipid hydroperoxides generated previously on photosensitization and free radical chain reactions are initiated. This pro-oxidant combination of iron and ascorbate was proved to enhance cytotoxicity of PDT on L1210 leukaemia cells and human squamous carcinoma cells as well.

The other therapeutic strategy is based on selective inhibition of iron-containing enzymes [260]. The inhibition of ferrochelatase by iron chelating agent 1,2-diethyl-3-hydroxy-4-pyridione increases the level of its substrate protoporphyrin IX, which acts as an endogenous photosensitizer formed during PDT from 5-aminolevulinic acid [5, 261].

17.3.5 Combination of Polypyrrolic Photosensitizers and Metallopharmaceuticals

The high binding affinity of cisplatin towards DNA has led to its popularity as an anticancer agent [17]. However, due to cumulative drug resistance, low tumour selectivity, and toxic side effects of this classical anticancer drug, researchers explore other related metallodrugs. As has been established before, a combination of chemotherapy and PDT has a significantly additive anti-tumour effect [262, 263], which suggests a new way for classic anti-tumour drugs modification.

Studied systems involve porphyrin moieties linked to a peripheral platinum moiety (see Figure 17.5c) [264–268]. Both synthetic (sulphonated pyridinetriphenylporphyrin) and natural (haematoporphyrin) PDT photosensitizers were successfully coupled with complexes structurally similar to already used classic anti-tumour drugs such as cisplatin and carboplatin. The presence of the light absorber allows excitation of these systems, which can then act as photodynamic and chemotherapeutic agents at the same time. Solubility, reactivity, and selectivity of these complexes can be modulated by changing the first Pt^{II} coordination sphere or peripheral substituents at both the porphyrin ring and the chelating ligands [264–269].

17.3.6 Recent PDT Development

Although the original observation of photodynamic cell killing is over a century old, and the first patient treatment in the modern era took place in 1968, this concept is still evolving rapidly. The new approaches are driven by new optical technologies, advances in photosensitizer design and synthesis, and molecularly targeted drug delivery, by application of tools and concepts from molecular biology, and the identification of new potential applications.

New advances in PDT include: (1) metronomic PDT, in which the photosensitizer and light are delivered at a very low dose rate over extended periods; (2) two-photon PDT, with activation being realized either through simultaneous absorption of two low-energy photons (near infrared) or by a sequential absorption (excitation from the ground-state of photosensitizer followed by a second excitation from the triplet state); (3) PDT 'beacons' in which the photosensitizer is coupled to a quenching molecule; this molecule remains inactive until separation of the components by, for example, a specific enzymatic cleavage or hybridization to a target molecule; and (4) nanoparticle-based PDT where nanoparticles (NPs) are used as tools for improved photosensitizer delivery, photosensitizers themselves, or light energy transducers through a Förster resonant energy transfer (FRET). A development of new methods for monitoring cell and tissue responses to PDT treatment has been observed, particularly of those that are photonics based [140].

Metronomic PDT

PDT is given conventionally as a single treatment using the highest photosensitizer dose that avoids systematic toxicity and skin photosensitivity, and the highest light fluency rate that avoids tissue heating and photochemical depletion of oxygen. There have been a few studies of 'fractionated' PDT, particularly for tumours, using ALA-PpIX, which demonstrate improved responses compared with single doses [140, 270]. It has also been shown that, using ALA-PpIX, it is possible to apply the drug and light at low rates, avoiding tissue necrosis in either normal brain or tumour tissue while producing apoptotic death specifically in tumour cells [271]. By analogy to chemotherapy at low drug delivery rates, this approach has been termed metronomic – *m*PDT. Although there are still a lot of problems to be solved, this approach seems to be promising in inducing a specific mechanism of cellular death [140].

Two-photon PDT

Two-photon PDT can be realized in two ways: simultaneous (resonant) and two-photon (two-colour) PDT. In the first case exposure of the photosensitizer to a short pulse of NIR light results in the absorption of two photons by the photosensitizer (Figure 17.8a) [140]. The light pulse must provide sufficient photon flux to ensure effective two-photon absorption. The total energy absorbed is then equal to that of one-photon absorption at $\lambda/2$ and the resultant photochemical and photophysical processes are the same. In principle this has two potential advantages [140]. The first comes from the fact that the penetration of tissue by NIR light is better than that by visible light due to reduced absorption and scattering. The second potential advantage is that the quadratic dependence of two-photon absorption probability can be exploited to give an exquisite spatial confinement of the PDT effect by high numerical aperture focusing of a femtosecond activation laser beam. The second class of two-photon PDT (Figure 17.8b) involves two separate short laser pulses, the first of which activates the photosensitizer to a singlet excited state. After a short time interval this transforms to the triplet state. A second laser pulse, at a wavelength efficiently absorbed by the triplet state, generates higher excited triplet states, which can interact directly with target biomolecules. The advantage of this philosophy is the independence from oxygen of the effects [272], so the desired endpoint can be achieved even under hypoxic conditions, which is particularly relevant to PDT for solid tumours. Again technical challenges to enable this clinical treatment are formidable [140]. First, photosensitizers with high absorption coefficients in the

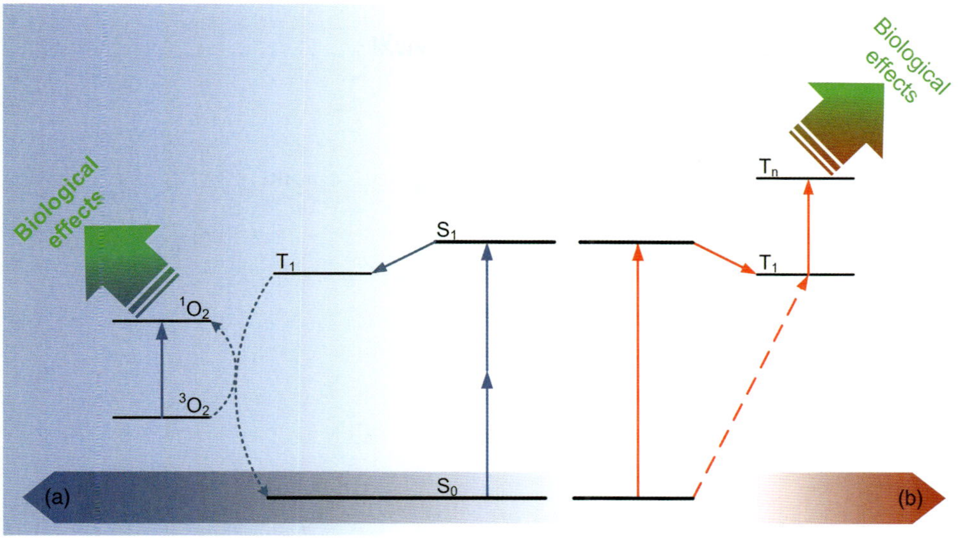

Figure 17.8 *Two-photon photodynamic therapy (PDT): (a) simultaneous absorption of two near-infrared (NIR) photons raises the sensitizer to the S_1 excited state; (b) sequential absorption of two photons, one from S_0 and the other from T_1, raises the sensitizer to a higher excited state*

ground and excited triplet states at reasonably long wavelengths are required, otherwise the depth of the treatment would be very limited. Recently it has been reported that Cu-substituted phthalocyanines absorbing at 670 and 514 nm in the ground and excited triplet states, respectively, could be considered for such applications [273]. Second, appropriate laser sources are needed [140].

PDT Molecular Beacons
The concept of this strategy [274] is shown in Figure 17.9. By analogy to the use of beacons for target-specific in vivo fluorescence imaging [274], PDT molecular beacons (PMBs) comprise a photosensitizer and a quencher linked by a target-labile linker, which keeps the photosensitizer and quencher in a close proximity so that FRET prevents activation of the photosensitizer. On interaction with the target, the linker is either broken or opened, enabling separation of the photosensitizer from quencher and allowing the PDT action to occur. This has been demonstrated in solution, in cells in vitro, and recently also in vivo in animal tumour models, using enzyme-cleavable peptide linkers [275].

Nanoparticle-Based PDT
Recent advances in nanotechnology and nanomaterials provides several different novel approaches to PDT [140, 276]. First, NPs can improve the delivery of photosensitizers, as they are postulated to do with many other imaging contrast agents or therapeutic compounds [140, 276–278]. Second, some NPs are photochemically active. Thus, for example, porous silicon NPs, with a huge surface-to-volume ratio,

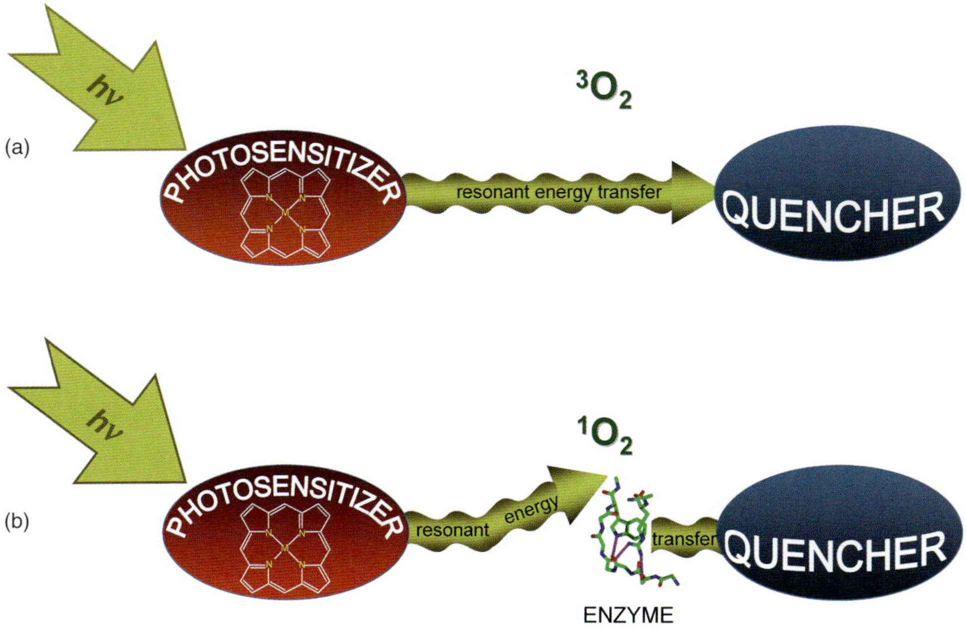

Figure 17.9 Concept of photodynamic molecular beacons: (a) the singlet oxygen formation via energy transfer from photosensitizer is prevented due to transfer of the energy to the nearby quencher; (b) singlet oxygen formation after the spatial separation of the photosensitizer and quencher

are able to generate singlet oxygen on light absorption as a result of direct energy transfer [140, 276].

The third novel class of PDT NPs is the use of nanoparticle-photosensitizer (NP-PS) conjugates, where the NP serves as the primary light absorber and then, through FRET, activates the photosensitizer. This principle has been demonstrated [140, 276] and the potential advantage is that the NP can be selected to have a very high photon cross-section at an optimal wavelength for the specific application, whereas the photosensitizer can be selected taking into account its high singlet oxygen formation quantum yield.

An issue with all strategies involving NPs is the delivery of the NPs or NP-PS conjugates to target cells or tissues. However, the potential toxicity of NPs should be considered.

17.4 Nanomedical Methods

Applications of nanotechnology and nanomaterials for treatment, diagnosis, monitoring, and control of human biological systems are referred to as nanomedicine [276, 279, 280].

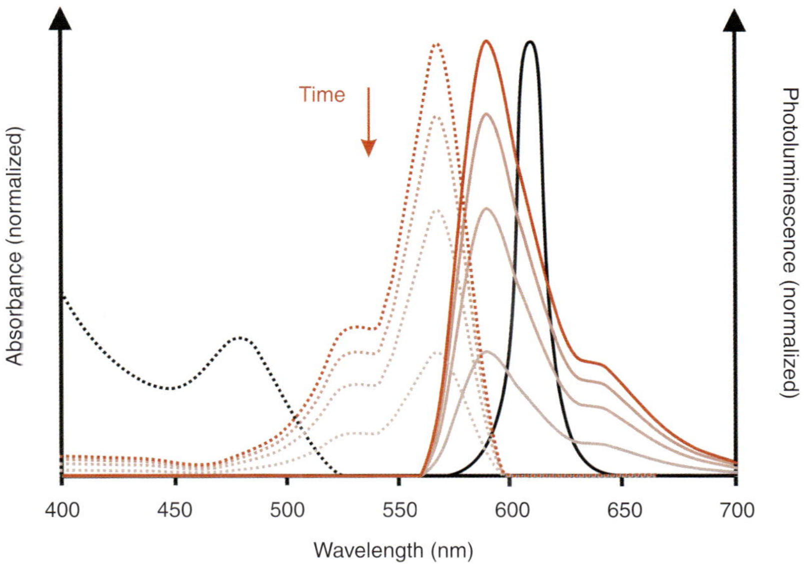

Figure 17.10 *Comparison of the schematic excitation (dotted line) and emission (solid line) spectra of the quantum dot (black) and an organic photosensitizer (red)*

Research into the rational delivery and targeting of pharmaceutical, therapeutic, and diagnostic agents is at the forefront of projects in nanomedicine. These involve the identification of precise targets (cells and receptors) related to specific clinical conditions and choice of the appropriate nanocarriers to achieve the required responses while minimizing the side effects. Mononuclear phagocytes, dendritic cells, endothelial cells, and cancers (tumour cells, as well as tumour neovasculature) are key targets [280].

It has been observed that NPs are promising tools for the advancement of drug delivery and medical imaging, and as diagnostic sensors. The size of NPs endows them with properties that can be very useful in diagnostics (particularly in imaging) and therapy [281, 282]. Quantum dots with their size-tunable light emission, when used together with MRI (magnetic resonance imaging), can produce exceptional images of tumour sites. These NPs are much brighter than organic dyes and need only one light source for excitation (Figure 17.10). This means that the use of fluorescent quantum dots could produce a higher contrast image at lower cost than organic dyes used today as contrast agents [283, 284].

Other properties of nanostructures, high surface area: volume ratio, allow many functional groups to be attached to an NP, which can seek out and bind to certain target cells. In addition, the small size of NPs (10–100 nm) allows them to accumulate preferentially at tumour sites (because tumours lack an effective lymphatic drainage system), which can make NPs especially useful in oncology [285, 286]. A very exciting research question arises: how to make these imaging NPs more useful for anticancer therapy? For instance, is it possible to manufacture multifunctional NPs that would detect, image, and then proceed to treat a tumour [287]?

Use of lasers, over the previous decades, has emerged to be highly promising to cancer therapy modalities, most commonly the PTT method, which employs light-absorbing dyes to achieve the photothermal damage of tumours, and PDT, which employs chemical photosensitizers generating singlet oxygen or other ROSs capable of tumour destruction [4, 279].

PDT and bionanotechnology show striking similarities in clinical design and mechanism. The PDT paradigm of photosensitizer application, light activation, and singlet oxygen generation does in fact occur on the nanoscale level as the resultant outcomes do. Nanotechnology has the potential to explain as well as modify each of the critical steps of PDT, particularly photosensitizer design and delivery, light source miniaturization, and greater insight into dosimetry and mechanisms of action [276, 288]. Nanomaterials are extensively used in life sciences, eg in histological studies, diagnostic assays, drug delivery systems, and separation techniques. As NPs can effectively penetrate tissues and are generally absorbed by cells, they are considered mainly as carriers of photosensitizers. However, some of them, eg Si, TiO_2 (see above), ZnO, and fullerenes, are capable of generating singlet oxygen themselves. For PDT purposes ceramic-based, metallic (eg gold), and biodegradable nanoparticles with phthalocyanines were tested in order to improve hydrophilicity and increase yield of singlet oxygen generation, as well as enhance cell affinity of the photosensitizer [289].

Recent advances in the field of nanoscience have shown the emergence of noble metal nanostructures with unique photophysical properties, including large optical field enhancements resulting in the strong scattering and absorption of light [290–294]. Noble metal nanoparticles, on account of the phenomenon of surface plasmon resonance (LSPR), possess strongly enhanced visible and NIR light absorption, several orders of magnitude more intense than conventional laser phototherapy agents. The use of plasmonic nanoparticles as highly enhanced photoabsorbing agents has thus introduced a much more selective and efficient strategy of cancer therapy, namely plasmonic photothermal therapy [295, 296] (Figure 17.11).

The ability to integrate metal NPs into biological systems is important for their biomedical applications. Gold NPs conjugated to antibodies can be selectively targeted to cancer cells without significant binding to healthy cells [297, 298]. Irradiation of the cancer cells selectively labelled with the NPs, with a laser of frequency overlapping with the LSPR absorption maximum of the NPs, results in selective heating and destruction of cancer cells at much lower powers than those required to destroy healthy cells to which NPs do not bind specifically. Gold nanoparticles (10–50 nm) offer five or more orders of magnitude larger absorption coefficients than conventional dyes, so much lower laser energies can be used for cell destruction [296–298]. The surface plasmon resonance peaks of gold nanostructures can be tuned from visible to NIR region by controlling the shape and structure (solid vs hollow). A combination of this optical tunability with the inertness of gold makes nanostructures well suited to various biomedical applications. The strong LSPR scattering of gold nanoparticles conjugated with specific targeting molecules allows the cancer cells specific imaging and diagnosis [296, 299, 300]. For gold nanorods or nanoshells, the LSPR can be tuned to their NIR region, making in vivo imaging and therapy possible [298, 299, 301].

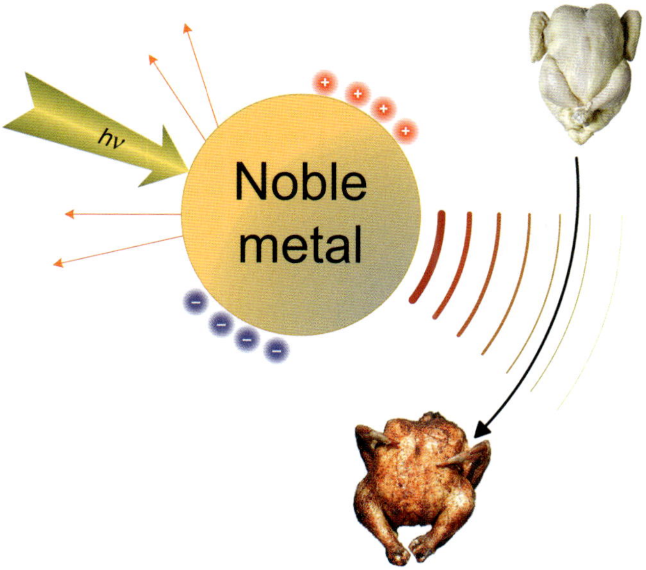

Figure 17.11 *Schematic view of plasmonic photothermal action*

In addition metal nanostructures have higher photostability, and they do not suffer from photobleaching. Currently, gold nanospheres [279, 297, 302–304], gold nanorods [305, 306], gold nanoshells [307, 308], gold nanocages [309, 310], and carbon nanotubes [311] are the chief nanostructures that have been demonstrated in photothermal therapeutics due to their strongly enhanced absorption in the visible and NIR regions on account of their surface plasmon resonance oscillations. Of these structures, the first three are especially promising because of their ease of preparation, ready multi-functionalization, and tunable optical properties.

NPs that consist of a plasmonic layer and an iron oxide moiety could provide a promising platform for development of multimodal imaging and therapy approaches in future medicine. Gold-coated iron oxide hybrid NPs have been synthesized and suggested for combined molecular-specific MRI/optical imaging and PTT of cancer cells [312]. The gold layer exhibits a surface plasmon resonance providing an optical contrast due to light scattering in the visible region and also presents a convenient surface for conjugating targeting moieties, while the iron oxide cores give strong spin–spin relaxation time contrast. The strong optical absorption of the plasmonic gold layer also makes these nanoparticles a promising agent for PTT [296]. It has been shown that the receptor-mediated aggregation of specific hybrid NPs allows highly selective, molecular-specific destruction of highly proliferative cancer cells using pulsed NIR laser radiation. The addition of MRI and optical contrast results in the possibility of obtaining the location and concentration of administered doses of the hybrid agent to help guide laser therapy [312]. Designing and producing nanometre scale multifunctional entities that can diagnose, deliver therapeutic agents, and monitor cancer treatment is one of the future nanomedical strategies.

Inorganic nanostructures are highly versatile and tunable materials and therefore they are very interesting for this and other bioapplications, including biophysical studies [296, 313, 314], biological sensing [296, 315, 316], imaging [296, 299, 317], medical diagnostics [281, 296], and therapy [296, 297, 305, 307, 308, 318–320].

References

1. Milgrom LR. *The Colours of Life: An introduction to the chemistry of porphyrins and related compounds*. Oxford: Oxford University Press, 1997.
2. Wolken JJ. *Light and Life Processes*. New York: Van Nostrand Reinhold Co., 1998.
3. Nalwa HS, ed. Photobiology. In: *Handbook of Photochemistry and Photobiology,* Vol 4. *Photobiology*. Stevenson Ranch: American Scientific Publishers, 2003:1–158.
4. Vo-Dinh T. Biomedical Photonics Handbook. New York: CRC Press, 2003.
5. Szaciłowski K, Macyk W, Drzewiecka-Matuszek A, Brindell M, Stochel G. Bioinorganic photochemistry: frontiers and mechanisms. *Chem Rev* 2005;**105**:2647–94.
6. Vo-Dinh T. Biomedical Photonics: a revolution at the interface of science and technology. In: Vo-Dinh T, ed. *Biomedical Photonics Handbook*. New York: CRC Press, 2003.
7. Bonnett R. Photosensitizers of the porphyrin and phthalocyanine series for photodynamic therapy. *Chem Soc Rev* 1995;**24**:19.
8. Jesionek A, von Tappeiner H. On the treatment of skin cancers with fluorescent substances. *Arch Klin Med* 1905;**82**:223–7.
9. Lippard SJ, Berg JM. *Principles of Bioinorganic Chemistry*. Mill Walley, CA: University Science Books, 1994.
10. Kaim W, Schwederski B. *Bioinorganic Chemistry: Inorganic elements in the chemistry of life*. Amsterdam: John Wiley & Sons, 1994.
11. Cowan JA. *Inorganic Biochemistry*, 2nd edn. Oxford: Oxford University Press 1996.
12. de Silva Frausto JJR, Williams RJP. *The Biological Chemistry of the Elements: The inorganic chemistry of life*, 2nd edn. Oxford: Oxford University Press, 2001.
13. Farrell N. Biomedical uses and applications of inorganic chemistry. *Coord Chem Rev* 2002;**232**:1–4.
14. Farrell N. Metal complexes as drugs and chemotherapeutic agents. In: McCleverty JA, Meyer TJ, eds. *Comprehensive Coordination Chemistry II*. Amsterdam: Elsevier, 2003:809.
15. Williams RJP, Frausto Da Silva JJR. The distribution of elements in cells. *Coord Chem Rev* 2000;**200–202**:247–348.
16. Ali H, van Lier JE. Metal complexes as photo- and radiosensitizers. *Chem Rev* 1999;**99**:2379–450.
17. Guo Z, Sadler PJ. Medicinal inorganic chemistry. *Adv Inorg Chem* 1999;**49**:183.
18. Frausto Da Silva JJR, Williams RJP. *The Biological Chemistry of the Elements. The inorganic chemistry of life*. Oxford: Oxford University Press, 2001.
19. Bonnett R. Metal complexes for photodynamic therapy. In: McCleverty JA, Meyer TJ, eds. *Comprehensive Coordination Chemistry II*. Amsterdam: Elsevier 2003:945.
20. Faulkner S, Matthews JL. Fluorescent complexes for biomedical applications. In: McCleverty JA, Meyer TJ, eds. *Comprehensive Coordination Chemistry II*. Amsterdam: Elsevier 2003:913.
21. Stochel G, Wanat A, Kuliś E, Stasicka Z. Light and metal complexes in medicine. *Coord Chem Rev* 1998;**171**:203–20.
22. Stochel G, Drzewiecka A, Kuliś E, et al. Photochemistry and photophysics of coordination compounds in medicine. In: Melnik M, Sirota A, eds. *Challenges for Coordination Chemistry in the New Century*. Bratislava: Slovak Technical University Press, 2001:301–6.

23. Tuner J, Hode L. *Low Lever Laser Therapy: Clinical practice and scientific background*. Goengesberg: Prima Books, 1999.
24. Karu T. Low-power laser therapy. In: Vo-Dinh T, ed. *Biomedical Photonics Handbook*. New York: CRC Press, 2003:48.1–48.25.
25. Baxter GD. *Therapeutic Lasers: Theory and practice*. London: Churchill Livingstone, 1994.
26. Simunovic Z. *Laser in Medicine and Dentistry*. Rijeka: Vitgraf, 2000.
27. Vo-Dinh T. Basic instrumentation in photonics. In: Vo-Dinh T, ed. *Biomedical Photonics Handbook*. New York: CRC Press, 2003.
28. Palumbo G, Pratesi R. *Laser and Current Optical Techniques in Biology*. Cambridge: Royal Scoiety of Chemistry, 2004.
29. Hartman KM. Action spectroscopy. In: Lohmann W, Marke H, Ziegler H, eds. *Biophysics*. Heidelberg: Spinger-Verlag, 1983:115.
30. Lipson ED. Action spectroscopy: methodology. In: Horspool WH, Song P-S, eds. *Handbook of Organic Chemistry and Photobiology*. Boca Raton, FL: CRC Press, 1995:1257.
31. Smith KS. The photophysical basis of low-level laser radiation therapy. *Laser Ther* 1991;**3**:19–25.
32. Karu T. Primary and secondary mechanisms of action of visible to near-IR radiation on cells. *J Photochem Photobiol B: Biol* 1999;**49**:1–17.
33. Davies MJ. Singlet oxygen-mediated damage to proteins and its consequences. *Biochem Biophys Res Commun* 2003;**305**:761–70.
34. Endo I, Odaka M, Yohda M. An enzyme controlled by light: the molecular mechanism of photoreactivity in nitrile hydratase. *Trends Biotechnol* 1999;**17**:244–8.
35. Pieroni O, Fissi A, Angelini N, Lenci F. Photoresponsive photopeptides. *Acc Chem Res* 2001;**34**:9–17.
36. Lion-Dagan M, Willner I. Nitrospiran-modified alfa-chymotrypsin, a photostimulated biocatalyst in an organic solvent: effects of bioimprinting. *J Photochem Photobiol A: Chem* 1997;**108**:247–52.
37. Shipway AN, Willner I. Electronically transduced molecular mechanical and information functions on surfaces. *Acc Chem Res* 2001;**34**:421–32.
38. Willner I. Photoswitchable biomaterials: en route to optoelectronic systems. *Acc Chem Res* 1997;**30**:347–56.
39. Harvey AJ, Abell AD. Azobenzene-containing, peptidyl alfa-ketoesters as photobiological switches of alfa-chymotrypsin. *Tetrahedron* 2000;**59**:9763–71.
40. Fiamengo R, Crego-Calama M, Reinhoudt DN. Synthetic self-assembled models with biomimetic functions. *Curr Opin Chem Biol* 2001;**5**:660–73.
41. Chain C-K, Hofrichter J, Eaton WA. Optical triggers of protein folding. *Science* 1996;**274**:628–9.
42. Pascher T, Chesick JP, Winkler JR, Gray HB. Protein folding triggered by electron transfer. *Science* 1996;**271**:1558–60.
43. Jori G, Spikes JD. Photothermal sensitizers: possible use in tumor therapy. *J Photochem Photobiol B: Biol* 1990;**6**:93–101.
44. Karu TI. Photobiology of low-power laser effects. *Health Phys* 1989;**56**:691–704.
45. Vladimirov YA, Osipov AN, Klebanov GI. Review: Photobiological principles of therapeutic applications of laser radiation. *Biokhimiya* 2004;**69**(1):103–13.
46. Anderson RR, Parrish JA. Selective photothermolysis: precise microsurgery by selective absorption of pulsed radiation. *Science* 1983;**220**:524–7.
47. Dierick CC, Casparian JM, Venugopalan V, Farinelli WA, Anderson RR. Thermal relaxation of port-wine stain vessels probed in vivo: the need for 1–10 millisecond laser pulse treatment. *J Invest Dermatol* 1995;**105**:709–14.
48. Ho WS, Ying SY, Chan PC, Chan HH. Treatment of port wine stains with intense pulsed light. *Dermatol Surg* 2004;**30**:887–90.
49. Katugampola GA, Rees AM, Lanigan SW. Laser treatment of psoriasis. *Br J Dermatol* 1995;**133**:909–13.
50. Vladimirov Y, Borisenko G, Boriskina N, Kazarinov K, Osipov A. NO-hemoglobin may be a light-sensitive source of nitric oxide both in solution and in red blood cells. *J Photochem Photobiol B: Biol* 2000;**59**:115–22.

51. Borishenko GG, Osipov AN, Kazarinov KD, Vladimirov Y. Photochemical reactions of nitrosyl hemoglobin during exposure to low-power laser irradiation. *Biochemistry (Moscow)* 1997;**62**:661–6.
52. Karu TI, Kalendo GS, Letokhov VS. Control of RNA synthesis rate in tumor cells HeLa by action of low-intensity visible light of copper laser. *Lett Nuov Cim* 1981;**32**:55.
53. Giese AC. Photosensitization of organisms with special reference to natural photosensitizers. In: Hillenkampf F, Pratesi R, Sacchi C, eds, *Lasers in Biology and Medicine*. New York: Plenum Press, 1980:299.
54. Karu TI, Kalendo GS, Letokhov VS, Lobko VV. Biostimulation of HeLa cells by low intensity visible light. *Nuov Cim D* 1982;**1**:828.
55. Karu TI. *Photobiology of Low-Power Laser Therapy*. London: Harwood Academic Publishers, 1989.
56. Karu T. *The Science of Low Power Laser Therapy*. London: Gordon & Breach, 1998.
57. Karu TI. Molecular mechanism of the therapeutic effect of low-intensity laser radiation. *Lasers Life Sci* 1988;**2**:53.
58. Karu TI, Afanasyeva NI. Cytochrome oxidase as primary photoacceptor for cultured cells in visible and near IR regions. *Dokl Akad Nauk (Moscow)* 1995;**342**:693.
59. Brown GC. Nitric oxide and mitochondrial respiration. *Biochem Biophys Acta* 1999;**1411**:351.
60. Karu T, Andreichuk T, Ryabykh T. Changes in oxidative metabolism of murine spleen following diode laser (660–950 nm) irradiation: effect of cellular composition and radiation parameters. *Lasers Surg Med* 1993;**13**:453.
61. Forman NJ, Boveris A. Superoxide radical and hydrogen peroxide in mitochondria. In: Pryor A, ed., *Free Radicals in Biology*, Vol 5. New York: Academic Press, 1982:65.
62. Karu TI, Tiphlova OA, Matveyets YA, Yartsev AP, Letokhov VS. Comparison of the effects of visible femtosecond laser pulses and continuous wave laser radiation of low average intensity on the clonogenicity of *Escherichia coli*. *J Photochem Photobiol B: Biol* 1991;**10**:339.
63. Karu TI. Local pulsed heating of absorbing chromophores as a possible primary mechanism of low-power laser effects. In: Galetti G, Bolognani L, Ussia G (eds), *Laser Applications in Medicine and Surgery*. Bologna: Monduzzi Editore, 1992.
64. Lakowicz JR. Radiative decay engineering: biophysical and biomedical applications. *Anal Biochem* 2001;**298**:1–24.
65. Jori G, Schindl L, Schindl A, Polo L. Novel approaches towards a detailed control of the mechanism and efficiency of photosensitized processes in vivo. *J Photochem Photobiol A: Chem* 1996;**102**:101–7.
66. Banerjee A, George J. Bronchoscopic photodynamic diagnosis and therapy for lung cancer. *Curr Opin Pulm Med* 2000;**6**:378–83.
67. Wolfsen HC, Ward EM. Clinical use of photodynamic therapy for treatment and photodetection in gastroenterology. In: Nalwa HS, ed. *Handbook of Photochemistry and Photobiology*. New York: American Scientific Publishers, 2003:1–39.
68. Mycek MA, Pogue B. *Handbook of Biomedical Fluorescence*. New York: CRC Press, 2003.
69. Dougherty TJ, Gomer CJ, Henderson BW, et al. Photodynamic therapy: A review. *Proc Natl Cancer Inst* 1998;**90**:889–902.
70. Pandey R. Recent advances in photodynamic therapy. *J Porphyrins Phthalocyanines* 2000;**4**:368–73.
71. Wainwright M. Local treatment of viral disease using photodynamic therapy. *Int J Antimicrob Agents* 2003;**21**:510–20.
72. Hamblin MR, Hasan T. Photodynamic therapy: a new antimicrobial approach to infectious disease? *Photochem Photobiol Sci* 2004;**3**:436–50.
73. Ignarro LJ. Nitric oxide: a unique endogenous signalling molecule in vascular biology. *Angew Chem Int Ed* 1999;**38**:1882–92.
74. Ignarro LJ. Nitric oxide as a unique signalling molecule in the vascular system: a historical overview. *J Physiol Pharm* 2002;**53**:503–14.

75. Murad F. Discovery of the biological effects of nitric oxide and its role in cell signalling. *Angew Chem Int Ed* 1999;**38**:1856–68.
76. O'Bryan MK, Zini A, Yan Cheng C, Schlegel PN. Human sperm endothelial nitric oxide synthase expression: correlation with sperm motility. *Fertil Steril* 1998;**70**: 1143–7.
77. Xie K, Huang S. Contribution of nitric oxide-mediated apoptosis to cancer metastasis inefficiency. *Free Rad Biol Med* 2003;**34**:969–86.
78. Adams DR, Brochwicz-Lewinski M, Butler AR. Nitric oxide: Physiological roles, biosynthesis and medical uses. *Progr Chem Org Nat Prod* 1999;**76**:1–186.
79. Methods in Nitric Oxide Research. Chichester: John Wiley & Sons 1996.
80. Dorman T, Thompson DA, Breslow MJ, Lipsett PA, Rosenfeld BA. Nicardipine versus nitroprusside for breakthrough hypertension following carotid endarterectomy. *J Clin Anesth* 2001;**13**:16.
81. Macyk W, Franke A, Stochel G. Metal compounds and small molecules activation – case studies. *Coord Chem Rev* 2005;**249**:2547–57.
82. Szaciłowski K, Chmura A, Stasicka Z. Interplay between iron complexes, nitric oxide and sulfur ligands: Structure, (photo)reactivity and biological importance. *Coord Chem Rev* 2005;**249**:2408–36.
83. Macyk W, Stochel G. From coordination chemistry to photocatalysts and drug. *GIT Lab J* 2005;**9**:44–5.
84. Szaciłowski K, Macyk W, Stochel G, Sostero S, Traverso O, Stasicka Z. Ligand and medium controlled photochemistry of iron and ruthenium mixed ligand complexes: prospecting for versatile systems. *Coord Chem Rev* 2000;**208**:277–97.
85. Flitney FW, Megson IL, Flitney DE, Butler AR. Iron-sulphur cluster nitrosyls, a novel class of nitric oxide generator: mechanism of vasodilator action on rat isolated tail artery. *Br J Pharmacol* 1992;**107**:842.
86. Flitney FW, Megson IL, Thomson JLM, Kennovin GD, Butler AR. Vasodilator responses of rat isolated tail artery enhanced by oxygen-dependent, photochemical release of nitric oxide from iron-sulphur nitrosyls. *Br J Pharmacol* 1996;**117**:1549.
87. Ford PC, Bourassa J, Miranda K, et al. Photochemistry of metal nitrosyl complexes. Delivery of nitric oxide to biological targets. *Coord Chem Rev* 1998;**171**(F5):185–202.
88. Jańczyk A, Wolnicka-Glubisz A, Chmura A, et al. NO dependent phototoxicity of Roussin's black salt against cancer cells. *Nitric Oxide* 2004;**10**:42–50.
89. Oszajca J, Stochel G, Wasielewska E, et al. Cyanonitrosylmetallates as potential NO-donors. *J Inorg Biochem* 1998;**69**:121–7.
90. Matthews EK, Seaton ED, Forsyth MJ, Humphrey PPA. Photon pharmacology of an iron-sulphur cluster nitrosym compound acting on smooth muscle. *Br J Pharmacol* 1994;**113**:87.
91. Mascharak PK, Papaefthymiou GC, Frankel RB, Holm RH. Evidence for the localized Fe(III)/Fe(II) oxidation state configuration as an intrinsic property of $[Fe_2S_2(SR)_4]^{3-}$ clusters. *J Am Chem Soc* 1981;**103**(F6):6110–16.
92. Tfouni E, Krieger M, McGarvey BR, Franco DW. Structure, chemical and photochemical reactivity and biological activity of some ruthenium amine nitrosyl complexes. *Coord Chem Rev* 2003;**236**:57–69.
93. Serli B, Zangrando E, Gianferrara T, Yellowlees L, Alessio E. Coordination and release of NO by ruthenium-dimethylsulfoxide complexes – implications for antimetastases activity. *Coord Chem Rev* 2003;**245**:73–83.
94. Marcondes FG, Ferro AA, Sousa-Torsoni A, et al. In vivo effects of the controlled NO donor/scavenger ruthenium cyclam complexes on blood pressure. *Life Sci* 2002;**70**:2735–52.
95. Cox AB, Wallace RM. Photolysis of nitrosylruthenium chloro complexes. *Inorg Nucl Chem Lett* 1971;**7**:1191.
96. Komozin PN, Kozakova VM, Miroshnichenko IV, Smitsyn NM. An EPR study of the photolysis of nitrosyl compounds of ruthenium. *Russ J Inorg Chem* 1983;**28**:1806.

97. Works CF, Ford PC. Photoreactivity of the ruthenium nitrosyl complex, ru(salen)(cl)(no). solvent effects on the back reaction of NO with the Lewis acid Ru^{III}(salen)(Cl). *J Am Chem Soc* 2000;**122**:7592.
98. Lang DR, Davies JA, Lopes LGF, et al. A controlled NO-releasing compound: synthesis, molecular structure, spectroscopy, electrochemistry, and chemical reactivity of *R,R,S,S-trans*-$[RuCl(NO)(cyclam)]^{2+}$ (1,4,8,11-tetraazacyclotetradecane). *Inorg Chem* 2000;**39**:2294.
99. DeLeo MA, Ford PC. Photoreactions of coordinated nitrite ion. Reversible nitric oxide labilization from the chromium(III) complex [*trans*-$Cr(cyclam)(ONO)_2]^+$. *Coord Chem Rev* 2000;**208**:47.
100. Carlos RM, Ferro AA, Silva HAS, et al. Photochemical reactions of *trans*-$[Ru(NH_3)_4L(NO)]^{3+}$ complexes. *Inorg Chim Acta* 2004;**357**:1381–8.
101. Sauaia MG, de Souza Oliveira F, Tedesco AC, Santana a Silva R. Control of NO release by light irradiation from nitrosyl-ruthenium complexes containing polypyridyl ligands. *Inorg Chim Acta* 2003;**355**:191–6.
102. de Souza Oliveira F, Togniolo V, et al. Nitrosyl ruthenium complex as nitric oxide deliver agent: synthesis, characterization and photochemical properties. *Inorg Chem Commun* 2004;**7**:160–4.
103. Clarke MJ. Ruthenium metallopharmaceuticals. *Coord Chem Rev* 2003;**236**:209–33.
104. Megson IL. Nitric oxide donor drugs. *Drugs Future* 2000;**25**:701.
105. Rakova OA, Sanina NA, Aldoshin SM, et al. Synthesis and characterization of potential NO donors: novel iron-sulfur nitrosyls containing the N-C-S skeleton. *Inorg Chem Commun* 2003;**6**:145–8.
106. Sanina NA, Rakova OA, Aldoshin SM, Chuev II, Atovmyan EG, Ovanesyan NS. Synthesis and X-ray and spectral study of the compounds $[Q_4N]_2[Fe_2(S_2O_{3)2}(NO)_4]$ (Q = Me, Et, *n*-Pr, *n*-Bu). *Russ J Coord Chem* 2001;**27**:179–83.
107. Wolak M, Stochel G, van Eldik R. Aquacobalamin (Vitamin B_{12a}) does not bind NO in aqueous solution. Nitrite impurities account for observed reaction. *Inorg Chem* 2000;**39**:2018–9.
108. Wolak M, Stochel G, van Eldik R. Mechanistic studies on the interaction of reduced cobalamin (vitamin B_{12r}) with nitroprusside. *J Am Chem Soc* 2003;**125**:1334–51.
109. Wolak M, Zahl A, Schneppensieper T, Stochel G, van Eldik R. Kinetics and mechanism of the reversible binding of nitric oxide to reduced cobalamin. *J Am Chem Soc* 2001;**123**:9780–91.
110. Wanat A, Gdula-Argasińska J, Rutkowska-Żbik D, Witko M, Stochel G, van Eldik R. Nitrite binding to metmyoglobin and methemoglobin in comparison to nitric oxide binding. *J Biol Inorg Chem* 2002;**7**:165–76.
111. Ford E, Laverman L, Lorkovic I. Reaction mechanisms of nitric oxide with biologically relevant metal centers. *Adv Inorg Chem* 2003;**54**:203–57.
112. Traylor TG, Magde D, Marsters J, Jongeward K, Wu GZ, Walda K. Geminate processes in the reaction of nitric oxide with 1-methylimidazole-iron(II) porphyrin complexes. Steric, solvent polarity, and viscosity effects. *J Am Chem Soc* 1993;**115**:4808.
113. Hoshino M, Kogure M. Photochemistry of nitrosyl porphyrins in the temperature range 180–300 K and the effects of pyridine on photodenitrosylation of nitrosyliron tetraphenylporphyrin. *J Phys Chem B* 1989;**93**:5478.
114. Hoffman BM, Gibson QH. On the photosensitivity of liganded hemoproteins and their metal-substituted analogues. *Proc Natl Acad Sci USA* 1978;**75**:21.
115. Gibson K, Kernohan W. Lasers in medicine – a review. *J Med Eng Technol* 1993;**17**(2):51–7.
116. Gibson QH, Hoffman BM, Crepeau RJH, Edelstein SJ, Bull C. Manganese hemoglobin: Allosteric effects in stopped flow flash photolysis and sedimentation measurements. *Biochem Biophys Res Commun* 1974;**59**:146.
117. Hoshino M, Arai S, Yamaji M, Hama Y. Laser photolysis studies of nitric oxide adducts of cobalt(II) porphyrins. Photoinduced denitrosylation at the temperature range 160–300 K. *J Phys Chem* 1986;**90**:2109.

118. Yamaji M, Hama Y, Miyazaki Y, Hoshino M. Photochemical formation of oxochromium (IV) tetraphenylporphyrin from nitritochromium(III) tetraphenylporphyrin in benzene. *Inorg Chem* 1992;**31**:932.
119. Hoshino M, Laverman L, Ford PC. Nitric oxide complexes of metalloporphyrins: an overview of some mechanistic studies. *Coord Chem Rev* 1999;**187**:75–102.
120. Morlino EA, Rodgers MAJ. $Ca_6Li_xAl_{23-x}$, $Sr_9Li_{7+x}Al_{36-x}$, and $Ba_2Li_{3+x}Al_{6-x}$: New ternary intermetallic compounds linking close-packed metal structures and zintl phases. *J Am Chem Soc* 1996;**118**:11789.
121. Morlino EA, Rodgers MAJ. The photoinduced dissociation of nitric oxide from nitrosyl metalloporphyrins. *Progr React Kin Mech* 1998;**23**:91.
122. Miranda KM, Bu X, Lorkovic IM, Ford PC. Synthesis and structural characterization of several ruthenium porphyrin nitrosyl complexes. *Inorg Chem* 1997;**36**:4838.
123. Lorkovic IM, Miranda K, Lee B, Bernard S, Schoonover J, Ford PC. Flash photolysis studies of the ruthenium(II) porphyrins Ru(P)(NO)(ONO). Multiple pathways involving reactions of intermediates with nitric oxide. *J Am Chem Soc* 1998;**120**:11674.
124. Denninger JW, Marletta MA. Guanylate cyclase and the NO/cGMP signalling pathway. *Biochem Biophys Acta* 1999;**1411**:334.
125. Kröncke KD, Fehsel K, Kolb-Bachofen V. Nitric oxide: cytotoxicity versus cytoprotection – how, why, when and where? *Nitric Oxide* 1997;**1**:107–20.
126. Adams DR, Brochwicz-Lewinski M, Butler AR. Nitric oxide: Physiological roles, biosynthesis and medical uses. *Prog Chem Org Nat Prod* 1999;**76**:1.
127. Wink DA, Mitchell JB. Chemical biology of nitric oxide: insights into regulatory, cytotoxic and cytoprotective mechanisms of nitric oxide. *Free Rad Biol Med* 1998;**25**: 434.
128. Cooper CE. Nitric oxide and iron proteins. *Biochem Biophys Acta* 1999;**1411**:290–309.
129. Franke A, Jung C, Stochel G, van Eldik R. Substrate binding favors enhanced NO binding to P450cam. *J Am Chem Soc* 2004;**126**:4181–91.
130. Wanat A, Wolak M, Orzeł Ł, Brindell M, van Eldik R, Stochel G. Laser flash photolysis as tool in the elucidation of the nitric oxide binding mechanism to metallobiomolecules. *Coord Chem Rev* 2002;**229**:37–49.
131. Laverman L, Wanat A, Oszajca J, Stochel G, Ford PC, van Eldik R. Mechanistic studies on the reversible binging of nitric oxide to metmyoglobin. *J Am Chem Soc* 2001;**123**:285–93.
132. Noguchi T, Nojiri M, Takei K-i, Odaka M, Kamiya N. Nitrile hydratase. *Biochemistry* 2003;**42**:11642–50.
133. Popescu V-C, Muenck E, Fox BG, et al. Nitrile hydratase. *Biochemistry* 2001;**40**: 7984–91.
134. Kovacs JA. Synthetic analogues of cysteine-ligated non-heme iron and non corrinoid cobalt enzymes. *Chem Rev* 2004;**104**:825–48.
135. Zayat L, Calero C, Albores P, Baraldo L, Etchenique R. A new strategy for neurochemical photodelivery: Metal-lined heterolytic cleavage. *J Am Chem Soc* 2003;**125**:882–3.
136. Tuchin V. Light–tissue interaction. In: Vo-Dinh T, ed. *Biomedical Photonics Handbook*. New York: CRC Press, 2003:3.1–3.26.
137. Letokhov VS. Laser light in biomedicine and the life sciences: From the present to the future. In: Vo-Dinh T, ed. *Biomedical Photonics Handbook*. New York: CRC Press, 2003.
138. Calzavara-Pinton PG, Szeimies RM, Ortel B. *Photodynamic Therapy and Fluorescence Diagnosis In Dermatology*. Amsterdam: Elsevier, 2001.
139. Steiner R. New laser technology and future applications. *Medical Laser Application* 2006;**21**:131–40.
140. Wilson B, Patterson MS. The physics, biophysics and technology of photodynamic therapy. *Phys Med Biol* 2008;**53**:R61–R109.
141. Nikolenko V, Youstic R, Zayat L, Baraldo LM, Etchenique R. Two-photon uncaging of neurochemicals using inorganic metal complexes. *Chem Commun* 2005;1752–4.

142. Wecksler SR, Mikhailovsky A, Korystov D, Buller F. Single- and two-photon properties of a dye-derivatized Roussin's red salt ester ($Fe_2(\mu\text{-}RS)_2(NO)_4$) with a large TPA cross section. *Inorg Chem* 2007;**46**:395–402.
143. Kratochwil NA, Zabel M, Range K-J, Bednarski PJ. Synthesis and X-ray crystal structure of *trans,cis*-[$Pt(OAc)_2I_2(en)$]: A novel type of cisplatin analog that can be photolyzed by visible light to DNA-binding and cytotoxic species in vitro. *J Med Chem* 1996;**39**: 2499–507.
144. Kratochwil NA, Bednarski PJ, Mrozek H, Vogler A, Nagle JK. Photolysis of an iodoplatinum(IV) diamine complex to cytotoxic species by visible light. *Anti-Cancer Drug Res* 1996;**11**:155–71.
145. Kratochwil NA, Parkison JA, Bednarski PJ, Sadler PJ. Nucleotide platination induced by visible light. *Angew Chem Int Ed Eng* 1999;**38**:1460–3.
146. Kasparkowa J, Mackay FS, Brabec V, Sadler PJ. Formation of platination GG cross-links on DNA by photoactivation of a platinum(IV) azide complex. *J Biol Inorg Chem* 2003;**8**:741–5.
147. Muller P, Schroder B, Coxall JA, Parkin A, Parson P, Sadler PJ. Nucleotide cross-linking induced by photoreactions of platinum(IV)-azide complexes. *Angew Chem Int Ed Eng* 2003;**42**:335–9.
148. Bednarski P, Grunert R, Zielzki M, Wellner A, Mackay F, Sadler P. Light-activated destruction of cancer cell nuclei by platinum diazide complexes. *Chem Biol* 2006;**13**(1):61–7.
149. Brindell M, Piotrowska D, Shoukry AA, Stochel G, van Eldik R. Kinetics and mechanism of the reduction of (ImH)[trans-RuC_{14}(dmso)(Im)] by ascorbic acid in acidic aqueous solution. *J Biol Inorg Chem* 2007;**12**:809–18.
150. Brindell M, Stochel G, Bertolasi V, Boaretto R, Sostero S. Photochemistry of *trans-* and *cis*-[$RuCl_2(dmso)_4$] in aqueous and nonaqueous solutions. *Eur J Inorg Chem* 2007:2353–9.
151. Brindell M, Kuliś E, Elmroth SCK, Urbanska K, Stochel G. Light-induced anticancer activity of [$RuCl_2(DMSO)_4$] complexes. *J Med Chem* 2005;**23**:7298–304.
152. Menon EL, Perera R, Navarro M, Kuhn RJ, Morrison H. Phototoxicity against tumor cells and Sindbis Virus by an octahedral rhodium bisbipyridyl complexes and evidence for the genome as a target in viral photoinactivation. *Inorg Chem* 2004;**43**: 5373–81.
153. Magennis SW, Habtemariam A, Novakova O, et al. Dual triggering of DNA binding and fluorescence via photoactivation of dinuclear ruthenium(II) arene complexes. *Inorg Chem* 2007;**46**:5059–68.
154. Pauly M, Kayser I, Schmitz M, et al. In vitro inhibition of gene transcription by novel photo-activated polyazaaromatic ruthenium(II) complexes. *J Chem Soc Chem Commun* 2002:1086–7.
155. Odom DT, Parker CS, Barton JK. Site-specific inhibition of transcription factor binding to DNA by a metallointercalator. *Biochemistry* 1999;**38**:5155–63.
156. Lentzen O, Defrancq E, Constant J-F, et al. Determination of DNA guanine sites forming photo-adducts with Ru(II)-labeled oligonucleotides; DNA polymerase inhibition by resulting photo-crosslinking. *J Biol Inorg Chem* 2004;**9**:100–8.
157. Garcia-Fresnadillo D, Boutonnet N, Schumm S, et al. Luminescence quenching of Ru-labeled oligonucleotide by target complementary strands. *Biophys J* 2002;**82**: 978–87.
158. Lentzen O, Constant J-F, Defrancq E, et al. Photocrosslinking in ruthenium-labelled duplex oligonucleotides. *ChemBioChem* 2003;**4**:195–202.
159. Crean CW, Kavanagh YT, O'Keeffe CM, et al. Targeting of photooxidative damage on single stranded DNA representing the *bcr-abl* chimeric gene using oligonucleotide-conjugates containing $[Ru(phen)_3]^{2+}$-like photosensitizer group. *Photochem Photobiol Sci* 2002;**1**:1024–33.
160. Ortmans I, Content S, Boutonnet N, et al. Ru-Labeled oligonucleotide for photoinduced reactions on targeted DNA guanines. *Chem Eur J* 1999;**5**:2712–21.

161. Lentzen O, Constant J-F, Defrancq E, Moucheron C, Dumy P, Kirsch-De Mesmaeker A. Photoadduct leading to crosslinking in Ru^{II}-derivatized oligonucleotides. *Nucleosides, Nucleotides Nucleic Acid* 2003;**22**:1487–9.
162. Blasius R, Moucheron C, Kirsch-De Mesmaeker A. Photoadducts of metallic compounds with nucleic acids – Role played by the photoelectron transfer process and by the TAP and HAT ligands in the Ru^{II} complexes. *Eur J Inorg Chem* 2004: 3971–9.
163. Ossipov D, Gohil S, Chattopadhyaya J. Synthesis of the DNA-$[Ru(tpy)(dppz)(CH_3CN)]^{2+}$ and their photo cross-linking studies with the complementary DNA strands. *J Am Chem Soc* 2002;**124**:13416–33.
164. Angeles-Boza AM, Bradley PM, Fu PK-L, et al. Photocytotoxicity of a new Rh_2(II,II) complex: increase in cytotoxicity upon irradiation similar to that of PDT agent hematoporphyrin. *Inorg Chem* 2005;**44**:7262–4.
165. Angeles-Boza AM, Bradley PM, Fu PK-L, et al. DNA binding and photocleavage in vitro by new dirhodium(II) dppz complexes: correlation to cytotoxicity and photocytotoxicity. *Inorg Chem* 2004;**43**:8510–19.
166. Lutterman DA, Fu PK-L, Turro C. cis-$[Rh_2(\mu\text{-}O_2CCH_3)_2(CH_3CN)_6]^{2+}$ as a photoactivated cisplatin analog. *J Am Chem Soc* 2005;**128**:738–9.
167. Grossweiner L, Grossweiner J, Rogers G, Jones L. *The Science of Phototherapy: An introduction*. New York: Springer, 2005.
168. Renger G. *Primary Processes of Photosynthesis – Part 1*. Cambridge: Royal Society of Chemistry, 2007.
169. Renger G. *Primary Processes of Photosynthesis – Part 2*. Cambridge: Royal Society of Chemistry, 2008.
170. Giacomoni PU. *Biophysical and Physiological Effects of Solar Radiation on Human Skin*. Cambridge: Royal Society of Chemistry, 2007.
171. Raab O. The effect of fluorescent substances on infusoria. *Z Biol* 1900;**39**:524–46.
172. Jesionek A, Tappeiner VH. Therapeutische Versuche mit fluoreszierenden Stoffen. *Muench Med Wochneshr* 1903;**47**:2042–4.
173. Hausman W. Die sensibilisierende wirkung deshematoporphyrins. *Biochem Z* 1911;**30**:276–86.
174. Lipson RL, Baldes EJ. The photodynamic properties of a particular hematoporphyrin derivative. *Arch Dermatol* 1960;**82**:508–16.
175. Figge FHJ, Weiland GS, Manganiello LOJ. Affinity of neoplastic embryonic and traumatized tissue for metalloporphyrins. *Proc Soc Exp Biol Med* 1948;**68**:640–1.
176. Vincente MGH. Porphyrin-based sensitizers in the detection and treatment of cancer: recent progress. *Curr Med Chem Anti-Cancer Agents* 2001;**1**:175–94.
177. Nalwa HS, ed. *Handbook of Photochemistry and Photobiology*, Vol 1. *Inorganic Photochemistry*. Stevenson Ranch: American Scientific Publishers, 2003.
178. Wu F, Chen WZ, Bai J, et al. Pathological changes in human malignant carcinoma treated with high-intensity focused ultrasound. *Ultrasound Med Biol* 2001;**27**:1099–106.
179. Ara G, Anderson R, Mandel K, Oseroff AR. Absorption of ns photoradiation of melanosomes generates acoustic waves and induced pigmented melanoma cell toxicity. *Photochem Photobiol* 1988;**47**:37S-40S.
180. Chen WR, Adams RL, Heaton E, Dickey ET, Bartels KE, Nordquist RE. Chromophore-enhanced laser tumor tissue photothermal interaction using an 808 nm diode laser. *Cancer Lett* 1995;**88**:15–9.
181. Jori G, Spikes JD. Photothermal sensitizers: possible use in tumor therapy. *J Photochem Photobiol B: Biol* 1990;**6**:93–101.
182. Oleinick NL, Morris RL, Belichenko I. The role of apoptosis in response to photodynamic therapy: what, where, why and how. *Photochem Photobiol Sci* 2001;**1**:1–21.
183 Henderson B, Dougherty TJ. How does photodynamic therapy work. *Photochem Photobiol* 1992;**55**:145–57.
184. Sobolev AS, Jans DA, Rosenkranz AA. Targeted intracellular delivery of photosensitizers. *Progr Biophys Mol Biol* 2000;**73**:51–90.

185. Bonnett R, Martinez G. Photobleaching of sensitisers used in photodynamic therapy. *Tetrahedron* 2001;**57**:9513–47.
186. Korbelik M. Induction of tumor immunity to photodynamic therapy. *J Clin Laser Med Surg* 1996;**14**:329–34.
187. Ochsner M. Photophysical and photobiological processes in the photodynamic therapy of tumours. *J Photochem Photobiol B: Biol* 1997;**39**:1–18.
188. Oleinick NL, Evans HH. The photobiology of photodynamic therapy. Cellular targets and mechanisms. *Radiat Res* 1998;**150**:S146–S56.
189. Nyman ES, Hynninen PH. Research advances in the use of tetrapyrrolic photosensitizers for photodynamic therapy. *J Photochem Photobiol B: Biol* 2004;**73**:1–28.
190. DeRosa MC, Crutchley RJ. Photosensitized singlet oxygen and its applications. *Coord Chem Rev* 2002;**233–234**:351–71.
191. Detty MR, Gibson SJ, Wagner SL. Current clinical and preclinical photosensitizers for use in photodynamic therapy. *J Med Chem* 2004;**47**:389–3915.
192. Dougherty T, Hicks C, Maletta A, Fan J, Rutenberg I, Gafney HD. Excited-state coordination chemistry: a new quenching mechanism. *J Am Chem Soc* 1998;**120**:4226–7.
193. Graczykowa A. *Fotodynamiczna metoda rozpoznawania i leczenia nowotworów.* [*Photodynamic Method of Diagnosis and Therapy of Tumors (in Polish).*] Warsaw: Bellona, 1999.
194. Boulnois J-L. Photophysical processes in recent medical laser developments. A review. *Lasers Med Sci* 1986;**1**:47.
195. Allison RR, Downie GH, Cuenca R, Hu X-H, Childs CJH, Sibata CH. Photosensitizers in clinical PDT. *Photodiagnosis Photodynamic Ther* 2004;**1**:27–42.
196. Sharman WM, Allen CM, van Lier JE. Photodynamic therapeutics: basic principles and clinical applications. *Drug Disc Today* 1999;**4**:507–17.
197. Drzewiecka A, Urbanska K, Matuszak Z, et al. Tritolylporphyrin dimer as a new potent hydrophobic sensitizer for photodynamic therapy of melanoma. *Acta Biochim Pol* 2001;**48**:277–82.
198. Lang K, Mosinger J, Wagnerova DM. Photophysical properties of porphyrinoid photosensitizers non-covalently bound to host molecules; models for photodynamic therapy. *Coord Chem Rev* 2004;**248**:321–50.
199. Moucheron C, Kirsch-De Mesmaeker A, Choua S. Photophysics of $Ru(phen)_2(PHEHAT)^{2+}$: a novel 'light switch' for DNA and photooxidant for mononucleotides. *Inorg Chem* 1997;**36**:584–92.
200. Moucheron C, Kirsch-De Mesmaeker A, Kelly JM. Photoreactions of ruthenium and osmium complexes with deoxyribonucleic acid. *J Photochem Photobiol B: Biol* 1997;**40**:91–106.
201. Ishii K, Kobayashi N. The photophysical properties of phthalocyanines and related compounds. In: Kadish KM, Smith KM, Guilard R, eds. *The Porphyrin Handbook*. New York: Academic Press, 2003:1–40.
202. Lecomte C, Rohmer MM, Benard M. Theoretical and physical characterisation. In: Kadish KM, Smith KM, Guilard R, eds. *The Porphyrin Handbook*. New York: Academic Press, 2003:39–75.
203. Buchler JW. Static coordination chemistry of metalloporphyrins. In: Smith KC, ed. *Porphyrins and Metalloporphyrins*. Amsterdam: Elsevier, 1975:157–231.
204. Scheer H, ed. *Chlorophylls*. Boca Raton, FL: CRC Press, 1991.
205. Felton RH. Primary redox properties of metalloporphyrins. In: Dolphin D, ed. *The Porphyrins*. New York: Academic Press, 1978:53–125.
206. Tanielian C, Schweitzer C, Mechin R, Wolff C. Quantum yield of singlet oxygen production by monomeric and aggregated forms of hematoporphyrin derivative. *Free Rad Biol Med* 2001;**30**:208–12.
207. van Leengoed HLLM, Van Der Veen N, Versteeg AAC, Oullet R, Van Lier JE, Star WM. In vivo fluorescence kinetics of phthalocyanines in a skin-fold observation chamber model: role of central metal ion and degree of sulfonation. *Photochem Photobiol* 1993;**58**:233–7.

208. Falk JE. *Porphyrins and Metalloporphyrins. Their general, physical and coordination chemistry, and laboratory methods*. Amsterdam: Elsevier, 1964.
209. Garbo GM. Purpurins and benzochlorins as sensitizers for photodynamic therapy. *J Photochem Photobiol B: Biol* 1996;**34**:109–16.
210. Hartwich G, Fiedor L, Simonin I, et al. Metal-substituted bacteriochlorophylls. 1. Preparation and influence of metal and coordination on spectra. *J Am Chem Soc* 1998;**120**: 3675–83.
211. Grossweiner LI, Bilgin MD, Berdusis P, Mody TD. Singlet oxygen generation by metallotexaphyrins. *Photochem Photobiol* 1999;**70**:138–45.
212. Sessler JL, Miller RA. Texaphyrins. New drugs with diverse clinical applications in radiation and photodynamic therapy. *Biochem Pharmacol* 2000;**59**:733–9.
213. Baldo MA, O'Brien DF, You Y, et al. Highly efficient phosphorescent emission from organic electroluminescent devices. *Nature* 1998;**395**:151–4.
214. Musewald C, Hartwich G, Pöllinger-Dammer F, Lossau H, Scheer H, Michel-Beyerle ME. Time-resolved spectral investigation of bacteriochlorophyll a and its transmetalated derivatives [Zn]-bacteriochlorophyll a and [Pd]-bacteriochlorophyll a. *J Phys Chem B* 1998;**102**:8336–42.
215. Schreiber S, Gross S, Brandis A, et al. Local photodynamic therapy (PDT) of rat C6 glioma xenografts with Pd-bacteriopheophorbide leads to decreased metastases and increase of animal cure compared with surgery. *Int J Cancer* 2002;**99**:279–85.
216. Musewald C, Hartwich G, Lossau H, et al. Ultrafast photophysics and photochemistry of [Ni]-bacteriochlorophyll a. *J Phys Chem B* 1999;**103**:7055–60.
217. Kaliya OL, Lukyanets EA, Vorozhtsov GN. Catalysis and photocatalysis by phthalocyanines for technology, ecology and medicine. *J Porphyrins Phthalocyanines* 1999;**3**:592–610.
218. Wojaczyński J, Latos-Grażyński L. Poly- and oligometalloporphyrins associated through coordination. *Coord Chem Rev* 2000;**204**:113–71.
219. D'Amico A, Di Natale C, Paolesse R, Macagnano A, Mantini A. Metalloporphyrins as basic material for volatile sensitive sensors. *Sens Actuators* 2000;B**65**:209–15.
220. Bucking M, Dickson EFG, Farahani M, Fischer F, Holmes D, Jori G, et al. Quantification of the selective retention of palladium octabutoxynaphthalocyanine, a potential photothermal drug, in mouse tissues. *J Photochem Photobiol B: Biol* 2000;**58**:87–93.
221. Soncin M, Busetti A, Fusi F, Jori G, Rodgers MAJ. Irradiation of amelanotic melanoma cells with 532 nm high peak power pulsed laser radiation in the presence of the photothermal sensitizer Cu(II)-hematoporphyrin: a new approach to cell photoinactivation. *Photochem Photobiol* 1999;**69**:708–12.
222. Busetti A, Soncin M, Reddi E, Rodgers MAJ, Kenney ME, Jori G. Photothermal sensitization of amelanotic melanoma cells by Ni(II)-octabutoxy-naphthalocyanine. *J Photochem Photobiol B: Biol* 1999;**53**:103–9.
223. Fujishima A, Hashimoto K, Watanabe T. *TiO_2 Photocatalysis. Fundamentals and applications*. Tokyo: BKC Inc., 1999.
224. Fujishima A, Rao TN, Tryk DA. Titanium dioxide photocatalysis. *J Photochem Photobiol C: Photochem Rev* 2000;**1**:1–21.
225. Nakagawa Y, Wakuri S, Sakamoto K, Tanaka N. The photogenotoxicity of titanium dioxide particles. *Mutat Res* 1997;**394**:125–32.
226. Uchino T, Tokunaga H, Ando M, Utsumi H. Quantitative determination of OH radical generation and its cytotoxicity induced by TiO_2 – UVA treatment. *Toxicol in Vitro* 2002;**16**:629–35.
227. Huang N, Xu M, Yuan C, Yu R. The study of the photokilling effect and mechanism of ultrafine TiO_2 particles on U937 cells. *J Photochem Photobiol A: Chem* 1997;**108**: 229–33.
228. Wamer WG, Yin J-J, Wei RR. Oxidative damage to nucleic acids photosensitized by titanium dioxide. *Free Rad Biol Med* 1997;**23**:851–8.
229. Lu P-J, Ho I-C, Lee T-C. Induction of sister chromatid exchanges and micronuclei by titanium dioxide in Chinese hamster ovary-K1 cells. *Mutat Res* 1998;**414**:15–20.

230. Xu M, Huang N, Xiao Z, Lu Z. Photoexcited TiO_2 nanoparticles through OH-radicals induced malignant cells to necrosis. *Supramol Sci* 1998;**5**:449–51.
231. Mills A, Le Hunte S. An overview of semiconductor photocatalysis. *J Photochem Photobiol A: Chem* 1997;**108**(1):1–35.
232. Blake DM, Maness P-C, Huang Z, Wolfrum EJ, Huang J, Jacoby WA. Application of the photocatalytic chemistry of titanium dioxide to disinfection and the killing of cancer cells. *Separation Purification Methods* 1999;**28**(1):1–50.
233. Kubota Y, Niwa C, Ohnuma T, et al. Protective effect of TiO_2 particles on UV light induced pyrimidine dimer formation. *J Photochem Photobiol A: Chem* 2001;**141**: 225–30.
234. Sakai H, Ito E, Cai RX, et al. Intracellular Ca^{2+} concentration change of T-24 cell under irradiation in the presence of TiO_2 ultrafine particles. *Biochim Biophys Acta* 1994;**1201**:259.
235. Kubota Y, Shuin T, Kawasaki C, et al. Photokilling of T-24 human bladder cancer cells with titanium dioxide. *Br J Cancer* 1994;**70**:1107.
236. Sakai H, Baba R, Hashimoto K, Kubota Y, Fujishima A. Selective killing of a single cancerous T24 cell with TiO_2 semiconducting microelectrode under irradiation. *Chem Lett* 1995:185.
237. Hirakawa K, Mori M, Yoshida M, Oikawa S, Kawanishi S. Photo-irradiated titanium dioxide catalyzes site specific DNA damage via generation of hydrogen peroxide. *Free Rad Res* 2004;**38**:439–47.
238. Sökmen M, Candan F, Sümer Z. Disinfection of *E. coli* by the Ag-TiO_2/UV system: lipid peroxidation. *J Photochem Photobiol A: Chem* 2001;**143**:241–4.
239. Ashikaga T, Wada M, Kobayashi H, et al. Effect of photocatalytic activity of TiO_2 on plasmid DNA. *Mutat Res* 2000;**466**:1–7.
240. Muszkat L, Feigelson L, Bir L, Muszkat KA. Titanium dioxide photocatalyzed oxidation of proteins in biocontaminated waters. *J Photochem Photobiol B: Biol* 2001;**60**:32–6.
241. Hancock-Chen T, Scaiano JC. Enzyme inactivation by TiO_2 photosensitization. *J Photochem Photobiol B: Biol* 2000;**57**:193–6.
242. Sakai H, Ito E, Cai R, et al. Intracellular Ca^{2+} concentration change of T24 cell under irradiation in the presence of TiO_2 ultrafine particles. *Biochim Biophys Acta* 1994;**1200**:259–65.
243. Matsunaga T, Tomoda T, Nakajima T, Wake H. Photochemical sterilization of microbial cells by semiconductor powder. *FEMS Microbiol Lett* 1985;**29**:211–4.
244. Kikuchi Y, Sunada K, Iyoda T, Hashimoto K, Fujishima A. Photocatalytic bactericidal effect of TiO_2 thin films: dynamic view of the active oxygen species responsible for the effect. *J Photochem Photobiol A: Chem* 1997;**106**:51–6.
245. Sunada K, Watanabe T, Hashimoto K. Studies on photokilling of bacteria on TiO_2 thin film. *J Photochem Photobiol A: Chem* 2003;**156**:227–33.
246. Huang Z, Maness P-C, Blake DM, Wolfrum EJ, Smolinski SL, Jacoby WA. Bactericidal mode of titanium dioxide photocatalysis. *J Photochem Photobiol A: Chem* 2000;**130**:163–70.
247. Mitoraj D, Jańczyk A, Strus M, et al. Visible light inactivation of bacteria and fungi by modified titanium dioxide. *Photochem Photobiol Sci* 2007;**6**:642–8.
248. Yu JC, Xie Y, Tang HY, Zhang L, Chan HC, Zhao J. Visible light-assisted bactericidal effect of metal phthalocyanine-sensitized titanium dioxide films. *J Photochem Photobiol A: Chem* 2003;**156**:235–41.
249. Shieh K-J, Li M, Lee Y-H, Sheu S-D, Liu Y-T, Wang Y-C. Antibacterial performance of photocatalyst thin film fabricated by defection effect in visible light. *Nanomed Nanotechnol Biol Med* 2006;**2**:121–6.
250. Jańczyk A, Wolnicka-Glubisz A, Urbanska A, Stochel G, Macyk W. Photocytotoxicity of platinum(IV)-chloride surface modified TiO_2 irradiated with visible light against murine macrophages. *J Photochem Photobiol B: Biol* 2008:**92**:54–8.

251. Jańczyk A, Wolnicka-Głubisz A, Urbańska K, Kisch H, Stochel G, Macyk W. Photodynamic activity of platinum(IV)-chloride surface modified TiO_2 irradiated with visible light. *Free Rad Biol Med* 2008;**44**:1120–30.
252. Kisch H, Macyk W. Visible-light photocatalysis by modified titania. *ChemPhysChem* 2002;**3**:399–400.
253. Macyk W, Burgeth G, Kisch H. Photoelectrochemical properties of platinum(IV)-chloride surface modified TiO_2. *Photochem Photobiol Sci* 2003;**2**:322.
254. Macyk W, Kisch H. Photosensitization of crystalline and amorphous titanium dioxide by platinum(IV) chloride surface complexes. *Chem Eur J* 2001;**7**:1862–7.
255. Kisch H, Burgeth G, Macyk W. Visible Light photocatalysis by a titania transition metal complex. *Adv Inorg Chem* 2004;**56**:241–59.
256. Konaka R, Kasahara E, Dunlap WC, Yamamoto Y, Chien KC, Inoue M. Irradiation of titanium dioxide generates both singlet oxygen and superoxide anion. *Free Rad Biol Med* 1999;**27**:294–300.
257. Jańczyk A, Krakowska E, Stochel G, Macyk W. Singlet oxygen photogeneration at surface modified titanium dioxide. *J Am Chem Soc* 2006;**128**:15574–5.
258. Buettner GR, Kelley EE, Burns CP. Membrane lipid free radicals produced from L1210 murine leukemia cells by photofrin photosensitization: an electron paramagnetic resonance spin trapping study. *Cancer Res* 1993;**53**:3670–3.
259. Kelley EE, Domann FE, Buettner GR, Oberley LW, Burns CP. Increased efficacy in vitro Photofrin photosensitization of human oral squamous cell carcinoma by iron and ascorbate. *J Photochem Photobiol B: Biol* 1997;**40**:273–7.
260. Liu ZD, Hider RC. Design of iron chelators with therapeutic application. *Coord Chem Rev* 2002;**232**:151–71.
261. Curnov A, McIlroy BW, Postle-Hacon MJ, Porter JB, MacRobert AJ, Bown SG. Enhancement of 5-aminolaevulinic acid induced photodynamic therapy using hydroxypyridinone iron chelating agents. *Br J Cancer* 1998;**78**:1278.
262. Canti G, Nicolin A, Cubeddu R, Taroni P, Bandieramonte G, Valentini G. Antitumor efficacy of the combination of photodynamic therapy and chemotherapy in murine tumors. *Cancer Lett* 1998;**125**(1–2):39–44.
263. Nonaka M, Ikeda H, Inokuchi T. Effect of combined photodynamic and chemotherapeutic treatment on lymphoma cells in vitro. *Cancer Lett* 2002;**184**:171–8.
264. Lottner C, Knuechel R, Bernhardt G, Brunner H. Combined chemotherapeutic and photodynamic treatment on human bladder cells by hematoporphyrin–platinum(II) conjugates. *Cancer Lett* 2004;**203**:171–80.
265. Brunner H, Arndt MR, Treittinger B. Porphyrin platinum conjugates – new aims. *Inorg Chim Acta* 2004;**357**:1649–69.
266. Brunner H, Schellerer K-M. New porphyrin platinum conjugates for the cytostatic and photodynamic tumor therapy. *Inorg Chim Acta* 2003;**350**:39–48.
267. Song R, Kim Y-S, Sohn YS. Synthesis and selective tumor targeting properties of water soluble porphyrin–Pt(II) conjugates. *J Inorg Biochem* 2002;**83**:83–8.
268. Kim Y-S, Song R, Kim DH, Jun MJ, Sohn YS. Synthesis, biodistribution and antitumor activity of hematoporphyrin–platinum(II) conjugates. *Bioorg Med Chem* 2003;**11**:1753–60.
269. Kim YS, Song R, Lee CO, Sohn YS. Synthesis and biological activity of novel platinum(II) complexes of glutamate to hydrophilic hematoporphyrin derivatives. *Bioorg Med Chem Lett* 2004;**14**:2889–92.
270. de Haas ERM, De Vijlder HC, Sterenborg HJCM, Neumann HAM, Robinson DJ. Fractionated aminolevulinic acid-photodynamic therapy provides additional evidence for the use of PDT for non-melanoma skin cancer. *J Eur Acad Dermatol Venereol* 2008;**22**:426–30.
271. Lilge L, Portnoy M, Wilson BC. Apoptosis induced in vivo by photodynamic therapy in normal brain and intracranial tumor tissue. *Br J Cancer* 2000;**83**:1110–7.

272. Smith G, McGimpsey WG, Lynch MC, Kochevar IE, Redmond R. An efficient oxygen independent 2-photon photosensitization mechanism. *Photochem Photobiol* 1994;**59**: 135–9.
273. Mir Y, Houde D, Van Lier JE. Two-photon absorption of copper tetrasulfophthalocyanine induces phototoxicity towards jurkat cells in vitro *Photochem Photobiol Sci* 2006;**5**:1024–30.
274. Juan C, Stefflova K, Niedre K, et al. Protease-triggered photosensitizing beacon based on singlet oxygen quenching and activation. *J Am Chem Soc* 2004;**126**:11450–1.
275. Zheng G, Chen J, Stefflova K, Jarvi M, Li H, Wilson BC. Photodynamic molecular beacon as an activatable photosensitizer based on protease-controlled singlet oxygen quenching and activation. *Proc Natl Acad Sci USA* 2007;**104**:8989–94.
276. Allison RR, Mota HC, Bagnato VS, Sibata CH. Bio-nanotechnology and photodynamic therapy – State of the art review. *Photodiagnosis and Photodynamic Therapy* 2008;**5**:19–28.
277. Koo YE, Fan W, Hah H. Photonic explorers based on multifunctional nanoplatforms for biosensing and photodynamic therapy. *Appl Opt* 2007;**46**:1924–30.
278. Roy I, Ohulchanskyy TY, Pudvar HE. Ceramic-based nanoparticles entrapping water-insoluble photosensitizing anticancer drugs: a novel drug carrier system for photodynamic therapy. *J Am Chem Soc* 2003;**125**:7860–5.
279. Huang X, Prashant KJ, El-Sayed IH, El-Sayed MA. Plasmonic photothermal therapy (PPTT) using gold nanoparticles. *Lasers Med Sci* 2008;**23**:217–28.
280. Moghimi SM, Hunter AC, Murray JC. Nanomedicine: current status and future prospects. *FASEB J* 2005;**19**:311–30.
281. Rosi NL, Mirkin CA. Nanostructures in biodiagnostics. *Chem Rev* 2005;**105**:1547–62.
282. Jain KK. Applications of nanobiotechnology in clinical diagnostics. *Clin Chem* 2007;**53**:2002–9.
283. Mulder W, Koole R, Brandwijk R, et al. Quantum dots with paramagnetic coating as a bimodal molecular imaging probe. *Nano Lett* 2006;**6**:1–6.
284. Warren CWC, Maxwell DJ, Gao X, Bailey RE, Han M, Nie S. Luminescent quantum dots for multiplexed biological detection and imagining. *Curr Opin Biotechnol* 2002;**13**:40–6.
285. West JL, Halas NJ. Engineered nanomaterials for biophotonics application: improving sensing, imaging, and therapeutics. *Annu Rev Biomed Eng* 2003;**5**:285–92.
286. Hu M, Chen J, Li Z-Y, Au L, et al. Gold nanostructures: engineering their plasmonic properties for biomedicinal applications. *Chem Soc Rev* 2006;**35**:1084–94.
287. Philipp CM, Berlien HP. The future of biophotonics in medicine – A proposal. *Medical Laser Application* 2006;**21**:115–22.
288. Rosenkranz AA, Jans DA, Sobolev AS. Targeted intracellular delivery of photosensitizers to enhance photodynamic efficiency. *Immunol Cell Biol* 2000;**78**:452–64.
289. Wang X-L, Chao H, Li H, et al. DNA interactions of cobalt(III) mixed-polypyridyl complexes containing asymmetric ligands. *J Inorg Biochem* 2004;**98**:1143–50.
290. Stewart ME, Anderton CR, Thompson LB, et al. Nanostructured plasmonic sensors. *Chem Rev* 2008;**108**:494–521.
291. Jain PK, Lee KS, El-Sayed I, El-Sayed MA. Calculated absorption and scattering properties of gold nanoparticles of different size, shape and composition: Application in biological imaging and biomedicine. *J Phys Chem B* 2006;**110**:7238–48.
292. Kelly KL, Coronado E, Zhao LL, Schartz GC. The optical properties of metal nanoparticles: The influence of size, shape and dielectric environment. *J Phys Chem B* 2003;**107**:668–77.
293. Kreibig U, Vollmer M. *Optical Properties of Metal Clusters*. Berlin: Springer, 1995.
294. Link S, El-Sayed MA. Optical properties and ultrafast dynamics of metallic nanocrystals. *Annu Rev Phys Chem* 2003;**54**:331–66.
295. Jain PK, Huang X, El-Sayed IH, El-Sayed MA. Review of some interesting surface plasmon resonance-enhanced properties of noble metal nanoparticles and their applications to biosystems. *Plasmonics* 2007;**2**:107–18.

296. Prasanant KJ, Huang X, El-Sayed IH, El-Sayed MA. Noble metals on the nanoscale: Optical and photothermal properties and some applications in imaging, sensing, biology and medicine. *Acc Chem Res* 2008;in press.
297. El-Sayed IH, Huang X, El-Sayed MA. Selective laser photo-thermal therapy of epithelial carcinoma using anti-EGFR antibody conjugated gold nanoparticles. *Cancer Lett* 2006;**239**:129–35.
298. Huang X, Jain PK, El-Sayed IH, El-Sayed AM. Selective photothermal destruction of cancer cells with the use of immunotargeted gold nanoparticles. *Photochem Photobiol* 2006;**82**:412–17.
299. Sokolov K, Follen M, Aaron J, et al. Real-time vital optical imaging of pancancer using anti-epidermal growth factor receptor antibodies conjugated gold nanoparticles. *Cancer Res* 2003;**63**:1999–2004.
300. El-Sayed IH, Huang X, El-Sayed MA. Surface plasmon resonance scattering and absorption of anti-EGFR antibody conjugated gold nanoparticles in cancer diagnostics: applications in oral cancer. *Nano Lett* 2005;**5**:829–34.
301. Yiwen W, Xie X, Wang X, et al. Photoacoustic tomography of nanoshell contrast agent in the in vivo rat brain. *Nano Lett* 2004;**4**:1689–92.
302. Zharov VP, Galitovsky V, Viegas M. Photothermal detection of local thermal effects during selective nanophotothermolysis. *Appl Phys Lett* 2003;**83**:4897–9.
303. Hainfeld JF, Slatkin DN, Smilowitz HM. The use of gold nanoparticles to enhance radiotherapy in mice. *Phys Med Biol* 2004;**49**:N309–M15.
304. Khlebtsov B, Zharov V, Melnikov A, Tuchin V, Khlebtsov N. Optical amplification of photothermal therapy with gold nanoparticles and nanoclusters. *Nanotechnology* 2006;**17**:5167–79.
305. Huang X, El-Sayed IH, El-Sayed AM. Cancer cell imaging and photothermal therapy in the near-infrared region by using gold nanorods. *J Am Chem Soc* 2006;**126**:2115–20.
306. Takahashi H, Niidome T, Nariai A, Niidome Y, Yamada S. Photothermal reshaping of gold nanorods on tumor cells. *Nanotechnology* 2006;**17**:4431–5.
307. Hirsch LR, Stafford RJ, Bankson JA, et al. Nanoshell-mediated near infrared thermal therapy of tumors under MR Guidance. *Proc Natl Acad Sci USA* 2003;**100**:13549–54.
308. O'Neal DP, Hirsch LR, Halas NJ, Payne JD, West JL. Photothermal tumor ablation in mice using near infrared absorbing nanoparticles. *Cancer Lett* 2004;**209**:171–6.
309. Chen J, Wiley B, Li ZY, et al. Gold nanocages: engineering their structure for biomedical applications. *Adv Mater* 2005;**17**:2255–61.
310. Hu M, Petrova H, Chen J, et al. Ultrafast laser studies of the photothermal properties of gold nanocages. *J Phys Chem B* 2006;**110**:1520–4.
311. Shi Kam NW, O'Connell M, Wisdom JA, Dai H. Carbon nanotubes as multifunctional biological transporters and near-infrared agents for selective cancer cell destruction. *Proc Natl Acad Sci USA* 2005;**102**:11600–5.
312. Larson TA, Bankson J, Aaron J, Sokolov K. Hybrid plasmonic magnetic nanoparticles as molecular specific agents for MIRI/optical imaging and photothermal therapy of cancer cells. *Nanotechnology* 2007;**18**:325101–9.
313. Reinhard B, Sheikholeslami S, Mastroianni A, Alivisatos AP, Liphardt J. Use of plasmon coupling to reveal the dynamics of DNA bending and cleavage by single EcoRV restriction enzymes. *Proc Natl Acad Sci USA* 2007;**104**:2667–72.
314. Sonnichsen C, Reinhard BM, Liphardt J, Alivisatos AP. A molecular ruler based on plasmon coupling of single gold and silver nanoparticles. *Nat Biotechnol* 2005;**23**: 741–5.
315. Elghanian R, Storhoff JJ, Mucic RC, Letsinger RL, Mirkin CA. Selective colorimetric detection of polynucleotides based on the distance-dependent optical properties of gold nanoparticles. *Science* 1997;**277**:1078–81.
316. Haes AJ, van Duyne RP. A nanoscale optical biosensor: Sensitivity and selectivity of an approach based on the localized surface plasmon resonance spectroscopy of triangular silver nanoparticles. *J Am Chem Soc* 2002;**124**:10596–604.
317. Alivisator P. The use of nanocrystals in biodetection. *Nature Biotech* 2004;**22**:47–52.

318. Loo CA, Lowery A, Halas N, West JL, Drezek R. Immunotargeted nanoshells for integrated cancer imaging and therapy. *Nano Lett* 2005;**5**:709–11.
319. Jain PK, El-Sayed I, El-Sayed MA. Au nanoparticles target cancer. *Nano Today* 2007;**2**:18–29.
320. Huang X, Jain PK, El-Sayed IH, El-Sayed AM. Gold nanoparticles: Interesting optical properties and recent applications in cancer diagnostics and therapy. *Nanomedicine* 2007;**2**:681–93.

18

Photodynamic Inactivation of Microorganisms

Nur Gesundung ist noch schöner als Gesundheit.
[Only recovery is more beautiful than health.]
Gerhard Kocher

The problem of microorganism inactivation is important for both clinical treatment and environmental protection.

The field of antimicrobial chemotherapy (ACT) is one of the constant challenges, particularly in view of the rapid evolutionary changes and wide variety of pathogens encountered. Despite major advances in medicine in the last century, microbiologically based diseases continue to present enormous global health problems. The development and widespread use of antibiotics to treat bacterial infections represents one of the most revolutionary advances ever made in scientific medicine. The development of good water supplies and hygiene-based procedures for a whole range of human activities has reduced the likelihood of transmission of microbiological diseases.

It might have been expected that microbiologically based diseases would have been reduced to a level that no longer had a serious impact on human health. But this is not the case. Resistance to antibiotics, which were previously highly effective, has developed. A multidrug resistance and superinfection via effects on normal flora starts to be an enormous problem. Medical science has failed to find a comprehensive therapy to combat viral pathogens and new agents such as prions have emerged. Frequent travelling and lack of adoption of good disinfection methods in poor countries has caused transmission of microorganisms to become a global problem [1, 2].

New antimicrobial approaches that are effective, affordable, and widely applicable are urgently needed. The photodynamic method has been suggested as a possible new strategy for microorganism inactivation [1, 2].

Photodynamic ACT (PACT), similar to photodynamic therapy (PDT) described in Chapter 17, utilizes photosensitizers and visible or UV light in order to induce a phototoxic response, usually via oxidative damage. For some time the disinfection of blood products, particularly for viral inactivation, has been the major use of PACT, although more and more clinically based protocols are being developed, eg in the treatment of oral infections. The technique has been shown to be effective in vitro against bacteria (including drug resistance strains), yeasts, viruses, and parasites.

The ability of a light–drug combination to kill microorganisms is not a new idea and has been known for over 100 years [3]. In fact the concept of PDT was initiated in 1900 when Raab described the antimicrobial action of acridine and light on *Paramecium* species [4]. However, only recently has the search for alternative treatments for antibiotic-resistant pathogens stimulated the PACT development.

Photosensitizers that can target non-economic cells (pathogenic, microbes, tumours, etc) can thus be activated *in situ* resulting in cell inactivation. The antitumour applications of this strategy, usually known as a PDT, have been in clinical use for over 20 years whereas antimicrobial applications, photoantimicrobial chemotherapy – also called antimicrobial photodynamic therapy (APDT) or photodynamic inactivation of microorganisms (PDI) – is not yet in widespread use [5, 6]. A wide range of photosensitizers, both natural and synthetic, has been tested for PACT. Although in cancer PDT porphyrin sensitizers are the most commonly studied compounds, the development of PACT focused mainly on the phenothiazinium dye methylene blue, which was first introduced as an antimalarial agent over a century ago [7].

Progressive development of a large number of synthetic and natural photosensitizers for PACT has occurred [1, 6, 8]. More than a century ago the results first published described the uses of acridine hydrochloride for the photoinactivation of microorganisms [4, 9]. The phenothiazine dyes such as methylene blue and toluidine blue showed promising activity against bacteria and fungi [1]. Several cyanine dyes, such as pyrvinium and merocyanines, were studied for photoinactivation of microbes in plasma and serum [10, 11]. The photosensitizers with highly conjugated molecules, such as porphyrins and phthalocyanines, appeared especially promising [12]. In this context photobactericidal activity of the natural hematoporphyrin precursor, 5-aminolevulinic acid (ALA), has been tested [13]. The photodynamic and accumulation properties of porphyrins, chlorins, and phthalocyanines can be modulated by various substituents, eg metal ions [13, 14]. The binding of some aluminium phthalocyanines to phospholipid membranes has been suggested to drive the interaction of central metal ions with phosphate groups of phospholipids [14].

Microbial cells exhibit a large variety of size, subcellular architecture, biochemical composition, and therefore also susceptibility to externally added chemical agents. Despite such a great diversity all types of microorganisms have been shown to be susceptible to photodynamic action, although specific demands on both photosensitizers and light characteristics have to be expected.

18.1 Bacteria

Gram-positive bacteria can easily take up photosensitizer molecules and therefore be readily photoinactivated. This is not the case for Gram-negative bacteria, which are relatively impermeable to neutral or anionic drugs due to their highly negatively charged surface [6, 15]. With neutral or negatively charged photosensitizers the coadministration of disrupting cationic agents, such as polymixin, could help [2]. Cationic porphyrins and cationic phthalocyanines are active for both Gram-negative and -positive bacteria [16, 17].

Positively charged photosensitizers, particularly cationic metallophthalocyanines, have been proved to be most efficient in photodynamic inactivation of both Gram-negative and -positive bacteria. The reason is believed to lie in the electrostatic interaction of cationic photosensitizer with negatively charged sites at the outer surface of the bacterial cell wall, which facilitates the photosensitizer molecule binding to bacterial cells.

It has been shown that Gram-positive bacteria, such as *Escherichia coli*, are susceptible to the photosensitizing action of a variety of photosensitizers, among them amphiphilic metalloporphyrins (Pd^{II}) [16], cationic metallophthalocyanines (Zn^{II}) [15, 17], or water-soluble derivatives of chlorophylls [18].

Titanium dioxide is a particularly interesting photosensitizer capable of bacterial [19–25] inactivation. Several papers deal with cytotoxicity of reactive oxygen species (ROS) photogenerated at a TiO_2 surface upon UV light irradiation [23, 26, 27]. This activity of TiO_2 can be used to produce sterile films protecting various surfaces. Matsunaga and his coworkers made the first attempts to elucidate the mechanism of titania photocytotoxicity demonstrated by inactivation of *Saccharomyces cerevisiae* [28]. As a conclusion to their studies a possible mechanism of photoinduced cell death involves oxidation of coenzyme A (CoA). However, the photogenerated ROS may attack the microorganism from outside, oxidizing cell membrane at the start (mainly by lipid peroxidation) and then destroying nucleic acids, proteins (enzymes deactivation), etc. A cooperative effect of various oxidative species (involving hydroxyl radicals, superoxide anions, and H_2O_2 produced from photogenerated superoxide anions) explains the bacterial inactivation [26, 29]. The attack of these species leads to the destruction of three layers of cellular wall: outer membrane, peptidoglycan, and cytoplasmic membrane [23].

Hydroxyl radicals are probably the most toxic for microorganisms [30]. They promote peroxidation of polyunsaturated phospholipid components of the lipid membrane and induce disorder in the cell membrane [31]. The damage of the outer membrane increases the permeability to ROSs. This process is possible thanks to a sufficient lifetime of ROSs generated at the TiO_2 surface. ROS diffusion was studied by Fujishima et al. Their experiments demonstrated the bactericidal effect of irradiated TiO_2 film on *E. coli* even at the distance of 50 μm from the film [21]. Furthermore, oxidative perforation of the cellular membrane allows the photocatalyst nanoparticles to penetrate the interior of the cell, causing severe, efficient oxidation of the cell content [27, 32]. ROSs are responsible for oxidation of amino acids, peptides [33], enzymes [34], and nucleic acids [32, 35–37]. Destruction of nucleic acids results in pyrimidine dimer formation [38], DNA structural changes [37], decrease of DNA molecular weight [32], or hydroxylation of guanine bases [35].

An additional biocidal effect can be obtained on TiO_2 loading with some metals possessing bactericidal properties (eg Ag, Cu). Tests with silver-doped TiO_2 (Ag/TiO_2) with 1% w/w silver content showed increased efficiency of *E. coli* photoinactivation [19]. The process of lipid peroxidation in this system was significantly faster than the system with neat titania. The same effect was achieved with TiO_2 with deposited copper [20, 22]. These properties of selected M^0/TiO_2 photocatalysts have been used to prepare bactericidal ceramic tiles. Production of such materials is relatively simple – the titania layers can be impregnated with metal salts solutions followed by ion photoreduction to form metal nanocrystals [20]. The activity of such surfaces is twofold: (1) bacteria killing by metal (in dark) and/or by photogenerated ROSs; and (2) photodecomposition (photomineralization) of biofilm, in particular endotoxins, lipids, etc. The second property of M^0/TiO_2 materials is especially important because removal of a biofilm from the surface prevents it from the growth of new microorganism generation due to the lack of available nutrients.

Recently a few reports on TiO_2-based photocatalysts showing photodynamic activity on visible light irradiation appeared. Yu et al. demonstrated bactericidal effect of metallophthalocyanine-modified TiO_2 films under visible light irradiation [41]. In the presence of thin films fabricated from nitrogen-doped TiO_2 Shieh et al. reached 99.99% deactivation of *E. coli* under visible light irradiation [42]. Mitoraj et al. have proven a similar photoactivity of carbon-doped TiO_2 (mechanism shown in Figure 18.1) [39]. The possibility of various reactive species generation at photosensitized TiO_2 as a result of both charge and energy transfer processes (see Chapter 7) are summarized in Figure 18.2.

18.2 Viruses

Viruses represent a completely different type of targets for PDI, although similarly to bacteria, viruses must be destroyed without causing unacceptable damage to the host cells. The design and mode of action of the photosensitizer are critical in this field and must be linked to the structures and function of the particular viral target. In the decontamination of blood products phthalocyanines have shown a significant potential [43]. Enveloped viruses such as HIV are generally amenable to photoinactivation, in contrast to the non-enveloped viruses, indicating that the viral envelope may be a target for the phthalocyanine photosensitization rather than nucleic acid [44, 45]. The use of a free base, aluminium, and silicon phthalocyanines against viruses in blood has been reported to show high efficiencies of singlet oxygen production [43, 46]. The current and future use of the photodynamic approach in the treatment of viral diseases and the decontamination of blood and other biological products constitutes an area of a very active research.

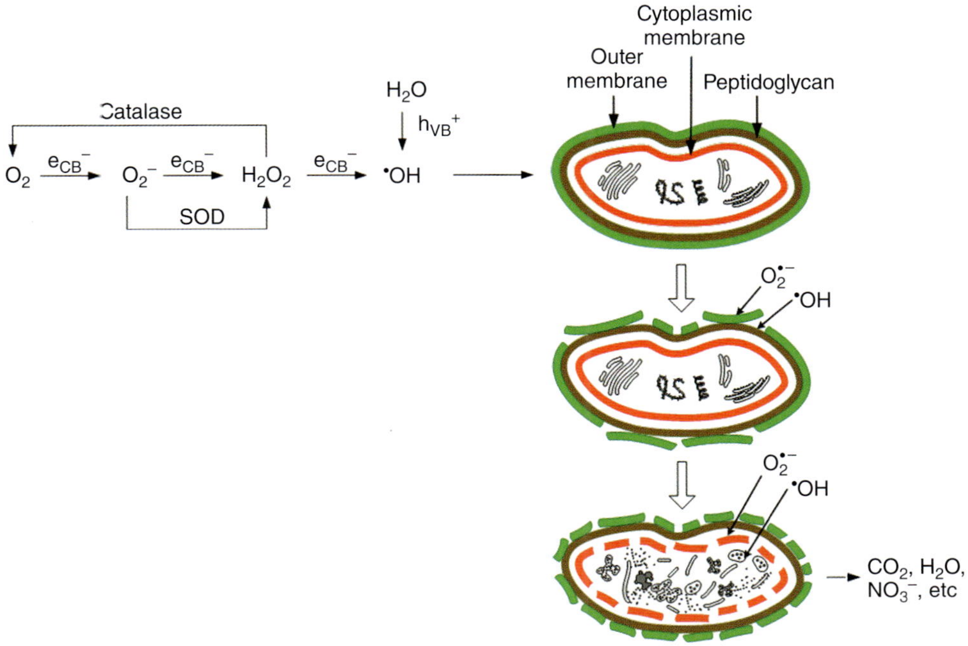

Figure 18.1 *Selected processes occurring during the photocatalytic inactivation of* E. coli *in the presence of semiconductor photosensitizer (SOD – superoxide dismutase;* e_{CB}^-*,* h_{VB}^+ *– photogenerated electrons and holes in conduction and valence band, respectively). Adapted from Mitoraj et al. [39], Szaciłowski et al. [40] and Sunada et al. [23]*

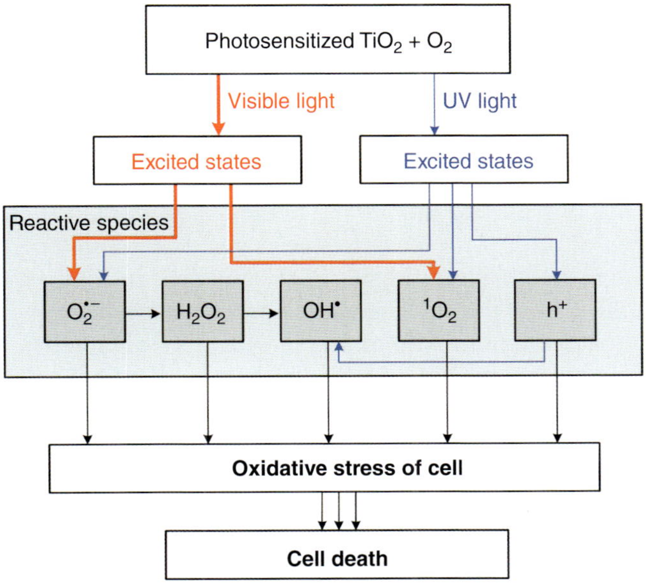

Figure 18.2 *Schematic diagram illustrating origin and fate of reactive species photogenerated at the surface of photosensitized* TiO_2 *irradiated with visible or UV light*

18.3 Fungi

In comparison to antimicrobial therapy, antifungal treatment is limited to a very small number of drug substances.

Superficial skin mycosis, caused by either *Candida* species or dermatophytes, is one of the most frequent diseases in human beings and animals. The search for new therapeutic approaches is stimulated by the fact that standard drug treatments are prolonged and expensive, and the appearance of drug-resistant strains is more and more frequent in high-risk groups of patients, such as immunocompromised individuals. The photodynamic effects on the *Candida* pathogens of different photosensitizers, mainly belonging to three chemical groups – phenothiazine dyes, porphyrins, and phthalocyanines – have been shown. Unlike mammalian cells, fungi are surrounded by a rigid cell composed principally of soluble and insoluble polysaccharide polymers. The cellular uptake is negatively influenced by the lipophilicity of the molecules and positively by the hydrophilicity and presence of charged groups. It has recently been shown that the presence of positive charges is necessary for inactivation of *Candida albicans*, but an excess of charges, especially if homogeneously distributed, causes a decrease of activity. On the basis of these findings a tetracationic Zn-phthalocyanine, bearing four amino-alkylated peripheral substituents capable of inactivating even strains of multi-drug resistant *Candida albicans*, has been developed [13].

Photodynamic inactivation of fungi can also be achieved in the presence of irradiated TiO_2. Maneerat and Hayata demonstrated antifungal activity of TiO_2 films on *Penicillium expansum* [47]. Titanium and zinc oxides can photoinactivate *Aspergillus niger* and *Candida albicans* [48]. Various wild fungi belonging to the *Fusarium* genus (*F. equiseti, F. oxysporum, F. anthophilum, F. verticillioides*, and *F. solani*) have been inactivated by photoexcited TiO_2 [49].

Fungi present a much more complex target than bacteria, eg yeasts, which constitute a large group of rather disparate eukaryotic organisms, are enveloped by a mixture of glucan, mannan, chitin, and lipoproteins, and separated from the plasma membrane by a periplasmic space.

The presence of a nuclear membrane in the yeasts, the larger cell size, and the reduced number of targets for singlet oxygen per unit volume of cell have been suggested as the main reason for higher PDT resistance in comparison to bacteria. However, the efficiency of the photosensitized process is markedly more pronounced by the conjugation of the porphyrin molecule with glycosyl moieties, which enhance their penetration into inner cellular areas [50].

18.4 Parasites

Several physical and chemical approaches are used to inactivate parasites; however, the level of disinfection is often limited. In addition, some conventional treatments may also generate harmful effects on the environment [51]. Recent findings indicate that photodynamic technology can be successfully applied for the inactivation of parasitic pathogens. Cationic porphyrins and phthalocyanines have been found to be phototoxic on parasitic and free living protozoa in both the cystic and vegetative stages of their development [52, 53].

18.5 Perspectives

Future potential of photodynamic technology in the antimicrobial application looks very encouraging. In therapeutic treatment there is a range of dermatological conditions such as infected ulcers, infected burn wounds, and skin disease involving microorganisms. This would require topical application of photosensitizers selective for the microorganisms, which would not cause unacceptable damage to the host tissue. Applications to the dental field have already been investigated and are likely to be established as soon as improved sensitizers are developed. There are many other possible therapeutic applications.

Sterilisation of blood products and other organ transplant tissues opens another area for photodynamic effect applications. Photodynamic activity of some inorganic sensitizers together with their photocatalytic activity can be very useful in water, air, and surface detoxification (see Chapter 21). APDT can be a new approach for safe food. The reduction of several Gram-positive bacteria and yeasts has been observed in tests with visible light and haematoporphyrin or chlorophylls (natural constituents of food) as photosensitizers [54].

References

1. Wainwright M. Photodynamic antimicrobial chemotherapy (PACT). *J Antimicrob Chemother* 1998;**42**:13–28.
2. Malik Z, Hanania J, Nitzan Y. Bactericidal effects of photoactivated porphyrins: an alternative approach to antimicrobial drugs. *J Photochem Photobiol B: Biol* 1990;**5**: 281–93.
3. Moan J, Peng Q. An outline of the hundred-year history of PDT. *Anticancer Res* 2003;**23**:3591–600.
4. Raab O. Ueber die Wirkung fluorescirender Stoffe auf Infusorien (Action of fluorescent materials on infusorial substances). *Z Biol* 1900;**39**:524–46.
5. Wainwright M. Dyes in the development of drugs and pharmaceuticals. *Dyes Pigments* 2008;**76**:582–9.
6. Jori G, Brown SB. Photosensitized inactivation of microorganisms. *Photochem Photobiol Sci* 2004;**3**:403–5.
7. Gutman P, Ehlich P. Uber die Wirkung der Methylen blau. *Berlin Klinische Wochenschr* 1891 **39**:903–55.
8. Maisch T, Jori G, Abels C. Antibacterial photodynamic therapy in dermatology. *Photochem Photobiol Sci* 2004;**3**:907–17.
9. Tappeiner HJ. Therapeutische Versuche mit fluoreszierenden Stoffen. *Munch Med Wochenschr* 1903;**50**:2041.
10. Anderson GS, Gunter WHH, Searle R. Chemistry of biologically important synthetic organoselenium compounds. *Photochem Photobiol* 1996;**64**:683.

11. Neyndorff H, Barel LD, Turafo F, Levy GJ. Development of a model to demonstrate photosensitizer-mediated viral inactivation in blood. *Transfusion* 1990;**30**:485.
12. Sternberg ED, Dolphin D. Porphyrin-based photosensitizers for use in photodynamic therapy. *Tetrahedron* 1998;**54**:4151–202.
13. Calzavara-Pinton PG, Venturini M, Sala R. A comprehensive overview of photodynamic therapy in the treatment of superficial fungal infections of the skin. *J Photochem Photobiol B* 2005;**78**:1–6.
14. Pashkovskaya AA, Maizlish VE, Shaposhnikov GP, Kotova EA, Antonenko YN. Role of electrostatics in the binding of charged metallophthalocyanines to neutral and charged phospholipid membranes. *Biochim Biophys Acta* 2008;**1778**:541–8.
15. Mantareva V, Kussovski V, Angelov I, et al. Photodynamic activity of water-soluble phthalocyanine zinc(II) complexes against pathogenic microorganisms. *Bioorgan Med Chem* 2007;**15**:4829–35.
16. Merchat M, Berttoloni G, Giacomoni P, Villanueva A, Jori G. Meso-substituted cationic porphyrins as efficient photosensitisers of Gram-positive and Gram-negative bacteria. *J Photochem Photobiol B: Biol* 1996;**32**:159–64.
17. Minnock A, Vernon DI, Schofield J, Griffiths J, Parish JH, Brown SB. Photoinactivation of bacteria. Use of a cationic water-soluble zinc phthalocyanine to photoinactivate both Gram-negative and Gram-positive bacteria. *J Photochem Photobiol B: Biol* 1996;**32**: 159–64.
18. Soukos NS, Hamblin MR, Hasan T. The effect of charge on cellular uptake and phototoxicity of poly-lysine chlorine e6 conjugates. *Photochem Photobiol* 1997;**65**:723–9.
19. Sökmen M, Candan F, Sümer Z. Disinfection of *E. coli* by the Ag-TiO_2/UV system: lipid peroxidation. *J Photochem Photobiol A: Chem* 2001;**143**:241–4.
20. Fujishima A, Hashimoto K, Watanabe T. *TiO_2 Photocatalysis. Fundamentals and Applications*. Tokyo: BKC Inc., 1999.
21. Kikuchi Y, Sunada K, Iyoda T, Hashimoto K, Fujishima A. Photocatalytic bactericidal effect of TiO_2 thin films: dynamic view of the active oxygen species responsible for the effect. *J Photochem Photobiol A: Chem* 1997;**106**:51–6.
22. Amezaga-Madrid P, Nevarez-Moorillon GV, Orrantia-Borunda E, Miki-Yoshida M. Photoinduced bactericidal activity against *Pseudomonas aeruginosa* by TiO_2 based thin films. *FEMS Microbiol Lett* 2002;**211**:183–8.
23. Sunada K, Watanabe T, Hashimoto K. Studies on photokilling of bacteria on TiO_2 thin film. *J Photochem Photobiol A: Chem* 2003;**156**:227–33.
24. Kühn KP, Chaberny IF, Massholder K, et al. Disinfection of surfaces by photocatalytic oxidation with titanium dioxide and UVA light. *Chemosphere* 2003;**53**:71–7.
25. Yoshinari M, Oda Y, Kato T, Okuda K. Influence of surface modifications to titanium on antibacterial activity in vitro. *Biomaterials* 2001;**22**:2043–8.
26. Kikuchi Y, Sunada K, Iyoda T, Hashimoto K, Fujishima A. Photocatalytic bactericidal effect of TiO_2 thin film: dynamic view of the active oxygen species responsible for the effect. *J Photochem Photobiol A: Chem* 1997;**106**:51–6.
27. Huang Z, Maness P-C, Blade DM, Wolfram EJ, Smolinski AL, Jacoby WJ. Bactericidal mode of titanium dioxide photocatalysis. *J Photochem Photobiol A: Chem* 2000;**130**:163–70.
28. Matsunaga T, Tomoda T, Nakajima T, Wake H. Photochemical sterilization of microbial cells by semiconductor powder. *FEMS Microbiol Lett* 1985;**29**:211–4.
29. Saito T, Iwase T, Morioka T. Mode of photocatalytic bactericidal action of powdered semiconductor TiO_2 on *Streptococcus mutans*. *J Photochem Photobiol B: Biol* 1992; **14**:369–79.
30. Srinivasan C, Somasundaram N. Bactericidal and detoxification effects of irradiated semiconductor catalyst, TiO_2. *Curr Sci* 2003;**85**(10):25.
31. Maness P-C, Smolinski S, Blake D, Huang Z, Wolfrum AJ, Jacoby WA. Bactericidal activity of photocatalytic TiO_2 reaction: toward an understanding of its killing mechanism. *Appl Environ Microbiol* 1999:4094–8.
32. Huang N, Xu M, Yuan C, Yu R. The study of the photokilling effect and mechanism of ultrafine TiO_2 particles on U937 cells. *J Photochem Photobiol A: Chem* 1997;**108**:229–33.

33. Muszkat L, Feigelson L, Bir L, Muszkat KA. Titanium dioxide photocatalyzed oxidation of proteins in biocontaminated waters. *J Photochem Photobiol B: Biol* 2001;**60**:32–6.
34. Hancock-Chen T, Scaiano JC. Enzyme inactivation by TiO_2 photosensitization. *J Photochem Photobiol B: Biol* 2000;**57**:193–6.
35. Wamer WG, Yin J-J, Wei RR. Oxidative damage to nucleic acids photosensitized by titanium dioxide. *Free Rad Biol Med* 1997;**23**:851–8.
36. Nakagawa Y, Wakuri S, Sakamoto K, Tanaka N. The photogenotoxicity of titanium dioxide particles. *Mutat Res* 1997;**394**:125–32.
37. Ashikaga T, Wada M, Kobayashi H, et al. Effect of photocatalytic activity of TiO_2 on plasmid DNA. *Mutat Res* 2000;**466**:1–7.
38. Kubota Y, Niwa C, Ohnuma T, et al. Protective effect of TiO_2 particles on UV light induced pyrimidine dimer formation. *J Photochem Photobiol A: Chem* 2001;**141**:225–30.
39. Mitoraj D, Jańczyk A, Strus M, et al. Visible light inactivation of bacteria and fungi by modified titanium dioxide. *Photochem Photobiol Sci* 2007;**6**:642–8.
40. Szaciłowski K, Macyk W, Drzewiecka-Matuszek A, Brindell M, Stochel G. Bioinorganic Photochemistry: Frontiers and Mechanisms. *Chem Rev* 2005;**105**:2647–94.
41. Yu JC, Xie Y, Tang HY, Zhang L, Chan HC, Zhao J. Visible light-assisted bactericidal effect of metalphthalocyanine-sensitized titanium dioxide films. *J Photochem Photobiol A: Chem* 2003;**156**:235–41.
42. Shieh K-J, Li M, Lee Y-H, Sheu S-D, Liu Y-T, Wang Y-C. Antibacterial performance of photocatalyst thin film fabricated by defection effect in visible light. *Nanomed Nanotechnol Biol Med* 2006;**2**:121–6.
43. Rywkin S, Ben-Hur E, Malik Z, et al. New phthalocyanines for photodynamic virus inactivation. *Photochem Photobiol* 1994;**60**:165–70.
44. Moor AC, Wagenaars-van Gompel AE, Brand A, Dubbelman MA, vanSteveninck J. Primary targets for photoinactivation of vesicular stomatitis virus by AIPcS4 or Pc4 and red light. *Photochem Photobiol* 1997;**65**:465–70.
45. Wainwright M. Photoinactivation of viruses. *Photochem Photobiol Sci* 2004;**3**:406–11.
46. Horowitz B, Williams B, Rywkin S, et al. Inactivation of viruses in blood with aluminum phthalocyanine derivatives. *Transfusion* 1991;**31**:102–8.
47. Maneerat C, Hayata Y. Antifungal activity of TiO_2 photocatalysis against *Penicillium expansum* in vitro and in fruit tests. *Int J Food Microbiol* 2006;**107**:99–103.
48. Seven O, Dindar B, Aydemir S, Metin D, Ozinel MA, Icli S. Solar photocatalytic disinfection of a group bacteria and fungi aqueous suspensions with TiO_2, ZnO and Sahara desert dust. *J Photochem Photobiol A: Chem* 2004;**165**:103–7.
49. Sichel C, de Cara M, Tello J, Blanco J, Fernandez-Ibanez P. Solar photocatalytic disinfection of agricultural pathogenic fungi: *Fusarium* species. *Appl Catal B: Environ* 2007; **74**:152–60.
50. Carré V, Gaud O, Sylvain I, et al. Fungicidal properties of meso-arylglycosyl-porphyrins: influence of the sugar substituents on photoinduced damage in the yeast *Saccharomyces cerevisiae*. *Photochem Photobiol* 1999;**48**:57–62.
51. Wagner A. Water resource management in an interdisciplinary perspective. In: Kayamanidou M, ed. *Euro-Mediterranean Sciences and Technology Cooperation*. London: Ballière Tindall, 1997:44–46.
52. Kassab K, Ben Amor T, Jori G, Coppellotti O. Photosensitisation of *Colpoda inflata* by meso-substituted cationic porphyrins. *Photochem Photobiol Sci* 2002;**1**:560–4.
53. Lustigman S, Berthus E. Photosensitised inactivation of *Plasmodium falciparum* in human red cells by phthalocyanins. *Transfusion* 1996;**28**:643–8.
54. Kreitner M, Wagner K-H, Alth G, Ebermann R, Foißy H, Elmadfa I. Haematoporphyrin- and sodium chlorophyll-induced phototoxicity towards bacteria and yeasts: a new approach for safe foods. *Food Control* 2001;**12**:529–33.

19

Photodelivery and Phototargeting

Mehr Licht! [More light]
Johann Wolfgang von Goethe

In medicinal chemistry there is a constant challenge to minimize unwanted biological side effects occurring during drug application [1–7]. Light-assisted drug delivery and phototargeting give a possibility of drug action triggering at the desired site, which helps to avoid unwanted side effects [8, 9]. The pharmaceutical can be packed and transported in the form of a thermally stable compound and released at the target site by point irradiation (using laser and optical fibres). The carrier molecule should be easy to prepare, have minimal toxicity in the dark, should undergo efficient photochemical reaction leading to the release of the pharmaceutical in the appropriate form and generate only non-toxic, easily excreted by-products. There are two main approaches to photodelivery. They are based on photodissociation of the pro-drug complex or photoinduced drug release from vesicle structures [8].

Vitamin B_{12} and its derivatives were considered as possible carrier molecules, since they meet most of the aforementioned requirements. Since the photochemical reactions of the cobalamins involve the bond breaking between the cobalt centre and the axial ligand [10–21], the model assemblies of metallocobalamins with a second metal complex (such as $[Pt(CN)_4]^{2-}$ or $[Au(CN)_2]^{-}$) [14, 22, 23] as the sixth ligand have been synthesized. Upon irradiation these cobalamin derivatives undergo photosolvation with the concomitant metal complex release.

Nitric oxide (NO) photodissociation from metallonitrosyl complexes, described in Chapter 18, can be regarded as another example of light-assisted drug delivery. Research on compounds capable of NO delivery has undergone tremendous growth in recent years [8, 11, 24, 25]. Although a number of NO photogenerators appeared in the literature [8, 26–29], only a few examples addressed their integration in appropriate materials. In a very small area NO photorelease has in fact been demonstrated in the case of polymer films [30], hydrogels [31] and self-assembled

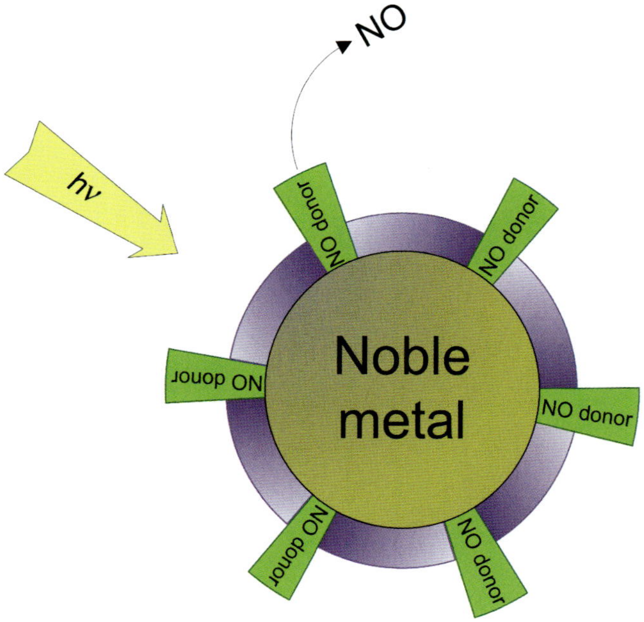

Figure 19.1 *Schematic view of the NO-photoreleasing noble metal nanoparticle*

monolayers on two-dimensional gold surfaces [32]. A new paradigm for increasing the utility of chemical NO release strategies is based on coupling NO donors to nanoparticles. Thiol-stabilized metal nanoparticles, commonly referred to as monolayer-protected clusters (MPCs), are fascinating hybrid systems with intriguing properties and multifaceted applications in various fields [33]. Water-soluble gold monolayer-protected clusters functionalized with diazeniumdiolate may release NO due to a proton-initiated diazeniumdiolate decomposition in solution [34, 35]. Recently the first example of water-soluble monolayer-protected platinum nanoparticles able to deliver NO on visible light irradiation has been reported (Figure 19.1) [36]. Noble metal nanoparticles functionalized with NO donors are very promising three-dimensional NO donors [34, 36]. Construction of these three-dimensional nanostructures is expected to offer additional advantages compared with the two-dimensional self-assembled monolayers on noble metals, ie improved light absorption as well as reduced quenching of the excited states of the photoactive units [37]. These hybrid nanosystems exhibit a combination of indispensable properties for biological applications, such as ultra-small sizes (about 1 nm), water solubility and thermal stability under physiological conditions, photoactivation with low-energy light, and photogeneration of non-toxic intermediates [36].

Activation by light is an exceptionally promising method for triggering release of vesicle contents because it provides a variety of adjustable parameters (photosensitizer, vesicle-forming components, irradiation wavelength, irradiation time, etc), which can be optimized for biological compliance. Light may induce phase transitions within lipid bilayers, resulting in vesicle opening, or can lead to direct

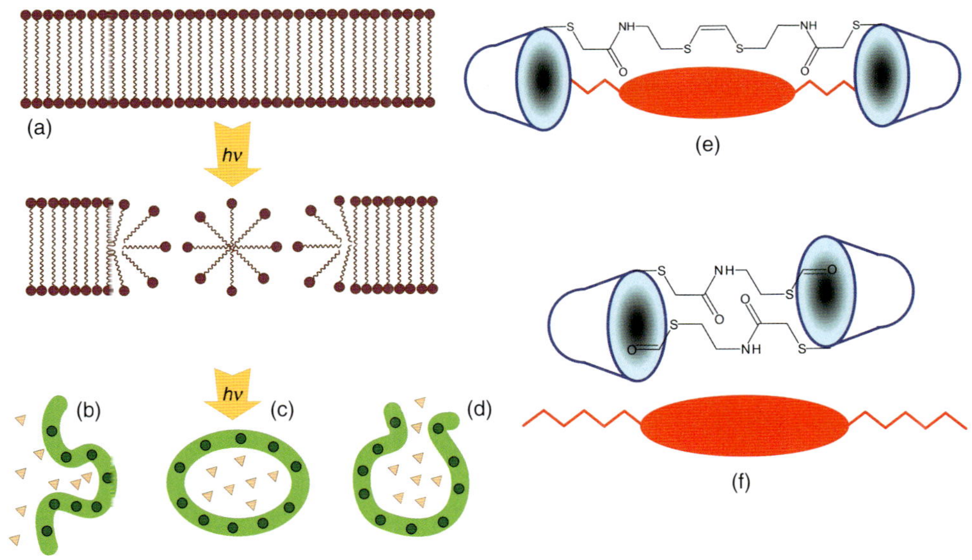

Figure 19.2 Schematic view of drug delivery with photosensitive liposomes and cyclodextrin cages: (a) photoinduced lamellar-to-micellar phase transition in the liposome bilayer, (b) encapsulation of the drug (triangles) inside the liposome-containing photosensitizer (circles), (c) irradiation of the liposome, (d) drug release from the opened liposome, (e) hydrophobic photosensitizer encapsulated in a water-soluble cyclodextrin cage, and (f) liberation of the photosensitizer due to cleavage of the linker

interactions of liposomes with target sites [38, 39]. Lamellar-to-micellar phase transitions may be induced by photoisomerization of liposome-bound photoreceptors or by oxidation of some membrane components with singlet oxygen or other reactive oxygen species (Figure 19.2a). The drug can be encapsulated in light-sensitive liposomes containing a photosensitizer (aluminum phthalocyanines, bacteriochlorophylls, etc) incorporated into membranes (Figure 19.2b). After liposome administration, point irradiation leads to photodynamic opening of lipid membranes and subsequently to topical drug release (Figure 19.2c, d). In some cases liposomes can be opened with the strategies characteristic for a photothermal therapy (PTT) approach [38, 39].

A multifunctional micellar nanoassembly with porphyrins for the simultaneous photodelivery of NO and singlet oxygen (Figure 19.3) has recently been reported [36]. This micellar nanoassembly, able simultaneously to supply, in the same region of space and under visible light control, two species playing a key role in anticancer therapy, in addition to being characterized by considerable fluorescence and reduced size (about 40 nm), seems to be a promising system for biomedical applications [36].

Another strategy for drug photodelivery was proposed by Breslow et al. [40]. His strategy involves encapsulation of the hydrophobic phthalocyanine photosensitizer in cyclodextrin dimer with a sulphur-containing olephin linker (Figure 19.2e). The supramolecular complex shows a good solubility, but irradiation generates

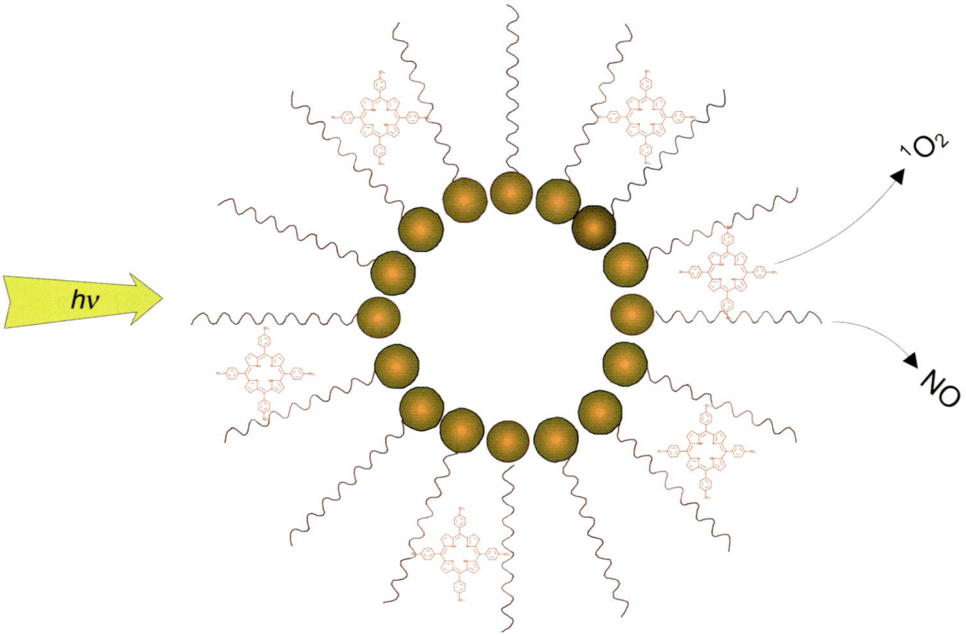

Figure 19.3 *Schematic view of the multifunctional nanostructure photoreleasing NO and $^1O_2$*

singlet oxygen and thus induces liberation of the hydrophobic photosensitizer (Figure 19.2f), which can effectively bind to neighbouring cells or other biological structures.

Light-assisted delivery and phototargeting of bioactive molecules over the last few years have also been established as a promising technique in the biological sciences, especially in the field of neurophysiology [41, 42]. Caged compounds are powerful experimental tools in physiology because they provide means for rapid and localized delivery of bioactive substances [41–46]. In the last few years a number of photolabile protecting groups has been developed for caging of a variety of biomolecules [44,45]. Chemical bonds established between bioactive ligands and proper protecting groups should be strong enough to be water stable and weak enough to be broken with low-energy photons. Most of the available caged compounds are based on the use of nitrobenzols or nitrophenols as protecting groups [45]. These caged compounds need UV light (<350 nm) to damage living tissue. Use of metal complexes as photolabile protecting groups for bioactive molecules offers the lower energies needed for photodissociation. The advantage of coordination versus organic compounds as caged compounds relies also on the diversity of biomolecules that can be coordinated to the metal centre. The potassium channel blocker caged compound, $[Ru(bpy)L_2]^{2+}$ (L = 4-aminopyridine), can photodissociate ligand on irradiation with visible light (around 470 nm) [41]. Some chelating compounds have been used as caged metal ions (Ca^{2+}, La^{3+}, etc) [47]. The release of radical NO from nitrosyl complexes is the sole example of the use of photochemical properties of coordination compounds for this purpose [8].

Stimulation of the neuron response by neurochemical photorelease from a coordination compound has been used in recent studies in neurophysiology [41, 42]. Application of coordination compounds presents several important advantages compared with organic analogues. The metal–organic ligand bond is normally weaker than a covalent σ bond and therefore can be broken using a lower energy radiation. Moreover, as the coordination sphere of the metal can be changed by means of ligand replacement or modification, absorption bands can be tuned to utilize various colours for delivery of different biomolecules. Redox potential, size, lipophilicity, charge, and almost any chemical property can be tuned by means of ligand modification.

References

1. Patrick GL. *An Introduction to Medical Chemistry*. Oxford: Oxford University Press, 2007.
2. Nogrady T, Weaver DF. *Medicinal Chemistry: A molecular and biochemical approach*. Oxford: Oxford University Press, 2005.
3. Wermuth CG. *The Practice of Medicinal Chemistry*. London: Academic Press, 2007.
4. Krogsgaard-Larsen P, Liljefors T, Madsen U. *Textbook of Drug Design and Discovery*, Basingstoke: Taylor & Francis, 2002.
5. Bundgaard H. *Design of Prodrugs*. Oxford: Elsevier Science Publishers BV (Biomedical Division), 1985.
6. Bladon CM. *Pharmaceutical Chemistry Therapeutic Aspects of Biomacromolecules*. Chichester: John Wiley & Sons, 2002.
7. Crowe J, Bradshaw T, Monk P. *Chemistry for the Biosciences: The essential concepts*. Oxford: Oxford University Press, 2006.
8. Szaciłowski K, Macyk W, Drzewiecka-Matuszek A, Brindell M, Stochel G. Bioinorganic photochemistry: frontiers and mechanisms. *Chem Rev* 2005;**105**:2647–94.
9. Stochel G, Wanat A, Kuliś E, Stasicka Z. Light and metal complexes in medicine. *Coord Chem Rev* 1998;**171**:203–20.
10. Pratt JM. *Inorganic Chemistry of Vitamin B_{12}*. New York: Academic Press, 1972.
11. Brown DG. The chemistry of vitamin B_{12} and related inorganic model system. *Prog Inorg Chem* 1973;**18**:177.
12. Shrauzer GN. New developments in the field of vitamin B_{12}: reactions of the cobalt atom in corrins and in vitamin B_{12} model compounds. *Angew Chem Int Ed Eng* 1976;**15**: 417–26.
13. Toscano PJ, Marzili LG. B_{12} and related organocobalt chemistry: formation and cleavage of cobalt-carbon bonds. *Prog Inorg Chem* 1984;**31**:105–204.
14. Kunkely H, Pavlovski V, Vogler A. Photochemistry of vitamin B_{12} derivatives. New observations and conclusions. in: contributions to development of coordination chemistry. In: Ondrejovic G, Sirota A, eds. *Proceedings of the 14th International Conference on Coordination Chemistry*. Bratislava: Slovak Technical University Press, 1993: 363.
15. Jarett JT, Drennan CL, Amaratunga M, Scholten JD, Ludwig ML, Matthews RG. A protein radical cage slows photolysis of methylcobalamin in methionine synthase from *Escherichia coli*. *Bioorg Med Chem* 1996;**4**:1237–46.
16. Chagovetz AM, Grissom CB. Magnetic field effects in adenosylcob(III)alamin photolysis: relevance to B_{12} enzymes. *J Am Chem Soc* 1993;**115**:12152–7.
17. Lott WB, Chagovetz AM, Grissom CB. Alkyl radical geometry controls geminate cage recombination in alkylcobalamins. *J Am Chem Soc* 1995;**117**:12194–201.

18. Schrauzer GN, Lee LP, Sibert JW. Alkylcobalamins and alkylcobaloximes. Electronic structure, spectra, and mechanism of photodealkylation. *J Am Chem Soc* 1970; **92**:2997–3005.
19. Krautler B. Photochemical reactions of vitamin B_{12} derivatives. *Coord Chem Rev* 1991;**111**:215–20.
20. Chen E, Chance MR. Continuous-wave quantum yields of various cobalamins are influenced by competition between geminate recombination and cage escape. *Biochemistry* 1993;**32**:1480–7.
21. Pratt JM, Whitear BRD. Photolysis of methylcobalamin. *J Chem Soc A* 1971:252–5.
22. Stochel G, van Eldik R, Kunkely H, Vogler A. Kinetics and mechanism of the anation of aquocobalamin (vitamin B_{12a}) by cyanoferrates. Isolation and identification of a cyano-bridged product and mechanistic information from pressure effects. *Inorg Chem* 1989;**24**:4314–18.
23. Stochel G, Wanat A, Kuliś E, Stasicka Z. Light and metal complexes in medicine. *Coord Chem Rev* 1998;**171**:203–20.
24. Macyk W, Franke A, Stochel G. Metal compounds and small molecules activation – case studies. *Coord Chem Rev* 2005:2437–57.
25. Wang PG, Xian M, Tang X, et al. Nitric oxide donors: chemical activities and biological applications. *Chem Rev* 2002;**102**:1091–134.
26. Caruso EB, Cicciarella E, Sortino S. A multifunctional nanoassembly of mesogen bearing amphiphiles and porphyrins for the simultaneous photodelivery of nitric oxide and singlet oxygen. *Chem Commun* 2007:5028–30.
27. Baurassa J, Ford PC. Flash and continuous photolysis studies of Roussin's red salt dianion $Fe_2S_2(NO)_4^{2-}$ in solution. *Coord Chem Rev* 2000;**200**:887–900.
28. Sexton DJ, Muruganandam A, McKenney DJ, Mutus B. Visible light photochemical release of nitric oxide from *S*-nitrosoglutathione: potential photochemotherapeutic application. *Photochem Photobiol* 1994;**59**:463–7.
29. Suzuku T, Nagae O, Kato Y, Nakagawa H, Fukuhara K, Miyata N. Photoinduced nitric oxide release from nitrobenzene derivatives. *J Am Chem Soc* 2005;**127**:11720–6.
30. Frost MC, Meyerhoff ME. Controlled photoinitiated release of nitric oxide from polymer films containing *S*-nitroso-*N*-acetyl-dl-penicillamine derivatized fumed silica filler. *J Am Chem Soc* 2004;**126**:1348–9.
31. Shishido SM, Seabra AB, Loh W, de Oliveira MG. Thermal and photochemical nitric oxide release from *S*-nitrosothiols incorporated in Pluronic F127 gel: potential uses for local and controlled nitric oxide release. *Biomaterials* 2003;**24**:3543–53.
32. Etchenique R, Furman M, Olabe JA. Photodelivery of nitric oxide from a nitrosothiol-derivatized surface. *J Am Chem Soc* 2000;**122**:3967.
33. Templeton AC, Wuelfing WP, Murray RW. Monolayer-protected cluster molecules. *Acc Chem Res* 2000;**33**:27–36.
34. Polizzi MA, Stasko NA, Schoenfisch MH. Water soluble nitric oxide-releasing gold nanoparticles. *Langmuir* 2007;**23**:4938–43.
35. Rothrock AR, Donkers RL, Schoenfisch MH. Synthesis of nitric oxide-releasing gold nanoparticles. *J Am Chem Soc* 2005;**127**:9362–3.
36. Caruso EB, Petralia S, Conoci S, Giuffrida S, Sortino S. Photodelivery of nitric oxide from water-soluble platinum nanoparticles. *J Am Chem Soc* 2007;**129**:480–1.
37. Imahori H, Arimura M, Hanada T, et al. Photoactive three-dimensional monolayers: porphyrin-alkanethiolate-stabilized gold clusters. *J Am Chem Soc* 2001;**123**:335–6.
38. Gerasimov OV, Boomer JA, Qualls MM, Thompson DH. Cytosolic drug delivery using pH and light sensitive liposomes. *Adv Drug Deliv Rev* 1999;**38**:317–38.
39. Shum P, Kim J-M, Thompson DT. Phototriggering of liposomal drug delivery systems. *Adv Drug Deliv Rev* 2001;**53**:273–84.
40. Breslow R, Belvedere S, Gershell L, Leung D. The chelate effect in binding, catalysis, and chemotherapy. *Pure Appl Chem* 2000;**72**:333–42.
41. Zayat L, Calero C, Albores P, Baraldo L, Etchenique R. A new strategy for neurochemical photodelivery: Metal-lined heterolytic cleavage. *J Am Chem Soc* 2003;**125**:882–3.

42. Zayat L, Salierno M, Etchenique R. Ruthenium(II) bipyridyl complexes as photolabile caging groups for amines. *Inorg Chem* 2006;**45**:1728–31.
43. Bose M, Groff D, Xie J, Brustad E, Schultz PG. The incorporation of a photoisomerizable amino acid into proteins in *E. coli*. *J Am Chem Soc* 2005;**128**:388–9.
44. Shembekar VR, Chen Y, Carpenter BK, Hess GP. A protecting group for carboxylic acids that can be photolyzed by visible light. *Biochemistry* 2005;**44**:7107–14.
45. Banerjee A, Grewer C, Ramakrishnan L, et al. Toward the development of new photolabile protecting groups that can rapidly release bioactive compounds upon photolysis with visible light. *J Org Chem* 2003;**68**:8361–7.
46. Salierno M, Fameli C, Etchenique R. Caged amino acids for visible-light photodelivery. *Europ J Inorg Chem* 2008;1125–8.
47. Velez P, Gyorke S, Escobar AL, Vergara J, Fill M. Adaptation of single cardiac ryanodine receptor channels. *Biophys J* 1997;**72**:691–7.

20

Phototoxicity and Photoprotection

The doctor is to be feared more than the disease.
Latin proverb

20.1 Chemical and Physical Photoprotection

Phototoxicity may be caused by light interaction with both endogenous and exogenous chromophores. In medicinal photochemistry phototoxicons as a group do not differ from the group of phototherapeutics with regard to eventual biological effects. Nor is this the case for the kind of photoreactions that they can undergo and the biomolecules involved. Also the kind of cellular and tissue damage is the same. The difference between the two groups is found in appreciation of the photobiological effects: unwanted versus desired.

Many organic and inorganic drugs are apt to undergo photochemical reactions both in vivo and ex vivo. Ex vivo photochemistry may lead to the abatement of physiological reactivity or generation of harmful and toxic photoproducts, whereas in vivo photochemistry may also induce photoallergies and undesired phototoxic effects. A drug is phototoxic if light represents an essential condition causing deleterious biological effect, ranging from photoirritation to photocarcinogenicity [1–4]. Photoirritation is a light-induced, non-immunological skin response to a photoreactive chemical, applied directly to the skin or via the circulatory systems after systemic administration. Photoallergy is an immunological response to a photoproduct itself or adduct of a photoproduct with proteins or nucleic acids. Photogenotoxicity is a genotoxic response observed after exposure to a chemical photoactivated by UV or visible light. Photocarcinogenicity focuses on the potential of a drug to induce skin tumors in combination with UV. This may be either an indirect enhancement of UV-induced carcinogenic effects or a carcinogenic effect of a drug photoactivated under UV irradiation (also termed 'photochemical carcinogenesis'). It was also found that some photosensitizers enhance UV-associated skin carcinogenesis. Alteration in optical properties of skin is also responsible for photocarcinogenicity. Various phototoxic effects may be associated with several photochemical processes: direct interaction of excited molecule with biological targets, fragmentation of drug molecule into toxic species, photogeneration of reactive forms of oxygen (oxygen-centred radicals and singlet oxygen), etc [1, 2, 5, 6].

Drug photoprotection includes both physical protection against light (proper materials for immediate and market packages such as bottles, syringes, boxes), as well as additives preventing photochemical reactions by switching the photoreactivity of the excited state to the alternative pathways or changing the solution viscosity. The third possibility includes addition of chemical species acting as scavengers for toxic photoproducts (drug reformulation) [7]. The role of various metal compounds is significant in all three of the above-mentioned approaches [8].

Most of the drugs are packed in plastic or glass containers; however, any of these materials does not ensure itself sufficient protection against UV and visible irradiation. One of the possibilities includes formation of proper optical filters directly on the surface of the container. The filter can be composed of a silicate xerogel doped with titania or zirconia and organic pigments [9]. Optical and mechanical properties of these coatings make them well suited for drug packages, vials, syringes, etc. Other strategies assume changes in the composition of the drug preparation. The simplest strategy involves addition of dyes and pigments, acting as inner filters, to the pharmaceutical preparations. The most important additives are iron oxides and titanium dioxide. Also addition of quenchers may successfully prevent photoreactions by efficient energy transfer from the excited drug molecule to the quencher. In the case of photoisomerizable drugs, cyclodextrins are good 'molecular containers' that enforce preservation of molecular geometry/conformation and in this way help to maintain the desired physiological activity of the drug. Increase of the solution viscosity constitutes an efficient way to photostabilize some drugs, especially those that undergo photodissociation [1, 8].

One of the best known examples of metallodrugs showing photochemical reactivity ex vivo, which leads to generation of toxic photoproducts and abatement of physiological reactivity, is sodium nitroprusside [10, 11]. Its phototoxicity is well recognized and photoprotection strategies well developed. Extensive photolysis of the nitroprusside complex first leads to liberation of nitric oxide, and subsequently free cyanide is released in secondary photochemical steps. Highly energetic UV irradiation induces photoreduction of the complex yielding $[Fe(CN)_5NO]^{3-}$, which in turn spontaneously liberates axial cyanide in the fast thermal process. Other common metallopharmaceuticals that undergo photodecomposition on prolonged exposure to UV or visible light are cisplatin and carboplatin. Irradiation of these species in an aqueous solution leads to photoaquation [1]. In the case of sodium nitroprusside, simultaneous administration of vitamin B_{12}, which is a powerful cyanide scavenger, prevents cyanide poisoning [12]. An increase of the drug solution viscosity (addition of glycerol, glucose, etc) also prevents photodissociation due to disturbed diffusion of primary photoproducts from the solvent cage [10].

Phototoxicity, however, does not concern drugs only. Numerous natural (endogenous) chromophores may act as photosensitizers (generating reactive oxygen species [ROS] on irradiation) or may undergo photoreactions yielding deleterious products. Nature has evolved complex systems for free radical scavenging, frequently based on transition metal complexes. To prevent an overload of free radicals and peroxides aerobic organisms have elaborated defence mechanisms against ROSs. The effectiveness of the system can be modulated by the presence of redox transition metal ions and their complexes. The Cu^{2+}, Mn^{2+}, and Co^{2+} ions can act as redox scavengers of photochemically generated radicals [13]. Therefore these ions and their complexes exert an antioxidant activity and a radical blocking action, which to some extent protect from drug photoinduced DNA cleavage and lipid peroxidation. As the result, the photoinduced damage of a cell is reduced. However, high concentrations of these ions lead to toxic effects [1, 14, 15]. Another approach to increase photoprotective potential of a cell is the use of complexes, which can mimic enzyme superoxide dismutase (SOD), catalyzing the reduction of the superoxide anion to the less toxic H_2O_2. Various types of SOD enzymes are known: Cu, Zn-SOD in cytoplasm of eukaryotic cells [16], Mn-SOD in mitochondria, and Mn-SOD and Fe-SOD in prokaryotic cells. Therefore some Mn^{II}, Fe^{II}, and Cu^{II} complexes with oligopeptides, salicylates, and polyazamacrocycles were tested as efficient SOD models [17]. Use of many copper(II) complexes in vivo is limited by dissociation of Cu^{II} and binding to natural ligands such as albumins [18]. Some Mn^{II}–macrocycle complexes are biologically active [19]; however, Mn^{II} complexes are much less stable than analogous Cu^{II}–macrocycle complexes [17].

20.2 Inorganic Sunscreens

Today, topical sunscreens are divided into two broad categories: organic and inorganic. Formerly two other groups were considered: physical (associated with light reflection and scattering) and chemical (acting as light absorbers transforming photons energy to less harmful heat). Inorganic agents, mainly titanium dioxide and zinc oxide, form a film constituting an opaque barrier reflecting and scattering, but mainly absorbing UV (both UV-A and UV-B) radiation. These oxides are photostable (under certain conditions ZnO may undergo a partial photocorrosion with Zn formation), do not react with organic components of suntans [20], and have less variability in the photoprotective effect when compared with organic sunscreens [21]. Opaque inorganic sunscreens might give some protection against visible light-induced photosensitivity diseases [22]. However, inorganic sunscreens are often cosmetically unacceptable because of their opaque quality and occlusiveness. Recently modern pharmaceutical approaches such as micronization and encapsulation have allowed the development of high-quality inorganic sunscreens. Decreasing particle size to 10–50 nm (micronized form) results in less scattering of visible light, leading to a more cosmetically acceptable product. Micronization shifts the protective spectrum towards shorter wavelengths as a result of changed absorption properties [23]. Microfine titanium dioxide and zinc oxide have been found to be highly protective against the harmful UV-C rays, and offer good protection against short-term UV-B-induced immunomodulation in human trials [20, 24, 25]. Neither zinc oxide nor titanium dioxide has relevant skin-irritating properties [26].

Absorption properties of both titanium dioxide and zinc oxide are perfectly fitted to their applications as sunscreen agents [27–29] as their absorption onset around 400 nm determines the frontier between UV and visible light. However, the photocatalytic properties of these materials, described in Chapter 7, may be responsible for harmful ROSs generation. ROSs may be responsible for oxidation of organic components of suntan lotions but also for skin ageing and destruction. Scaiano et al. reported oxidation of commonly used UV-B organic sunscreens in the presence of TiO_2 [30]. Dunford et al. tested photocatalytic activity of TiO_2 isolated from commercial suntan lotions and found that all of them photocatalyzed phenol oxidation and caused DNA damage. A similar activity of ZnO was also confirmed [31]. Therefore inorganic sunscreens commonly used in cosmetic industry should be carefully tested for their photoactivity and phototoxicity. Sunscreens based on titanium and zinc oxides should contain additional antioxidants, eg α-tocopherol or β-carotene. Even more efficient protection against unwanted ROSs can be achieved by a smart titanium dioxide modification that would block its photocatalytic activity. Recently Lee et al. described a hydrophobic multicomponent polymer coating for nanocrystalline TiO_2 containing natural antioxidants extracted from grape seeds [32]. Such composite material shows good screening properties against UV, protecting DNA from photoinduced damage.

Although there is some risk of ROS photogeneration by TiO_2 the screening properties of TiO_2 have been proved in various experiments. DNA damage, observed as a pyrimidine dimer formation caused by UV irradiation [33], may be suppressed in the presence of TiO_2 [34], as demonstrated on normal human cells (TIG-1) and human cancer cells (T24). Under experimental conditions photogenerated ROSs did not manage to penetrate efficiently inside the cell.

Concerns related to the penetration of inorganic agents into the skin have been discussed. Some reports confirmed ZnO [35], TiO_2 [36, 37], and micronized TiO_2 [38] penetration. Contrary to these results other in vivo and in vitro studies found no evidence of penetration of TiO_2 and limited penetration only for ZnO, which is slightly soluble [26]. Microparticles tend to agglomerate and aggregate due to electrostatic effects, resulting in potentially greater loss in efficacy, so the micropigments have to be coated and kept in dispersion, which is still a major challenge for the cosmetic industry [39].

Suntan lotions containing inorganic sunscreens alone are generally recommended for children, because of their lack of penetration and subsequent degradation in the body [40], absence of photorelated effects (ie photoallergy), and no evidence of photogenotoxicity in vivo [26]. Recent controversy surrounding efficacy and safety of inorganic sunscreens [21, 41–43] has stimulated a reassessment of their use and properties. In this context wearing of sun-protective clothes and reducing sun exposure to a minimum must be preferred as the safest solution.

References

1. Albini A, ed. *Drugs. Photochemistry and Photostability*. Cambridge: Royal Society of Chemistry, 1998.
2. Albini A, Fasani E, eds. Photochemistry of drugs. In: Nalwa HS (ed.), *Handbook of Photochemistry and Photobiology*, Vol. **2**. *Organic Photochemistry*. Stevenson Ranch: American Scientific Publishers, 2003:143–93.
3. Miranda MA. Photosensitization by drugs. *Pure Appl Chem* 2001;**73**:481–6.
4. Cosa G. Photodegradation and photosensitization in pharmaceutical products: Assessing drug phototoxicity. *Pure Appl Chem* 2004;**76**:263–75.
5. Verma K, Agrawal N, Misra RB, Farooq M, Hans RK. Phototoxicity assessment of drugs and cosmetic products using *E. coli*. *Toxicology in Vitro* 2008;**22**:249–53.
6. Hans RK, Agrawal N, Verma K, Misra RB, Ray RS, Farooq M. Assessment of the phototoxic potential of cosmetic products. *Food and Chemical* 2008;**46**:1653–8.
7. Hubbard WK. International Conference on Harmonisation: Guideline for the photostability testing of new drug substances and products. *Fed Reg* 1997;**62**:27116–22.
8. Szaciłowski K, Macyk W, Drzewiecka-Matuszek A, Brindell M, Stochel G. Bioinorganic photochemistry: frontiers and mechanisms. *Chem Rev* 2005;**105**:2647–94.
9. Sanchez C, Soler-Illia GJDAA, Ribot F, Grosso D. Design of functional nano-structured materials through the use of controlled hybrid organic–inorganic interfaces. *C R Chimie* 2003;**6**:1131–51.
10. Stochel G, Wanat A, Kuliś E, Stasicka Z. Light and metal complexes in medicine. *Coord Chem Rev* 1998;**171**:203.
11. Oszajca J, Stochel G, Wasielewska E, et al. Cyanonitrosylmetallates as potential NO-donors. *J Inorg Biochem* 1998;**69**:121–7.
12. Leeuwenkamp OR, van der Mark EJ, van Bennekom WP, Bult A. Investigation of the photochemical and thermal degradation of aqueous nitroprusside solution using liquid chromatography. *Int J Pharm* 1985;**24**:27–41.
13. Giuffrida S, De Guidi G, Miano P, Sortino S, Condorelli G, Costanzo LL. Molecular mechanism of drug photosensitization: VIII effect of inorganic ions on membrane damage photosensitized by naproxen. *J Inorg Biochem* 1996;**63**:253–63.
14. Ledo A, Frade J, Barbosa RM, Laranjinha J. Nitric oxide in brain: diffusion, targets and concentration dynamics in hippocampal subregions. *Mol Asp Med* 2004;**25**:75.
15. Nelson EJ, Connolly J, McArthur P. Nitric oxide and *S*-nitrosylation: excitotoxic and cell signaling mechanism. *Biol Cell* 2003;**95**:3.
16. Vessey DA. The cutaneous antioxidant system. *Clin Dermatol* 1993;**8**:81–103.
17. Guo Z, Sadler PJ. Medicinal inorganic chemistry. *Adv Inorg Chem* 1999;**49**:183.
18. Nagano T, Hirano T, Hirobe M. Superoxide dismutase mimics based on iron in vivo. *J Biol Chem* 1989;**264**:9243–9.
19. Riley DP, Weiss RH. Manganese macrocyclic ligand complexes as mimics of superoxide dismutase. *J Am Chem Soc* 1994;**116**:387–8.
20. Mitchnick MA, Fairhurst D, Pinnell SR. Microfine zinc oxide (Z-cote) as a photostable UVA/UVB sunblock agent. *J Am Acad Dermatol* 1999;**40**:85–90.
21. Lautenschlager S, Wulf HC, Pittelkow MR. Photoprotection. *Lancet* 2007;**370**: 528–37.
22. Moseley H, Cameron H, MacLeod T, Clark C, Dawe R, Ferguson J. New sunscreens confer improved protection for photosensitive patients in blue light region. *Br J Dermatol* 2001;**145**:789–94.
23. Roelandts R. Shedding light on sunscreens. *Clin Exp Dermatol* 1998;**23**:147–57.
24. Gelis C, Girard S, Mavon A, Delverdier M, Paillous N, Vicendo P. Assessment of the skin photoprotective capacities of an organo-mineral broad-spectrum sunblock on two ex vivo skin models. *Photodermatol Photoimmunol Photomed* 2003;**19**:242–53.
25. van der Molen RG, Hursk HM. Efficacy of micronized titanium dioxide-containing compounds in protection against UVB-induced immunosuppression in humans in vivo. *J Photochem Photobiol B: Biol* 1998;**44**:143–50.

26. Nash JF. Human safety and efficacy of ultraviolet filters and sunscreen products. *Dermatol Clin* 2006;**24**:35–51.
27. West LE. Sun protection. *Dermatol Trends* 2004;**2**:81–95.
28. Bestak R, Barnetson RSC, Nearn MR, Halliday GM. Sunscreen protection of contact hypersensitivity responses from chronic solar-stimulated ultraviolet irradiation correlates with the absorption spectrum of sunscreen. *J Invest Dermatol* 1995;**105**:345–51.
29. van der Molen RG, Hurks HMH, Out-Luiting C, et al. Efficacy of micronized titanium dioxide-containing compounds in protection against UVB-induced immunosuppression in humans in vivo. *J Photochem Photobiol B: Biol* 1998;**44**:143–50.
30. Ricci A, Chretien MN, Maretti L, Scaiano JC. TiO_2-promoted mineralization of organic sunscreens in water suspension and sodium dodecyl sulfate micelles. *Photochem Photobiol Sci* 2003;**2**:487–92.
31. Dunford R, Salinaro A, Cai L, et al. Chemical oxidation of DNA damage catalysed by inorganic sunscreen ingredients. *FEBS Lett* 1997;**418**:87–90.
32. Lee WA, Pernodet N, Li B, Lin CH, Hatchwell E, Rafailovich MH. Multicomponent polymer coating to block photocatalytic activity of TiO_2 nanoparticles. *Chem Commun* 2007:4815–17.
33. Mitchell D. The relative cytotoxicity of (6–4) photoproducts and cyclobutane dimers in mammalian cells. *Photochem Photobiol* 1988;**48**:51–7.
34. Kubota Y, Niwa C, Ohnuma T, et al. Protective effect of TiO_2 particles on UV light induced pyrimidine dimer formation. *J Photochem Photobiol A: Chem* 2001;**141**:225–30.
35. Agren MS. Percutaneous absorption of zinc from zinc oxide applied topically to intact skin in man. *Dermatologica* 1990;**180**:36–9.
36. Dupre A, Touron P, Daste J, Lassere J, Bonafe JL, Viraben R. Titanium pigmentation. An electron probe microanalysis study. *Arch Dermatol* 1985;**121**:656–8.
37. Moran CA, Mullick FG, Ishak KG, Johnson FB, Hummer WB. Identification of titanium in human tissues: probable role in pathologic processes. *Hum Pathol* 1991;**22**:450–4.
38. Tan M-H, Commens CA, Burnett L, Snitch PJ. A pilot study on the percutaneous absorption of microfine titanium dioxide from sunscreens. *Australas J Dermatol* 1996;**37**: 185–7.
39. Lademann J, Weigmann H, Schafer H, Muller G, Sterry W. Investigation of the stability of coated titanium microparticles used in sunscreens. *Skin Pharmacol Appl Skin Physiol* 2000;**13**:258–64.
40. Janjua NR, Mogensen B, Andersson AM. Systemic absorption of the sunscreens benzophenone-3, octyl-methoxycinnamate, and 3-(4-methyl-benzylidene) camphor after whole-body topical application and reproductive hormone levels in humans. *J Invest Dermatol* 2004;**123**:57–61.
41. Westerdahl J, Ingvar C, Masback A, Olsson H. Sunscreen use and malignant melanoma. *Int J Cancer* 2000;**87**:145–50.
42. Schlumpf M, Cotton B, Conscience M, Haller V, Steinmann B, Lichtensteiger W. In vitro and in vivo estrogenicity of UV screens. *Environ Health Perspect* 2001;**109**:239–44.
43. Adam J. Sun-protective clothing. *J Cutan Med Surg* 1997;**3**:50–3.

21

Photocatalysis in Environmental Protection

'I am a spirit of no common rate'.
Titania,* Act III scene 1, *A Midsummer Night's Dream*, W. Shakespeare

21.1 Development of Homo- and Heterogeneous Methods

Nowadays, the main goal of research and development in air purification and waste water treatment is use of a technique that offers an advantage of destroying the pollutants, in contrast to conventional techniques, such as activated carbon or air stripping, that only transfer the contaminants from one phase to another. In this way, organic and inorganic compounds, and even microorganisms, are degraded or transformed into less harmful substances. The methods, called advanced oxidation processes (AOPs) or techniques (AOTs), are of considerable interest for treatment of wastewaters, contaminated surfaces, and ground waters, and for production of ultra-pure water [1–4].

The increased concern for the use of the AOPs may be explained by the need to seek new alternatives to the conventional ones, and environmentally friendly technologies. Under this term the scientific community refers to technologies with effectiveness based on the production of $OH^{\bullet}$ radicals, one of the most powerful oxidant reagents. They can easily attack organic molecules, leading to the production of organic peroxide radicals and their final conversion to CO_2, H_2O, and inorganic salts. With the exception of electron injection, AOPs rely entirely on oxidative degradation reactions, where organic radicals are generated on photolysis of the organic substrate or by reaction with hydroxyl radicals. These radical intermediates are

* Titania: (1) titanium dioxide; (2) the largest moon of Uranus named after Titania, queen of fairies, wife of Oberon in W. Shakespeare's *A Midsummer Night's Dream*.

subsequently trapped by dissolved molecular oxygen and lead via peroxyl radicals and peroxides to enhancement of the overall degradation process and finally to complete mineralization:

$$RH + OH^{\bullet} \rightarrow R^{\bullet} + H_2O \tag{21.1}$$

$$R^{\bullet} + O_2 \rightarrow RO_2^{\bullet} \tag{21.2}$$

$$RO_2^{\bullet} \rightarrow CO_2 + \text{other oxidation products} \tag{21.3}$$

Among these methods, homogeneous and heterogeneous solar photocatalytic detoxification methods (Fe^{3+}/H_2O_2, TiO_2/H_2O_2) have recently attracted great attention for the treatment of industrial wastewater, groundwater, and contaminated air, allowing the contribution of renewable sources of energy (solar energy) to the process of cleaning and restoring the environment [5–9].

21.2 Homogeneous Photocatalysis

One of the most friendly, attractive and economically justified techniques used in homogeneous systems consists of the photo-Fenton method. The photo-Fenton reagent ($Fe^{3+}/Fe^{2+}/H_2O_2$) is an efficient and cheap reagent, which in a very simple way produces $OH^{\bullet}$ radicals for wastewater treatment, due to the fact that iron is a very abundant and non-toxic element and hydrogen peroxide is easy to handle and environmentally safe. Iron in its ferrous and ferric forms acts as a photocatalyst and requires a working pH ≤4. At higher pH values, iron precipitates as a hydroxide. The Fenton reaction, which usually operates at or near ambient temperature and pressure:

$$Fe^{2+} + H_2O_2 \rightarrow Fe^{3+} + OH^{\bullet} + OH^{-} \tag{21.4}$$

with $k = 40–60\ mol^{-1}\,dm^3\,s^{-1}$ [10] is considerably enhanced by UV–Vis light, artificial or natural, producing additional $OH^{\bullet}$ radicals, and leading to Fe^{3+} reduction and thereby regeneration of the catalyst (photo-Fenton reagent):

$$Fe^{3+} + OH^{-} \xrightarrow{h\nu\ (\lambda < 350\,nm)} Fe^{2+} + OH^{\bullet} \tag{21.5}$$

The illumination therefore results in the formation of more $OH^{\bullet}$ radicals, regeneration of the catalyst and decarboxylation and/or complete oxidation of most of the organic compounds.

The degradation of aromatics, pesticides, herbicides, and other bio-recalcitrant compounds in industrial wastewaters can be efficiently achieved by means of the oxidative radicals generated in the Fenton and photo-assisted Fenton reaction [11]. Among others, the photo-Fenton method has proved to be effective in the degradation of phenol and its halogenated derivatives [12–14], dioxins [15], nitroaniline [16], various dyes, especially azo-dyes [10, 17–19], alcohols [9], carboxylic acids (citric [20] and formic acid [21–23]), some pharmaceuticals, such as diclofenac [24] and tetracycline [25], pesticides [26], herbicides [27, 28], and insecticides [29].

The photo-Fenton method of wastewater treatment is developed in two main fields: fitting of the usable radiation range to the solar light and immobilization of the photo-Fenton catalyst.

The feasibility of applying solar radiation as a source of UV–visible radiation has made the photo-Fenton system an economical and competitive process. Within this context an alternative method has been developed based on solar photocatalytic oxidation and natural processes of wastewater treatment [5], as well as sunlight-driven degradations of many compounds, such as EDTA [30], phenols [7, 13], pesticides [31–33], surfactants [34], diclofenac [24], formic acid [22], azo-dyes [19], non-biodegradable chlorinated solvents [35], nitroaniline [16], and other organic compounds [21, 36].

During the homogeneous Fenton reactions, the pH has to be adjusted twice, first to an acidic pH <4 to carry out the Fenton pre-treatment, and then back to a neutral pH. In this way, it is possible to proceed with the lower-cost second-stage biological destruction of the organic compounds once the biocompatibility has been attained during the Fenton pre-treatment. This is why supported Fenton-immobilized catalysts have recently been developed to overcome the costly pH adjustment on Nafion membranes [10, 18, 37], silica-structured surfaces [38], Nafion silica surfaces [8, 17], polyethylene copolymers, and alginate microcapsules. These supported catalysts have been developed over the last few years with the intention of (1) withstanding the corrosion due to the oxidative radicals produced in solution, (2) fixing the iron ions on a support in a stable way so as not to allow them to leach out during the reaction, and (3) allowing the Fenton-mediated catalysis to proceed at acceptable kinetic rates [10].

Pollutant degradation by photo-Fenton reagent was compared with effects of photocatalysis over TiO_2. The technical feasibility and performance of photocatalytic degradation of four water-soluble pesticides (diuron, imidacloprid, formetanate, and methomyl) have been studied at a pilot scale in two well-defined systems driven by natural solar UV light: heterogeneous photocatalysis with TiO_2 and homogeneous photocatalysis by the photo-Fenton method. The comparison of the degree of mineralization and toxicity achieved, as well as the transformation products generated en route to mineralization, showed that with the photo-Fenton system mineralization is faster and toxic oxygen compounds disappear faster than with TiO_2 [39, 40]. In the case of non-biodegradable chlorinated solvents, such as dichloroethane, dichloromethane, and trichloromethane dissolved in water, the photo-Fenton method was found to be the more effective treatment [35].

The study of the possible cooperation of the photo-Fenton reagent with a semiconductor photocatalyst showed that the Fenton catalyst immobilized in TiO_2 reveals enhanced photocatalytic activity and is more effective in pollutant degradation; this was revealed in the case of such electron donors as citric acid [20], maleic acid [41], chlorinated solvents [42], resorcinol [43], quinoline [44], herbicides [45, 46], and pesticides [47]. The optimization of the combined photocatalytic systems was obtained when each photocatalyst plays a specific role: Fe^{III} as a main $OH^{\bullet}$ radical source and TiO_2 as an oxidizing agent of Fe^{II} [45].

Similar to Fenton reactions, which proceed in a mixture of hydrogen peroxide and ferrous salts the 'Fenton-like reactions' are analogous reactions of peroxides with metal complexes, M_nL_m, in their low-valent oxidation states (e.g. Fe^{II}, Cu^{I}, Ti^{III}, Cr^{II}, Co^{II}). The hybrid photocatalysts composed of semiconductor and transition metal complex were announced to be effective in visible light detoxification of azo-dyes (TiO_2 modified by copper tetracarbonyl phthalocyanine) [48], in copper ion-catalyzed aerobic oxidation of ascorbic acid [49], or in diverse processes of solar energy conversion [50].

'Fenton-like' reactions play an important role in a variety of catalytic and biological processes. In biology these reactions are believed to be the main source of reactive oxygen species (ROS) in the cell causing a variety of diseases, eg cancer, arteriosclerosis, essential hypertension, Alzheimer's disease, amyloidosis, osteoarthritis [9].

Disinfection is the ultimate phase preceding water supply to customers. Nowadays, the most common disinfectant in use is chlorine. Despite its major advantages, chlorine has the undisputed disadvantage of forming chloro-organic compounds in the presence of organic matter naturally present in water. These compounds are carcinogenic and mutagenic. To minimize the risk to humans, modifications of conventional treatment and other alternative methods of disinfection, including the removal of chloro-organic compounds, have been proposed. Moreover, standard water disinfection techniques are often too expensive. In this respect, the use of solar energy as an alternative to chlorination has recently been explored because this technology could be economically viable for communities in regions with a high degree of sunlight radiation [51].

The disinfecting effect of solar light can be enhanced by addition of a catalyst and such generated systems belong to the group of advanced oxidation processes. The generated $OH^{\bullet}$ radical is highly toxic towards microorganisms and very reactive in the oxidation of organic substances. Therefore, a solar photocatalytic treatment could become an efficient method to simultaneously inactivate bacteria and degrade organic matter. Moreover, Fenton reactions can naturally occur in biological systems, especially where H_2O_2 is formed during the course of normal cell functioning. There are studies indicating the toxic nature of iron and copper and their role in ageing of biological systems. Both copper and iron combined with H_2O_2 have been investigated as a substitute for conventional disinfectants. Fenton-based disinfection using H_2O_2 is considered more efficient, with essentially no side effects. The Fenton reaction has been found to be the key reaction in the oxidation of membrane lipids and amino acids, and in reactions where biological reduction agents are present, such as ascorbic acid or thiols. Its occurrence is also supposed in heart diseases, such as ischaemia and reperfusion. Investigations of the Fenton reaction in cell biology are in progress. The addition of TiO_2, TiO_2/Fe^{3+}, or Fe^{3+}/H_2O_2 to the water accelerates the bactericidal action of sunlight, leading to total disinfection by heliophotocatalysis [52].

To maintain our surroundings biologically clean, in recent years TiO_2 photosterilization has attracted increasing attention because of its stable and strong oxidative reactions resulting in biocidal activity. Recently, it was found that the coexistence of Cu^{2+} and H_2O_2 in liquid phase together with the TiO_2 film exerted a synergistic effect on killing *Escherichia coli* cells, though these chemical species were individually at non-toxic levels to the cells [53]. Cu^{II} was also reported to enhance the protein-damaging potential of photosensitized tetracycline hydrochloride assessed using serum albumin as a model protein [54].

Transition metal compounds were also suggested as biogenic photocatalysts, which could catalyze low-temperature reactions with the help of solar light quanta. Species that might serve as the photocatalysts include cations of some transition metals (Cu^{2+}, Mn^{2+}, $Fe^{2+/3+}$, etc), dissolved in water basins or droplets, solid oxide-based aerosols of both semiconducting (TiO_2, Fe_2O_3, ZnO) and insulating (MgO, CaO) nature, Fe and Ti oxides containing sands, soot particles, etc [55].

21.3 Heterogeneous Photocatalysis

A tremendous increase in the number of photocatalytically active products introduced to the market has been observed for last few years. The main fields of their applications include photomineralization of organics, photosterilization, photodemisting, and photovoltaics. Other applications did not reach the level allowing commercialization and can be considered as perspectives (eg photocatalytic CO_2 reduction and fixation, H_2 production, synthesis).

Fundamentals of heterogeneous photocatalysis have been described in Chapter 7. The photodynamic activity of TiO_2 (based on its photocatalytic activity and high efficiency in ROS generation) can find applications in photodynamic therapy (PDT) and photodynamic inactivation of microorganisms (PDI), described in Chapters 17 and 18, respectively. Selected environmental aspects of heterogeneous photocatalysis are described below.

21.3.1 Water and Air Detoxification

Application of heterogeneous photocatalysis to water and air purification appears a particularly convenient method in the cases of easily accessible UV light (mainly solar, but also artificial) and diluted pollutants. TiO_2-based photocatalysis offers unique possibilities of mineralization of most organic compounds and, in addition, sterilization. This type of AOP useful for water and air purification has been studied and developed extensively over the last few decades. UV light, especially as a part of solar radiation, must be used to activate this process. In addition oxygen and water are indispensable.

Photocatalytic systems became a subject of many detailed studies and improvements. A need to compare photocatalytic activity of various materials enforced selection of a 'model pollutant'. Perhaps the most commonly chosen is 4-chlorophenol (4-CP) [56, 57]. The route of 4-CP oxidation in the presence of excited TiO_2 was described in many original and review papers [58–64]. The oxidative pathway includes attack of hydroxyl radicals or direct oxidation with a hole leading to hydrogen atom abstraction or introduction of additional –OH groups to the aromatic ring. Further steps include the C–Cl bond cleavage, introduction of the next hydroxyl groups, and ring opening. Decarboxylation of carboxylic acids and subsequent oxidation reactions lead finally to complete mineralization.

Up to now hundreds of organic and inorganic compounds have been successfully degraded in the photomineralization process [60, 65]. Photodegradation processes may be useful for either water or air purification. There are several reports also on photocatalyzed removal of selected inorganic ions such as CN^- [66], NO_3^- [67], NO_2^- [67], etc. Several pilot plants for water and air cleaning were built and successfully operated, especially in regions having large number of sunny days. Nevertheless, the efficiency of the photocatalytic process is still insufficient. Only about 3% of solar light energy, available as UV radiation, can be utilized by TiO_2. Therefore two main points should sustain main objects of research and development: sensitization towards visible light and improvement of quantum yield of the photo-oxidation process.

Depending on the nature of pollutants (specific light absorption properties, specific stability) an appropriate photocatalyst or its modification should be selected. Crystal structure of TiO_2 plays a crucial role in its activity. In general TiO_2 in the form of anatase is more photoactive than rutile or amorphous material. However, mixtures of anatase and rutile appear to be especially active which was often confirmed in comparative studies comprising commercial TiO_2 sample P25 produced by *Degussa* (about 70% of anatase and 30% of rutile, specific surface area $50\,m^2\,g^{-1}$ [68]) [57, 61, 64, 69–71]. This material is often regarded as a standard TiO_2 photocatalyst [57, 71]. Nevertheless TiO_2 in the form of rutile, as well as an amorphous material, shows a certain activity. Recently the third TiO_2 phase, brookite, has also attracted some interest [72].

An improvement of charge separation within the excited photocatalyst resulting in decreased recombination efficiency can be achieved by a combination of TiO_2 with other semiconducting oxides. Improved photocatalytic performances have been reported for TiO_2/WO_3 [73], TiO_2/ZrO_2 [74, 75], TiO_2/SnO_2 [76], TiO_2/Nb_2O_5 [77]. Interparticle charge transfer prevents fast electron-hole recombination. The same effect is probably responsible for improved performance of anatase/rutile mixtures.

Metal deposition on TiO_2, usually achieved by photochemical reduction of appropriate metal ions in the presence of alcohols as reducing agents, can be used for photocatalytic recovery of noble metals from diluted solutions [78]. The presence of Pt^0, Ag^0, Au^0, and other metallic nanoparticles on TiO_2 particles, playing a role of electron sinks, also influences photocatalytic processes [79, 80]. In this case a better charge separation and longer lifetime of electron/hole pairs are achieved enhancing efficiency of the IFET processes [80]. Presence of metal at the catalyst surface can also influence a possible products distribution.

The mechanisms of TiO_2 photosensitization have been described in Chapter 7. A novel version of sensitization was reported for TiO_2 containing up to 3% of Pt^{IV}, Rh^{III}, and Au^{III} chlorides in the bulk [81–83]. Visible light irradiation causes a complete mineralization of 4-CP in the presence of these photocatalysts. The surface modification of TiO_2 with $[PtCl_6]^{2-}$ complexes gave even more efficient catalysts [84–87]. Upon irradiation the excited platinum complex undergoes homolytic metal–chloride bond cleavage, yielding Pt^{III} transient species and an adsorbed chlorine atom. Subsequent electron transfers from the former to titania and from 4-CP to the chlorine atom reform the sensitizer.

Although the dye sensitization method can be very efficient, its application in photocatalytic water purification is usually very limited due to a low stability of such systems.

21.3.2 Photocatalytic CO_2 Reduction

Photocatalytic CO_2 reduction over simple oxide semiconductors, although characterized by low efficiencies, might contribute to natural processes occurring in atmosphere of young Earth (see section 10.3, Chapter 10). More sophisticated systems developed in future may be useful for CO_2 utilization as a substrate for production of chemicals. Until now yields of these processes, usually allowing formation of micromoles of products, are still not sufficient. Selected examples of CO_2 reduction in heterogeneous photocatalytic systems are described below.

Experiments with ZrO_2 loaded with Pt, Au, or RuO_2 revealed the production of hydrogen, oxygen, and carbon monoxide in the solution of $NaHCO_3$ and $KHCO_3$ [88]. Simultaneous hydrogen evolution was more then 100 times faster than CO_2 reduction for not loaded zirconium oxide, but when semiconductor was doped with about 1% of copper H_2 evolution was suppressed and the rate of hydrogen, oxygen, and carbon monoxide production amounted to 19.5, 10.8, and 2.5 µmol h^{-1}, respectively. The activity of the system was stable for more than 150 hours. In the case of not loaded ZrO_2 rates were higher, but also water splitting was preferential: 309 µmol h^{-1} (H_2), 167 µmol h^{-1} (O_2), and 3.0 µmol h^{-1} (CO) in $NaHCO_3$ solution.

Methanol and methane production from CO_2 was observed when aqueous suspension of ZnO was irradiated with visible light [89]. Experiments with ZnS carried out by Yoneyama et al. [90, 91] or recently by Martin et al. [92] showed high efficiencies in carbon dioxide reduction to formate.

The carbon dioxide photoreduction upon visible light has been performed using Pt-CdS-RuO_2 powder (particles <1 µm) and K[Ru(H-edta)Cl] · $2H_2O$ in CO_2 saturated water solution [93]. Formic acid, formaldehyde as the main products and trace amounts of methanol, methane, and carbon monoxide have been identified. Also hydrogen and oxygen as products of water decomposition were evolved. The proposed mechanism of the process is given below (L = H-edta):

$$\mathrm{CdS} \xrightarrow{h\nu} \mathrm{CdS}(\mathrm{h}^+, \mathrm{e}^-) \tag{21.6}$$

$$\mathrm{H_2O} + 2\mathrm{h}^+ \rightarrow 2\mathrm{H}^+ + 1/2\,\mathrm{O_2} \tag{21.7}$$

$$[\mathrm{LRu^{III}(H_2O)}] + \mathrm{e}^- \rightarrow [\mathrm{LRu^{II}(H_2O)}]^- \tag{21.8}$$

$$[\mathrm{LRu^{II}(H_2O)}]^- + \mathrm{CO_2} \rightleftharpoons [\mathrm{LRu^{II}(CO_2)}]^- + \mathrm{H_2O} \tag{21.9}$$

$$[\mathrm{LRu^{II}(CO_2)}]^- + 2\,\mathrm{H}^+ + 2\,\mathrm{e}^- \rightarrow [\mathrm{LRu^{II}(H_2O)}]^- + \mathrm{HCOOH} \tag{21.10}$$

$$[\mathrm{LRu^{II}(CO_2)}]^- + 4\,\mathrm{H}^+ + 4\,\mathrm{e}^- \rightarrow [\mathrm{LRu^{II}(H_2O)}]^- + \mathrm{HCHO} + \mathrm{H_2O} \tag{21.11}$$

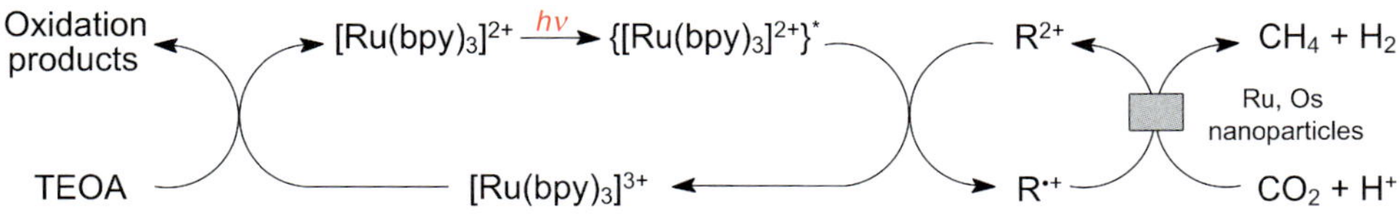

Figure 21.1 *Photosensitized electron transfer cycle for the reduction of carbon dioxide in aqueous photosystem consisting of* $[Ru(bpy)_3]^{2+}/R^{2+}$*/TEOA/Ru- or Os-colloid. TEOA, triethanolamine. (Adapted from Maidan and Willner [94] and Willner et al. [95])*

Formed holes oxidize water giving molecular oxygen and protons. The Ru^{III} complex is reduced to the Ru^{II} one, which reacts with CO_2. Formic acid and formaldehyde are formed as a result of the reduction process.

It was found, that also Ru and Os colloids can act as catalysts for the photoreduction of carbon dioxide to methane [94, 95]. $[Ru(bpy)_3]^{2+}$ plays a role of a photosensitizer, triethanolamine (TEOA) works as an electron donor, while bipyridinium electron relays (R^{2+}) mediate the electron transfer process. The production of hydrogen, methane, and small amounts of ethylene may be observed in such a system (Figure 21.1). Excited $[Ru(bpy)_3]^{2+}$ is oxidized by bipyridinium salts, whereas formed $[Ru(bpy)_3]^{3+}$ is reduced back to $[Ru(bpy)_3]^{2+}$ by TEOA. The reduced bipyridinium salt $R^{\bullet+}$ reduces hydrogen and CO_2 in the presence of metal colloids. Recombination of surface-bound H atoms competes with a multi-electron CO_2 reduction. More selective reduction of CO_2 to CH_4, ethylene, and ethane was obtained using ruthenium(II)-trisbipyrazine, $[Ru(bpz)_3]^{2+}$/TEOA/Ru colloid system. The elimination of hydrogen evolution is thought to be caused by a kinetic barrier towards H_2 evolution in the presence of $[Ru(bpz)_3]^{2+}$ and noble metal catalysts [96].

Photochemical reduction of CO_2 was also achieved in the presence of the *p*-type semiconductor (copper oxide) or silicon carbide electrodes [97]. Irradiation of this system generates methanol and methane as the main products in the case of CuO electrode whereas hydrogen (with efficiency about 80%), methanol (16%), methane, and carbon monoxide in the case of SiC electrode. Also TiO_2/CuO systems appeared relatively efficient (up to 19.2% quantum yield) in photocatalytic CO_2 to CH_3OH reduction [98].

In addition to described heterogeneous systems a homogeneous photocatalytic CO_2 reduction was tested using heteroleptic rhenium complexes [99, 100] and supramolecular ruthenium and rhenium bi- and tetranuclear complexes; their excited states were quenched by 1-benzyl-1,4-dihydronicotinamide (BNAH) and CO_2 was reduced by the electron donor intermediate species [101].

21.3.3 Other Applications of Photocatalysis

Several other photocatalytic environmentally friendly processes may also be considered. Photocatalytic synthesis of chemicals may serve as a clean alternative route to traditional synthetic methods (see section 6.5). Palmisano et al. reviewed various types of organic reactions that could be achieved by heterogeneous photocatalysis

[102]. They include aromatic hydroxylation, hydrocarbon and alcohol oxidation, alkene epoxidation, nitro-aromatic reduction, dehydrogenation, carbonylation, cyclization, heterocycle functionalization, etc.

Photosensitized generation of singlet oxygen in heterogeneous systems may be interesting for medical applications. Neat TiO_2 is not a good photosensitizer capable of singlet oxygen generation; however, modifications of its surface with selected inorganic or organic species may improve the efficiency of $^1O_2$ formation significantly. As demonstrated in the process of cyanuric acid degradation singlet oxygen is very probably formed in the presence of fluorinated TiO_2, silylated TiO_2, and TiO_2 modified with hexachloroplatinate(IV) ions [103].

Photodemisting properties of TiO_2 films result from high hydrophilicity of irradiated TiO_2 surface. This effect is maintained also in darkness for several hours after irradiation, therefore covering with thin TiO_2 films became an easy method of production of self-demisting glass and mirrors.

21.4 New Ideas in Pollution Abatement

The development of pollution abatement methods is directed towards new techniques and new ways of upgrading the quality of life.

21.4.1 New Emerging Techniques

With regard to the new techniques, what seems to keep the secret of future advances is photocatalysis in supercritical fluids. Recently, homogeneous catalysis in supercritical fluids has been developed considerably from a new and essentially unexplored field into a mature field with an appreciable body of information [104]. Supercritical fluids (SCFs), which are pure compounds or mixtures heated and pressurized beyond their critical points, have many advantages as solvents for homogeneous catalysis. A few of these advantages are true for all SCF solvents and essentially all reactions: mass transfer is very rapid, the solvent is completely miscible with gaseous reactants, and the solvent is easy to remove from the product. As these advantages are also relevant to homogeneous photocatalysis the fast development of this discipline may be anticipated.

Other perspectives of creating a new technique in pollution abatement are offered by an organized reaction microenvironment, which is employed to constrain and pattern initial reagents (precursors) and to perform photoinduced heterogeneous or homogeneous synthesis (decomposition) in a controllable and pre-selected way. Mesoporous oxides and hybrid photocatalysts based on molecularly organized assemblies offer new possibilities of concentrating low reactant levels, controlling the reaction environment, separating photocatalyst and reaction products, and realizing host–guest catalysis of light-driven reactions. This allows one to organize catalytic photolysis in an especially efficient way, to increase its selectivity and ensure another set of products compared with photocatalysis in a suspension of dispersed photocatalyst. Further increase in selectivity can be attained through functionalization of spatially confined photoreactors to impart them biological or chemical recognition properties, to couple different photocatalysts and redox catalysts, and to create complicated hybrid microphotoreactors including those that mimic natural photosynthetic centres [105].

Another idea worth considering is the concept of templated photochemistry, ie photochemical reaction control in homogeneous solutions by templates. Shielding of prochiral faces, topochemical reaction control, templating, and aggregate or inclusion complex formation are strategies that have been found to enhance and control reaction efficiency and regio- or stereoselectivity. However, the number of truly catalytic examples and their efficiency remain limited. Development of catalytic templates controlling the stereochemistry of photoreactions with high precision and activity will therefore be a future challenge in the field [106].

To improve the photoprocess performance, diverse combinations of heterogeneous photocatalysis with chemical and physical operations have been proposed, including among others, photo-Fenton, ozonization, biological or electrochemical treatment, and ultrasonic irradiation; these attempts were recently reviewed and analysed [107].

21.4.2 Renewable Energy Resources

A great technological challenge facing our global future is the development of renewable energy resources such as artificial photosynthetic systems and fuel cells free of environmental pollution, since global warming is closely related to tremendous consumption of fossil fuels and emission of carbon dioxide. A spectacular new goal of photocatalysis in pollution abatement is therefore the contribution to renewable sources of energy, which could guarantee a more sustainable development. To realize this task, solar energy should be converted into useful energy, via reduction catalyses indispensable for creating new energy systems, especially in terms of molecular catalysts.

The potential substrates should be cheap and freely available and products should not pollute the environment. Taking these prerequisites into account, water and carbon dioxide should be regarded as the best candidates for the substrates. In both cases, however, the reactive excited states cannot be reached simply by absorption of solar light. Fortunately, solar energy can be transferred by photosensitizers or photocatalysts.

Examples of the molecular catalysts for CO_2, H^+, and O_2 reductions were reviewed recently and a role for molecular aggregates composed of a simple metal complex and a functional polymer was emphasized. When using molecular aggregates as a catalyst, efficient catalysis by the complex via a multi-electron transfer reduction often takes place [108, 109].

The efficient utilization of CO_2 has attracted considerable attention from fundamental research to industrial application in recent years. Heterogeneous catalysis, electrocatalysis, and photocatalysis are presently the three predominant chemical methods for converting CO_2 into some useful chemicals, such as methanol, formic acid, formaldehyde. These methods need high temperature and pressure or additional electric or luminous energy. Examples of photocatalytic systems capable of CO_2 reduction have been described in section 21.3.2 and in section 10.3, Chapter 10. However, photocatalytic processes of CO_2 reduction are characterized by low selectivity and yields. From this perspective biological and enzymatic CO_2 fixation systems have also received much attention [110, 111].

Molecular hydrogen is known as a clean-burning fuel free of CO_2 emission and is now considered a promising candidate to solve the current energy problems. Photochemical hydrogen production is potentially one of the most fascinating ways for solar energy conversion and storage. Since 1977, several homogeneous, quasi-homogeneous, or microheterogeneous model systems of hydrogen or oxygen generation from water by visible light irradiation have been proposed. These photo-systems are based on different approaches: (1) multimolecular systems, (2) systems involving supramolecular structure of polyad structure, and (3) systems incorporated in organized and constrained or confined media [112].

The tris(2,2′-bipyridine)ruthenium(II), $[Ru(bpy)_3]^{2+}$, has attracted considerable attention for many years due to its potential application as a photosensitizer promoting visible-light-induced splitting of H_2O into H_2 and O_2 (see sections 6.4 and 6.5, Chapter 6). The fundamental concepts on the application of this photosensitizer have been well demonstrated by the observation that the photochemical hydrogen production is effectively promoted on visible-light irradiation of an aqueous solution containing a photosensitizer, such as $[Ru(bpy)_3]^{2+}$, an electron relay, such as methyl viologen (*N*,*N*′-dimethyl-4,4′-bipyridinium, abbreviated MV^{2+}), a catalyst evolving H_2 (eg Pt, Pt^{II}, or Rh^{I} complexes) and a sacrificial electron donor, such as EDTA [113].

Hydrogen production by molecular photocatalysis from different substrates, using different photosensitizers, sacrificial donors, and catalysts, was recently extensively reviewed by Esswein and Nocera [114]. Among others, platinum(II) [115], di-iron(III)-μ-oxo-bisporphyrins [116], dimeric μ-oxo-bridged ruthenium diimine complexes [117], rhodium(II)/(III) mixed valence complexes [118], and iridium(III) complexes were suggested as better photosensitizers than those of Ru^{II}, because they enable two-electron water photolysis [119]. Screening of 32 heteroleptic iridium complexes showed that 6 of them produced considerably more hydrogen than commonly tested photosensitizers and had relative quantum efficiencies of hydrogen production up to 37 times greater than that of the $[Ru(bpy)_3]^{2+}$ complex [120]. A unique, homogeneous photocatalyst for H_2 production containing ruthenium light absorbers and a rhodium electron collector was reported to produce photocatalytically hydrogen from water. The supramolecular complex used, $[\{(bpy)_2Ru(dpp)\}_2RhCl_2]^{5+}$ (dpp = 2,3-bis(2-pyridyl)pyrazine), possesses a central Rh^{III} core that serves as an electron collector for multi-electron photochemistry [121].

Various alternatives to the conventional one-electron photochemistry were reviewed recently [116, 122]. A rational framework for advancing the multi-electron photochemistry of new metal complexes has been developed using two-electron mixed-valence complexes as the redox platform. Two-electron mixed valence is a useful design concept for hydrogen and oxygen photocatalysis. As single-electron mixed-valence compounds react in one-electron steps, two-electron mixed-valence compounds may react in two-electron steps at the constituent redox sites, whether they are metal or ligand based. With redox centres working in concert on photoexcitation, two- and four-electron transformations are promoted along excited-state pathways. Such a strategy is ideally suited to the activation of small molecules. The photochemistry for hydrogen and oxygen activation was described with two-electron mixed-valence complexes of three different classes: (1) M^n-M^{n+2} bimetallic complexes, (2) tetrapyrrole macrocycles, and (3) externally bridged di-iron(III) μ-oxo-porphyrin dimers. The excited-state chemistry of these compounds offers an interesting alternative to the conventional one-electron photochemistry. To date, hydrogen and oxygen were successfully activated, which made it possible to construct photocycles for H_2 production from protic media, as well as efficient oxygen atom transfer (OAT) photochemistry using O_2 as the terminal oxidant and oxygen atom source [116].

Production of H_2 fuel from water by solar energy may involve H_2 complexes of transition metals, at least as intermediates, and H_2 complexes have been implicated in solar energy conversion schemes based on photoreduction of water. Industrially important water gas shift and related H_2-producing reactions undoubtedly proceed via transient H_2 complexes [116, 123].

Since the discovery of the first water-splitting system based on TiO_2 and Pt in 1972 by Fujishima and Honda, over 130 inorganic materials have been discovered as catalysts for this reaction [124]. These materials are mostly oxides and sulphides of the transition metals, with TiO_2 as the most promising catalyst for both photocatalytic and photoelectrochemical generation of hydrogen through water splitting using solar energy [125–128]. Platinum deposited at TiO_2 surface enhances hydrogen evolution whereas RuO_2 particles enhance oxygen evolution in the process of photocatalyzed decomposition of water on a RuO_2/Pt/TiO_2 photocatalyst [129]. In this system, which can be considered as a short-circuited micro-photoelectrochemical cell, platinum plays the role of the cathode, whereas RuO_2 is the anode. Deposited platinum and ruthenium oxide reduce the over-potentials for H_2 and O_2 formation.

References

1. Legrini O, Oliveros E, Braun AM. Photochemical processes for water treatment. *Chem Rev* 1993;**93**:671–98.
2. Braun AM, Oliveros E. How to evaluate photochemical methods for water treatment. *Wat Sci Tech* 1997;**35**:17–23.
3. Litter MI. Heterogeneous photocatalysis. Transition metal ions in photocatalytic systems. *Appl Catal B: Environ* 1999;**23**:89–114.
4. Andreozzi R, Caprio V, Insola A, Marotta R. Advanced oxidation processes (AOP) for water purification and recovery. *Catal Today* 1999;**53**:51–9.
5. Antoniadis A, Takavakoglou V, Zalidis G, Poulios I. Development and evaluation of an alternative method for municipal wastewater treatment using homogeneous photocatalysis and constructed wetlands. *Catal Today* 2007;**124**:260–5.
6. Peralta-Hernandez JM, Meas-Vong Y, Rodriguez FJ, Chapman TW, Maldonado MI, Godinez LA. In situ electrochemical and photo-electrochemical generation of the

Fenton reagent: a potentially important new water treatment technology. *Water Res* 2006;**40**:1754–62.

7. Rodriguez M, Malato S, Pulgarin C, et al. Optimizing the solar photo-Fenton process in the treatment of contaminated water. Determination of intrinsic kinetic constants for scale-up. *Solar Energy* 2005;**79**:360–8.
8. Bozzi A, Yuranova T, Mielczarski E, et al. Superior biodegradability mediated by immobilized Fe-fabrics of waste waters compared to Fenton homogeneous reactions. *Appl Catal B: Environ* 2003;**42**:289–303.
9. Masarwa A, Rachmilovich-Calis S, Meyerstein N, Meyerstein D. Oxidation of organic substrates in aerated aqueous solutions by the Fenton reagent. *Coord Chem Rev* 2005;**249**:1937–43.
10. Parra S, Guasaquillo I, Enea O, et al. Abatement of an azo dye on structured C-Nafion/Fe-ion surfaces by photo-Fenton reactions leading to carboxylate intermediates with a remarkable biodegradability increase of the treated solution. *J Phys Chem B* 2003;**107**:7026–35.
11. Pignatello JJ, Liu D, Huston P. Evidence for an additional oxidant in the photoassisted Fenton reaction. *Environ Sci Technol* 1999;**33**:1832–9.
12. Benitez FJ, Beltran-Heredia J, Acero JL, Rubio FJ. Chemical decomposition of 2,4,6-trichlorophenol by ozone, Fenton's reagent, and UV radiation. *Ind Eng Chem Res* 1999;**38**:1341–9.
13. Gernjak W, Krutzler T, Glaser A, et al. Photo-Fenton treatment of water containing natural phenolic pollutants. *Chemosphere* 2003;**50**:71–8.
14. Ji H, Song W, Chen C, Yuan H, Ma W, Zhao J. Anchored oxygen-donor coordination to iron for photodegradation of organic pollutants. *Environ Sci Technol* 2007;**41**:5103–7.
15. Katsumata H, Kaneco S, Suzuki T, Ohta K, Yobiko Y. Degradation of polychlorinated dibenzo-*p*-dioxins in aqueous solution by Fe(II)/H_2O_2/UV system. *Chemosphere* 2006;**63**:592–9.
16. Sun J-H, Sun S-P, Fan M-H, Guo H-Q, Lee Y-F, Sun R-X. Oxidative decomposition of *p*-nitroaniline in water by solar photo-Fenton advanced oxidation process. *J Hazard Mater* 2008;**153**:187–93.
17. Gumy D, Fernandez-Ibanez P, Malato S, Pulgarin C, Enea O, Kiwi J. Supported Fe/C and Fe/Nafion/C catalysts for the photo-Fenton degradation of Orange II under solar irradiation. *Catal Today* 2005;**101**:375–82.
18. Kiwi J, Denisov N, Gak Y, et al. Catalytic Fe^{3+} clusters and complexes in Nafion active in photo-Fenton processes. High-resolution electron microscopy and femtosecond studies. *Langmuir* 2002;**18**:9054–66.
19. Chacon JM, Leal MT, Sanchez M, Bandala ER. Solar photocatalytic degradation of azo-dyes by photo-Fenton process. *Dyes Pigments* 2006;**69**:144–50.
20. Quici N, Morgada ME, Gettar RT, Bolte M. Photocatalytic degradation of citric acid under different conditions: TiO_2 heterogeneous photocatalysis against homogeneous photolytic processes promoted by Fe(III) and H_2O_2. *Appl Catal B: Environ* 2007;**71**:117–24.
21. Farias J, Rossetti GH, Albizzati ED, Alfano OM. Solar degradation of formic acid: temperature effects on the photo-Fenton reaction. *Ind Eng Chem Res* 2007;**46**:7580–6.
22. Rossetti GH, Albizzati ED, Alfano OM. Modeling of a flat-plate solar reactor. Degradation of formic acid by the photo-Fenton reaction. *Solar Energy* 2004;**77**:461–70.
23. Rossetti GH, Albizzati ED, Alfano OM. Decomposition of formic acid in a water solution employing the photo-Fenton reaction. *Ind Eng Chem Res* 2002;**41**:1436–44.
24. Perez-Estrada LA, Malato S, Gernjak W, et al. Diclofenac: identification of main intermediates and degradation pathway. *Environ Sci Technol* 2005;**39**:8300–6.
25. Bautitz IR, Nogueira RFP. Degradation of tetracycline by photo-Fenton process – Solar irradiation and matrix effects. *J Photochem Photobiol A: Chem* 2007;**187**:33–9.
26. Catastini C, Sarakha M, Mailhot G, Bolte M. Iron(III) aquacomplexes as effective photocatalysts for the degradation of pesticides in homogeneous aqueous solutions. *Sci Total Environ* 2002;**298**:219–28.

27. Katsumata H, Kaneco S, Suzuki T, Ohta K, Yobiko Y. Photo-Fenton degradation of alachlor in the presence of citrate solution. *J Photochem Photobiol A: Chem* 2006;**180**:38–45.
28. Poulain L, Mailhot G, Wong-Wah-Chung P, Bolte M. Photodegradation of chlortoluron sensitised by iron(III) aquacomplexes. *J Photochem Photobiol A: Chem* 2003;**159**:81–8.
29. Evgenidou E, Konstantinou I, Fytianos K, Poulios I. Oxidation of two organophosphorous insecticides by the photo-assisted Fenton reaction. *Water Res* 2007;**41**:2015–27.
30. Emilio CA, Jardim WF, Litter MI, Mansilla HD. EDTA destruction using the solar ferrioxalate advanced oxidation technology (AOT). Comparison with solar photo-Fenton treatment. *J Photochem Photobiol A: Chem* 2002;**151**:121–8.
31. Perez HM, Penuela G, Maldonado MI, et al. Degradation of pesticides in water using solar advanced oxidation processes. *Appl Catal B: Environ* 2006;**64**:272–7.
32. Malato S, Blanco J, Vidal A, et al. Applied studies in solar photocatalytic detoxification: an overview. *Solar Energy* 2003;**75**:329–42.
33. Maldonado Rubio MI, Gernjak W, Oller Alberola I, Blanco Galvez J, Fernandez-Ibanez P, Malato Rodriguez S. Photo-Fenton degradation of alachlor, atrazine, chlorfenvinphos, diuron, isoproturon and pentachlorophenol at solar pilot plant. *Int J Environ Pollut* 2006;**27**:135–46.
34. Amat AM, Arques A, Miranda MA, Segui S. Photo-Fenton reaction for abatement of commercial surfactants in a solar pilot plant. *Solar Energy* 2004;**77**:559–66.
35. Rodriguez SMR, Galvez JB, Rubio MIM, Ibanez PF, Gernjak W, Alberola IO. Treatment of chlorinated solvents by TiO_2 photocatalysis and photo-Fenton: influence of operating conditions in a solar pilot plant. *Chemosphere* 2005;**58**:391–8.
36. Nogueira RFP, Silva MRA, Trovo AG. Influence of the iron source on the solar photo-Fenton degradation of different classes of organic compounds. *Solar Energy* 2005;**79**:384–91.
37. Parra S, Henao L, Mielczarski E, et al. Synthesis, testing, and characterization of a novel Nafion membrane with superior performance in photoassisted immobilized Fenton catalysis. *Langmuir* 2004;**20**:5621–9.
38. Feng J, Hu X, Yue PL, Zhu HY, Lu GQ. Degradation of azo-dye Orange II by a photoassisted Fenton reaction using a novel composite of iron oxide and silicate nanoparticles as a catalyst. *Ind Eng Chem Res* 2003;**42**:2058–66.
39. Malato S, Caceres J, Aguera A, et al. Degradation of imidacloprid in water by photo-Fenton and TiO_2 photocatalysis at a solar pilot plant: a comparative study. *Environ Sci Technol* 2001;**35**:4359–66.
40. Malato S, Blanco J, Cáceres J, Fernández-Alba AR, Agüera A, Rodriguez A. Photocatalytic treatment of water-soluble pesticides by photo-Fenton and TiO_2 using solar energy. *Catal Today* 2002;**76**:209–20.
41. Franch MI, Ayllon JA, Peral J, Domenech X. Enhanced photocatalytic degradation of maleic acid by Fe(III) adsorption onto the TiO_2 surface. *Catal Today* 2005;**101**:245–52.
42. Yang P, Lu C, Hua N, Du Y. Titanium dioxide nanoparticles co-doped with Fe^{3+} and Eu^{3+} ions for photocatalysis. *Mater Lett* 2002;**57**:794–801.
43. Lam SW, Chiang K, Lim TM, Amal R, Low GK-C. The role of ferric ion in the photochemical and photocatalytic oxidation of resorcinol. *J Catal* 2005;**234**:292–9.
44. Cermenati L, Pichat P, Guillard C, Albini A. Probing the TiO_2 photocatalytic mechanisms in water purification by use of quinoline, photo-Fenton generated OH radicals and superoxide dismutase. *J Phys Chem B* 1997;**101**:2650–8.
45. Mestankova H, Mailhot G, Jirkovsky J, Krysa J, Bolte M. Mechanistic approach of the combined (iron–TiO_2) photocatalytic system for the degradation of pollutants in aqueous solution: an attempt of rationalisation. *Appl Catal B: Environmental* 2005;**57**:257–65.
46. Mestankova H, Krýsa J, Jirkovský J, Mailhot G, Bolte M. The influence of Fe(III) speciation on supported TiO_2 efficiency: example of monuron photocatalytic degradation. *Appl Catal B: Environmental* 2005;**58**:185–91.
47. Devipriya S, Yesodharan S. Photocatalytic degradation of pesticide contaminants in water. *Solar Energy Materials Solar Cells* 2005;**86**:309–48.

48. Chen F, Deng Z, Li X, Zhang J, Zhao J. Visible light detoxification by 2,9,16,23-tetracarboxyl phthalocyanine copper modified amorphous titania. *Chem Phys Lett* 2005;**415**: 85–8
49. Song B, Wang G, Yuan J. Measurement and characterization of singlet oxygen production in copper ion-catalyzed aerobic oxidation of ascorbic acid. *Talanta* 2007;**72**:231–6.
50. Garcia CG, de Lima JF, Iha NYM. Energy conversion: from the ligand field photochemistry to solar cells. *Coord Chem Rev* 2000;**196**:219–47.
51. Rincon A-G, Pulgarin C. Comparative evaluation of Fe^{3+} and TiO_2 photoassisted processes in solar photocatalytic disinfection of water. *Appl Catal B: Environ* 2006;**63**: 222–31.
52. Rincon A-G, Pulgarin C. Fe^{3+} and TiO_2 solar-light-assisted inactivation of *E. coli* at field scale Implications in solar disinfection at low temperature of large quantities of water. *Catal Today* 2007;**122**:128–36.
53. Sato T, Taya M. Copper-aided photosterilization of microbial cells on TiO_2 film under irradiation from a white light fluorescent lamp. *Biochem Eng J* 2006;**30**:199–204.
54. Khan MA, Musarrat J. Tetracycline-Cu(II) photo-induced fragmentation of serum albumin. *Comp Biochem Physiol C* 2002;**131**:439–46.
55. Parmon VN. Abiogenic catalysis in Nature. *Colloids and Surfaces A: Physicochemical and Engineering Aspects* 1999;**151**:351–65.
56. Mills A, Davies RH, Worsley D. Water purification by semiconductor photocatalysis. *Chem Soc Rev* 1993:417–25.
57. Serpone N, Sauvé G, Koch R, et al. Standardization protocol of process efficiencies and activation parameters in heterogeneous photocatalysis: relative photonic efficiencies. *J Photochem Photobiol A: Chem* 1996;**94**:191–203.
58. Barbeni M, Pramauro E, Pelizzetti E, Borgarello E, Grätzel M, Serpone N. Photodegradation of 4-chlorophenol catalyzed by titanium dioxide particles. *Nouv J Chim* 1984;**3**:547–50.
59. Al-Ekabi H, Serpone N, Pelizzetti E, Minero C, Fox MA, Draper RB. Kinetic studies in heterogeneous photocatalysis. 2. TiO_2-mediated degradation of 4-chlorophenol alone and in a three-component mixture of 4-chlorophenol, 2,4-dichlorophenol, and 2,4,5-trichlorophenol in air-equilibrated aqueous media. *Langmuir* 1989;**5**:250–5.
60. Hoffmann MR, Martin ST, Choi W, Bahnemann DW. Environmental applications of semiconductor photocatalysis. *Chem Rev* 1995;**95**:69–96.
61. Mills A, Morris S, Davies R. Photomineralisation of 4-chlorophenol sensitised by titanium dioxide: a study of the intermediates. *J Photochem Photobiol A: Chem* 1993;**70**:183–91.
62. Stafford U, Gray KA, Kamat PV. Radiolytic and TiO_2-assisted photocatalytic degradation of 4-chlorophenol. A comparative study. *J Phys Chem* 1994;**98**:6343–51.
63. Cunningham J, Al-Sayyed G. Factors influencing efficiencies of TiO_2-sensitised photodegradation. Part 1 – Substituted benzoic acids: discrepancies with dark-adsorption parameters. *J Chem Soc Faraday Trans* 1990;**86**:3935–42.
64. Theurich J, Lindner M, Bahnemann DW. Photocatalytic degradation of 4-chlorophenol in aerated aqueous titanium dioxide suspensions: a kinetic and mechanistic study. *Langmuir* 1996;**12**:6368–76.
65. Fox MA, Dulay MT. Heterogeneous photocatalysis. *Chem Rev* 1993;**93**:341–57.
66. Augugliaro V, Loddo V, Marcè G, Palmisano L, López-Munoz MJ. Photocatalytic oxidation of cyanides in aqueous titanium dioxide suspensions. *J Catal* 1997;**166**: 272–83.
67. Ranjit KT, Viswanathan B. Photocatalytic reduction of nitrite and nitrate ions over doped TiO_2 catalyst. *J Photochem Photobiol A: Chem* 1997;**107**:215–20.
68. Ettlinger M. Highly dispersed metallic oxides produced by AEROSIL® process. *Tech Bull Pigments Degussa* 1984;**56**:12.
69. Lichtin NN, Avudaithai M. TiO_2-photocatalyzed oxidative degradation of CH_3CN, CH_3OH, $CHCl_3$, and CH_2Cl_2 supplied as vapors and in aqueous solution under similar conditions. *Environ Sci Technol* 1996;**30**:2014–20.

70. Cunningham J, Sedlák P. Interrelationships between pollutant concentration, extent of adsorption, TiO_2-sensitized removal, photon flux and levels of electron or hole trapping additives. 1. Aqueous monochlorophenol-TiO_2 (P25) suspensions. *J Photochem Photobiol A: Chem* 1994;**77**:255–63.
71. Ahuja S, Kutty TRN. Nanoparticles of $SrTiO_3$ prepared by gel to crystallite conversion and their photocatalytic activity in the mineralization of phenol. *J Photochem Photobiol A: Chem* 1996;**97**:99–107.
72. Agrios AG, Pichat P. State of the art and perspectives on materials and applications of photocatalysis over TiO_2. *J Appl Electrochem* 2005;**35**:655–63.
73. Papp J, Soled S, Dwight K, Wold A. Surface acidity and photocatalytic activity of TiO_2, WO_3/TiO_2, and MoO_3/TiO_2 photocatalysts. *Chem Mater* 1994;**6**:496–500.
74. Zorn ME, Tompkins DT, Zeltner WA, Anderson MA. Photocatalytic oxidation of acetone vapor on TiO_2/ZrO_2 thin films. *Appl Catal B: Environ* 1999;**23**:1–8.
75. Moret MP, Zallen R, Vijay DP, Desu SB. Brookite-rich titania films made by pulsed laser deposition. *Thin Solid Films* 2000;**366**:8–10.
76. Vinodgopal K, Kamat PV. Enhanced rates of photocatalytic degradation of an azo dye using SnO_2/TiO_2 coupled semiconductor thin films. *Environ Sci Technol* 1995;**29**:841–5.
77. Cui H, Dwight K, Soled S, Wold A. Surface acidity and photocatalytic activity of Nb_2O_5/TiO_2 photocatalysts. *J Solid State Chem* 1994;**115**:187–91.
78. Borgarello E, Serpone N, Emo G, Harris R, Pelizzetti E, Minero C. Light-induced reduction of rhodium(III) and palladium(II) on titanium dioxide dispersions and the selective photochemical separation and recovery of gold(III), platinum(IV), and rhodium(III) in chloride media. *Inorg Chem* 1986;**25**:4499–503.
79. Linsebigler AL, Lu G, Yates JT Jr. Photocatalysis on TiO_2 surfaces: principles, mechanisms, and selected results. *Chem Rev* 1995;**95**:735–58.
80. Kamat PV, Meisel D. Nanoscience opportunities in environmental remediation. *C R Chimie* 2003;**6**:999–1007.
81. Zang L, Lange C, Maier WF, Abraham I, Storck S, Kisch H. Amorphous microporous titania modified with platinum(IV) chloride – a new type of hybrid photocatalyst for visible light detoxification. *J Phys Chem B* 1998;**102**:10765.
82. Kisch H, Zang L, Lange C, Maier WF, Antonius C, Meissner D. Modifiziertes, amorphes Titandioxid – ein Hybrid-Photohalbleiter zur Detoxifikation und Stromerzeugung mit sichtbarem Licht. *Angew Chem* 1998;**110**:3201.
83. Zang L, Macyk W, Lange C, et al. Visible light detoxification and charge generation by transition metal chloride modified titania. *Chem Eur J* 2000;**6**:379–84.
84. Burgeth G, Kisch H. Photocatalytic and photoelectrochemical properties of titania-chloroplatinate(IV). *Coord Chem Rev* 2002;**230**:40.
85. Kisch H, Macyk W. Visible-light photocatalysis by modified titania. *Chem Phys Chem* 2002;**3**:399–400.
86. Macyk W, Burgeth G, Kisch H. Photoelectrochemical properties of platinum(IV)-chloride surface modified TiO_2. *Photochem Photobiol Sci* 2003;**2**:322–8.
87. Kisch H, Burgeth G, Macyk W. Visible light photocatalysis by a titania transition metal complex. *Adv Inorg Chem* 2004;**56**:241–59.
88. Sayama K, Arakawa H. Photocatalytic decomposition of water and photocatalytic reduction of carbon dioxide over ZrO_2 catalyst. *J Phys Chem* 1993;**97**:531–3.
89. Watanabe M. Photosynthesis of methanol and methane from CO_2 and H_2O molecules on a ZnO surface. *Surf Sci* 1992;**279**:L236–42.
90. Inoue H, Torimoto T, Sakata T, Mori H, Yoneyama H. Photoreduction of carbon dioxide on quantized zinc sulfide. *Chem Lett* 1990:1483–6.
91. Kuwabata S, Nishida K, Tsuda R, Inoue H, Yoneyama H. Photochemical reduction of carbon dioxide to methanol using ZnS microcrystallite as a photocatalyst in the presence of methanol dehydrogenase. *J Electrochem Soc* 1994;**141**:1498–503.
92. Zhang XV, Ellery SP, Friend CM, et al. Photodriven reduction and oxidation reactions on colloidal semiconductor particles: Implications for prebiotic synthesis. *J Photochem Photobiol A: Chem* 2007;**185**:301–11.

93. Khan MMT, Rao NN, Chatterjee D. A novel photosynthetic mimic reaction catalysed by K[Ru(H-EDTA)Cl]·$2H_2O$; reduction of carbon dioxide to formate and formaldehyde in the presence of an aqueous suspension of Pt–CdS–RuO_2. *J Photochem Photobiol A: Chem* 1991;**60**:311–18.
94. Maidan R, Willner I. Photoreduction of CO_2 to CH_4 in aqueous solutions using visible light. *J Am Chem Soc* 1986;**108**:8100–1.
95. Willner I, Maidan R, Mandler D, Durr H, Dorr G, Zengerle K. Photosensitized reduction of CO_2 to CH_4 and H_2-evolution in the presence of Ru and Os-colloids: strategies to design selectivity of products distribution. *J Am Chem Soc* 1987;**109**:6080–6.
96. Durr H, Dorr G, Zengerle K, Mayer E, Curchod J-M, Braun AM. Possibilities et limites des diazacomplexes de Ru^{2+} en tant que sensibilisateurs induisant la photolyse de l'eau. *Nouv J Chim* 1985;**9**:717–20.
97. Halmann MM, Steinberg M. *Greenhouse Gas Carbon Dioxide Mitigation: Science and technology*. New York: Lewis Publishers, 1998: 520 and references cited herein.
98. Yahaya AH, Gondal MA, Hameed A. Selective laser enhanced photocatalytic conversion of CO_2 into methanol. *Chem Phys Lett* 2004;**400**:206–12.
99. Hori H, Johnson FPA, Koike K, Ishatani O, Ibusuki T. Efficient photocatalytic CO_2 reduction using $[Re(bpy)(CO)_3P(OEt)_3]^+$. *J Photochem Photobiol A: Chem* 1996;**96**:171–4.
100. Hori H, Ishatani O, Koike K, Johnson FPA, Ibusuki T. Efficient carbon dioxide photoreduction by novel metal complexes and its reaction mechanisms. *Energy Conversion* 1995;**36**:621–4.
101. Gholamkhass B, Mametsuka H, Koike K, Tanabe T, Furue M, Ishitani O. Architecture of supramolecular metal complexes for photocatalytic CO_2 reduction: ruthenium-rhenium bi- and tetranuclear complexes. *Inorg Chem* 2005;**44**:2326–36.
102. Palmisano G, Augugliaro V, Pagliaro M, Palmisano L. Photocatalysis: a promising route for 21st century organic chemistry. *Chem Commun* 2007:3425–37.
103. Jańczyk A, Krakowska E, Stochel G, Macyk W. Singlet oxygen photogeneration at surface modified titanium dioxide. *J Am Chem Soc* 2006;**128**:15574–5.
104. Jessop PG. Homogeneous catalysis using supercritical fluids: Recent trends and systems studied. *J Supercrit Fluids* 2006;**38**:211–31.
105. Shchukin D, Sviridov D. Photocatalytic processes in spatially confined micro- and nanoreactors. *J Photochem Photobiol C: Photochem Rev* 2006;**7**:23–39.
106. Svoboda J, Konig B. Templated photochemistry: toward catalysts enhancing the efficiency and selectivity of photoreactions in homogeneous solutions. *Chem Rev* 2006;**106**:5413–30.
107. Augugliaro V, Litter M, Palmisano L, Soria J. The combination of heterogeneous photocatalysis with chemical and physical operations: A tool for improving the photoprocess performance. *J Photochem Photobiol C: Photochem Rev* 2006;**7**:127–44.
108. Abe T, Kaneko M. Reduction catalysis by metal complexes confined in a polymer matrix. *Prog Polym Sci* 2003;**28**:1441–88.
109. Yagi M, Kaneko M. Molecular catalysts for water oxidation. *Chem Rev* 2001;**101**:21–35.
110. Lu Y, Jiang Z, Xu S, Wu H. Efficient conversion of CO_2 to formic acid by formate dehydrogenase immobilized in a novel alginate-silica hybrid gel. *Catal Today* 2006;**115**: 263–8.
111. Tsujisho I, Toyoda M, Amao Y. Photochemical and enzymatic synthesis of formic acid from CO_2 with chlorophyll and dehydrogenase system. *Catalysis Commun* 2006;**7**:173–6.
112. Amouyal E. Photochemical production of hydrogen and oxygen from water: A review and state of art. *Solar Energy Materials Solar Cells* 1995;**38**:249–76.
113. Ozawa H, Haga M, Sakai K. A photo-hydrogen-evolving molecular device driving visible-light-induced EDTA-reduction of water into molecular hydrogen. *J Am Chem Soc* 2006;**128**:4926–7.
114. Esswein AJ, Nocera DG. Hydrogen production by molecular photocatalysis. *Chem Rev* 2007;**107**:4022–47.

115. Zhang D, Wu L-Z, Zhou L, et al. Photocatalytic hydrogen production from Hantzsch 1,4-dihydropyridines by platinum(II) terpyridyl complexes in homogeneous solution. *J Am Chem Soc* 2004;**126**:3440–1.
116. Rosenthal J, Bachman J, Dempsey JL, et al. Oxygen and hydrogen photocatalysis by two-electron mixed-valence coordination compounds. *Coord Chem Rev* 2005;**249**: 1316–26.
117. Hurst JK. Water oxidation catalyzed by dimeric μ-oxo bridged ruthenium diimine complexes. *Coord Chem Rev* 2005;**249**:313–28.
118. Esswein AJ, Veige AS, Nocera DG. A photocycle for hydrogen production from two-electron mixed-valence complexes. *J Am Chem Soc* 2005;**127**:16641–51.
119. Lowry MS, Goldsmith JI, Slinker JD, et al. Single-layer electroluminescent devices and photoinduced hydrogen production from an ionic iridium(III) complex. *Chem Mater* 2005;**17**:5712–9.
120. Goldsmith JI, Hudson WR, Lowry MS, Anderson TH, Bernhard S. Discovery and high-throughput screening of heteroleptic iridium complexes for photoinduced hydrogen production. *J Am Chem Soc* 2005;**127**:7502–10.
121. Elvington M, Brown J, Arachchige SM, Brewer KJ. Photocatalytic hydrogen production from water employing a Ru, Rh, Ru molecular device for photoinitiated electron collection. *J Am Chem Soc* 2007;**129**:10644–5.
122. Dempsey JL, Esswein AJ, Manke DR, Rosenthal J, Soper JD, Nocera DG. Molecular chemistry of consequence to renewable energy. *Inorg Chem* 2005;**44**:6879–92.
123. Kubas GJ. Fundamentals of H_2 binding and reactivity on transition metals underlying hydrogenase function and H_2 production and storage. *Chem Rev* 2007;**107**:4152–205.
124. Osterloh FE. Inorganic materials as catalysts for photochemical splitting of water. *Chem Mater* 2008;**20**:35–54.
125. Nowotny J, Sorrell CC, Sheppard LR, Bak T. Solar-hydrogen: Environmentally safe fuel for the future. *Int J Hydr Energy* 2005;**30**:521–44.
126. Nowotny J, Bak T, Nowotny MK, Sheppard LR. Titanium dioxide for solar-hydrogen I. Functional properties. *Int J Hydr Energy* 2007;**32**:2609–29.
127. Gondal MA, Hameed A, Yamani ZH, Suwaiyan A. Production of hydrogen and oxygen by water splitting using laser induced photo-catalysis over Fe_2O_3. *Appl Catal A: General* 2004;**268**:159–67.
128. Aroutiounian VM, Arakelyan VM, Shahnazaryan GE. Metal oxide photoelectrodes for hydrogen generation using solar radiation-driven water splitting. *Solar Energy* 2005;**78**:581–92.
129. Duonghong D, Borgarello E, Grätzel M. Dynamics of light-induced water cleavage in colloidal systems. *J Am Chem Soc* 1981;**103**:4685–90.

Index

